COMPUTATIONAL METHODS
FOR PROTEIN FOLDING

A SPECIAL VOLUME OF ADVANCES IN CHEMICAL PHYSICS

VOLUME 120

EDITORIAL BOARD

COMPUTATIONAL METHODS FOR PROTEIN FOLDING

ADVANCES IN CHEMICAL PHYSICS
VOLUME 120

Edited by

RICHARD A. FRIESNER

Columbia University, New York, NY

Series Editors

I. PRIGOGINE

Center for Studies in Statistical Mechanics
and Complex Systems
The University of Texas
Austin, Texas
and
International Solvay Institutes
Universite Libre de Bruxelles
Brussels, Belgium

and

STUART A. RICE

Department of Chemistry
and
The James Franck Institute
The University of Chicago
Chicago, Illinois

AN INTERSCIENCE PUBLICATION
A JOHN WILEY & SONS, INC. PUBLICATION

For ordering and customer service, call 1-800-CALL-WILEY

Library of Congress Catalog Number: 58-9935

ISBN 0-471-20955-4

Printed in the United States of America.

10 9 8 7 6 5 4 3 2 1

CONTRIBUTORS TO VOLUME 120

BENOIT CROMP, Département de Chimie, Université de Montréal, Montréal, Québec, Canada; Centre de Recherche en Calcul Appliqué, Montréal, Québec, Canada; and Protein Engineering Network of Centers of Excellence, Edmonton, Alberta, Canada

R. I. DIMA, Institute for Physical Science and Technology and Department of Chemistry and Biochemistry, University of Maryland, College Park, MD, U.S.A.

AARON R. DINNER, New Chemistry Laboratory, University of Oxford, Oxford, U.K.

RON ELBER, Department of Computer Science, Cornell University, Ithaca, NY, U.S.A.

VOLKER A. EYRICH, Department of Chemistry and Center for Biomolecular Simulation, Columbia University, New York, NY, U.S.A.

ANTHONY K. FELTS, Department of Chemistry, Rutgers University, Wright-Rieman Laboratories, Piscataway, NJ, U.S.A.

CHRISTODOULOS A. FLOUDAS, Department of Chemical Engineering, Princeton University, Princeton, NJ, U.S.A.

RICHARD A. FRIESNER, Department of Chemistry and Center for Biomolecular Simulation, Columbia University, New York, NY, U.S.A.

EMILIO GALLICCHIO, Department of Chemistry, Rutgers University, Wright-Rieman Laboratories, Piscataway, NJ, U.S.A.

JOHN GUNN, Schrödinger, Inc., New York, NY, U.S.A.; Centre de Recherche en Calcul Appliqué, Montréal, Québec, Canada; and Protein Engineering Network of Centers of Excellence, Edmonton, Alberta, Canada

PIERRE-JEAN L'HEUREUX, Département de Chimie, Université de Montréal, Montréal, Québec, Canada; Centre de Recherche en Calcul Appliqué, Montréal, Québec, Canada; and Protein Engineering Network of Centers of Excellence, Edmonton, Alberta, Canada

MARTIN KARPLUS, New Chemistry Laboratory University of Oxford, Oxford, U.K.; Department of Chemistry and Chemical Biology, Harvard University, Cambridge, MA, U.S.A.; and Laboratoire de Chimie Biophysique, Institut le Bel, Université Louis Pasteur, Strasbourg, France

JOHN L. KLEPEIS, Department of Chemical Engineering, Princeton University, Princeton, NJ, U.S.A.

D. K. KLIMOV, Institute for Physical Science and Technology and Department of Chemistry and Biochemistry, University of Maryland, College Park, MD, U.S.A.

ANDRZEJ KOLINSKI, Laboratory of Computational Genomics, Danforth Plant Science Center, Creve Coeur, MO, U.S.A.; and Department of Chemistry, University of Warsaw, Warsaw, Poland

RONALD M. LEVY, Department of Chemistry, Rutgers University, Wright-Rieman Laboratories, Piscataway, NJ, U.S.A.

ÉRIC MARTINEAU, Département de Chimie, Université de Montréal, Montréal, Québec, Canada; Centre de Recherche en Calcul Appliqué, Montréal, Québec, Canada; and Protein Engineering Network of Centers of Excellence, Edmonton, Alberta, Canada

JAROSLAW MELLER, Department of Computer Science, Cornell University, Ithaca, NY, U.S.A.; and Department of Computer Methods, Nicholas Copernicus University, Torun, Poland

HEATHER D. SCHAFROTH, Department of Chemical Engineering, Princeton University, Princeton, NJ, U.S.A.

JEFFREY SKOLNICK, Laboratory of Computational Genomics, Danforth Plant Science Center, Creve Coeur, MO, U.S.A.

SUNG-SAU SO, Hoffman-La Roche, Inc., Discovery Chemistry, Nutley, NJ, U.S.A.

DARON M. STANDLEY, Schrödinger Inc., New York, NY, U.S.A.

D. THIRUMALAI, Institute for Physical Science and Technology and Department of Chemistry and Biochemistry, University of Maryland, College Park, MD, U.S.A.

ANDERS WALLQVIST, Department of Chemistry, Rutgers University, Wright-Rieman Laboratories, Piscataway, NJ, U.S.A.

KARL M. WESTERBERG, Department of Chemical Engineering, Princton University, Princeton, NJ, U.S.A.

INTRODUCTION

Few of us can any longer keep up with the flood of scientific literature, even in specialized subfields. Any attempt to do more and be broadly educated with respect to a large domain of science has the appearance of tilting at windmills. Yet the synthesis of ideas drawn from different subjects into new, powerful, general concepts is as valuable as ever, and the desire to remain educated persists in all scientists. This series, *Advances in Chemical Physics*, is devoted to helping the reader obtain general information about a wide variety of topics in chemical physics, a field that we interpret very broadly. Our intent is to have experts present comprehensive analyses of subjects of interest and to encourage the expression of individual points of view. We hope that this approach to the presentation of an overview of a subject will both stimulate new research and serve as a personalized learning text for beginners in a field.

I. Prigogine
Stuart A. Rice

PREFACE

The first attempts to model proteins on the computer began almost 30 years ago. Over the past three decades, our understanding of protein structure and dynamics has dramatically increased as a result of rapid advances in both theory and experiment. The Protein Data Bank (PDB) now contains more than 10,000 high-resolution protein structures. The human genome project and related efforts have generated an order of magnitude more protein sequences, for which we do not yet know the structure. Spectroscopic measurement techniques continue to increase in resolution and sensitivity, allowing a wealth of information to be obtained with regard to the kinetics of protein folding and unfolding, complementing the detailed structural picture of the folded state. In parallel to these efforts, algorithms, software, and computational hardware have progressed to the point where both structural and kinetic problems may be studied with a fair degree of realism.

Despite these advances, many major challenges remain in understanding protein folding at both a conceptual and practical level. There is still significant debate about the role of various underlying physical forces in stabilizing a unique native structure. Efforts to translate physical principles into practical protein structure prediction algorithms are still at an early stage; most successful prediction algorithms employ knowledge-based approaches that rely on examples of existing protein structures in the PDB, as well as on techniques of computer science and statistics. Theoretical modeling of the dynamics of protein folding faces additional difficulties; there is a much smaller body of experimental data, which is typically at relatively low resolution; carrying out computations over long time scales requires either very large amounts of computer time or the use of highly approximate models; and the use of statistical methods to analyze the data is still in its infancy.

The importance of the protein folding problem—underscored by the recent completion of the human genome sequence—has led to an explosion of theoretical work in areas of both protein structure prediction and kinetic modeling. An exceptionally wide variety of computational models and techniques are being applied to the problem, due in part to the participation of scientists from so many different disciplines: chemistry, physics, molecular biology, computer science, and statistics, to name a few. This has made the field very exciting for those of us working in it, but it also poses a challenge; how can the key issues in state of the art research be communicated to different audiences, given the interdisciplinary nature of the task at hand and the methods being brought to bear on it?

The objective of this volume of *Advances in Chemical Physics* is to discuss recent advances in the computational modeling of protein folding for an audience of physicists, chemists, and chemical physicists. Many of the contributors to this volume have their roots in chemical physics but have committed a significant fraction of their resources to studying biological systems. The chapters thus address the target audience but incorporate approaches from other areas because they are relevant to the methods that the various authors have developed in their laboratories. While some of the chapters contain review sections, the principal focus is on the authors' own research and recent results.

When modeling protein folding the key questions are (a) the nature of the physical model to be used and (b) the questions that the calculations are aimed at answering. It is impossible in a single volume to cover all of the different approaches that are currently being used in research on protein folding. Nevertheless, a reasonably broad spectrum of computational methods is represented here, as is briefly described below. The volume is organized so as to group together contributions in which similar approaches are adopted.

The simplest models of proteins involve representations of the amino acids as beads on a chain (typically taken to be hydrophobic or hydrophilic, depending upon the identity of the amino acid) embedded in a lattice. Primitive models of this type employ a simple lattice such as a cubic lattice, and they use a single center to represent each amino acid. These models are very fast computationally, but lack a level of detail (both structurally and in their potential energy function) to permit prediction of protein structure from the amino acid sequence. On the other hand, they can be extremely valuable in providing conceptual insight into the general thermodynamic and kinetic issues as to why and how proteins fold into a unique native state; they can also be profitably used to model folding kinetics, as well as to make testable predictions for such kinetics that can be compared with experimental data. The contributions of Thirumulai et al. and Dinner et al. discuss models of this type, presenting both conceptual insights into the basis of protein folding and results for modeling of specific protein folding events.

Reduced models of proteins (i.e., models not containing complete atomic detail) can be used to make structural predictions, either by allowing assessment of the fitness of a protein structure already in the PDB as a model for an unknown sequence ("threading") or by carrying out Monte Carlo simulations using the model and a suitable potential energy function. The contribution by Meller and Elber describes a classical threading approach in which the amino acid sequence is "threaded" in an optimal fashion onto a set of candidate template structures using dynamic programming techniques, and the suitability of the template is evaluated by a potential energy function. These authors have worked out new methods for optimizing such functions, which are discussed in detail in their chapter.

If a reduced (or other) model is used to predict protein structure via simulation, without direct reference to structures in the PDB, this is referred to as "*ab initio* protein" structure prediction. Potential energy functions for *ab initio* prediction can be derived either from physical chemical principles or from a "knowledge-based" approach based on statistics from the PDB (e.g., the probability of observing a residue–residue distance for a given pair of amino acids). For reduced models, the use of knowledge-based potential of some sort is mandated. The contributions of Eyrich et al., Skolnick and Kolinsiki, and L'Heureux et al. derive originally from an *ab initio* approach using reduced models. However, all of these groups have in the past several years increasingly incorporated empirical elements from threading and other such approaches, so that what is described in these contributions is more of an attempt to integrate reduced model simulations with additional information and techniques that can improve practical structure prediction results. Several of these research groups have entered the CASP (Critical Assessment of Protein Structure Prediction) blind test experiments, which allow a comparative evaluation of the prediction accuracy of the different methods employed by the participants; results from the most recent such experiment, CASP4 (not reported in this volume because the results were available subsequent to submission of most of the chapters), were encouraging with regard to the ability of these hybrid methods to provide improvement in many cases over methods not incorporating simulations.

The use of models employing an atomic level of detail (e.g. a molecular mechanics potential function) in addressing the protein folding problem presents significant difficulties for two reasons: (1) A large expenditure of computation time is required to evaluate the model energy at each configuration; (2) the quality of the potential energy functions and solvation model are critical in being able to accurate compare the stability of alternative structures. The contribution by Klepeis et al. discusses both algorithms designed to reduce the required computational effort by sampling phase space more efficiently and a wide variety of applications of atomic level models using these more efficient sampling techniques. The contribution from Wallqvist et al. is more narrowly focused on a single problem: the use of detailed atomic potential functions in conjunction with a continuum solvation model to distinguish native and "native-like" protein structures from "decoys"—alternative structures generated by various means and intended to challenge the model's accuracy. Both of these contributions demonstrate that considerable progress is being made in the application of atomic level models with regard to improving both accuracy and efficiency.

In the end, a thorough description of all aspects of protein folding will require the use of the full range of models and methods discussed in this volume. In the simplest hierarchical picture, one can imagine using inexpensive reduced models to generate low-resolution structures that can then be refined using more detailed (and computationally expensive) approaches. Although progress will undoubtedly continue in the development of physical chemical models, empirical information and phenomenological approaches will always provide additional speed and reliability if practical results are desired. How to best combine all of these elements represents one of the principal issues facing those working in the field; it also exemplifies the need for new ideas and approaches.

Columbia University   RICHARD A. FRIESNER
New York, New York

CONTENTS

STATISTICAL ANALYSIS OF PROTEIN FOLDING KINETICS

AARON R. DINNER

New Chemistry Laboratory, University of Oxford, Oxford, U.K.

SUNG-SAU SO

Hoffmann-La Roche Inc., Discovery Chemistry, Nutley, NJ, U.S.A.

MARTIN KARPLUS

New Chemistry Laboratory, University of Oxford, Oxford, U.K.; Department of Chemistry and Chemical Biology, Harvard University, Cambridge, MA, U.S.A.; and Laboratoire de Chimie Biophysique, Institut le Bel, Université Louis Pasteur, Strasbourg, France

CONTENTS

Computational Methods for Protein Folding: A Special Volume of Advances in Chemical Physics, Volume 120, Edited by Richard A. Friesner. Series Editors I. Prigogine and Stuart A. Rice. ISBN 0-471-20955-4. © 2002 John Wiley & Sons, Inc.

I.   INTRODUCTION

Experimental and theoretical studies have led to the emergence of a unified general mechanism for protein folding that serves as a framework for the design and interpretation of research in this area [1]. This is not to suggest that the details of the folding process are the same for all proteins. Indeed, one of the most striking computational results is that a single model can yield qualitatively different behavior depending on the choice of parameters [1–3]. Consequently, it remains to determine the behavior of individual sequences under given environmental conditions and to identify the specific factors that lead to the manifestation of one folding scenario rather than another. Although doing so requires investigation of the kinetics of particular proteins at the level of individual residues, for which protein engineering [4] and nuclear magnetic resonance (NMR) [5] experiments are very useful, complementary information about the roles played by the sequence and the structure can also be obtained by a statistical analysis of the folding rates of a series of proteins.

Statistical methods have been applied for many years in attempts to predict the structures of proteins (for a review of progress in this area, see the chapter by Meller and Elber, this volume), but their use in the analysis of folding kinetics is relatively recent. The first such investigations focused on "toy" protein models in which the polypeptide chain is represented by a string of beads restricted to sites on a lattice. It was found that the ability of a sequence to fold correlates strongly with measures of the stability of its native (ground) state (such as the Z-score or the gap between the ground and first excited compact states) [6–9], but the native structure also plays an important role for longer chains [10,11]. While lattice models are limited in their ability to capture the structural features of proteins, they have the important advantage that the results of statistical analyses can be compared with calculated folding trajectories to determine the physical bases of observed correlations. Consequently, studies based on such models are particularly useful for the quantitation of observed effects, the generalization from individual sequences, the identification of subtle relationships, and ultimately the design of additional sequences that fold at a given rate.

Analogous statistical analyses of experimentally measured folding kinetics of proteins were hindered by the fact that complex multiphasic behavior was exhibited by most of the proteins for which data were available (e.g., barnase and lysozyme). In recent years, an increasing number of proteins that lack significantly populated folding intermediates and thus exhibit two-state folding kinetics have been identified, and a range of data have been tabulated for them [12–14]. The initial linear analyses of such proteins indicated that their folding rates are determined primarily by their native structures [12,14]. More recently, a nonlinear, multiple-descriptor approach revealed that there is a significant dependence on the stability as well [15]. These and related studies are discussed in Section IV.A, after an overview of the statistical methods employed in this area (Section II) and a review of the results from lattice models (Section III).

An in-depth analysis of a database of 33 proteins that fold with two- or weakly three-state kinetics is presented in Sections IV.B through V. We explore one-, two-, and three-descriptor nonlinear models. A structurally based cross-validation scheme is introduced. Its use in conjunction with tests of statistical significance is important, particularly for multiple-descriptor models, due to the limited size of the database. Consistent with the initial linear studies [12,14], it is found that the contact order and several other measures of the native structure are most strongly related to the folding rate. However, the analysis makes clear that the folding rate depends significantly on the size and stability as well. Due to the importance ascribed to the stability by analytic [16–18] and simulation [2,3,6–11] studies, as well as its recent use in one-dimensional models for fitting and interpreting experimental data [19,20], we examine its connection to the folding rate in more detail. The unfolding rate, which correlates more strongly with stability, is considered briefly. The relation of the statistical results to experiments and the model studies is discussed in Sections VI and VII.

II. STATISTICAL METHODS

Before reviewing the results for specific systems, we introduce the statistical methods that have been used to analyze folding kinetics. Perhaps the simplest such method is to group sequences; here, one categorizes each sequence in a database according to one or more of its native properties ("descriptors") and its folding behavior. Visualization can be used to identify patterns, and averages and higher moments of the distributions of descriptors can be used to quantitate differences between groups. For properties on which the folding kinetics depend strongly, such as the energy gap in lattice models, this type of analysis has proven effective [6].

However, simple grouping is often insufficient to identify weaker but still significant trends and makes it difficult to determine the relative importance of relationships. Consequently, more quantitative methods are necessary. One statistic that is employed widely is the Pearson linear correlation coefficient ($r_{x,y}$):

$$r_{x,y} = \frac{\sigma_{xy}^2}{\sigma_x \sigma_y} = \frac{\sum_i (x_i - \bar{x})(y_i - \bar{y})}{\sqrt{\sum_i (x_i - \bar{x})^2 \sum_i (y_i - \bar{y})^2}} \tag{1}$$

Typically, the x_i are a set of values of a particular descriptor, such as the sequence length, and the y_i are a set of values for a measure of the folding kinetics, such as the logarithm of the folding rate constant (log k_f) [9,10,12]. The magnitude of $r_{x,y}$ determines its significance, and its sign indicates whether x_i and y_i vary in the same or opposite manner: $r_{x,y} = 1$ corresponds to a perfect correlation, $r_{x,y} = -1$ to a perfect anticorrelation, and $r_{x,y} = 0$ to no correlation. In spite of its popularity, this statistic has several shortcomings when used by itself. It is limited to the identification of linear relationships between pairs of properties; it is not straightforward to test or cross-validate those relationships, which is important, as discussed below; and it cannot be used directly to predict the behavior of additional sequences.

These limitations can be overcome by constructing models to predict folding behavior and then quantifying their accuracy. For the latter step, the Pearson linear correlation coefficient can be used with x_i as the observed values and y_i as the predicted ones (for which we introduce the shorthand notations r_{trn}, r_{jck}, and r_{cv}, described below). Alternatively, one can calculate the root-mean-square error or the closely related fraction of unexplained variance:

$$q^2 = 1 - \frac{\sum_i (y_i - x_i)^2}{\sum_i (x_i - \bar{x})^2} \tag{2}$$

Again, x_i (y_i) are the observed (predicted) values. Typically, r and q^2 behave consistently. The latter is useful for quantitating the improvement obtained upon extending a model with N descriptors to one with $N + 1$ with Wold's statistic: $E = (1 - q^2_{N+1})/(1 - q^2_N)$ [21,22]. A value of less than 1.0 for the latter shows that q^2 increases upon adding a descriptor. The statistical significance of a particular value of E depends on the specific data, but $E = 0.4$ has been suggested to correspond typically to the 95% confidence interval [23].

For constructing the models themselves, linear regression (on one or more descriptors) is attractive in that the best fit for a set of data can be determined analytically, but, as its name implies, it is limited to detecting linear relationships. While fits with higher-order polynomials are possible, a general and flexible alternative is to use neural networks (NNs). The latter are computational tools for model-free mapping that take their name from the fact that they are based on simple models of learning in biological systems [24,25]. Neural networks have been used extensively to derive quantitative structure–property relationships in medicinal chemistry (for a review, see Ref. 26) and were first used to analyze folding kinetics in Ref. 11. A schematic diagram of a neural network is shown in Fig. 1. In this example, there are three inputs (indicated by the rectangles on the left); in the present study these would each contain the value of a descriptor, such as the free energy of unfolding or the fraction of

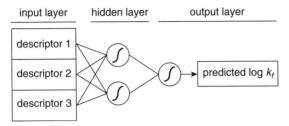

Figure 1. Schematic of a neural network.

helical contacts. The circles represent sigmoidal functions (nodes). There are many possible choices for the specific form of these functions; we use

$$f = \frac{1}{1 + \exp\left(-\theta - \sum_i w_i p_i\right)} \tag{3}$$

where the sum ranges over the previous layer (to the left in the diagram), p_i are the values of the elements of that layer, w_i are the weights for each of those elements (represented by the connecting lines in the diagram), θ is an arbitrary constant, and the data are assumed to be normalized for clarity. Thus, to "fire" the network in Fig. 1, a weighted sum over the three inputs to each hidden node is made, the resulting sums are used to calculate the values of the sigmoidal functions associated with those nodes, a weighted sum of those values is then made, and the final sigmoidal function of the output node is calculated. To fit data, the w_i are initialized to random values and adjusted with standard optimization techniques to maximize the accuracy of the output for the (training) set. In the present study, we varied the weights with the scaled conjugate gradient method [27].

When one wishes to test many different possible descriptors, the number of possible NN input combinations can be very large. One can avoid making an exhaustive search by using a genetic algorithm (GA) to select the descriptors to test. This tool is also biologically motivated—in this case, by evolution. A population is created in which each individual consists of a particular set of descriptors. Repeatedly, each such set (a "parent") is duplicated ("asexual repro-duction"), the new copy (a "child") is changed by one descriptor ("mutated"), and then only the best ("fittest") individuals in the combined pool of parents and children are kept. Here, "best" means that a linear regression or NN model employing those descriptors yields the greatest accuracy for the training set. Alternative schemes that involve combining features from different individuals ("sexual reproduction") also exist but are not employed here; for a compre-hensive review of the use of GAs in medicinal chemistry see Ref. 28. In the present study, we used 40 individuals with 20 genetic cycles; a few trials with 200 individuals and 50 cycles did not yield significantly different results.

An important point concerning neural networks, and indeed any multiple parameter model, is that it is possible to overfit the data. For small sample sizes (here, a small number of proteins), even relatively simple neural networks can memorize the examples in the training set at the expense of learning more general rules. Thus, it is important to test a model on novel data not used during the fitting process. One approach is cross-validation, in which one partitions the existing data into a series of training and test sets. In the special case of jackknife cross-validation, all possible combinations are formed in which a single protein is used to test the network and the remainder are used to train it. While jackknife cross-validation is straightforward to automate, it is not appropriate if any members of the database are significantly related (e.g., homologous proteins) because the inclusion of the similar data in the training set can bias the test. A structurally based partitioning scheme is presented in Section IV.B. Throughout, care is taken to distinguish statistics (r and q^2) for *fits* of the entire (training) set (denoted "trn") from those for *predictions* obtained with either jackknife or structurally based cross-validation (denoted "jck" and "cv," respectively).

III. LATTICE MODELS

The first study in which a large number of unrelated sequences were analyzed to identify the factors that determine their folding kinetics was based on a 27-residue chain of beads subject to Monte Carlo dynamics on a simple cubic lattice [6]. In this and the subsequent studies of 125-residue sequences [10,11], folding rate constants were calculated for only a few sequences due to the large number of trajectories required to obtain accurate results. Folding "ability" was measured by either (a) the fraction of Monte Carlo trials that reached the native state within the allotted simulation time or (b) the average fraction of native contacts in the lowest energy states sampled. When the results for the 27-residue sequences were grouped according to the former, it was found that the stability of the native (ground) state is the only feature that distinguishes those that folded repeatedly within the simulation time from those that did not. If the native state is maximally compact, the stability criterion can be simplified to a consideration of the difference in energy between the ground state and the first fully compact ($3 \times 3 \times 3$) excited state [6]. These criteria have been used in the design of fast folding sequences [29] and are consistent with similar studies which focus on exhaustive enumeration of folding paths for two-dimensional chains [7,30] or on the ratio of the folding and the "glass" transition temperatures for the (three-dimensional) 27-residue model [8].

In a number of subsequent studies of the 27-residue model, it was argued that the kinetic folding behavior is determined by factors other than the energy gap

[31–33]. Unger and Moult [31] suggested that the dependence on the energy gap derived from the variation in the simulation temperature in Ref. 6 and identified the structure of the ground state as the primary determinant of the folding kinetics of this system. However, in a study of 15- and 27-residue three-dimensional chains that employed the Pearson linear correlation coefficient to quantitate the relationships between various sequence factors and the logarithm of the mean first passage time, the correlation with the Z-score was robust to use of a single temperature [9]. Examination of Ref. 31 showed that sequences were designed to have strong short-range contacts without mandating a certain fraction of long-range contacts, so that the resulting ground states were more appropriate for modeling a helix-coil transition than protein folding. Nevertheless, as will be discussed below, native structure does play a role for certain lattice models [10,11] as it does for proteins [12,14,15]. Klimov and Thirumalai [32,33] introduced the parameter $\sigma = 1 - T_f/T_\theta$, where T_f is the temperature at which the fluctuation of the order parameter is at its maximum and T_θ is the temperature at which the specific heat is at its maximum. They found that σ is positively correlated with the logarithm of the mean first passage time (i.e., small sigma gives fast folding). However, the interpretation of T_θ as the collapse transition temperature is not correct in general, and the correlation described above arises from the fact that σ is related to the energy gap [9]. These statistical studies of short chains are discussed in detail in Ref. 9.

The correlation of the folding time with the energy gap can be understood in terms of its effect on the energy surface. For random 27-residue sequences, folding proceeds by a fast collapse to a semicompact disordered globule, followed by a slow, nondirected search through the relatively small number of semicompact structures for one of the many transition states that lead rapidly to the native conformation [2]. A large energy gap results in a native-like transition state that is stable at a temperature high enough for the folding polypeptide chain to overcome barriers between random semicompact states. As the energy gap increases to the levels obtainable in designed sequences, the model exhibits Hammond behavior [34] in that there is a decrease in the fraction of native contacts required in the transition state from which the chain folds rapidly to the native state. Random sequences with relatively small gaps must form about 80% of the native contacts [2], whereas designed sequences with large gaps need form only about 20% [35]. This shift increases the ratio of the number of transition states to the number of semicompact states and results in a nucleation mechanism [35].

The first study to employ the Pearson linear correlation coefficients between various individual sequence properties and measures of folding ability concerned the analysis of 125-residue lattice model simulations [10]. It revealed that, in addition to the stability, the native structure plays an important role in determining folding ability for chain lengths comparable to that typical of certain well-studied proteins (e.g., barnase and lysozyme); that is, a strong correlation was observed between the frequency of reaching the native state within the simulation time and the number of native contacts in tight turns or antiparallel sheets. On the lattice, these are the cooperative secondary structural elements that have the shortest sequential separations between contacts; lattice "helices," which typically consist only of $i, i + 3$ contacts, are noncooperative and thus do not accelerate folding. The physical basis of the relation between structure and kinetics in lattice models and in proteins is discussed in Section IV.E.

The initial linear analysis of the 125-residue model also made clear that one descriptor can compensate for others, so that it is necessary to consider more than one simultaneously [10]. Accordingly, the functional dependence of the folding ability on sets of sequence properties was derived with an artificial neural network, and a genetic algorithm was used to select the sets that maximize the accuracy of the predictions. Not only did the nonlinear, multiple-descriptor method increase the correlation coefficients between the observed folding abilities and the cross-validated predictions from about 0.5 to greater than 0.8, but it revealed (in addition to the strong dependences on the stability and structure of the native state) a role for the spatial distribution of strong and weak pairwise interactions within the native structure. Sequences with native structures that have more labile contacts between surface residues were found to fold faster in general because misfolded subdomains are less likely to form and lead to off-pathway traps [10,11,36]. This observation indicates that, as one goes to longer sequences, the relationship between the folding rate and the native state descriptors becomes more complex.

The genetic neural network (GNN) method was further validated by use of one of the resulting quantitative structure–property relationships (QSPRs) to design additional fast-folding 125-residue sequences [37]. The target native structure and the pairwise interaction energies were varied to maximize the output of a network trained on the original set of sequences to predict the average fraction of native contacts in the lowest energy structure sampled in each of 10 Monte Carlo simulations [10,11]. The specific descriptors employed were the number of contacts in antiparallel sheets, the estimated gap in energy between the native state and the lower limit of the quasi-continuous spectrum [38], and the total energy of the contacts between surface residues. On average, the designed sequences folded more rapidly than those for which only the stability of the native state was optimized [29,39]. The studies of the 125-residue lattice models thus make clear that simultaneous consideration of multiple descriptors can improve our understanding of protein folding and our ability to extrapolate from the analysis to predict the behavior of novel sequences. The utility of the statistical approach for obtaining a better understanding of the folding rates of proteins is described in the following section.

IV. FOLDING RATES OF PROTEINS

In this section we describe statistical analyses of measured rates of protein folding. Earlier studies are reviewed and an analysis of currently available experimental data is presented. The physical bases of the results are then discussed.

A. Review

As mentioned in the Introduction, statistical analyses of the folding kinetics of proteins were delayed until a sufficient number of proteins that fold with two-state kinetics overall were identified [12,13]. Plaxco et al. [12] carried out an analysis much like the initial 125-mer lattice model study mentioned above [10] for a set of 12 two-state proteins (extended to 24 proteins in Ref. 14); that is, they calculated linear correlation coefficients between several individual sequence properties and the logarithm of the measured folding rate constants ($\log k_f$). The only descriptor examined that exhibited a high correlation ($r_{c/n,\log k_f} = 0.81$) was the structure of the native state as measured by the normalized contact order (c/n), the average sequential residue separation of atoms in contact divided by the length of the sequence (see the footnote to Table III for the exact definition of c/n employed here). It is important to note that the contact order does not include any information about the energies of the interactions in the native state; it is only a measure of the structure (we use the term "structure" rather than "topology" [12,14] because, according to the standard mathematical meaning of the latter, all proteins that lack disulfide bonds have the same topology).

We used a neural network to carry out a nonlinear, two-descriptor analysis of the database of 33 proteins described in Section IV.B [15] and demonstrated that the stability contributes significantly to determining folding rates for a given contact order. Moreover, for 14 slow-folding proteins with high contact orders (mixed-α/β and β-sheet proteins), the free energy of unfolding can be used by itself to predict folding rates. By contrast, the folding rates of α-helical proteins show essentially no dependence on the stability. The variation in behavior observed for the structural classes suggests that, although there is a general mechanism of folding (see the Introduction), its expression for individual proteins can lead to very different behavior.

A number of simple physically motivated one-dimensional models have been introduced recently to fit and interpret data on peptide and protein folding [19, 20,40–42]. These models, which use only native state data, have elements in common with earlier theoretical treatments by Zwanzig, Wolynes, and their co-workers [16,17,43]. The conformation of a protein is represented by a series of binary variables (based on one or two residues), each of which can be either native or random coil. Pairwise interactions (which are assumed to be entirely favorable, as in a Gō model [44,45]) are counted if and only if both the sequence positions involved are native. Often, an additional approximation is made in which the formation of the native structure is limited to one or two sequential segments [46]. Independent of this assumption, the one-dimensional character of these models and the choice of energy functions typically force the native structure to propagate in an essentially sequential manner. By adjusting parameters, one of these models was shown to fit log k_f with an accuracy of $0.83 \leq r_{trn} \leq 0.87$ for 18 proteins [20]. The fact that this correlation is somewhat higher than that obtained using only the contact order (Table I and Refs. 12,14, and 20) has been used as evidence for the physical basis of the model; that is, it provides an "explanation" of the empirical relationship between the folding rate and the contact order. However, the improvement appears to be due to the incorporation of the protein stabilities into the model. These were introduced by adjusting the pairwise interactions separately for each protein such that the model yielded free energies for folding that matched experimental ΔG values. Using the methods described in Section II and applied in Section IV.B, we were able to obtain $r_{trn} = 0.93$ with two descriptors (ΔG and q_a, described in Table I) and $r_{trn} = 0.98$ with three (ΔG, c, and b) for the same set of 18 proteins; for c/n, and $\Delta G/n$, $r_{trn} = 0.85$, which is very similar to the correlations reported in Ref. 20 ($0.83 \leq r_{trn} \leq 0.87$). Thus, further work is required to show that such simple phenomenological models can predict aspects of the folding reaction that go beyond the experimental data used in the fitting procedures. Although these model studies consider the prediction of ϕ values [4], it appears from the published results and statements in the text of Ref. 20 that the correlation is poor. This suggests that quantitative comparisons of predicted ϕ-values with the observed ones could serve as a meaningful test of such phenomenological models.

An alternative phenomenological model was developed by Debe and Goddard [47]. In essence, they assumed a sequence of events which is, in a certain sense, the reverse of the diffusion–collision model [48,49]: the correct overall (tertiary) structure is formed at low-resolution first by a random search and then local (secondary) refinement takes place within the manifold of states in that fold. Thus, the factor that determines the relative rate of folding for a series of proteins is the probability of randomly sampling a structure with the known native contacts (estimated by a Monte Carlo procedure); the distance at which a contact was counted was adjusted to optimize the fit. For mixed-α/β and β-sheet proteins, an accuracy of $r_{trn} = 0.78$ was obtained. This statistic is comparable to the correlation coefficients associated with the contact order (Table I and Refs. 12 and 14), which could suggest that this model is a rather complex procedure for reproducing the simple (essentially linear) dependence of log k_f on that descriptor. For α-helical proteins, the folding rates were considerably underestimated, which led Debe and Goddard to conclude that hose proteins must instead fold by a diffusion–collision mechanism [48,49]. The discussion in the present section shows that phenomenological models can be useful for interpreting the observed statistical correlations. However, it is important to keep in mind that the ability to fit a particular set of data is not sufficient to demonstrate that the folding *mechanism* on which the model is based is correct.

B. Database

To illustrate the methods described in Section II and to show that simultaneous consideration of multiple descriptors improves prediction of protein folding kinetics, we describe a detailed analysis of the available data for the folding rates of two- and weakly three-state proteins. The descriptors tested are listed in Table I and can be divided into several categories: native state stability (0 and 1), size (2 to 5), native structure (8 to 15), and the propensity for a given structure (16 to 23). Definitions and sources for the descriptors as well as the data themselves are given in Tables II and III. Although certain descriptors are significantly

TABLE I
Descriptors Tested as Inputs to the GNN and Their Correlations[a]

Index	Symbol	Description	$r_{x,\log k_f}$	r_{trn}	r_{cv}	q^2_{cv}
0	ΔG	Stability	0.29	0.40	0.06	-0.16
1	$\Delta G/n$	Normalized stability	0.37	0.42	-0.00	-0.13
2	m	Buried surface area	-0.04	0.38	-0.16	-0.40
3	m/n	Normalized surface area	-0.04	0.24	-0.29	-0.21
4	n	Sequence length	-0.10	0.35	-0.52	-0.19
5	n_c	Number of atomic contacts	-0.08	0.34	-0.32	-0.18
6	c	Contact order	-0.73	0.74	0.67	0.45
7	c/n	Normalized contact order	-0.79	0.83	0.74	0.54
8	h	α-Helix content	0.63	0.64	0.39	0.11
9	e	β-Sheet content	-0.67	0.71	0.59	0.34
10	t	H-bonded turn content	0.04	0.34	-0.07	-0.21
11	s	Bend content	-0.11	0.31	-0.25	-0.26
12	g	3_{10}-Helix content	-0.01	0.35	-0.47	-0.28
13	b	β-Bridge content	-0.15	0.30	-0.36	-0.32
14	o	Other 2° structure	-0.05	0.27	-0.32	-0.44
15	a	Total helix content $(h+g)$	0.63	0.67	0.28	-0.04
16	P_h	Predicted α-helix	0.47	0.49	0.05	-0.10
17	P_e	Predicted β-sheet	-0.48	0.57	0.29	0.01
18	P_o	Predicted other 2°	-0.27	0.43	-0.39	-0.32
19	p_h	α-Helix propensity	0.51	0.55	0.21	-0.03
20	p_e	β-Sheet propensity	-0.47	0.64	0.42	0.14
21	p_o	Other 2° propensity	-0.40	0.50	-0.20	-0.16
22	q_e	Expected 2° prediction accuracy	0.21	0.42	0.07	-0.14
23	q_a	Actual 2° prediction accuracy	0.40	0.45	-0.14	-0.45

[a]Here r_{trn} and r_{cv} are correlation coefficients between observed and calculated values of log k_f for training set fits and cross-validated predictions, respectively. Correlations are the maximum ones observed for 10 independent trials, each with a different random number generator seed. Statistics for linear regression are available in Table V.

TABLE II
Rate, Stability, and Size Descriptors[a]

Group	Protein	Reference	log k_f	log k_u	ΔG	$\Delta G/n$	m	m/n	n	n_c
SH3	1NYF		1.973128	−3.00382	6.0	0.0895522	1.40	0.0208955	67	378
	1PKS		−0.455932	−3.17335	3.4	0.0404762	2.30	0.0273810	84	710
	1SHG		0.612784	−2.55238	2.9	0.0467742	0.80	0.0129032	62	406
	1SRL		1.755875	−0.99982	4.1	0.0640625	1.60	0.0250000	64	389
β-Sandwich	1FNF-9		−0.397940	−1.30080	1.2	0.0133333	3.00	0.0333333	90	686
	1FNF-10	61	2.380211	−3.63762	9.4	0.1000000	6.50	0.0691489	94	648
	1HNG	61	1.255273	−2.76905	6.8	0.0693878	1.10	0.0112245	98	560
	1TEN		0.462398	−2.55238	4.8	0.0533333	1.30	0.0144444	90	600
	1TIT	61	1.505150	−3.30921	7.5	0.0842697	2.50	0.0280899	89	472
	1WT	61	0.176091	−3.55220	4.0	0.0430108	1.30	0.0139785	93	893
Acylphosphatase	1APS		−0.638272	−3.95789	5.4	0.0551020	1.25	0.0127551	98	833
	1HDN		1.173186	−2.67730	4.6	0.0541176	2.20	0.0258824	85	705
	1PBA		2.952792	−0.18705	4.1	0.0512500	1.00	0.0125000	80	589
	1URN		2.499687	−4.19990	9.3	0.0911765	2.30	0.0225490	102	749
	2HQI	63	0.079181		3.8	0.0527778	2.35	0.0326389	72	730
Cytochrome	1HRC		3.447158	−1.76923	6.9	0.0663462	2.40	0.0230769	104	828
	1HRC-oxidized		2.602060		17.7	0.1701923	3.30	0.0317308	104	828
	1YCC		4.176091		14.6	0.1417476	3.10	0.0300971	103	863
Cold shock	1CSP		3.029384	1.07899	3.0	0.0447761	0.76	0.0113433	67	346
	1MJC		2.274158	0.51842	2.9	0.0420290	0.57	0.0082609	69	400
λ-Repressor	1LMB		3.690196	1.47686	3.0	0.0375000	1.10	0.0137500	80	632
	1LMB-G46A/G48A	13,64	4.944483	1.55602	4.8	0.0600000	1.10	0.0137500	80	632
Ubiquitin	1UBQ		3.185259	−3.35891	7.1	0.0934211	1.90	0.0250000	76	510
	1UBQ-V26A		2.008600	−1.09671	3.9	0.0513158	2.00	0.0263158	76	510

Unique									
1COA		1.681241	−3.74405	7.0	0.1093750	1.80	0.028250	64	376
1DIV	65	2.857332	−0.12492	4.1	0.0725000	1.46	0.0260714	56	461
1FKB		0.633468	−3.76887	5.5	0.0514019	1.40	0.0130841	107	758
1IMQ	66	3.161368	−1.90623	6.3	0.0732558	1.10	0.0127907	86	936
2ABD		2.445604	−3.99928	7.1	0.0825581	3.00	0.0348837	86	1118
2AIT		1.826075	−4.34600	8.1	0.1094595	1.30	0.0175676	74	621
2PDD	67,68	4.255272	−2.36130	3.1	0.0720930	0.80	0.0186047	43	199
2PTL		1.778151	−1.69866	4.6	0.0741935	1.90	0.0306452	62	482
2VIK		2.954243	−1.21445	6.2	0.0492063	1.60	0.0126984	126	1089

[a]Unless otherwise noted, the stabilities, m values, and rates were taken from Ref. 13. Concerning the measured values of $\log k_f$ and $\log k_u$, it should be noted that the available data were obtained at different temperatures, and no correction for this variation was made. For CI2, for which data are available, the folding rate varies by about one order of magnitude over the full temperature range in the database (5°C to 37°C), but it changes by only a factor of about 1.5 over a range that includes most of the database (20°C to 25°C).

TABLE III
Structural and Structural Propensity Descriptors[a]

Set	Protein	c	c/n	h	e	t	s	g	b	o	a	P_h	P_e	P_o	P_a	P_e	P_o	q_e	q_a
SH3	1NYF	18.11	27.03	0.00	35.82	7.46	11.94	4.48	4.48	35.82	4.48	0.00	26.87	73.13	8.84	31.59	60.66	73.36	79.10
	1PKS	24.75	29.46	4.71	30.59	14.12	4.71	10.59	3.53	31.76	15.29	17.65	16.47	65.88	18.93	23.95	57.97	72.14	57.65
	1SHG	18.05	29.11	0.00	41.94	9.68	8.06	4.84	4.84	30.65	4.84	19.35	29.03	51.61	23.51	29.68	48.03	75.48	58.06
	1SRL	20.49	32.02	0.00	25.00	3.12	20.31	4.69	1.56	45.31	4.69	0.00	32.81	67.19	6.80	37.56	56.79	73.91	82.81
β-Sandwich	1FNF-9	27.97	31.08	0.00	54.44	6.67	11.11	0.00	0.00	27.78	0.00	0.00	24.44	75.56	4.10	32.12	64.65	78.54	65.56
	1FNF-10	26.53	28.22	0.00	50.00	8.51	9.57	0.00	1.06	30.85	0.00	0.00	42.55	57.45	2.88	43.56	54.40	79.10	79.79
	1HNG	31.65	32.30	0.00	53.06	16.33	7.14	3.06	0.00	20.41	3.06	21.43	19.39	59.18	24.28	24.81	51.63	70.25	55.10
	1TEN	27.42	30.47	0.00	53.33	11.11	7.78	0.00	0.00	27.78	0.00	4.44	27.78	67.78	8.98	31.87	59.96	73.39	70.00
	1TIT	28.78	32.34	0.00	35.96	6.74	17.98	0.00	1.12	38.20	0.00	17.98	31.46	50.56	16.87	35.86	48.08	71.95	64.04
	1WIT	30.25	32.53	0.00	51.61	13.98	5.38	3.23	3.23	22.58	3.23	15.05	31.18	53.76	15.79	32.00	52.92	66.71	64.52
Acylphosphatase	1APS	34.38	35.08	18.37	36.73	9.18	14.29	0.00	0.00	21.43	18.37	2.04	37.76	60.20	11.47	34.69	54.53	67.42	52.04
	1HDN	24.24	28.52	37.65	30.59	8.24	11.76	0.00	0.00	11.76	37.65	36.47	21.18	42.35	37.00	23.48	40.46	80.03	77.65
	1PBA	20.67	25.84	25.93	11.11	4.94	18.52	3.70	0.00	35.80	29.63	18.52	23.46	58.02	22.00	26.30	52.56	69.92	70.37
	1URN	26.88	26.35	27.45	23.53	10.78	9.80	5.88	0.00	22.55	33.33	44.12	14.71	41.18	40.63	16.67	43.44	75.22	68.63
	2HQI	22.35	31.04	26.39	27.78	18.06	11.11	0.00	0.00	16.67	26.39	23.61	30.56	45.83	23.32	30.52	47.13	69.92	72.22
Cold shock	1CSP	19.51	29.12	0.00	53.73	14.93	7.46	4.48	1.49	17.91	18.37	0.00	38.81	61.19	5.09	44.23	51.83	77.16	70.15
	1MJC	18.57	26.91	0.00	47.83	15.94	5.80	4.35	1.45	24.64	4.35	0.00	34.78	65.22	4.00	37.38	59.76	78.64	75.36
Cytochrome	1HRC	17.00	16.35	41.35	0.00	17.31	7.69	0.00	1.92	31.73	41.35	20.19	14.42	65.38	26.93	19.10	54.69	75.50	64.42
	1HRC-oxidized	17.00	16.35	41.35	0.00	17.31	7.69	0.00	1.92	31.73	41.35	20.19	14.42	65.38	26.93	19.10	54.69	75.50	64.42
λ-Repressor	1YCC	16.97	16.48	42.72	0.00	15.53	5.83	0.00	1.94	33.98	42.72	8.74	14.56	76.70	19.20	20.67	60.81	70.51	58.25
	1LMB	9.77	12.21	70.00	0.00	7.50	5.00	3.75	0.00	13.75	73.75	67.50	0.00	32.50	61.23	5.65	34.08	76.44	77.50
	1LMB-G46A/G48A	9.77	12.21	68.75	0.00	7.50	5.00	3.75	0.00	15.00	72.50	70.00	0.00	30.00	65.07	4.69	31.23	78.19	81.25

Ubiquitin	1UBQ	17.44	22.95	15.79	31.58	15.79	5.26	7.89	2.63	21.05	23.68	11.84	28.95	59.21	19.22	31.34	50.35	68.92	84.21
	1UBQ-V26A	17.44	22.95	15.79	31.58	15.79	5.26	7.89	2.63	21.05	23.68	13.16	28.95	57.89	22.34	29.45	49.14	70.87	85.53
Unique	1COA	16.07	25.11	17.19	21.88	15.62	4.69	6.25	29.69	21.88	17.19	29.69	53.12	22.52	32.99	45.66	75.26	76.56	
	1DIV	10.62	18.96	33.93	19.64	16.07	8.93	5.36	3.57	12.50	39.29	28.57	21.43	50.00	33.88	23.60	43.77	70.29	71.43
	1FKB	30.99	28.96	7.48	38.32	22.43	5.61	2.80	1.87	21.50	10.28	2.80	33.64	63.55	8.61	33.15	58.92	73.23	76.64
	1IMQ	16.14	18.77	52.33	0.00	11.63	11.63	0.00	0.00	24.42	52.33	40.70	5.81	53.49	38.60	10.73	51.59	79.38	82.56
	2ABD	17.03	19.80	56.98	0.00	10.47	8.14	3.49	0.00	20.93	60.47	54.65	0.00	45.35	50.22	8.33	42.34	76.71	81.40
	2AIT	27.62	37.32	0.00	40.54	8.11	16.22	0.00	2.70	32.43	0.00	12.16	36.49	51.35	15.25	36.12	49.67	77.62	67.57
	2PTL	20.53	33.11	19.35	38.71	9.68	14.52	0.00	1.61	16.13	19.35	24.19	41.94	33.87	24.46	38.45	38.30	75.24	85.48
	2PDD	6.45	15.00	44.19	0.00	16.28	9.30	0.00	0.00	30.23	44.19	34.88	16.28	48.84	32.75	23.04	45.88	71.79	74.42
	2VIK	25.87	20.53	19.05	23.81	15.08	16.67	2.38	0.00	23.02	21.43	24.60	23.02	52.38	24.16	27.98	48.43	71.23	77.78

[a]The secondary structure contents were obtained with the program (DSSP) [69], and the secondary structure predictions and propensities were obtained with the program PRED2ARY [70] (these descriptors are expressed as percentages of the total numbers of residues). Each of the mutations involved the substitution of an alanine into a helix; because such a change is likely to increase the propensity for forming a helix in that region, the contact orders and secondary structure content were taken to be the same as those of the wild types, and the secondary structure content and predictions were calculated with the modified sequences. Likewise, the structural data for the two forms of horse cytochrome c (1HRC) were taken to be the same. A contact was defined as two heavy atoms that are within 4 Å of each other and separated by at least two residues (i.e., $i, i+1$ and $i, i+2$ contacts are ignored). The (unnormalized) contact order is $c = \frac{1}{n_c} \sum_{i>j} \Delta(i,j)|s_i - s_j|$, where n_c is the total number of contacts, s_i is the sequence position of the residue containing atom i, and $\Delta(i,j)$ selects the atoms (i and j) that are in contact (as defined above). The normalized contact order (c/n) is multiplied by 100 for consistency with Refs. 12 and 13.

correlated with others (Table IV), consideration of all of them is useful because exhaustive enumeration or a genetic algorithm (GA) is employed to determine which to include for optimal fitting and prediction.

The database consists of 33 proteins. Twenty-four of these fall into six structurally related groups, and nine are structurally unique. The former are SH3 domains [1NYF (82 to 148), 1PKS, 1SHG, and 1SRL], Ig-like β-sandwiches [1FNF (1326 to 1415), 1FNF (1416 to 1509), 1HNG, 1TEN (802 to 891), 1TIT, and 1WIT], members of the acylphosphatase family (1APS, 1HDN, 1PBA, 1URN, and 2HQI), cytochromes (1HRC, 1HRC-oxidized, 1YCC), cold shock proteins [1CSP and 1MJC (2 to 70)], λ-repressor variants (1LMB wild type and G46A/G48A), and ubiquitin variants (1UBQ wild type and V26A). The remainder of the proteins are 1COA (20 to 83), 1DIV (1 to 56), 1FKB, 1IMQ, 2ABD, 2AIT, 2PDD, 2PTL (94 to 155), and 2VIK. Numbers in parentheses indicate the residue numbers of the domain or fragment studied.

To cross-validate the results, each group of structurally related proteins is left out of the training set in turn and used to test the network. Such a partitioning scheme (in contrast to a jackknife one, for example) minimizes the likelihood of biasing the results in favor of structural descriptors (see Section II). Its use yields true predictions (denoted "cv") in contrast to fits of the data, in which all the proteins are included during the training (denoted "trn"). The latter tend to yield inflated accuracy statistics, but we describe them here as well for comparison with earlier studies [12,13,20,47], which failed to cross-validate their results [however, it should be noted that the relationship in Ref. 12 has been used successfully for blind predictions (K. W. Plaxco and D. Baker, personal communication)].

C. Single-Descriptor Models

We begin by examining the relationship between $\log k_f$ and each individual descriptor.

1. Linear Correlations

The first column of statistics given in Table I contains the Pearson linear correlation coefficients between the descriptor values (x) and $\log k_f (r_{x,\log k_f})$. This is the statistical measure used by Plaxco et al. in their analysis of a subset of the descriptors considered here [12,14]. Consistent with their results, the two coefficients with the largest magnitudes are associated with the contact order (c and c/n). Several descriptors not examined by Plaxco et al. [12,14] exhibit $|r_{x,\log k_f}| > 0.5$ as well: the α-helix content and propensity (h and p_h), total helix content (a), and β-sheet content (e). Additional linear statistics are provided in Table V. Physical interpretations of the results are given in Section IV.E.

2. Neural Network Predictions

The second and third columns of statistics in Table I measure the ability of a single-input neural network to predict the folding rate. They contain Pearson

TABLE IV

Descriptor–Descriptor Pearson Linear Correlation Coefficients[a]

	ΔG	ΔG/n	m	m/n	n	n_c	c	c/n	h	e	t	s	g	b	o	a	P_h	P_e	P_o	p_h	p_e	p_o	q_e	q_a
0 ΔG		0.94	0.28	0.21	0.50	0.37	-0.02	-0.29	0.24	-0.41	0.26	-0.05	-0.34	0.02	0.28	0.20	0.01	-0.17	0.16	0.06	-0.18	0.10	-0.04	-0.17
1 ΔG/n	0.93		0.40	0.26	0.20	0.16	-0.16	-0.25	0.20	-0.39	0.19	0.03	-0.25	0.22	0.30	0.16	-0.01	-0.07	0.09	0.05	-0.08	0.00	-0.02	-0.02
2 m	0.28	0.43		0.98	0.39	0.31	0.12	-0.09	0.02	-0.06	-0.05	-0.01	-0.27	-0.08	0.22	-0.01	-0.11	0.04	0.14	-0.13	0.08	0.14	0.20	0.06
3 m/n	0.21	0.40	0.95		0.12	0.13	0.01	-0.04	-0.03	-0.02	-0.12	0.06	-0.16	0.06	0.20	-0.05	-0.12	0.13	0.06	-0.13	0.16	0.04	0.21	0.21
4 n	0.21	0.20	-0.10	-0.27		0.73	0.48	-0.10	0.17	-0.15	0.27	-0.17	-0.40	-0.47	0.07	0.12	0.02	0.13	0.23	0.02	0.16	0.04	-0.12	0.21
5 n_c	-0.19	-0.38	-0.20	-0.28	0.60		0.17	-0.28	0.50	-0.44	0.14	-0.15	-0.34	-0.42	-0.12	0.46	0.34	-0.52	-0.03	0.34	-0.55	0.02	0.00	-0.19
6 c	0.11	0.00	-0.34	-0.43	-0.34	-0.02		0.81	-0.63	0.66	-0.03	0.29	-0.28	-0.22	0.18	-0.67	-0.49	0.51	0.28	-0.56	0.50	0.44	-0.26	-0.46
7 c/n	-0.08	-0.06	-0.10	-0.08	-0.68	-0.54	0.43		-0.84	0.83	-0.22	0.47	-0.10	0.08	0.24	-0.86	-0.61	0.77	0.20	-0.68	0.77	0.35	-0.18	-0.26
8 h	-0.01	-0.01	0.13	0.12	0.08	0.39	-0.39	-0.84		-0.89	-0.01	-0.22	-0.11	-0.34	-0.41	0.99	0.83	-0.82	-0.51	0.88	-0.88	-0.60	0.23	0.24
9 e	-0.18	-0.12	-0.35	-0.36	-0.15	-0.39	0.44	0.45	-0.75		0.00	-0.01	0.02	0.12	0.00	-0.89	-0.64	0.77	0.26	-0.71	0.80	0.38	-0.11	-0.18
10 t	-0.15	-0.26	0.01	-0.12	0.29	0.34	-0.03	-0.27	0.08	0.00		-0.70	0.20	0.32	-0.27	0.01	-0.17	0.01	0.26	-0.11	0.00	0.21	-0.23	-0.14
11 s	0.30	0.34	0.01	0.06	-0.17	-0.15	0.29	0.47	-0.22	-0.01	-0.70		-0.38	-0.25	0.45	-0.27	-0.20	0.32	0.00	-0.23	0.31	0.06	0.01	0.02
12 g	-0.05	-0.15	-0.14	-0.12	0.17	0.19	-0.11	-0.28	-0.01	-0.31	0.02	-0.55		0.43	0.26	0.02	0.07	-0.09	-0.03	0.09	-0.06	-0.10	-0.29	0.02
13 b	-0.04	0.08	-0.05	-0.02	-0.26	0.00	0.03	-0.42	-0.47	0.09	-0.16	0.42	0.49		0.26	-0.29	-0.32	0.31	0.20	-0.27	0.32	0.13	-0.22	-0.03
14 o	0.26	0.33	0.29	0.28	0.05	-0.26	0.03	0.30	-0.47	0.09	-0.27	0.45	0.08	0.43		-0.42	-0.50	0.22	0.59	-0.50	0.31	0.59	-0.10	-0.23
15 a	-0.02	-0.10	0.29	0.28	-0.13	-0.42	0.08	0.23	0.97	-0.81	0.16	0.07	0.22	-0.34	-0.70		0.84	-0.84	-0.51	0.89	-0.89	-0.62	0.19	0.27
16 P_h	0.23	0.19	-0.20	-0.19	0.11	0.04	-0.45	-0.38	0.68	-0.66	0.05	0.11	0.33	-0.19	-0.41	0.74		-0.80	-0.80	0.99	-0.86	-0.83	0.27	0.23
17 P_e	0.22	0.32	0.31	0.33	0.04	0.13	0.31	0.43	-0.33	0.33	-0.13	0.25	-0.62	0.16	0.26	-0.46	-0.69		0.29	-0.83	0.98	0.38	-0.18	-0.03
18 P_o	-0.51	-0.53	0.01	-0.01	-0.17	0.13	0.35	0.16	-0.66	0.63	0.03	-0.36	0.07	0.13	0.35	-0.62	-0.79	0.10		-0.76	0.41	0.96	-0.25	-0.34
19 p_h	0.21	0.16	0.42	0.43	0.08	0.15	-0.36	-0.33	0.74	-0.67	0.10	0.10	0.32	-0.24	-0.51	0.79	0.98	-0.68	-0.77		-0.89	-0.82	0.21	0.22
20 p_e	0.12	0.22	0.42	0.43	0.06	-0.08	0.24	0.37	-0.55	0.48	-0.22	0.25	0.06	0.19	0.38	-0.67	-0.80	0.47	0.31	-0.83		0.48	-0.15	-0.06
21 p_o	-0.47	-0.48	0.06	0.04	0.05	-0.08	0.35	0.19	-0.69	0.63	0.09	-0.41	-0.22	0.21	0.12	-0.65	-0.84	0.21	0.97	-0.84	0.48		-0.21	-0.36
22 q_e	0.18	0.26	0.53	0.53	-0.21	-0.34	-0.42	-0.14	0.08	0.06	-0.48	0.14	-0.22	-0.13	0.12	0.03	0.03	-0.02	-0.02	0.00	0.06	-0.06		0.37
23 q_a	0.20	0.15	0.42	0.39	0.18	0.21	-0.55	-0.58	0.29	-0.20	0.09	0.14	-0.30	-0.15	-0.19	0.21	0.11	0.17	-0.30	0.05	0.16	-0.24	0.54	

[a]Data for all 33 proteins are above the diagonal, and data for the 14 proteins with $c > 21$ are below the diagonal.

17

TABLE V
Linear Regression Statistics for $\log k_f$

Index	Symbol	r_{trn}	r_{cv}	q_{cv}^2
0	ΔG	0.29	-0.02	-0.09
1	$\Delta G/n$	0.37	0.13	-0.05
2	m	0.04	-0.65	-0.19
3	m/n	0.04	-0.52	-0.20
4	n	0.10	-0.53	-0.27
5	n_c	0.08	-0.60	-0.24
6	c	0.73	0.70	0.48
7	c/n	0.79	0.77	0.59
8	h	0.63	0.55	0.30
9	e	0.67	0.59	0.34
10	t	0.04	-0.76	-0.23
11	s	0.11	-0.52	-0.19
12	g	0.01	-0.75	-0.41
13	b	0.15	-0.43	-0.26
14	o	0.05	-0.74	-0.31
15	a	0.63	0.57	0.32
16	P_h	0.47	0.29	0.06
17	P_e	0.48	0.31	0.08
18	P_o	0.27	-0.27	-0.28
19	p_h	0.51	0.37	0.13
20	p_e	0.47	0.28	0.05
21	p_o	0.40	0.07	-0.09
22	q_e	0.21	-0.21	-0.14
23	q_a	0.40	0.12	-0.07

linear correlation coefficients (r_{trn} and r_{cv}) between observed and calculated values of $\log k_f$; thus, only positive values of r are significant. Because there are only 24 different input possibilities, it is feasible to consider each one in turn, so that use of a genetic algorithm is not necessary at this stage. However, the NN weights depend on the random number generator seed through the training procedure. Consequently, for each descriptor, the network was trained independently with ten different seeds. The maximum correlation coefficient for each set of 10 networks corresponding to a particular descriptor is listed in Table I; the average standard deviation for a given descriptor was 0.03 for r_{trn} and 0.06 for r_{cv}.

As stated above, the coefficients denoted "trn" are for results obtained with networks trained on all 33 proteins; in other words, they are not true predictions since all the data are included in the training set. For descriptors that are linearly related to $\log k_f$, r_{trn} is expected to be comparable in magnitude to $r_{x,\log k_f}$ (in fact, for linear regression, $r_{trn} = |r_{x,\log k_f}|$), whereas, for ones that are non-linearly related, it should be higher. Thus, r_{trn} can be viewed as essentially a nonlinear version of the statistic employed in Ref. 12. Accordingly, most of the descriptors that exhibit high r_{trn} were included in the analysis of $r_{x,\log k_f}$.

The coefficients denoted "cv" are for the predictions obtained with the structurally based cross-validation scheme. Negative values of r_{cv} indicate that the accuracy of the network is lower than that which would be obtained from random guesses. If a network fails in this way when confronted with novel test data, it has derived a spurious relationship by memorizing the information in the training set at the expense of learning more general rules. The highest r_{cv} do correspond to the highest r_{trn}, but overall the cross-validated coefficients are much lower. The large differences between r_{trn} and r_{cv} in many cases (Table I) make clear that the former is a relatively indiscriminate statistic for such a small database. If linear regression is used, r_{trn} and r_{cv} are often closer due to the decreased flexibility of the fitting method (Table V). However, such an approach fails to identify nonlinear relationships and can hide complexities in the results.

In summary, the contact order yields relatively good prediction of log k_f but is not alone in doing so. Several measures of the propensity of the sequence for a given structure also exhibit significant relationships with the folding rate. Although r_{cv} values for the various descriptors obtained from the secondary structure prediction program (indices 16 to 21 in Table I) are lower than those for measures of the known native structure (indices 6 to 15), the former correlations may be sufficiently high that the calculated descriptors could be used to identify particularly fast or slow proteins without the need for high-resolution structures. The stability, which has been suggested to be of importance based on model studies, exhibits no clear relation to the folding rate. An essential additional point of the single-descriptor analysis is that large differences are observed between most of the values obtained with and without cross-validation. This highlights the need for care in assessing the significance of correlations when working with small numbers of sequences.

D. Multiple-Descriptor Models

We present results for two- and three-descriptor models; addition of a fourth descriptor yielded no significant improvement in predictive accuracy. In the two-descriptor case there are only 276 possible input combinations, so we examine each explicitly, whereas, in the three-descriptor case there are 2024, so we use the genetic algorithm (GA) to optimize the descriptor selection. Use of the GA in the two-descriptor case gives models of comparable quality to the exhaustive search, but this test of the algorithm is not very stringent because the space of input combinations is small. Because both the GA and the NN depend on the random number generator seed, several trials were performed in each case (as detailed in Section IV.D.2).

1. Two Descriptors

The best five two-descriptor models are shown in Table VI, and selected examples to illustrate the types of behavior that are observed are shown in Fig. 2.

There is a significant increase in fitting ability (training) and, more importantly, in predictive accuracy (cross-validation) upon adding a second descriptor. In Figure 2, we see that the squares (□) tend to be closer to the ideal line than the circles (○), particularly for lower log k_f (slower-folding proteins). To quantitate the improvement, we calculated Wold's E statistic from the q_{cv}^2 values (Table VI). While these figures suggested to us that the additional descriptors significantly improve the accuracies of the cross-validated predictions, general confidence limits are not straightforward to calculate. Consequently, we did the following. We shuffled the values of each secondary descriptor (other than c/n) 10 times and then trained neural networks to predict log k_f as for the actual data. Averages and standard deviations of the correlation coefficients are reported in Table VII. We see that, even though the r_{trn} values are comparable to those in Table VI, the

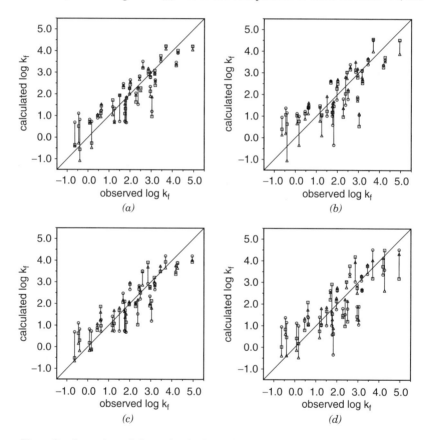

Figure 2. Comparison of observed and calculated values of log k_f for selected models. (a and b) $c/n(○)$; c/n and $\Delta G/n$ (□); and c/n, $\Delta G/n$ and $p_e(\triangle)$. (c and d) $c/n(○)$; c/n and n_c (□); and c/n, n_c, and $\Delta G(\triangle)$. (a and c) Training set fits. (b and d) Cross-validated predictions.

TABLE VI

The Best (as Measured by r_{cv}) Five Two-Descriptor Models Obtained by Examining All Possible Combinations for Ten Different Random Number Generator Seeds[a]

Descriptors		r_{trn}	r_{cv}	q^2_{cv}	E
c/n	$\Delta G/n$	0.89	0.81	0.66	0.74
c/n	P_h	0.87	0.80	0.63	0.81
c/n	n_c	0.89	0.79	0.62	0.82
c/n	p_h	0.86	0.77	0.57	0.93
c/n	q_a	0.84	0.77	0.59	0.89

[a]For the calculation of E, q^2_{cv} was compared with that for c/n. Statistics for linear regression and additional measures of the predictive accuracy are available in Tables VII and VIII.

r_{cv} values are close to that for c/n by itself (Table I); the NN ignores the randomized descriptor. The fact that the r_{cv} values for the actual data are two to four standard deviations above the average r_{cv} values for the randomized data demonstrates that the improvement is significant and is not due to the increase in the number of fitting parameters.

The best predictions are obtained with $\Delta G/n$ paired with c/n (ΔG with c is the sixth best set of inputs with $r_{cv} = 0.77$ and $E = 0.76$) This combination of input descriptors was investigated previously [15], but it is of interest that it ranks first in the exhaustive search performed here. To better understand the physical basis for the correlations, we show the dependence of log k_f on c/n and $\Delta G/n$ in Fig. 3a. When c/n is small ($c/n \leq 19$; mainly α-helical proteins), folding is always fast ($k_f > 400\,s^{-1}$), whereas when c/n is large ($c/n \geq 25$; either mixed-α/β or β-sheet proteins), the rate spans over three orders of magnitude. Thus, proteins with lower contact orders fold fast regardless of their stabilities, whereas for those with higher contact orders, the rate increases with $\Delta G/n$. As described in Ref. 15, a single-input neural network can be trained to predict log k_f from ΔG for the 14 proteins with $c > 21$ (Fig. 4); $r_{trn} = 0.81$, and $r_{cv} = 0.64$, which confirms that stability plays a significant role in determining the folding rates of mixed-α/β and β-sheet proteins. For these 14

TABLE VII

Randomization Tests for the Models in Table VI[a]

Descriptors		r_{trn}	r_{cv}	q^2_{cv}
c/n	$\Delta G/n$	0.83 ± 0.01	0.71 ± 0.03	0.49 ± 0.04
c/n	P_h	0.84 ± 0.03	0.68 ± 0.07	0.43 ± 0.12
c/n	n_c	0.87 ± 0.02	0.69 ± 0.04	0.46 ± 0.05
c/n	p_h	0.84 ± 0.02	0.68 ± 0.06	0.42 ± 0.10
c/n	q_a	0.84 ± 0.00	0.68 ± 0.07	0.44 ± 0.11

[a]In each case, the second descriptor listed was shuffled 10 times, and the networks were trained as for the original data. Values shown are averages for the 10 trials; ranges indicate standard deviations.

TABLE VIII
Linear Regression Statistics for the Models in Table VI

Descriptors		r_{trn}	r_{cv}	q_{cv}^2	E
c/n	$\Delta G/n$	0.81	0.72	0.47	1.27
c/n	P_h	0.79	0.75	0.57	1.04
c/n	n_c	0.82	0.79	0.62	0.92
c/n	p_h	0.79	0.75	0.56	1.05
c/n	q_a	0.80	0.77	0.60	0.97

proteins, $r_{\Delta G, \log k_f} = 0.80$ while $r_{c, \log k_f} = -0.22; E = (1 - q_{c,\Delta G}^2)/(1 - q_c^2)$ $= 0.23$.

Two of the other models in Table VI combine the contact order with a measure of the α-helical propensity: c/n with either P_h or p_h. These pairings essentially reflect the results of the previous section. The remaining model couples c/n with n_c, which reveals a secondary dependence on protein size. Consistent with the sign of $r_{n_c, \log k_f}$ (Table I), the functional dependences of $\log k_f$ on these descriptors for the models in Table VI indicate that shorter proteins fold faster than longer ones (Fig. 3b).

2. Three Descriptors

As mentioned above, there are 2024 possible combinations of three descriptors, so we use a GA to identify the inputs that are likely to yield the greatest predictive accuracy. Use of the GA requires selection of a particular measure of predictive accuracy to decide which models to keep at each cycle. Because we are interested primarily in cross-validated predictions, r_{cv} is a natural choice. However, the structurally based partitioning scheme is less straightforward to automate than a jackknife one. Consequently, for the GNN, we used the Pearson linear correlation coefficient for the jackknife cross-validated outputs (r_{jck}) and subsequently tested each selected combination of descriptors with the structurally based cross-validation scheme (r_{cv}). We performed five GNN trials, from each of which we saved the best 20 models. Of these 100 models, 46 were unique, and each of these was subjected to 10 trials with the structurally based cross-validation scheme.

In general, the GA combines the descriptors that were identified above by the two-dimensional exhaustive search ($c, c/n, \Delta G, \Delta G/n,$ and n_c) to further refine the predictions (Tables IX to XI and Fig. 2). The propensity for sheet structure (p_e) appears in two of the five models; not surprisingly, it is strongly anti-correlated with the propensity for helical structure, which appeared in Table VI ($r_{p_e, p_h} = -0.89$). In considering these results, it is necessary to keep in mind that the database is small, so that there is a danger of overfitting (but see Table X). Nevertheless, given this disclaimer, we see that simultaneous consideration of multiple descriptors improves prediction of the folding rate and that both the

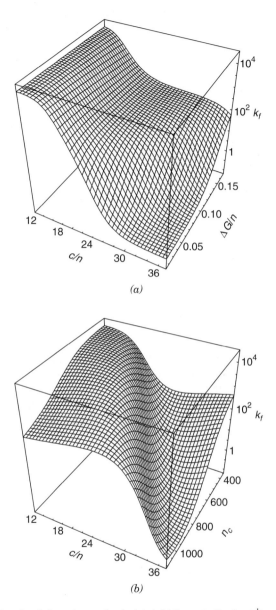

Figure 3. Functional dependence of calculated folding rate (k_f, in s^{-1}) on the normalized contact order (c/n) and either (a) the normalized stability ($\Delta G/n$ in kcal/mol) or (b) the total number of atomic contacts (n_c).

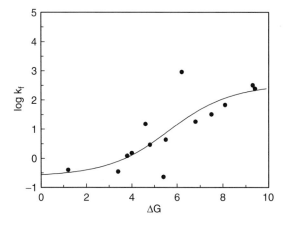

Figure 4. Observed (points) and calculated (line) log k_f as a function of the stability in kcal/mol for the 14 proteins in the database with $c > 21$.

size and the stability play significant secondary roles that could not have been anticipated from the single-descriptor analyses.

E. Physical Bases of the Observed Correlations

Consistent with earlier, single-descriptor linear analyses of protein folding [12,13,50], the primary determinants of the folding rate are measures that characterize the native structure; that is, proteins with more sequentially local interactions tend to fold faster. As discussed below, the equilibrium structure and the kinetics are connected by the fact that the structure of the transition state resembles that of the native state in many small proteins [50]. Thus, the kinetics and the underlying thermodynamics of the reaction are affected in a similar way, in accord with linear free energy relations.

The microscopic origin for the statistical dependence of the folding kinetics on the structure is the stochastic diffusive search that is required to find the

TABLE IX
The Best (as Measured by r_{cv}) Five Unique Three-Descriptor Models Obtained from the GNN Protocol for Ten Different Random Number Generator Seeds[a]

Descriptors			r_{trn}	r_{jck}	r_{cv}	q_{cv}^2	E
c/n	$\Delta G/n$	p_e	0.92	0.84	0.86	0.74	0.76
c/n	ΔG	n_c	0.93	0.84	0.84	0.70	0.80
c/n	$\Delta G/n$	n_c	0.92	0.81	0.83	0.67	0.97
c/n	ΔG	c	0.90	0.83	0.83	0.66	0.81
c/n	ΔG	p_e	0.91	0.80	0.83	0.67	0.72

[a]For the calculation of E, q_{cv}^2 was compared with the highest observed q_{cv}^2 of the six possible two-descriptor models that could be formed from the three selected inputs (corresponding to the unshuffled pair in Table X). Statistics for linear regression and additional measures of the predictive accuracy are available in Table X and XI.

TABLE X
Randomization Tests for the Models in Table IX

Descriptors			Randomized	r_{trn}	r_{cv}	q_{cv}^2
c/n	$\Delta G/n$	p_e	p_e	0.89 ± 0.02	0.80 ± 0.03	0.61 ± 0.07
c/n	ΔG	n_c	ΔG	0.88 ± 0.02	0.72 ± 0.05	0.48 ± 0.10
c/n	$\Delta G/n$	n_c	n_c	0.89 ± 0.01	0.74 ± 0.04	0.49 ± 0.09
c/n	ΔG	c	c/n	0.89 ± 0.01	0.71 ± 0.04	0.46 ± 0.08
c/n	ΔG	p_e	ΔG	0.88 ± 0.01	0.69 ± 0.06	0.41 ± 0.10

transition state. As described in the formulation of the "hydrophobic zipper hypothesis" [51,52] and in the statistical analyses of 125-residue lattice models [10,11], having sequentially short-range contacts in the transition state should increase the folding rate for two reasons. First, such contacts are found more readily because there are fewer conformations to search (the number grows exponentially with loop length). Second, making sequentially long-range contacts costs more entropy because they constrain the chain to a greater degree. These advantages correspond to different components of the macroscopic rate law $[k_f = A(T)\exp(-\Delta G/k_B T)]$. In this regard, it is necessary to keep in mind that the preexponential factor can be nontrivial for protein folding [53,54]. If $A(T)$ is sufficiently large, there is a separation of time scales; the protein reaches an effective equilibrium within the unfolded state rapidly, and the rate is dominated by the time required to surmount the barrier [55]. In this case, the observed statistical dependence on the structure implies that the barrier is entropic (as in Fig. 3a of Ref. 1 and Figs. 6 and 7 of Ref. 36). Based on these ideas, Fersht recently derived a simple relationship to show that changes in contact order are directly proportional to changes in log k_f [50]. On the other hand, if $A(T)$ is small, there is no separation of time scales. Because a dependence on the structure enters through the preexponential factor in this case, the barrier, if there is one, could be either entropic or energetic (as in Fig. 3b of Ref. 1).

Free energy surfaces for folding have now been determined for high-resolution (all-atom) models of several peptides and proteins [72–77]. For both α-helical and β-hairpin peptides, decomposition of the surfaces into contributions from the effective energies (which include the full solvent free

TABLE XI
Linear Regression Statistics for the Models in Table IX

Descriptors			r_{trn}	r_{cv}	q_{cv}^2	E
c/n	$\Delta G/n$	p_e	0.83	0.71	0.46	1.57
c/n	ΔG	n_c	0.84	0.73	0.46	1.42
c/n	$\Delta G/n$	n_c	0.84	0.76	0.55	1.29
c/n	ΔG	c	0.83	0.71	0.41	1.40
c/n	ΔG	p_e	0.82	0.69	0.38	1.34

energies) and configurational entropies indicated that the free energy barriers derive primarily from the fact that the entropy decreases more rapidly than the energy [75–77], as in Ref. 36 discussed above. However, consistent with the statistical analyses of proteins, differences in secondary structure content correspond to differences in the general shapes of the free energy surfaces. For α-helical sequences, the transition states tend to be less folded, and secondary and tertiary structure form concurrently [72,77]. For peptides and proteins which contain β-hairpins and β-sheets, a collapse to a native-like radius of gyration occurs first, and rearrangement to the native state follows wihout significant expansion [73–75]. At least for peptides at elevated temperatures [76,77], determination of the rate of diffusion on the free energy surfaces, which relates directly to the pre-exponential factor in the rate law [53], should now be possible but has not been done and would be of interest.

In connecting these ideas with earlier phenomenological models, it is not obvious how to reconcile the dependence of the rate on the structure with a nucleation mechanism, as in Ref. 50. The statistical relationship suggests that the transition state contains a considerable amount of native structure, while a nucleus, in the classic sense of the word, is a small part of the structure. However, it could be that a limited number of native contacts (i.e., those in the nucleus) are sufficient to confine the transition state ensemble to a native-like fold. This idea is supported by a recent analysis of the folding transition state of acylphosphatase in which key residues, as determined by a φ value analysis, play a critical role [56].

V. UNFOLDING RATES OF PROTEINS

To function, a protein must not only fold (kinetic criterion) but populate its native state for a significant fraction of the time (thermodynamic criterion). The unfolding rate (k_u) as well as k_f contribute to the equilibrium constant, which determines to what degree the latter condition is satisfied. To find the factors that affect the unfolding rate, we carried out an analysis for $\log k_u$. Rate data for unfolding in water were not available for three of the proteins (2HQI, 1YCC, and 1HRC-oxidized), so these were excluded from the analysis; the choice of descriptors was the same.

For single-descriptor models, the best cross-validated predictions are obtained with the contact order (c and c/n), the free energy of unfolding (ΔG and $\Delta G/n$), and the buried surface area (m) (Table XII). The strong dependence of the unfolding rate on the contact order for these proteins is somewhat surprising because no significant correlation was observed in a previous study of a database of 24 proteins [14], 19 of which are included here. For those 19 proteins we have $r_{\Delta G, \log k_u} = -0.61$, $r_{c, \log k_u} = -0.56$ and $r_{c/n, \log k_u} = -0.45$, whereas for the 11 additional proteins included in the present analysis of the unfolding rate we have $r_{\Delta G, \log k_u} = -0.64$, $r_{c, \log k_u} = -0.85$, and $r_{c/n, \log k_u} = -0.83$. The proteins

TABLE XII
Single-Input Correlations for Unfolding Rates

Index	Symbol	$r_{x, \log k_u}$	r_{trn}	r_{cv}	q_{cv}^2
0	ΔG	-0.64	0.69	0.53	0.21
1	$\Delta G/n$	-0.45	0.55	0.40	0.12
2	m	-0.41	0.61	0.45	0.14
3	m/n	-0.31	0.36	0.08	-0.11
4	n	-0.43	0.58	0.09	-0.09
5	n_c	-0.40	0.53	0.09	-0.05
6	c	-0.68	0.77	0.67	0.44
7	c/n	-0.58	0.69	0.52	0.20
8	h	0.40	0.49	-0.57	-0.86
9	e	-0.34	0.53	0.16	-0.06
10	t	-0.01	0.39	-0.25	-0.12
11	s	-0.08	0.26	-0.19	-0.24
12	g	0.03	0.36	-0.16	-0.32
13	b	-0.27	0.27	-0.19	-0.23
14	o	-0.20	0.55	0.15	-0.08
15	a	0.40	0.50	-0.27	-0.27
16	P_h	0.29	0.53	-0.64	-0.32
17	P_e	-0.28	0.30	-0.38	-0.47
18	P_o	-0.20	0.52	-0.22	-0.20
19	p_h	0.29	0.50	-0.31	-0.42
20	p_e	-0.23	0.50	-0.38	-0.40
21	p_o	-0.27	0.49	-0.56	-0.11
22	q_e	0.14	0.35	-0.11	-0.14
23	q_a	0.24	0.48	0.19	-0.06

that appear to be primarily responsible for decreasing the correlation with the free energy of unfolding and increasing the correlation with the contact order are the helical proteins—in particular, 2PDD and 1LMB. Because for the 30 proteins considered in this section there is no significant correlation between the contact order and either the free energy of unfolding ($r_{\Delta G,c} = 0.28$) or the amount of buried surface area ($r_{m,c} = 0.23$), higher predictive accuracy is obtained by combining these descriptors (Table XIII). Only a slight improvement was observed upon adding a third descriptor.

We end this section by noting that, for these 30 proteins, there is a significant correlation between the folding and unfolding rates ($r_{\log k_f, \log k_u} = 0.59$). At least in the case that k_f and k_u are determined by an entropic barrier (Section IV.E), this relationship can be understood in the following way. Because all the proteins are roughly the same size, the stability of the native state does not depend on contact order (for the overall reaction, $\Delta S \propto n$). Changes to c that raise or lower the free energy of the transition state (TS) relative to the fixed endpoints (U and F) will change Δ_{U-TS} and ΔG_{F-TS} in the same manner. This dependence of the activation free energies is the basis not only for the correlation of $\log k_u$ with $\log k_f$ but also that with c.

TABLE XIII
The Best (as Measured by r_{cv}) Five Two-Descriptor Models for the Unfolding Rates

Descriptors		r_{trn}	r_{cv}	q_{cv}^2	E
c	$\Delta G/n$	0.90	0.85	0.71	0.53
c	ΔG	0.88	0.81	0.66	0.62
c/n	ΔG	0.89	0.80	0.61	0.49
c	m	0.83	0.73	0.53	0.85
c	m/n	0.90	0.71	0.49	0.92

VI. HOMOLOGOUS PROTEINS

Information about the transition state of a protein can be obtained from protein engineering experiments in which one compares the effects of mutations on the folding rate to their effects on the overall stability (ϕ values). Several proteins have been mutated extensively, and their kinetics have been measured. The fact that proteins with related structures but low sequence homologies are found to have similar transition states has been taken to support the relation between native structure and folding behavior; this is the case for the transition states of the src [57] and α-spectrin [58] SH3 domains, which have 36% sequence homology. A particularly interesting transition state comparison involves acylphosphatase (AcP) [59] and procarboxypeptidase A2 [60]. These two proteins fold to sandwich structures with two α-helices packed against a five- or four-stranded antiparallel sheet, respectively. Although their sequences have only 13% identity, the average ϕ values for all elements of secondary structure (except one, β-strand 4) are almost the same. Moreover, it has been suggested that the reason that procarboxypeptidase A2 folds about 4000 times faster than AcP is that the transition state of the latter involves longer loops and secondary structure elements; consistent with this observation, there is a strong correlation between $\log k_f$ and the contact order for proteins with this fold [59].

The dependence of the folding rate on the stability can be evaluated by measuring the kinetics of a family of proteins with native states that have similar structures but different ΔG values. Such an analysis was made recently for a set of six immunoglobulin-like β-sandwich domains [61]. They have stabilities that are distributed relatively uniformly over the range $1.2 \leq \Delta G \leq 9.4$ kcal/mol (in contrast to the AcP family discussed above, for which four of the five members have $3.8 \leq \Delta G \leq 5.4$ kcal/mol). Although there is some variation in the detailed structures of these six proteins, using the definition of the contact order given in Section II, all of them have $c/n > 28$ (for these six, $28.22 \leq c/n \leq 32.53$; for the five members of the AcP family, $25.83 \leq c/n \leq 35.08$; for all 33 proteins, $12.21 \leq c/n \leq 37.32$). In accord with the functional dependence on ΔG shown in Figs. 3a and 4, a strong positive correlation between $\log k_f$ and ΔG was observed for this family ($r_{\Delta G,\log k_f} = 0.99$). The data

TABLE XIV
Relation Between Stability and Folding Rate for Six Two-State Proteins That
Have Been Mutated Extensively[a]

Protein	Reference	c	Number of Mutants	$r_{\Delta G, \log k_f}$	r_{trn}	r_{jck}
Acylphosphatase	59	34.4	25	0.614	0.667	0.386
Procarboxypeptidase A2	60	20.7	19	0.531	0.712	0.464
src SH3	57	20.5	58	0.552	0.556	0.408
α-Spectrin SH3	58	18.0	18	0.481	0.476	0.099
CI2	71	16.1	86	0.554	0.606	0.519
λ-Repressor	64	9.8	9	0.720	0.760	0.307

[a]The coefficients r_{trn} and r_{jck} are for single-input (ΔG) neural networks. The α-spectrin SH3 domain values are those for pH 7; the src SH3 domain values are for pH 6. The λ-repressor values are for 2 M urea.

suggest that for a given structural family with significant variation in ΔG, the folding rates of individual sequences are determined by their stabilities.

This conclusion is consistent with the fact that both log k_f and log k_u typically vary linearly with the stability of the native state as a protein is mutated. Such Brønsted behavior has been used in protein engineering studies to argue that fractional φ values derive from partial structure formation rather than multiple parallel folding pathways [62]. Correlation coefficients for published folding rates of mutants of six two-state proteins are given in Table XIV. For the most part, there is a strong, essentially linear relation that is reasonably robust to jackknife cross-validation. For all the sequences, increases in stability tend to accelerate folding. Similar behavior is obtained simply by varying the conditions to affect the stability of a protein (for example, see Fig. 2a of Ref. 14). This analysis thus confirms that the stability is an important secondary factor in determining folding rate. As described in Ref. 9, in accord with the Hammond postulate [34], stabilizing the native state of a protein in most cases also lowers the energy of its transition state relative to the unfolded state and thus increases the folding rate.

VII. RELATING PROTEIN AND LATTICE MODEL STUDIES

The fact that the folding (and unfolding) kinetics of relatively small, two-state proteins can be predicted with reasonable accuracy from global features of the native state like the contact order, stability, and number of contacts supports the idea that the details of protein structure are not required to capture the key features of protein folding, so that reduced representations should be adequate. However, the most widely used simple heteropolymer models, those restricted to a simple cubic lattice, predict that stability is more important than native structure, in contrast to the experimental data for proteins. In this section we seek to understand why lattice models differ from proteins in this regard. Doing so is of importance because complete details of the folding trajectories of such models can be obtained and used to test phenomenological models like those described in Section IV.A.

In the case of the 27-residue model described in Section III [6,9], it is likely that the chain length is too short for there to be contacts that are sufficiently long range to slow-folding. In the case of the 125-residue model, which is larger than all but one (2VIK) of the proteins considered in the present study, significant correlations between various measures that characterize the native structure and the folding behavior were observed [10,11] (it should be mentioned that, in contrast to the number of antiparallel sheet contacts discussed in Section III, the contact order is a poor measure for characterizing lattice model structure; $18.7 \leq c \leq 31.0$ for the 100 helical proteins in Refs. 10 and 11, whereas $17.2 \leq c \leq 32.0$ for the 100 sheet proteins). However, in the lattice model, the functional dependence of the folding stability is essentially the same regardless of the native structure; at a particular threshold value of the stability (which varies only slightly with the number of antiparallel sheet contacts), the folding ability rises rapidly and then levels off [11,37]. There are two likely reasons that the functional dependence is much simpler than that for proteins (Fig. 3a). First, the 125-residue sequences were energetically optimized to observe folding on the time scale of feasible simulations and are thus expected to correspond to the more stable region in Fig. 3a. Second, due to the highly restricted conformational space of the lattice and the choice of move set, helices that form in isolation cannot diffuse as semirigid units [49]; as a result, lattice models cannot correctly capture the lower contact order region of Fig. 3a. Once one restricts oneself to the remaining part of Fig. 3a, the behaviors observed in the lattice models and proteins are consistent; in both, the folding ability increases sigmoidally with the stability [compare Fig. 4 with Fig. 16 of Ref. 11 and Fig. 1 of Ref. 37]. It should be noted, however, that an exact correspondence is not expected because, in the lattice model [2,6–11] and related analytic [16–18] studies, the stability descriptors are calculated from effective energies that include solvent effects implicitly rather than from full free energies, while the experimental ΔG values include the protein configurational entropy as well. It would be useful in this regard to have experimental enthalpies of folding for the proteins considered.

VIII. CONCLUSIONS

In the present study a nonlinear, multiple-descriptor method was applied to the prediction of the logarithm of the folding rate constant for a set of 33 two- and weakly three-state proteins. With two (three) descriptors, the Pearson linear correlation coefficient between the observed values and the training set and cross-validated predictions reach 0.89 (0.93) and 0.81 (0.86), respectively. These results are to be compared with those obtained by using the contact order by itself: $r_{trn} = 0.83$ and $r_{cv} = 0.74$. In addition to the contact order, some measures of the propensity of the sequence for a given structure also exhibited significant relationships with the folding rate; for example, $r_{cv} = 0.42$ for p_e. Although the propensity correlations are somewhat lower than those for measures obtained from the observed native structure, the sequence-based predictions may be sufficient to identify fast- or slow-folding proteins without the need for high-resolution structures. For example, using n and p_e, the folding rates for all 33 proteins, which range over almost six orders of magnitude, are predicted within a factor of 200; these (cross-validated) predictions are to be compared with those based on n_c and c/n, which are accurate within a factor of 60. In addition to the contact order, the size and stability play significant roles and are selected frequently for two- and three-descriptor models. Of particular interest is the finding that, for mixed-α/β and β-sheet proteins with higher contact orders ($c > 21$), the stability not only significantly improves the accuracy of multiple-descriptor models but gives excellent predictions by itself. The explicit or implicit inclusion of the stability in phenomenological models accounts for recent improvements in fitting experimental kinetic data [19,20,42]. Given the high quality of predictions that are obtained with the present analysis, further investigation of such correlations and their physical origins appear worthwhile, as has been suggested elsewhere [50].

Acknowledgments

A.R.D. is a Burroughs Wellcome Fund Hitchings-Elion Postdoctoral Fellow, and M.K. is the Eastman Visiting Professor at the Oxford Centre for Molecular Sciences. They would like to thank W. Graham Richards and Christopher M. Dobson for the hospitality that has been extended to them during their time in Oxford, John-Marc Chandonia for helpful discussions concerning the PRED2ARY program, and Jane Clarke for providing unpublished data for the proteins in Ref. 61. This work was supported in part by a grant from the National Science Foundation.

References

1. A. R. Dinner, A. Šali, L. J. Smith, C. M. Dobson, and M. Karplus, Understanding protein folding via free energy surfaces from theory and experiment. *Trends Biochem. Sci.* **25**, 331–339 (2000).

2. A. Šali, E. Shakhnovich, and M. Karplus, How does a protein fold? *Nature* **369**, 248–251 (1994).

3. Y. Zhou and M. Karplus, Interpreting the folding kinetics of helical proteins. *Nature* **401**, 400–403 (1999).

4. A. Fersht, *Structure and Mechanism in Protein Science: A Guide to Enzyme Catalysis and Protein Folding*, W. H. Freeman, New York, 1999.

5. C. M. Dobson, P. A. Evans, and S. E. Radford, Understanding how proteins fold: The lysozyme story so far. *Trends Biochem. Sci.* **19**, 31–37 (1994).

6. A. Šali, E. Shakhnovich, and M. Karplus, Kinetics of protein folding: A lattice model study of the requirements for folding to the native state. *J. Mol. Biol.* **235**, 1614–1636 (1994).

7. H. S. Chan and K. A. Dill, Transition states and folding dynamics of proteins and heteropolymers. *J. Chem. Phys.* **100**, 9238–9257 (1994).

8. N. D. Socci and J. N. Onuchic, Folding kinetics of proteinlike heteropolymers. *J. Chem. Phys.* **101**, 1519–1528 (1994).

9. A. R. Dinner, V. Abkevich, E. Shakhnovich, and M. Karplus, Factors that affect the folding ability of proteins. *Proteins* **35**, 34–40 (1999).

10. A. R. Dinner, A. Šali, and M. Karplus, The folding mechanism of larger model proteins: Role of native structure. *Proc. Natl. Acad. Sci. USA* **93**, 8356–8361 (1996).

11. A. R. Dinner, S.-S. So, and M. Karplus, Use of quantitative structure–property relationships to predict the folding ability of model proteins. *Proteins* **33**, 177–203 (1998).

12. K. W. Plaxco, K. T. Simons, and D. Baker, Contact order, transition state placement and the refolding rates of single domain proteins. *J. Mol. Biol.* **277**, 985–994 (1998).

13. S. E. Jackson, How do small single-domain proteins fold? *Folding & Design* **3**, R81–R91 (1998).

14. K. W. Plaxco, K. T. Simons, I. Ruczinski, and D. Baker, Topology, stability, sequence, and length: Defining the determinants of two-state protein folding kinetics. *Biochemistry* **39**, 11177–11183 (2000).

15. A. R. Dinner and M. Karplus, The roles of stability and contact order in determining protein folding rates. *Nature Struct. Biol.* **8**, 21–22 (2001).

16. J. D. Bryngelson and P. G. Wolynes, Spin glasses and the statistical mechanics of protein folding. *Proc. Natl. Acad. Sci. USA* **84**, 7524–7528 (1987).

17. J. D. Bryngelson and P. G. Wolynes, Intermediates and barrier crossing in a random energy model (with applications to protein folding). *J. Phys. Chem.* **93**, 6902–6915 (1989).

18. E. I. Shakhnovich and A. M. Gutin, Formation of unique structure in polypeptide chains: Theoretical investigation with the aid of a replica approach. *Biophys. Chem.* **34**, 187–199 (1989).

19. E. Alm and D. Baker, Prediction of protein-folding mechanisms from free-energy landscapes derived from native structures. *Proc. Natl. Acad. Sci. USA* **96**, 11305–11310 (1999).

20. V. Muñoz and W. A. Eaton, A simple model for calculating the kinetics of protein folding from three-dimensional structures. *Proc. Natl. Acad. Sci. USA* **96**, 11311–11316 (1999).

21. S. Wold, Cross-validatory estimation of the number of components in factor and principal components models. *Technometrics* **20**, 397–405 (1978).

22. S. Wold, Validation of QSARs. *Quant. Struct.–Act. Relat.* **10**, 191–193 (1991).

23. J. A. Malpass, D. W. Salt, M. G. Ford, E. W. Wynn, and J. Livingstone, Continuum regression: A new algorithm for the prediction of biological activity, in *Methods and Principles of Medicinal Chemistry*, Vol. 3, H. van de Waterbeemd, ed., VCH Publishers, New York, 1994, pp. 163–189.

24. J. J. Hopfield, Neural networks and physical systems with emergent collective computational abilities. *Proc. Natl. Acad. Sci. USA* **79**, 2554–2558 (1982).

25. J. Hertz, A. Krogh, and R. G. Palmer. *Introduction to the Theory of Neural Computation*, Addison-Wesley, Redwood City, CA, City, 1991.

26. D. T. Manallack and D. J. Livingstone, Neural networks in drug discovery: Have they lived up to their promise? *Eur. J. Med. Chem.* **34**, 195–208 (1999).

27. M. F. Møller, A scaled conjugate gradient algorithm for fast supervised learning. *Neural Networks* **6**, 525–533 (1993).

28. D. E. Clark, ed., *Evolutionary Algorithms in Molecular Design*, Wiley-VCH, Cambridge, 2000.

29. E. I. Shakhnovich and A. M. Gutin, Engineering of stable and fast-folding sequences of model proteins. *Proc. Natl. Acad. Sci. USA* **90**, 7195–7199 (1993).

30. H. S. Chan and K. A. Dill, Protein folding in the landscape perspective: Chevron plots and non-Arrhenius kinetics. *Proteins* **30**, 2–33 (1998).

31. R. Unger and J. Moult, Local interactions dominate folding in a simple protein model. *J. Mol. Biol.* **259**, 988–994 (1996).

32. D. K. Klimov and D. Thirumalai, Criterion that determines the foldability of proteins. *Phys. Rev. Lett.* **76**, 4070–4073 (1996).

33. D. K. Klimov and D. Thirumalai, Factors governing the foldability of proteins. *Proteins* **26**, 411–441 (1996).

34. G. S. Hammond, A correlation of reaction rates. *J. Am. Chem. Soc.* **77**, 334–338 (1955).

35. V. I. Abkevich, A. M. Gutin, and E. I. Shakhnovich, Specific nucleus as the transition state for protein folding: Evidence from the lattice model. *Biochemistry* **33**, 10026–10036 (1994).

36. A. R. Dinner and M. Karplus, The thermodynamics and kinetics of protein folding: A lattice model analysis of multiple pathways with intermediates. *J. Phys. Chem. B* **103**, 7976–7994 (1999).

37. A. R. Dinner, E. Verosub, and M. Karplus, Use of a quantitative structure–property relationship to design larger model proteins that fold rapidly. *Prot. Eng.* **12**, 909–917 (1999).

38. E. I. Shakhnovich and A. M. Gutin, Implications of thermodynamics of protein folding for evolution of primary sequences. *Nature* **346**, 773–775 (1990).

39. E. I. Shakhnovich and A. M. Gutin, A new approach to the design of stable proteins. *Prot. Eng.* **8**, 793–800 (1993).

40. V. Muñoz, P. A. Thompson, J. Hofrichter, and W. A. Eaton, Folding dynamics and mechanism of β-hairpin formation. *Nature* **390**, 196–199 (1997).

41. V. Muñoz, E. R. Henry, J. Hofrichter, and W. A. Eaton, A statistical mechanical model for β-hairpin kinetics. *Proc. Natl. Acad. Sci. USA* **95**, 5872–5879 (1998).

42. O. V. Galzitskaya and A. V. Finkelstein, A theoretical search for folding/unfolding nuclei in three-dimensional protein structures. *Proc. Natl. Acad. Sci. USA* **96**, 11299–11304 (1999).

43. R. Zwanzig, A. Szabo, and B. Bagchi, Levinthal's paradox. *Proc. Natl. Acad. Sci. USA* **89**, 20–22 (1992).

44. S. Takada, Gō-ing for the prediction of protein folding mechanisms. *Proc. Natl. Acad. Sci. USA* **96**, 11698–11700 (1999).

45. N. Gō, Theoretical studies of protein folding. *Annu. Rev. Biophys. Bioeng.* **12**, 183–210 (1983).

46. J. A. Schellman, The factors affecting the stability of hydrogen-bonded polypeptide structures in solution. *J. Phys. Chem.* **62**, 1485–1494 (1958).

47. D. A. Debe and W. A. Goddard III, First principles prediction of protein folding rates. *J. Mol. Biol.* **294**, 619–625 (1999).

48. M. Karplus and D. L. Weaver, Protein-folding dynamics. *Nature* **260**, 404–406 (1976).

49. M. Karplus and D. L. Weaver, Protein folding dynamics: The diffusion-collision model and experimental data. *Prot. Sci.* **3**, 650–668 (1994).

50. A. R. Fersht, Transition-state structure as a unifying basis in protein-folding mechanisms: Contact order, chain topology, stability, and the extended nucleus mechanism. *Proc. Natl. Acad. Sci. USA* **97**, 1525–1529 (2000).

51. K. M. Fiebig and K. A. Dill, Protein core assembly processes. *J. Chem. Phys.* **98**, 3475–3487 (1993).

52. K. A. Dill, K. M. Fiebig, and H. S. Chan, Cooperativity in protein-folding kinetics. *Proc. Natl. Acad. Sci. USA* **90**, 1942–1946 (1993).

53. N. D. Socci, J. N. Onuchic, and P. G. Wolynes, Diffusive dynamics of the reaction coordinate for protein folding funnels. *J. Chem. Phys.* **104**, 5860–5868 (1996).

54. M. Karplus, Aspects of protein reaction dynamics: Deviations from simple behavior. *J. Phys. Chem. B* **104**, 11–27 (2000).

55. R. Zwanzig, Two-state models of protein folding kinetics. *Proc. Natl. Acad. Sci. USA* **94**, 148–150 (1997).

56. M. Vendruscolo, E. Paci, C. M. Dobson, and M. Karplus, Three key residuees form a critical contact network in a transition state for protein folding. *Nature* **409**, 641–645 (2001).

57. D. S. Riddle, V. P. Grantcharova, J. V. Santiago, E. Alm, I. Ruczinski, and D. Baker, Experiment and theory highlight role of native state topology in SH3 folding. *Nature Struct. Biol.* **6**, 1016–1024 (1999).

58. J. C. Martínez and L. Serrano, The folding transition state between SH3 domains is conformationally restricted and evolutionarily conserved. *Nature Struct. Biol.* **6**, 1010–1016 (1999).

59. F. Chiti, N. Taddei, P. M. White, M. Bucciantini, F. Magherini, M. Stefani, and C. M. Dobson, Mutational analysis of acylphosphatase suggests the importance of topology and contact order in protein folding. *Nature Struct. Biol.* **6**, 1005–1009 (1999).

60. V. Villegas, J. C. Martínez, F. X. Avilés, and L. Serrano, Structure of the transition state in the folding process of human procarboxypeptidase A2 activation domain. *J. Mol. Biol.* **283**, 1027–1036 (1998).

61. J. Clarke, E. Cota, S. B. Fowler, and S. J. Hamill, Folding studies of immunolobulin-like β-sandwich proteins suggest that they share a common folding pathway. *Structure Fold. Des.* **7**, 1145–1153 (1999).

62. A. R. Fersht, L. S. Itzhaki, N. F. El Masry, J. M. Matthews, and D. E. Otzen, Single versus parallel pathways of protein folding and fractional formation of structure in the transition state. *Proc. Natl. Acad. Sci. USA* **91**, 10426–10429 (1994).

63. G. Aronsson, A.-C. Brorsson, L. Sahlman, and B.-H. Jonsson, Remarkably slow folding of a small protein. *FEBS Lett.* **411**, 359–364 (1997).

64. R. E. Burton, B. S. Huang, M. A. Daugherty, T. L. Calderone, and T. G. Oas, The energy landscape of a fast-folding protein mapped by ala→gly substitutions. *Nature Struct. Biol.* **4**, 305–310 (1997).

65. S. Sato, B. Kuhlman, W.-J. Wu, and D. P. Raleigh, Folding of the multidomain ribosomal protein L9: The two domains fold independently with remarkably different rates. *Biochemistry* **38**, 5643–5650 (1999).

66. N. Ferguson, A. P. Capaldi, R. James, C. Kleanthous, and S. E. Radford, Rapid folding with and without populated intermediates in the homologous four-helix proteins Im7 and Im9. *J. Mol. Biol.* **286**, 1597–1608 (1999).

67. S. Spector, B. Kuhlman, R. Fairman, E. Wong, J. A. Boice, and D. P. Raleigh, Cooperative folding of a protein mini domain: The peripheral subunit-binding domain of the pyruvate dehydrogenase multienzyme complex. *J. Mol. Biol.* **276**, 479–489 (1998).

68. S. Spector and D. P. Raleigh, Submillisecond folding of the peripheral subunit-binding domain. *J. Mol. Biol.* **293**, 763–768 (1999).

69. W. Kabsch and C. Sander, Dictionary of protein secondary structure: Pattern recognition of hydrogen-bonded and geometrical features. *Biopolymers* **22**, 2577–2637 (1983).

70. J. M. Chandonia and M. Karplus, New methods for accurate prediction of protein secondary structure. *Proteins* **35**, 293–306 (1999).

71. L. S. Itzhaki, D. E. Otzen, and A. R. Fersht, The structure of the transition state for folding of chymotrypsin inhibitor 2 analysed by protein engineering methods: Evidence for a nucleation–condensation mechanism for protein folding. *J. Mol. Biol.* **254**, 260–288 (1995).

72. Z. Guo, C. L. Brooks III, and E. M. Boczko, Exploring the folding free energy surface of a three-helix bundle protein. *Proc. Natl. Acad Sci. USA* **94**, 10161–10166 (1997).

73. F. B. Sheinerman and C. L. Brooks III, Molecular picture of folding of a small α/β protein. *Proc. Natl. Acad. Sci. USA* **95**, 1562–1567 (1998).

74. B. D. Bursulaya and C. L. Brooks III, Folding free energy surface of a three-stranded β-sheet protein. *J. Am. Chem. Soc.* **121**, 9947–9951 (1999).

75. A. R. Dinner, T. Lazaridis, and M. Karplus, Understanding β-hairpin formation. *Proc. Natl. Acad. Sci. USA* **96**, 9068–9073 (1999).

76. P. Ferrara and A. Caflisch, Folding simulations of a three-stranded antiparallel β-sheet. *Proc. Natl. Acad. Sci. USA* **97**, 10780–10785 (2000).

77. A. Hiltpold, P. Ferrara, J. Gsponer, and A. Caflisch, Free energy surface of the helical peptide Y(MEARA)$_6$, *J. Phys. Chem. B* **104**, 10080–10086 (2000).

INSIGHTS INTO SPECIFIC PROBLEMS IN PROTEIN FOLDING USING SIMPLE CONCEPTS

D. THIRUMALAI, D. K. KLIMOV, AND R. I. DIMA

Institute for Physical Science and Technology and Department of Chemistry and Biochemistry, University of Maryland, College Park, MD, U.S.A.

CONTENTS

Computational Methods for Protein Folding: A Special Volume of Advances in Chemical Physics, Volume 120, Edited by Richard A. Friesner, Series Editors I. Prigogine and Stuart A. Rice. ISBN 0-471-20955-4. © 2002 John Wiley & Sons, Inc.

I. INTRODUCTION

Protein folding is a process by which a polypeptide chain made up of a linear sequence of amino acids adopts a well-defined three-dimensional native structure [1]. Single-domain proteins reach their biologically active native conformations on time scales that are typically on the order of 10–1000 milliseconds [2]. Since Anfinsen's pioneering experiments it has been known that protein folding is a self-assembly process in which the information needed to determine the three-dimensional native structure is contained in the primary sequence [3]. Given this, the next important question is how the native state is kinetically reached in such a short time scale [4]. This issue was first emphasized by Levinthal, who wondered how a protein of a reasonable length can navigate the astronomically large conformational space so efficiently [5]. Seeking to resolve the paradox, Levinthal suggested that certain preferred pathways must guide the chain to the native state. For years the Levinthal paradox has served as an intellectual impetus in our quest to understand the mechanisms by which a polypeptide chain reaches the native conformation.

The last decade has witnessed considerable advances in our understanding of how a polypeptide chain folds starting from an ensemble of denatured states [6–13]. In recent years, protein folding kinetics has become increasingly important, largely because misfolding (i.e., errors in refolding) has been implicated in a number of diseases [14]. As a result, several advances have been made to probe the factors that govern the normal folding of proteins. Fast-folding experiments [2,8,15–19] and single-molecule methods [20–24] are beginning to provide direct glimpse into the early events in the assembly of proteins. Protein engineering in conjunction with the Φ-value analysis has become the cornerstone technique in deciphering the structures of the elusive transition state ensemble of two-state folders [25–27]. Although these tools have helped us to understand folding of individual proteins, considerable progress still needs to be made before the complex processes in misfolding and assembly of proteins with increasing complexity are well understood. In particular, to translate the functional genomics efforts into practical applications, it is important to solve rapidly the proteomics problem, namely, the determination of protein structures.

These multifaceted activities have ushered all aspects of protein folding at the center stage of molecular biology.

The major focus has been in understanding the folding mechanisms of proteins that display two-state behavior [28]. A variety of factors that determine the plausible folding scenarios have been identified [6,9–12,29–36]. A number of distinct folding mechanisms emerge depending on the characteristics temperatures that determine the phases of the polypeptide chain [10,34]. These findings explicitly link the underlying thermodynamic properties of proteins and their folding mechanisms. Several studies have focused on the factors that determine the folding rates of two-state proteins. Plausible relationships between folding rates and the contact order [37] (which emphasizes the role of structures involving proximal residues), stability [34,38], and Z score [34] have been established. Because many of these conceptual ideas have been described in recent reviews [6,9–12,33,39–41], we will not discuss these here.

A variety of computational and phenomenological approaches have been employed to obtain the general principles that control the folding rates and mechanisms of single-domain globular proteins [6,10,33]. It may be naively thought that the computational protocol for describing protein folding is straightforward. Indeed, the folding dynamics is well-described by the classical Newton equations of motion, and folding may be directly monitored from an appropriately long trajectory. However, there are two drastic limitations that prevent this approach to study the folding of proteins. First, the force fields for such a complex system are not precisely known. As a result, one needs to rely on the transferability hypothesis that interactions derived for small molecules can be used in larger systems, such as proteins. The second problem is simple: the limitations of current CPU power. Repeated folding of even a single-domain protein requires generating of multiple trajectories in a millisecond time scale. Even creative use of massively parallel simulations does not entirely solve this severe numerical constraint [42].[1]

In light of these difficulties, various simplified models of proteins have been suggested [10,39,41]. Most of the insights from computations came from the systematic studies of folding using coarse-grained models. The main rationale for their use is that a detailed study of such models will reveal general principles, if any, that govern the folding of proteins [10,39,41,43,44]. Such an approach

---

[1]We have recently achieved extraordinary speed-up of folding simulations for several β-hairpin sequences using distributed computing. In collaboration with Parabon Computations Inc., we have shown that distinct folding scenarios emerge even in the formation of β-hairpins. For the hairpin taken from the C-terminal of the immunoglobulin binding protein (GB1), the folding mechanisms and the time scales depend on the location of the hydrophobic cluster (D. Klimov, D. Newfield, and D. Thirumalai, unpublised results).

has yielded considerable insights into the mechanisms, time scales, and pathways in the folding of polypeptide chains.

The purpose of this chapter is to describe applications of simple concepts and computations to three specific problems in protein folding: (i) Are the requirements that folded states of proteins be compact and have low energy sufficient to explain the emergence of the finite number of folds from a very dense sequence space? An affirmative answer to this question, at the conceptual level, can be given using lattice models of proteins [44]. (ii) Phenomenological theory and lattice model computations are used to clarify the role of disulfide bonds in protein folding. The theory based on the proximity rule [45] and the lattice models investigating disulfide bonds formation [46] provided clarifications of the expected pathways in the refolding of bovine pancreatic trypsin inhibitor (BPTI). Recent calculations have explained quantitatively the effect of intact S–S bonds on the folding and stability of barnase. (iii) We describe a simple model of chaperonin-assisted folding [47]. Specific predictions about the coupling between conformational change of the chaperone molecule and the folding of the substrate protein emerge from the calculations. These predictions were subsequently tested experimentally.

To make this chapter as self-contained as possible, we briefly describe lattice models and the commonly employed computational methods. This is followed by a brief description of how a monomeric protein folds. The contents of this section are important to better appreciate the role of chaperones in the rescue of proteins. The chapter is concluded with brief comments about the challenges we face in the straightforward all-atom simulations of protein folding.

II. LATTICE REPRESENTATIONS OF PROTEINS

A. Basic Assumptions

Lattice models (LM) of single chains have long been used in polymer physics to obtain a number of universal properties (scaling of the size of the polymer with N, distribution of end-to-end distances, etc.) of real homopolymer chains [48]. For these issues the universal properties are unaffected by the precise interactions between monomers as long as they are short-ranged. It is not clear *a priori* that lattice models can be used to investigate general features of folding (e.g., cooperativity of transition from unfolded U to native N states). Single-domain proteins are finite-sized with the number of amino acid residues, N, not typically exceeding much beyond 200. Specific interactions that leads to the unique architecture of the N state cannot be fully represented using LM. The dynamics of the folding process can clearly depend on the precise move sets, so that the correspondence between the Monte Carlo simulations and the kinetics in aqueous solution is ambiguous at best. Nevertheless, a series of studies from several groups have yielded a number of predictions many of which have been affirmed experimentally [6,39,41,47].

In the context of protein folding, lattice models were first introduced by Gō and co-workers [49]. The insights brought by Dill and Chan in the late 1980s have had a great influence on the development of LM for understanding protein-folding kinetics [50]. Dill and co-workers argued that protein folding can be studied using short enough chains so that exact enumeration of all allowed conformations becomes possible. Exact enumeration enables precise computations of thermodynamic characteristics. Monte Carlo (MC) simulations, based on physically motivated move sets, can be used to monitor folding kinetics.

In the simplest LM, amino acids are represented by a single atom (treated as a backbone α-carbon) and the side chains are not explicitly considered. As a result, only a few basic interactions found in real proteins can be modeled. In the most popular version of LM the polypeptide chain adopts a self-avoiding walk on a cubic lattice [32,51,52]. The heterogeneity of interactions in amino acids is mimicked by having several interaction energy scales between the beads of the chain. In general, only short-range interactions between nonbonded residues that are nearest neighbors on the cubic lattice are taken into account. Thus, a generic energy function for such a model includes three components: (i) connectivity of the chain is preserved through rigid bonding of successive beads; (ii) a self-avoidance condition is imposed by the restriction that a given lattice site can be occupied only once; (iii) the contact interactions between the side chain beads i and j $B_{ij}(|i-j| \geq 1)$ are given by pairwise potentials. The energy of a conformation is

$$E = \sum_{i<j} B_{ij} \delta_{|\vec{r}_i - \vec{r}_j|, a} \tag{1}$$

where $\delta_{r,a}$ is the Kronecker delta function and a ($=3.8$ Å) is the lattice spacing.

1. Contact Energies

There are several models for the interaction matrix elements B_{ij} which take into account the diversity of interactions between amino acids. Because these models are at best a simple representation of the potentials in real proteins, it is not *a priori* clear that any particular model is better than the other. In the literature several different interaction schemes have been utilized [32,34,39,47,51,52]. These include HP model [39,51], random bond (RB) model [32], and the pairwise potentials derived from the statistical analysis of contacts between different amino acids in the protein structures [53–55]. In what follows we give a brief description of these models.

2. HP Model

This model reduces the set of 20 naturally occurring amino acids to two kinds, namely, hydrophobic (H) and polar (P) [39,51]. A sequence is given by the nature of the amino acid residue at a given position. For example, HPHPH is a sequence with $N = 5$. There are 2^N total sequences for a given N. In the HP model the interactions are given by a 2×2 matrix, whose elements are $B_{HH} = -\epsilon$ and zeros, otherwise. Despite the simplicity of the model, it is not exactly solvable due to the chain connectivity and excluded volume effects. Because the HP model can lead to microphase separation, variations in the interaction energies have been introduced. Various aspects of folding observed in the HP model (two-letter code) have been investigated by Dill et al. [39] and others [51].

3. Random Bond Model

In the RB model [32] the interaction elements are drawn from the Gaussian distribution

$$P(B_{ij}) = \frac{1}{\sqrt{2\pi}B} \exp\left(-\frac{(B_{ij} - B_0)^2}{2B^2}\right) \tag{2}$$

where B_0 is the average interaction that specifies the strength of the drive toward forming compact structures at low temperatures, and the dispersion B gives the extent of diversity of the interactions among beads. Energy is measured in terms of B which is set to unity. The choice of $B_0 = -0.1$ [32] ensures that the fraction of hydrophobic residues in a sequence (specified by the interaction matrix elements B_{ij}) is about 0.55, which roughly coincides with the fraction of hydrophobic residues in real proteins. A sequence is specified by the matrix of contact energies B_{ij}.

4. Statistically Derived Pairwise Potentials

In this case, the energies B_{ij} are given by pairwise statistical potentials computed by analyzing the frequency of amino acids interactions in the experimentally determined protein structures. Several sets of such potentials are currently available. These includes the potentials calculated by Miyazawa and Jernigan (MJ) [53], Kolinski, Godzik, and Skolnick (KGS) [54], Mirny and Shakhnovich [56], Tobi and Elber [57], and Betancourt and Thirumalai [55]. The major advantage of the such potential sets is that the model lattice sequence may now be described in terms of "real" amino acid composition, assuming that the contact energies reproduce the nature of interactions between amino acids.

5. Gō Model

The Gō model does not directly introduce a new force field, but modifies the existing energy function by tuning it to the known native structure [58]. Specifically, the Gō model considers only the interactions between residues (beads on the lattice) that are present in the native (ground) state. In other words, only native contacts are taken into account. The major advantage of the Gō model is almost a complete elimination of frustration in a model protein and, as a result, a substantial increase in the folding rates [59]. The severe shortcoming is that the energy function and the native structure cannot be decoupled; consequently the Gō model, despite being topologically frustrated, is "foldable" by definition.

B. Lattice Models with Side Chains

The cubic lattice models described above is the simplest version of the coarse-grained model. One obvious way to make it more realistic is to incorporate the explicit representation of side chains [60]. In this case, a polypeptide chain is modeled by a sequence of N backbone beads, representing the C_α carbons of a protein backbone. Side-chain beads, which mimic amino acid residues, are attached to each backbone bead. In all, there are $2N$ beads in the model, all of which occupy the vertices of cubic lattice. The conformation of a protein is specified by $2N$ vectors $\vec{r}_{b,i}, \vec{r}_{s,i}, i = 1, 2, \ldots, N = 15$, where $\vec{r}_{b,i}$ and $\vec{r}_{s,i}$ are the positions of backbone and side-chain beads, respectively. The energy function used for the side-chain model is typically the same as employed in the model without side chains. These models provide a more realistic description of cooperativity of folding, because they include effects of side-chain packing [39].

C. Computational Methods

1. Exhaustive Enumeration

The conformational space of short lattice sequences can be exhaustively enumerated. All conformations for a polypeptide chains with $N \lesssim 20$ on a cubic lattice can be enumerated using the Martin algorithm [61]. This algorithm successively generates all self-avoiding conformations for a given N, which allows *exact* calculation of any thermodynamic quantity. In order to reduce the sixfold symmetry on the cubic lattice, the direction of the first monomeric bond may be fixed in all conformations. The remaining conformations are still related by the eightfold symmetry on the cubic lattice (excluding the cases when conformations are completely confined to a plane or straight line). To decrease further the number of conformations, the Martin algorithm may be modified to reject all conformations related by this symmetry [32]. For longer model sequences the CPU time required to enumerate all conformations becomes prohibitively long. With constant upgrade in computer power this limitation is being steadily overcome.

2. Monte Carlo Method

The standard method for studying thermodynamics and kinetics of folding in the context of lattice models is the Monte Carlo (MC) algorithm [62]. Several types of moves are commonly used [32]. These are (i) corner moves (a flip of the residue across the diagonal of the square formed by the neighboring bonds), (ii) crankshaft rotations (rotation of the beads $i + 1$ and $i + 2$, while keeping the adjacent beads i and $i + 3$ fixed), and (iii) rotation of the end beads. Although the precise choice of moves or their probabilities affects the local structural dynamics, it is commonly believed that the general thermodynamic properties and even kinetic characteristics remain unchanged as long as the moves are ergodic. Even with the choice of physically motivated move sets their influence on the results must be tested.

3. Multiple Histogram Technique

The thermodynamic quantities for longer chains may be effectively computed using the multiple histogram method [51,60,63]. The method is based on the collection of a set of histograms at different values of the external parameter and combining them by reweightening the contribution from individual histograms. The thermal average of any quantity may then be calculated. Technically, multiple slow-cooling MC trajectories, each starting from different conditions, are needed to obtained the histograms. Each trajectory starts at a high temperature ($T_h > T_\theta$) and ends at the temperature $T_l < T_F$, where T_θ and T_F are the collapse and folding temperatures, respectively. In the course of a trajectory the temperature is changed periodically by small decrements, and the portions of simulations at a given fixed temperature (after quick equilibration intervals) are used for histogram collection. Usually, histograms for the values of energy, number of native contacts, radius of gyration, and so on, are obtained. There is no general prescription for choosing the lengths of the trajectory and of the equilibration interval because they depend strongly on the sequence and on the temperature. The number of trajectories is determined by the condition that the thermodynamics of the system should not change significantly with subsequent increase in sampling. Thus, by using multiple histogram technique, one can completely characterize the thermodynamics of the system by calculating the average of any quantity as a function of external parameter as well as the free energy profiles. Using the histograms, we can generate free energy profiles, provided that a useful reaction coordinate is chosen.

4. Folding Kinetics

The kinetics of folding of a lattice sequence is obtained using multiple folding trajectories at a fixed temperature. Each trajectory starts from a different high-temperature conformation. After a sudden quench of the temperature to T_s, the chain kinetics is monitored. Typically, the folding kinetics is characterized by time dependence of folding probes averaged over the total number of trajectories considered. The first passage to the native structure τ_{1i} is also recorded. From the distribution of τ_{1i} P_{fp}, the fraction of trajectories that have not reached the native conformation at a time t is calculated using

$$P_u(t) = 1 - \int_0^t P_{fp}(s)\, ds \tag{3}$$

The integral of $P_u(t)$ determines the average passage time τ_F as

$$\tau_F = \int_0^\infty P_u(t)\, dt \tag{4}$$

Accurate results require generation of hundreds of folding events.

III. REDUCTION IN CONFORMATIONAL SPACE

A. Importance of Excluded Volume Interactions

The impetus to examine the size of the conformational space of proteins comes from Levinthal [5], who wondered how can a polypeptide chain, even though it is relatively small, navigate the vast number of allowed conformations in search of the unique native state? A popular resolution of this argument suggests that fundamental constraints, notably the excluded volume (EV) interactions between atoms, so vastly reduces the conformations that only a very limited number is ever sampled. This idea can be precisely tested using appropriate models.

The number of independent conformations for a chain with N beads on a cubic lattice is $C_{IND} = Z^N$, where $Z \; (=6)$ is the lattice coordination number. If excluded volume interactions (also referred to as steric clashes [64]) are taken into account, then the number of allowed conformations is

$$C_{EV} \simeq Z_{eff}^N N^{\gamma-1} \tag{5}$$

where the universal exponent $\gamma \approx 1.16$, and $Z_{eff} = 4.684$ in a cubic lattice. Both C_{IND} and C_{EV} scale exponentially with N. However, it might be argued that the finite size of the proteins might make the reduction, due to EV interactions, so significant that the "entropy price" to adopt native-like conformations is not very large. In a cubic lattice the entropy change, ΔS, upon going from S_{IND} to S_{EV} is $\Delta S/k_B \approx N \ln(Z/Z_{eff})$. For $N = 10, \Delta S \approx 12.8$ eu, which is substantial. However, the absolute entropy associated with S_{EV} is $N \ln Z_{eff}$. Neglecting logarithmic corrections we get $S_{EV} = k_B \ln C_{EV} \approx 15.4$ eu. Thus, considering steric clashes alone *does substantially reduce the size of the conformational space*. However, this reduction is not sufficiently large to solve the "search problem" envisioned by Levinthal.

In a recent interesting article, Pappu et al. [64] have reemphasized the importance of excluded volume interactions by enumerating the allowed conformations for blocked all-atom polyalanine chains, Al–(Ala)$_n$–N$'$-methylamide for $n \leq 7$. By coarse graining the (ϕ, ψ) angles, they showed that the conformational space due to EV interactions is less than it would be if the (ϕ, ψ) angles are considered independent as suggested by Flory. This result is in qualitative accord with the estimates for lattice models given above. As pointed out above, this reduction is not sufficient to provide a qualitative explanation of the central kinetic issue raised by Levinthal.

Pappu et al. [64] suggest that EV interactions or steric clashes "bias" the conformations so that even in the unfolded state there is a significant tendency to form local structures. This is certainly the case in off-lattice models of proteins [10]. Typically, these fluctuating structures are stabilized by additional interactions (say, hydrogen bonding). If the favorable biasing interactions are too strong (greater than 2–3 $k_B T$), then the local interactions would become incompatible with the tertiary interactions. This has been shown to increase the topological frustration [65] see below, which in turn can lead to the dominance of kinetic traps. Thus, arguments that are based solely on the reduction of conformational space of proteins cannot account for the global folding mechanisms. Harmony (or consistency) between local and nonlocal interactions is necessary for efficient folding of proteins.

If only EV interactions are included in polypeptide chains, the chain cannot undergo a "phase transition" to any specific conformation. The effective mean-field one-body potential describing EV interactions is known to be long-ranged (scaling as $r^{-4/3}$). Consequently the polypeptide chain would adopt a random coil state at all temperatures, if only EV interactions are included. However, the chain can be induced to adopt a preferred structure (native conformation), if an additional attractive energy $-\epsilon$ between residues (hydrogen bond interactions, for example) is introduced. This is the basis of the popular HP model for proteins [39]. In a model, which takes into account the EV and attractive interactions, a phase transition into a native-like structure can occur at T such that $T \approx C_N \epsilon / S_U$, where C_N is the number of favorable native interactions and S_U is the entropy of the unfolded state. Pappu et al. [64] showed that by including an attractive energy term to mimic backbone hydrogen bonding, an apparent two-state transition from a stretched state to a contracted state takes place (Fig. 1). This kind of apparent two-state transitions, similar to those found in proteins, has been observed in simple lattice models as well [6]. The interesting feature of the calculations by Pappu et al. [64] is that a realistic model of even a short polypeptide chain with only one attractive energy scale can exhibit protein-like behavior.

IV. EMERGENCE OF STRUCTURES FROM THE DENSE SEQUENCE SPACE

The sequence space of proteins is extremely dense as the number of possible sequences for proteins of length N scales as 20^N. However, not all these

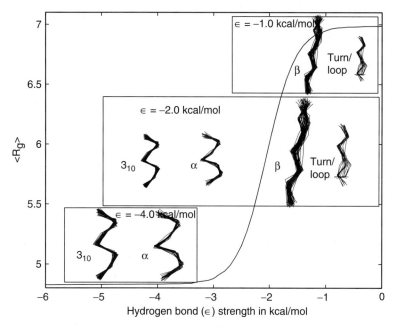

Figure 1. Dependence of the radius of gyration $\langle R_g \rangle$ for polyalanyl chain of length $n = 7$ [64] on the hydrogen bond length ϵ. As ϵ increases, compact conformations are populated preferentially. The transition from the extended conformations at higher values of ϵ to the contracted conformations occurs rather cooperatively in an apparent "two-state" manner. The radius of gyration is computed using a coarse-grained thermally weighted density of states (see Ref. 64 for details). Conformations that make the most significant contributions at different values of ϵ are also shown.

sequences encode for foldable protein structures, which for functional purposes are constrained to have specific physical characteristics. How do viable protein structures emerge from the dense sea of sequence space [66]? The extraordinary thinning of the sequence space as one gets to the structure space may be understood purely on the basis of accepted physical properties of proteins. To this end, two interrelated physical features of folded proteins must be taken into account. (i) Native proteins are compact. (ii) The interior of proteins consists mainly of hydrophobic residues, while the hydrophilic residues are typically found on the surface. This gives rise to a maximum number of favorable interactions making the native state very low in energy.

Lattice models are remarkably useful in answering the conceptual question posed above. To infer the sequence to structure mapping, we performed an exhaustive enumeration of all self-avoiding conformations for the sequences confined to cubic lattice with $N = 15$ [44]. The RB model has been used in the energy function with the parameters $B_0 = -0.1$ and $B = 1$. Protein-like structures are not only compact but also have low energy. We first computed the number of compact structures (CSs) for a given N, C_N, (CS). The number of CSs, in its most general form, is expected to scale as

$$C_N(\text{CS}) \simeq \bar{Z}^N Z_1^{N^{\frac{d-1}{d}}} N^{\gamma_c - 1} \tag{6}$$

where $\ln \bar{Z}$ is the conformational free energy (in units of $k_B T$), Z_1 is the surface fugacity, d is the spatial dimension, and γ_c measures possible logarithmic corrections to the free energy. The number of natural protein folds is limited (perhaps a few thousands), and their number is expected to grow at rates much smaller than those predicted in Eq. (2). To explore this we calculated by exact enumeration the number of minimum energy structures (MES), $C_N(\text{MES})$, as a function of N.

We define MES as those conformations whose energies lie within the energy interval Δ above the lowest energy E_0, corresponding to the native state. Several values for Δ (1.2 or 0.6) were used to ensure that no qualitative changes in the results are observed. We also tested another definition for $\Delta = 1.3|E_0 - tB_0|/N$, where t is the number of nearest-neighbor contacts in the ground state. It is worth noting that in the latter case Δ increases with N. Nevertheless, both definitions yield equivalent results. The computational technique involves exhaustive enumeration of all self-avoiding conformations for $N \leq 15$ on a cubic lattice. We calculated the energies of all conformations according to Eq. (1) and then determined the number of MES and CS. Each quantity, such as $C_N(\text{MES})$, $C_N(\text{CS})$, the lowest energy E_0, or the number of nearest-neighbor contacts t in the lowest energy structures, is averaged over 30 sequences. To test the reliability of the computational results, an additional sample of 30 RB sequences was generated. Note that in the case of $C(\text{MES})$ we computed the quenched average as $C_N(\text{MES}) = \exp\left[\overline{\ln\left[c(\text{MES})\right]}\right]$, where c is the number of MES for one sequence.

The number of MES $C(\text{MES})$ is plotted as a function of the number of residues N in Fig. 2 for $\Delta = 0.6$. A pair of squares for a given N represents $C(\text{MES})$ computed for two independent runs of 30 sequences each. For comparison, the number of self-avoiding walks $C(\text{SAW})$ and the number of CS $C(\text{CS})$ are also plotted in this figure (diamonds and triangles, respectively). As expected on general theoretical grounds, $C(\text{SAW})$ and $C(\text{CS})$ grow exponentially with N, whereas the number of MES $C(\text{MES})$ exhibits drastically different scaling behavior. There is no variation in $C_N(\text{MES})$ (normally, associated with the variation of shapes of compact structures) and its value remains steady within the entire interval of N starting with $N = 7$. We find (see Fig. 2) that $C_N(\text{MES}) \approx 10^1$. This result further validates our earlier finding for the two-dimensional model [46]. The results strongly suggest that $C_N(\text{MES})$ scales only as $\ln N$. Thus, the dual restriction of compactness and low energy of the native states may impose an upper bound on the number of distinct protein folds.

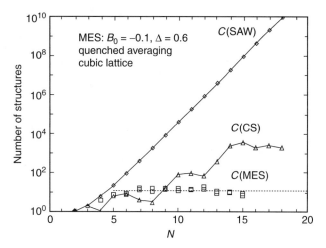

Figure 2. Scaling of the number of MES $C_N(\text{MES})$ (squares) on a cubic lattice. The data are obtained for $B_0 = 0.1$ and $\Delta = 0.6$. The pairs of squares for each N represent the quenched averages for different samples of 30 RB sequences. The number of compact structures $C_N(\text{CS})$ and self-avoiding conformations $C_N(\text{SAW})$ are plotted to highlight the dramatic difference in scaling behavior. It is clear that $C(\text{MES})$ remains practically flat; that is, it grows no faster than $\ln N$.

A.   Designability of Protein Folds

The computations described above indicate that minimal restrictions on the structures (compactness and low energies) make the structure space sparse. Consequently, each basin of attraction in the structure space must contain numerous sequences [66]. The way these sequences are distributed among the very slowly growing number (with respect to N) of conformations—that is, the density of sequences in structure space—is another important question. Li et al. [67] considered a three dimensional cubic lattice proteins with $N = 27$. By using the HP model and restricting themselves to only maximally compact structures as tentative candidates for protein native states, they showed that certain folds (i.e., structures) accommodate much larger number of sequences (see Fig. 3a) than the others. In one example, they found the NBA (the structure) that serves as a ground state for 3794(!) (when the total number is 2^{27}) sequences and, hence, was considered most designable. The precise distribution of sequences among NBAs is a function of the particular energy function.

An important conclusion of Li et al. [67] is that one can define, at least operationally, a designability index for every fold found in PDB. The structural characteristics of a given fold determine its designability. Several authors have suggested that if the fold has even an approximate symmetry, then it would be more designable [67,68]. This might explain the preponderance of TIM barrel

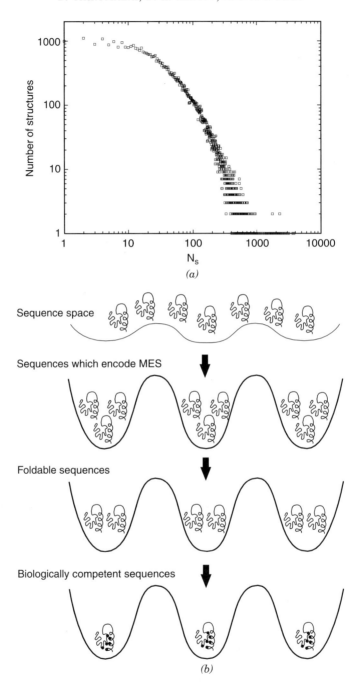

(a)

Sequence space

Sequences which encode MES

Foldable sequences

Biologically competent sequences

(b)

structures. If the symmetry argument is extended to RNAs, then we would conclude that certain symmetries should be hidden at the sequence level of mRNAs and ultimately the genes themselves encoding a given protein [44].

Because the number of NBA for the entire sequence space is very small, it is likely that proteins could have evolved randomly. Natural folds must correspond to one of the native basins of attraction in the structure space so that many sequences have these folds as the native conformations. In other words, natural protein folds, especially those with approximate symmetries, represent highly designable structures [67]. Further support for these ideas comes from the study of Lindgard and Bohr [69]. These authors showed that among maximally compact structures there are only very few folds that have protein-like characteristics. It was also estimated that the number of distinct protein folds is on the order of 10^3. Thus, each fold can be designed by many candidate sequences. However, there is also evolutionary pressure for sufficiently rapid folding to avoid aggregation. This kinetic requirement further restricts the possible sequences that can serve as biologically viable proteins (Fig. 3b).

V. PROTEIN FOLDING MECHANISM

Using lattice models with side chains we describe the most commonly found scenarios observed in protein folding. Because this topic has been subject of numerous reviews [6,9–12,41], we will stress a few points that are relevant in considering chaperonin-mediated protein folding that is discussed in Section VII.

A. Two-State Folders

Thermodynamics for the sequence with the native state shown in Fig. 4 with the contact interaction potentials B_{ij} taken from Table III of Ref. 54 reveals that it folds cooperatively in an apparent two-state manner. This is also reflected in the thermal distribution of the overlap function values $h(\chi)$ at the folding transition temperature T_F (Fig. 4). A nearly bimodal distribution of $h(\chi)$ with the peaks at $\chi \lesssim 0.2$ (NBA) and $\chi \sim 0.6$ (unfolded state) is observed. There is also nonnegligible contribution from the intermediate values of χ representing partially folded structures. Experiments that probe in more detail the thermal unfolding of proteins are beginning to reveal the possible importance of these

---

Figure 3. (a) A log–log plot of the histogram for number of structures with respect to the number of associated sequences N_s for 27-mer maximally compact cubic lattice conformations [67]. The plot illustrates a dramatic heterogeneity among structures in terms of their ability to encode protein sequences. (b) Schematic illustration of the mapping of vast sequence space onto the limited number of protein folds. This mapping involves drastic reduction in sequence space as polypeptide sequences evolve into functionally competent proteins.

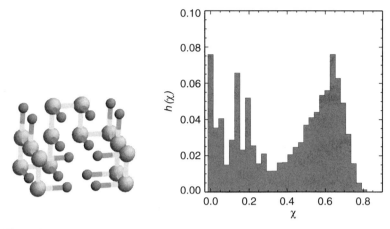

Figure 4. Native structures of sequences A generated using [126] is shown in the left panel. Backbone and side-chain beads are shown in light and dark gray, respectively. Native conformation is compact and has a well-defined hydrophobic core. The figure is generated using program RasMol [126]. The right panel displays the thermal distribution of states $h(\chi)$ calculated at $T \approx T_F$ for sequence A. $h(\chi)$ is approximately bimodal so that only NBA ($\chi \lesssim 0.2$) and unfolded state U($\chi \sim 0.6$) are significantly populated. Although small, the population of intermediate states nevertheless makes a sizable contribution to thermodynamics (affecting mainly cooperativity of folding).

conformations [70]. Due to substantial contribution from the partially folded structures, thermal unfolding cannot be quantitatively described as two-state.

The folding kinetics can be probed using the distribution of the first passage times, τ_{1i}. Several hundred (~ 600) folding events are used to obtain the distribution of τ_{1i}, from which the fraction of unfolded molecules $P_u(t)$ may be readily obtained. In addition to $P_u(t)$, we have computed the time dependence of the radius of gyration $\langle R_g(t) \rangle$, where the average is taken over 100 folding trajectories.

The sequence, whose native state is shown in Fig. 4, displays two-state kinetics for the temperatures $T \geq 0.8T_F$; that is, $P_u(t) \sim \exp(-\frac{t}{\tau_F})$, where τ_F is the folding time. To probe the sequence of events en route to the native conformation, we computed $\langle R_g(t) \rangle$, which reveals two stages in collapse. Initial rapid burst phase is followed by a gradual chain compaction (Fig. 5). The overall collapse time τ_c is associated with the second characteristic time. From the approach to the native conformation we draw the following general conclusions regarding two-state folders:

(a) The ratio τ_F/τ_c for two-state folders is typically less than 10. This is consistent with the fast-folding experiments on several two-state folders, which

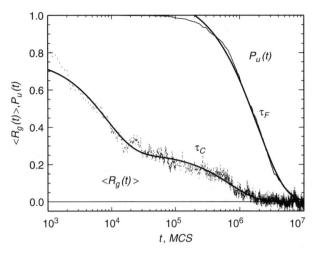

Figure 5. The time dependence of the normalized radius of gyration $\langle R_g(t) \rangle$ and the fraction of unfolded molecules $P_u(t)$ for sequence A at $T = 0.94T_F$. Data are averaged over 100 [for $\langle R_g(t) \rangle$] and 600 [for $P_u(t)$] trajectories. $P_u(t)$ decays exponentially with the time scale $\tau_F = 2.07 \times 10^6$ MCS. The approach of $\langle R_g(t) \rangle$ to equilibrium is biexponential with the times scales 0.083×10^6 MCS and 0.698×10^6 MCS. The first time scale is due to extremely rapid burst-phase partial collapse. The second time scale, which is associated with the collapse time τ_c, corresponds to the final compaction. The ratio τ_F / τ_c is approximately 3.0.

show that proteins rapidly collapse and reconfigure themselves to reach the native state. For the sequence in Fig. 4, $\tau_F / \tau_c \approx 3$. This ratio is in the range 5–10 for proteins.

(b) Analysis of the collapsed conformations shows that they are native-like; that is, the initial collapse in two-state folders is "specific" with very few nonnative interactions present. The overall scheme for reaching the NBA for two-state folders, which was predicted using theoretical arguments, is

$$\mathbf{U} \rightarrow \{\mathbf{I_N}\} \rightarrow \mathbf{N} \qquad (7)$$

where $\{\mathbf{I_N}\}$ is a collection of native-like structures. Fast-folding experiments on cyt-c and tendamistat [71] have been interpreted using this picture. Because the initial collapse is specific, the ensemble of native-like intermediates can be likened to an "on-pathway" intermediate. Lattice simulations (without side chains) using Gō model have come to a similar conclusion [72]. In the Gō model the only possible nonnative "interaction" comes from the topological entanglements, which are highly unlikely given the relatively small (48-mer) well-designed sequence.

B. Moderate Folders, Topological Frustration, and Kinetic Partitioning Mechanism

Many qualitative aspects of the folding kinetics of moderate folders can be understood in terms of the concept of topological frustration [10]. On average, about 55% of residues in proteins are hydrophobic, and their density along the sequence is roughly constant. As a result, on any local length scale there is a propensity for the hydrophobic residues to form tertiary contacts (structures) under folding conditions; that is, proximal residues adopt preferred structures. The assembly of the resulting structures would most likely be in conflict with the global native fold. The incompatibility of the low free-energy structures on local and global scales leads to a phenomenon called topological frustration. Topological frustration is an *intrinsic property of all foldable sequences* and arises due to the polymeric nature of proteins and the heterogeneity of amino acids. It follows that even the Gō model is topologically frustrated because residue connectivity can render certain favorable local structure incompatible with the global fold. An important physical outcome of topological frustration is that the free-energy folding landscape is rough, consisting of many minima that are separated by barriers of varying heights.

One of the principal consequences of topological frustration is that the folding kinetics follows the kinetic partitioning mechanism (KPM) [10]. Imagine an ensemble of unfolded molecules in search of the native conformation (Fig. 6). Due to the heterogeneity of folding pathways, a fraction of molecules, Φ, would reach the NBA (or N) rapidly without being kinetically trapped in the low-lying free-energy competing basins of attraction (CBA). The remaining fraction,

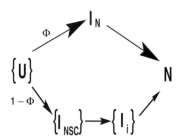

Figure 6. The sketch of the protein folding pathways. The fast (upper) folding pathway includes the formation of native-like collapsed states $\{I_N\}$, which rapidly convert into the native state N. The fraction of protein molecules, folding along this pathway, is Φ. For two-state folders, $\Phi \approx 1$. The lower track (followed by $1 - \Phi$ molecules) represents slow pathway(s), which fold by a three-stage kinetic mechanism. At the first stage, nonspecific collapse species I_{NSC} form, which later convert into a collection of discrete native-like intermediates $\{I_i\}$. The transition from $\{I_i\}$ to the native state is slow and represents the rate-limiting step in the slow pathway. The degree of heterogeneity in the folding pathways depends on the sequence and external conditions.

$(1-\Phi)$, would be trapped, and only on longer time scales would thermal fluctuations enable the chain to reach the NBA through an activated process. The value of the partition factor Φ depends on the sequence and external conditions. Thus, topological frustration leads to a separation of the initial ensemble of denatured molecules into fast- and slow-folding phases (Fig. 6). For two-state folders, which have a funnel-like free energy landscape, $\Phi = 1$.

According to the KPM $P_u(t)$ [see Eq. (3)] is given by

$$P_u(t) = \Phi \exp\left(-\frac{t}{\tau_N}\right) + \sum_k a_k \exp\left(-\frac{t}{\tau_k}\right) \tag{8}$$

where τ_N is the time scale for reaching the native state by the fast (direct) process (presumably by the nucleation-collapse), and τ_k is the time scale for indirect folding pathways, in which the native state is reached after escaping a local free-energy minimum (trap) k. Prefactors a_k are related to the "volumes" associated with the kth CBA. Thus, folding trajectories can be divided into those that reach the native conformation rapidly (their fraction or partition factor is Φ) and those that follow indirect off-pathway routes (Fig. 6).

The validity of the KPM has been demonstrated in several protein-like models beginning with the studies of Guo and Thirumalai [73]. More importantly, refolding experiments, on lysozyme [74] and large ribozymes [75] have confirmed the KPM. Using interrupted folding experiments, Kiefhaber [74] was the first to show that $\Phi \approx 0.15$ in lysozyme. Subsequent studies of lysozyme by Dobson and coworkers [76] show that $\Phi \approx 0.25$ in lysozyme. The difference is presumably due to changes in folding conditions. Perhaps the most direct demonstration of the validity of the KPM comes from the single-molecule FRET measurements on the L-21 Sca I ribozyme [77]. The results of these experiments analyzed by us showed that $\Phi \approx 0.06$, which is consistent with the estimates from ensemble measurements. These experiments show that KPM offers an unified picture of folding for a class of proteins and RNA [78].

VI. DISULFIDE BONDS IN FOLDING

A. Refolding of BPTI

Bovine pancreatic trypsin inhibitor (BPTI), a small protein with 56 amino acid residues (Fig. 7), is the first one for which a detailed map of the refolding pathways was deciphered. The native state of BPTI contains three disulfide (S–S) bonds formed between six Cys residues. Native state is specified by [30–51; 5–55;14–38] bonds. This notation indicates that Cys^{30} forms an S–S bond with Cys^{51}, and so on. Reduction of the S–S bonds unfolds BPTI. By using S–S bond formation as a "progress variable," Creighton [79–83] devised ingenious methods to trap the disulfide-bonded intermediates along the folding pathway.

Figure 7. See also color insert. The native-state conformation of the bovine pancreatic trypsin inhibitor (BPTI). The figure was produced with the program RasMol 2.7.1 [126] from the PDB entry 1bpi. There are three disulfide bonds in this protein: Cys5–Cys55 shown in red, Cys14–Cys38 shown in black, and Cys30–Cys51 shown in blue. The corresponding Cys residues are in the ball-and-stick representation and are labeled. The two helices (residues 2–7 and 47–56) are shown in green.

The refolding pathways were described in terms of the nature of the intermediates that accumulate during folding. There are 75 distinct intermediates containing one or more disulfide bonds that can be formed from six Cys residues. On the time scale of the experiments, Creighton discovered that only eight intermediates could be detected. These experiments were among the earliest to show that in the folding reaction only a small number of partially folded intermediates accumulates.

The most surprising discovery made by Creighton [79–83] was that in the refolding of BPTI, three non-native states—namely, the intermediates with disulfide bonds not present in the native state—are well-populated. More importantly, two of the non-native species, [30–51;5–14] and [30–51;5–38], are involved in the productive pathway; that is, folding proceeds through either of these two kinetically equivalent intermediates. The detection method employed by Creighton involves quenching the folding reaction using chemistry to stop the reaction. To isolate only the intermediate that would naturally occur in the refolding process, the quench rate must exceed rates of formation of other products. The chemistry of the quench method determines the time required to stop the reaction from progressing. Creighton's findings were challenged by

Weissman and Kim (WK) [84–87], who used pH changes (acid quenching) to disrupt the folding reaction. The most glaring difference between the two series of studies is that WK showed that, in the productive pathway, *only native intermediates* play a significant role. Non-native intermediates may only be involved as required by disulfide chemistry in the last stages of folding of BPTI; that is, they play a role in the formation of the precursor [30–51;5–55] from [30–51;14–38] (denoted by N_{sh}^{sh} and N', respectively).

In an attempt to resolve the apparent controversy between the findings of Creighton and WK, we introduced a phenomenological theory, referred to as the *proximity rule* [45], to predict the folding pathways in globular proteins. Our theory accounts for entropic effects analytically and energetic effects only approximately. The premise of the proximity rule is that local events, governed mainly by entropic considerations, dictate the initial events in protein folding. The importance of local events is the basis of the hierarchic mechanism of folding [11,12] and is also emphasized in the notion that contact order [37] is the primary determinant of folding rates of proteins. Just as in the applications of proximity rule, we expect that theories that rely largely on local events *can only* account for the early processes in folding. However, such theories often "work" in regimes for which they are not, in principle, applicable.

B. Proximity Rule

The major conformational changes in disulfide bonded proteins, such as BPTI and ribonuclease A [88], can be understood in terms of disulfide bond rearrangement. Thus, the conformations of the intermediates that determine the folding pathways are specified in terms of the S–S bonds. In such proteins the S–S bonds serve as a surrogate "reaction coordinate." These observations enable us to develop the proximity rule based on the following general principles.

1. Loop Formation Probability

We assume that the initial intramolecular disulfide bond rearrangement is a random process governed largely by entropic considerations. The probability of forming a disulfide bond under oxidizing conditions depends only on the loop length $l = |i - j|$, where i and j are the positions of the Cys residues along the polypeptide chain. The probability of simultaneous loop formation of lengths l_1 and l_2, $P(l_1, l_2)$, is assumed to be proportional to $P(l_1)P(l_2)$. The absence of correlation limits the theory to the prediction of only the earliest events in BPTI refolding. Similarly, theories that are based on local propensities alone can only describe the formation of secondary structures and initial tertiary structures in the folding of globular proteins. Despite this limitation, the utility of the proximity rule to predict the refolding pathways of BPTI was extended using parameters determined from experiments [45]. The loop formation probability $P(l)$ may be computed by modeling the polypeptide chain as a semiflexible chain.

2. *Folding Kinetics*

For slow-folding proteins, which require reconfiguration of partially folded structures, folding follows a three-stage kinetics [45]. These stages are as follows: (i) There is a rapid collapse of the chain to a set of compact conformations. At this stage, most of the free energy arises from a competition between hydrophobic forces and loop entropy. In BPTI this stage is characterized by the need to have proper loop contacts between Cys residues, so that a single S–S bond can form. At the end of this stage the most stable single disulfide species accumulate. (ii) The rearrangement of the single disulfide bonds leads to the formation of the native two-disulfide species. (iii) The rate-determining step involves the transition from the stable two-disulfide species to the native conformation. In this sequential progression bifurcations in the folding pathways are possible resulting in the parallel pathways to the native state.

Loop formation probability $P(l)$ may be obtained approximately using statistical mechanics of stiff chains [89]. Here, we provide the physical requirements. For chains with an effective persistence length l_p, we expect $P(l)$ to be negligible for $l < l_p$.[2] This is because the requisite self-avoidance criterion, bond angle, and dihedral angle constraints are violated for the loop lengths less than l_p. In the denatured conformations, excluded volume interactions are predominant; therefore for large enough l we expect $P(l)$ to decay as $\approx l^{-\theta_3}$ with $\theta_3 \approx 2.2$. Combining these requirements, we write $P(l)$ as

$$P(l) \approx \frac{1 - \exp(-l/l_p)}{l^{\theta_3}} \tag{9}$$

Experiments by Darby and Creighton [91], who measured the rates of formation of single disulfide intermediates in BPTI, can be understood using Eq. (9) for $P(l)$. The higher probability of forming loops between the ends of the chain is neglected in obtaining $P(l)$. This approximation should not have an effect in predicting the rates of single S–S bond formation in BPTI, but will be relevant in getting estimates of time scales for forming loops in polypeptide chains.

The general scheme described above has been applied to obtain approximately the refolding pathways in BPTI using experimentally determined rearrangement time τ_i for the transition from the single S–S intermediates to the double S–S species. Our results showed [45] that on a relatively long time scale, comparable to that used in the experiments by Creighton or WK, only native-like species should be populated. It may be that in the process of forming these native-like intermediates, certain non-native species identified by Creighton are transiently involved. Based on our estimate of τ_i, the transient

---

[2]In certain protein structures, loops with $l < l_p$ can form. However, such loops are stiff and often have very high strain energy [90].

population of non-native intermediates occurs on the time scales less than 30 seconds.

Because our theory is most accurate for predicting the ordering of single disulfide species, we focus on their rates and extents of accumulation. Considerations based on $P(l)$ suggest that only a small subset of the single disulfide intermediates can form. From $P(l)$ it follows that the probability of forming [14–38] is considerably greater than that of [5–55]. However using the kinetic constraints we have shown that although [14–38] forms rapidly and early in the folding process, its concentration decreases rapidly at subsequent times, whereas those of [30–51] and [5–55] increase. This specific prediction is one of the *striking outcomes* of the proximity rule [45]. The distinct kinetic behavior of the native [14–38] compared to the other two native single S–S intermediates is related to stability reasons [45,92]. The partially folded solvent-exposed state [14–38], which perhaps is the molten globule form of BPTI, can form without burying the hydrophobic core of the protein. On the other hand, the intermediates [5–55] and [30–51], in which the four Cys residues are in the interior, require the formation of the hydrophobic core of the protein (Fig. 7). The burial of hydrophobic residues that brings the Cys residues in proximity so those S–S bonds can form requires overcoming free energy barriers. This delays their formation compared to that of [14–38].

Proximity rule also predicts that the ratio of the maximum concentration of [30–51] to that of [5–55] is about $7:1$, whereas the concentration of [14–38] is negligible on the same time scale. This ratio is in excellent agreement with the experiments of WK, who found a ratio of $6:1$, and is in disagreement with Creighton's estimate of $20:1$.

The theoretical prediction that [14–38] should be the first intermediate to accumulate was *subsequently confirmed* by Dadlez and Kim [92]. Using oxidized glutathione (GSSH) to initiate disulfide bond formation and acid quenches to trap intermediates, they noted that the earliest intermediate that accumulates is [14–38]. The tenfold rearrangement of [14–38] compared to [30–51] or [5–55] was rationalized in terms of stability (see arguments given above). These findings are also consistent with the results for synthetic models, in which the Cys except at the positions 14 and 38 were replaced by α-amino-n-butyric acid (*Abu*) [93]. The folding of $[14–38]_{Abu}$ is similar to the formation of [14–38] in the wild type. This reinforces the notion that entropic considerations and overall hydrophobicity of BPTI rather than specific native interactions between the remaining cysteines, perhaps on the collapse time scale, determine the early formation of [14–38].

Despite being intensively studied, there are several major questions in the refolding of BPTI that are not understood. We mention two of them: (a) The *in vitro* folding pathways show that there are dead-end kinetic traps [84], which completely block the folding reaction. Weissman and Kim [87] showed that such kinetic traps are completely eliminated when the disulfide bonds rearran-gements are catalyzed by protein disulfide isomerase (PDI). The presence of PDI, which may be viewed as an intramolecular chaperone, enhances the folding rate by several thousands. The mechanism of action of PDI has not been elucidated. (b) In a beautiful experiment, Zhang and Goldenberg [94] showed that the dead-end kinetic traps in the wild-type BPTI are entirely eliminated by a single amino acid substitution. The mutant Y35L (tyrosine at position 35 is replaced by leucine) results in a rapid sequential pathway in which only native intermediates are populated. The simplistic explanation of this spectacular experiment is that the nonproductive intermediates in this mutant are destabilized. A fuller molecular explanation is required.

C.   Modeling the Role of S–S Bonds

A key disagreement between the early works [79–83] and the more recent studies WK on the refolding of BPTI is the role of non-native intermediates in directing the folding of BPTI. Creighton argued that not only were two non-native intermediates ([30–51;5–14] and [30–51;5–38]) accumulated substantially, but also they were equally involved in the productive folding pathways. WK showed that non-native states were not obligatory intermediates, and the only inter-mediates in the folding were native. Non-native intermediates may be involved in the transition state in the late stages of folding.

To clarify the relevance of non-native intermediates in the folding of proteins dictated by the formation of disulfide bonds Camacho and Thirumalai [45] used lattice models. While these models are merely caricatures of proteins, they contain the specific effects that can be studied in microscopic detail. We used a two-dimensional lattice sequence consisting of hydrophobic (H), polar (P), and Cys (C) residues. If two C beads are near neighbors on the lattice, they can form a S–S bond with an associated energy gain of $-\epsilon_s$ with $\epsilon_s > 0$. Thus, topological specificity is required for native S–S bond formation in this model. We have studied the folding kinetics of this model, which is perhaps the simplest model that can probe the characteristics of native and non-native disulfide bonded intermediates.

The sequence studied consists of $M = 23$ monomers, of which four represent C sites. The native conformation corresponds to [2–15;9–22] (Fig. 8a). The model sequence has six possible single and two disulfide intermediates including the native state. There are three *native* intermediates and two *non-native* inter-mediates. Even though the number of such intermediates are far less than the corresponding number in BPTI, it is sufficient to examine the crucial distinction between the roles played by native and non-native intermediates in the folding kinetics. Some of the questions that arise in the experimental studies of refolding of BPTI can be precisely answered using these simple models.

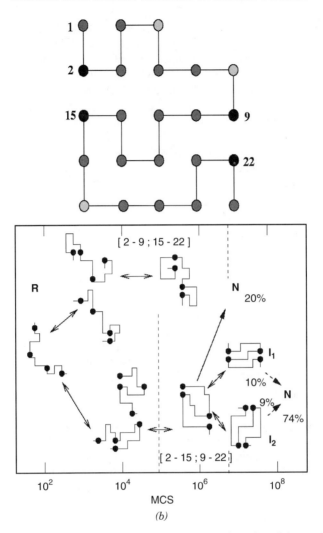

Figure 8. (See also color insert.) (a) The ground-state conformation of the two-dimensional model sequence with $M = 23$ beads and four covalent (S) sites. The red, green, and black circles represent, respectively, the hydrophobic (H), polar (P), and S sites. (b) Diagram of representative time snapshots along the main pathways of folding of the sequence in panel (a). The S sites are shown as black circles. Dotted lines delineate the three main folding regimes (random collapse, kinetics ordering and all-or-none). The arrows indicate the various transitions occurring in the system: the double-headed continuous arrows indicate backward and forward reactions where there is no substantial re-arrangement of the chain; the single-headed arrows indicate that the native-state is stable on the time scale of the simulations ($\sim 10^9$ MCS); the dashed arrows are for indirect transitions which occur by breaking the disulfide bonds and partial unfolding of the structure. The percentages indicate the concentration of the native and two native-like intermediates at the end of the second regime of kinetic ordering.

To probe the dynamical role played by the intermediates, we computed the time dependence of the concentration of the six species. The folding pathways are characterized in terms of the appearance of these intermediates (Fig. 8b). There are pathways that lead directly to **N** exclusively via native-like intermediates. In others, non-native intermediates are involved early in the folding process. For purposes of ascertaining the importance of the intermediates, all times are measured in terms of τ_F, the folding time. At the earliest time, $t < 10^{-5}\tau_F$ single disulfide species accumulate, whose probabilities of formation are determined by $P(l)$. At times that are roughly tenfold longer, the rearrangement of the nonnative single disulfide intermediates leads to the formation of two stable native single disulfide ([9–22] and [2–15]) species. These early intermediates act as seeds (nucleating sites) for subsequent formation of the native state [45]. At times on the order of about $10^{-4}\tau_F$, which coincide with the time at which native single disulfide species form, the concentration of these intermediates *cannot* be determined based on entropic considerations alone. Energetic considerations, such as favorable hydrophobic interactions, affect the formation of single disulfide intermediates.

In the second stage of the assembly we find that non-native two disulfide intermediate [2–9;15–22] can form transiently (Fig. 9b). Because this intermediate is unstable, it quickly rearranges to the more stable native **N** state. On relatively long time scales ($t \approx 0.01\tau_F$) we find that there are two native-like intermediates, in which the disulfides are in place but some other parts of the structure are not yet fully formed. This may be the analogue of the N_{sh}^{sh} state in BPTI which only needs the nearly solvent-exposed [14–38] bond to form. In the final stage of folding, structural fluctuations that transiently break the native S–S bonds enable the transition to **N**. This transition involves the transient formation of the non-native intermediate [2–9;15–22]. The two native-like intermediates I_1 and I_2 (Fig. 8b) rearrange almost exclusively via [2–9;15–22].

Even with an extremely simple model, several conclusions have been reached, which help clarify some of the issues in the refolding of BPTI. (1) Non-native species can form early in the folding process when bulk of the ordering is determined by entropic considerations. The current experiments on BPTI are far too slow to detect these early intermediates. On the time scale of collapse the more stable single disulfide species, which are native-like, form. (2) As the folding reaction progresses, native-like intermediates tend to form so that the productive pathways largely contain native-like intermediates. (3) The rate-determining step involves an activated transition from native-like species, via a high free-energy non-native transition state, to **N**. The transitions appear to involve rearrangement of the structure that does not involve the S–S bonds. These calculations suggest that although the folding pathways of BPTI can be described in terms of the disulfide intermediates, a complete description requires accounting for hydrophobic and charge effects as well. At present, these effects have not been completely examined experimentally or theoretically. The profound effect of point mutations [94] in altering the folding rates and the pathways of BPTI folding suggests that there are strong couplings between S–S bond formation and other forces that drive the native structure formation.

D. Engineering Disulfide Bonds in Barnase

To probe the folding pathways in BPTI the S–S bonds were initially reduced that results in unfolding. Refolding is initiated under oxidizing conditions that enable S–S bond formation and restoration of the native state. Alternatively, the impact of S–S bonds can be studied by engineering them at specific locations. With the S–S bonds intact, protein can be unfolded using denaturants such as urea. The folding kinetics can be initiated by diluting the denaturant. The latter procedure, which was first used by Clarke and Fersht [95], enables the study of the effect of intact S–S bonds on the stability and kinetics of folding. Clarke and Fersht used this procedure to engineer S–S bonds at two specific locations in barnase, whose folding without disulfide bonds has been well-characterized. This allows for a comparison of folding characteristics of proteins with and without disulfide bonds.

Two positions in barnase were constrained by S–S bonds that were left intact [95a]. One of them, between residues 85 and 102, connects two loops that apparently form early in the folding pathway of the wild type protein. A second disulfide between residues 43 and 80 connects two secondary structural elements. Barnase containing disulfide bonds is more stable than the wild type because the introduction of the S–S bond increases the free energy of the unfolded states. From the native state of barnase it is clear that the enhanced stability upon introduction of the disulfide bond between 43 and 80 cannot be accounted for solely by lowered entropy of the unfolded state compared to the WT. Using the Flory estimate we expect that stability of $[43–80]_{Bar}$ should be $1.5\,RT\ln 38 \approx 3.2\,\text{kcal/mol}$, whereas that of $[85–102]_{Bar}$ is $\approx 2.6\,\text{kcal/mol}$. These estimates do not compare favorably with the experimental values, which are 2.1 kcal/mol and 4.3 kcal/mol for $[43–80]_{Bar}$ and $[85–102]_{Bar}$, respectively. This suggests that the introduction of S–S bonds could also stabilize the native state to some extent.

Refolding kinetics of the mutated barnase depends strongly on the location of the S–S bond. Assuming that reduction in the conformation space leads to rate enhancement, we would predict that $[43–80]_{Bar}$ should fold faster than $[85–102]_{Bar}$. However, the opposite trend is found experimentally. The mutant with the shorter loop folds about five times more rapidly, whereas barnase with the disulfide between 43 and 80 folds two times slower than the wild type. Using lattice simulations, Abkevich and Shakhnovich [95b] argued that if S–S bonds are engineered into the regions highly structured in the transition state, refolding rates can be increased compared to the WT. The presence of S–S bonds elsewhere in the protein can either increase or decrease folding rates depending on the external conditions. Because the region containing residues 83 and 102 forms early in the folding process, it may be part of the folding nuclei. This explains the enhanced rate of folding of $[85–102]_{Bar}$ compared to WT. Because residues 43 and 80 are not part of the folding nuclei, the folding of $[43–80]_{Bar}$ is about 1.7 times slower than the WT. Thus, the simulations of Abkevich and Shakhnovich using simple lattice models are consistent with experiments.

VII. CHAPERONIN-FACILITATED PROTEIN FOLDING

According to the Anfinsen's hypothesis [88] natural proteins fold spontaneously to their lowest free energy states. By analyzing the weights of proteins and protein synthesis rates under glucose feeding conditions, Lorimer [96] estimated that in *Escherichia coli* more than 90% of proteins fold to their native states as envisaged by Anfinsen. This is remarkable because one might imagine that trafic (due to other macromolecules) in the crowded cellular environment might lead to strong intermolecular interactions which could potentially interfere with monomeric folding. Nevertheless, it appears that many proteins assemble spontaneously to their functionally competent states *in vivo* as envisioned by Anfinsen. However, there are some proteins that require the assistance of molecular chaperones to fold to the native conformation. The functions of the class I chaperonins belonging to heat shock protein family are the most extensively studied [97–100]. In this chapter we focus on insights into their function using simple lattice models [47,101,102].

The chaperonin family of proteins, namely GroEL and GroES, that function as a nanomachine by utilizing ATP, assist misfolded substrate proteins to reach their native states [100,103,104]. The crystal structures of GroEL [105], GroES [106], and the complex GroEL/GroES/ADP [107] have provided considerable insights into the chaperonin action. The chaperonin GroEL is a double-ringed oligomer consisting of two back-to-back stacked heptameric rings. It has an overall cylindrical structure divided into two nonconnected cavities, in which the substrate protein (SP) can be sequestered. Each subunit of the GroEL particle consists of three domains, namely, the equatorial domain, the inter-mediate domain, and the apical domain [100]. The heaviest of these is the equatorial domain, which contains more than half of the molecular weight of GroEL. We have argued that the concentration of dense inertial mass in the equatorial domain is necessary to generate the requisite force to peel the initially captured substrate protein (SP) from the apical domain. The concentric assembly of the subunits produces a ring structure having an architecture with an unusual sevenfold symmetry (Fig. 9a).

The co-chaperonin GroES, containing seven subunits [106], caps the GroEL particle as a dome. A remarkable feature, which has mechanistic implications, is that upon binding of GroES and ATP the volume of the cavity doubles [100]. This enhanced volume is accompanied by a series of concerted allosteric transition that the GroEL particle undergoes [108–110]. Because of the non-specificity of GroEL-SP interactions [111–113] and the plasticity of the architecture of the GroEL particle, this system acts as a "one size fits all" nanomachine.

Considerable progress in understanding the mechanism of this nanomachine has become possible due to a combination of an extraordinary body of experimental work [98,100] and some contributions from theoretical studies [114,115]. The hemicycle, which constitutes the fundamental functioning cycle of the GroEL machine [110], is schematically sketched in Fig. 9b. The process is initiated by the capture of the SP by the apical domain of the GroEL particle. To a first approximation, the mouth of the cavity can be thought of as a continuous hydrophobic surface formed by the helices in the apical domain. The nonspecific, but favorable, interaction between the SP and GroEL is due to the attraction between the exposed hydrophobic residues of the SP and the hydrophobic surface of the apical domain. Upon binding of ATP and GroES (in this specific order), significant concerted transitions occur in the GroEL particle. The series of transitions alters, in a fundamental way, the nature of interaction between GroEL and the SP [100]. Whereas in the process of capture the SP-GroEL interaction is attractive, the interaction is either neutral or even repulsive after encapsulation (step 2 in Fig. 9b). The surface remains hydrophilic until the restoration of GroEL to the initial state. This alteration between hydrophobic (H) and hydrophilic (P) surface enables this system to function as an annealing machine. The release of GroES and the encapsulated SP occurs when ATP and/or another SP molecule binds to the *trans*-ring [107].

Although the underpinnings of the cycle (Fig. 9b) are based on a number of experiments and theoretical arguments, several outstanding questions remain. A key issue is related to the coupling between the concerted allosteric transitions that the GroEL particle undergoes and the SP folding rate [47,116]. Consider the cycle displayed in Fig. 9b. Upon binding ATP to the upper ring, a cooperative transition $T \leftrightarrow R$ takes place. The terminology T and R are borrowed from the Monod–Wyman–Changeaux model [117] describing the binding of oxygen to hemoglobin. The tense state T has a higher affinity for ATP than the relaxed R state [109]. Upon binding ATP, the intermediate domain moves $25°$ toward the equatorial domain, which closes the ATP binding sites. Even with this relatively minor rigid body movement of the intermediate domain, the interaction between the SP and the walls (the apical domain) are weakened [108,109]. The weakened interaction is sufficient to enable the SP protein to unfold at least partially [118]. Subsequent binding of ATP and GroES leads to much larger domain movements in the GroEL particle. In particular, the apical domain moves upward by $60°$ and twists, with respect to the equatorial domain, by 90° [100]. This large segmental motion, which results in the encapsulation of the SP, doubles the volume of the cavity. Upon encapsulation, the interaction between the SP and the walls is either neural or repulsive depending on the size of SP. At least five independent rate constants are required to describe these large-scale concerted allosteric transitions in the GroEL particle [110]. This makes the description of the coupling between allostery of GroEL and the SP folding rate very difficult.

To examine the coupling between the allosteric transitions and SP folding rates, a model system may be considered in which the action of GroEL and ATP

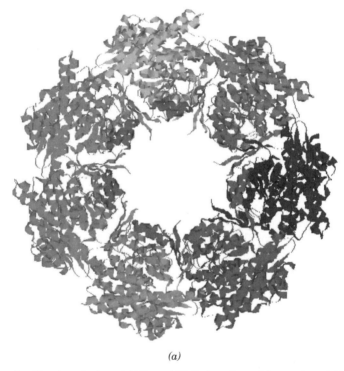

(a)

Figure 9. (See also color insert.) (a) Rasmol [126] view of one of the two rings of GroEL from the PDB file 1oel. The seven chains are indicated by different colors. The amino acid residues forming the binding site of the apical domain of each chain (199–204, helix H: 229–244 and helix I: 256–268) are shown in red. The most exposed hydrophobic amino acids that are facing the cavity and are implicated in the binding of the substrate as indicated by mutagenesis experiments [112, 127] are : Tyr199, Tyr203, Phe204, Leu234, Leu237, Leu259, Val263, and Val264. (b) A schematic sketch of the hemicycle in the GroEL–GroES-mediated folding of proteins. In step 1 the substrate protein is captured into the GroEL cavity. The ATPs and GroES are added in step 2, which results in doubling the volume, in which the substrate protein is confined. The hydrolysis of ATP in the *cis*-ring occurs in a quantified fashion (step 3). After binding ATP to the *trans*-ring, GroES and the substrate protein are released that completes the cycle (step 4).

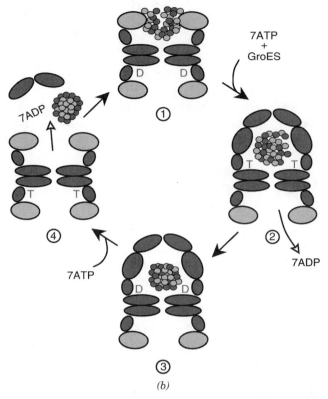

7ATP
+
GroES

7ADP

①

②

7ADP

③

④

7ATP

(b)

Figure 9 *(Continued)*

on SP can be investigated without the complication of the GroES interaction [116]. Many *in vitro* studies on the interaction between the GroEL and the SP have used only this subset [98]. In this reduced model system the equilibrium constant between T and R states and the time constants characterizing the SP folding are the only relevant parameters [110,116]. To examine the coupling in the reduced nanomachine, we modeled the central cavity as a cubic box, and a lattice model representation of the polypeptide chain was employed [47]. This, of course, is a highly simplified representation the GroEL–SP system. However, qualitative testable predictions of the coupling between allostery and the SP can be made using this caricature. Initially the interior walls of the GroEL particle (in the T state) are assumed to be hydrophobic. This description is reasonable, because in the T state the arrangement of the apical domain offers the SP a continuous lining of hydrophobic residues (Fig. 9a). We vary the wall hydrophobicity of GroEL by letting one particular residue (leucine) describe the wall character. Thus, the interactions between the wall and the SP protein is

$$E_c = \sum_i h E_{wi} \qquad (10)$$

where h $(0 \leq h \leq 1)$ gives the strength of the interaction, and E_{wi} is the contact interaction between the ith residue of the SP and the wall. The total energy of the encapsulated SP is given by the sum of Eq. (10) and the "internal" energy of the SP [Eq. (1)].

The key annealing action of the GroEL particle arises due to the changes in the hydrophobicity of its inner walls during the hemicycle [47]. In other words, during a single turnover the cavity changes from being able to capture the SP to that is which binding is not favored. This change in the wall character is accompanied by the allosteric transition of GroEL, resulting in the encapsulation of the SP. The effect of changing hydrophobicity is mimicked in our simplified model by letting the hydrophobicity of the confining cavity vary during the turnover time, τ_i. We divide τ_i into two subintervals. During a period τ_P the wall remains hydrophilic (P), and for the remainder $\tau_i - \tau_P$ the cavity is hydrophobic. Because the model does not include GroES, the situation we consider may serve as a model for the coupling between $T \leftrightarrow R$ transition and the SP folding. Here we have $\tau_P/(\tau_i - \tau_P) \sim L$, where L is the equilibrium constant between T and R [116]. By examining the effect of changing values of L on the rates of the SP folding the dependence of the SP folding rate on the allosteric equilibrium transitions can be examined. Simulations of the simplified lattice representation of the GroEL–SP system shows that there is an inverse correlation between the extent of the $T \leftrightarrow R$ transition and the folding rate of the SP. In other words, as the cooperativity of the $T \leftrightarrow R$ transition increases (higher values of L), the slower is the SP folding rate.

A. Unfolding Activity of GroEL

Although it is accepted that the GroEL nanomachine rescues the SP by stochastically enabling it to sample the rough free-energy landscape, the microscopic action on the SP has only recently become clear. A few experiments have shown that upon change in the wall characteristics of GroEL the SP unfolds partially, if not globally. Using hydrogen exchange labeling, Zahn et al. [118] showed that GroEL accelerates the exchange of highly protected amide protons. Because highly protected protons (high protection factor) are typically buried in the core of the SP, it follows that the SP unfolds in the presence of GroEL.

Nieba-Axmann et al. [119] also examined the plausible structural fluctuations in GroEL-bound cyclophilin A (CypA) using amide-proton exchange measurements. In the absence of nucleotides and GroES, folding of CypA is extremely sluggish. Upon addition of ADP, the rate increases by a factor of about 2.5, whereas the addition of ATP leads to a threefold enhancement in the folding rate. However, when GroES is added, the rate increases by a factor of 14 at 6°C and by nearly 30-fold at $T = 30°C$. The near independence of the refolding on nucleotides suggests that the full recovery of CypA occurs within a single turnover.

Upon binding of ATP and GroES, the domain moves upward and twists by 90° about the equatorial domain [100]. This results in the weakening of the interaction between the SP and the walls of the cavity. The simple lattice model described above, in which the character of the wall changes from H to P, can model the chaperonin-assisted folding provided the folding reaction is complete in a single turnover. To examine the structure of the polypeptide chain due to alterations in the wall character, we computed the inherent structures of the chain in the on state (hydrophobic wall) and in the off state (hydrophilic wall). According to the iterative annealing mechanism [103,104], upon going from the on state to the off state the polypeptide chain should undergo kinetic partitioning [Eq. (8)]. The inherent structures prior to and immediately following the change in the cavity characteristics allows us to compute the degree of commitment of the SP to folding. Because the GroEL machine operates stochastically, there ought to be a distribution of states of the SP that are populated as the on–off transition takes place. The simulations show (see Fig. 9 of Ref. 47) that before the transition in the cavity, a fraction of molecules is committed to folding, while most of the conformations fall into basins of attraction corresponding to misfolded or unfolded states. After the transition to the off state the chain is largely unfolded. This shows that upon weakening of the SP–GroEL interaction, which occurs as the GroEL particle undergoes the allosteric transitions, the polypeptide chain globally unfolds. In other words, the chief mechanism operative in the GroEL-mediated folding is that chaperonins *help fold proteins by globally unfolding them!* This is consistent with the predictions of IAM and is also affirmed by several experiments [118–120].

The simulations using simplified models are entirely consistent with several experiments including the one reported by Nieba-Axmann et al. [119], who noted that amide protons that are highly protected from hydrogen exchange in the native state of CypA in the absence of GroEL become much less protected when bound to the chaperonin. The protection factor decreases by nearly two orders of magnitude upon binding to GroEL. Thus binding to GroEL shifts the equilibrium from compact native-like states to globally unfolded conformations. In this dynamical picture of GroEL action, as opposed to the static Anfinsen cage model, chaperonins unfold the SP. It also follows that efficient folding can be induced by repeated unfolding of the chain.

B. Unfolding by Stretching

If, in the course of the allosteric transitions of the GroEL particle, the SP is unfolded, a natural question is, what is the mechanism of the GroEL-SP interaction that globally unfolds the protein? Lorimer and co-workers [120] have explored this issue using hydrogen–tritium exchange experiments in chaperonin-assisted folding of RUBISCO. They observed that within the time scale of a single turnover (approximately 13 seconds), complete exchange of tritiums takes place. This shows that RUBISCO unfolds at least partially, if not globally. From the crystal structures of GroEL and the GroEL–ADP–GroES complex, it is known that upon undergoing a series of concerted allosteric transitions, two adjacent subunits that are about 25 Å apart in the T state are about 33 Å apart in the R'' state [107,120]. This large-scale movement is presumed to generate force on the SP [99]. Recent pulling experiments on several proteins [121] show that the native state can be fully unfolded if a force exceeding a threshold value is applied. The magnitude of the threshold force depends on the SP [122].

To estimate the value of the force imparted to the SP, it is necessary to obtain the interaction energy between the SP and the apical domain of the GroEL particle. The SP–GroEL interaction energy must exceed $\frac{3}{2}k_B T S_{mis}$, where S_{mis} is the translational entropy of the misfolded chain molecule for capture by GroEL to occur. Assuming that the subunits move apart by about 0.2 nm, we estimate that the minimum force required to peel off the SP from the apical domain is about 35 pN. A more precise estimate of the interaction energy between the SP and GroEL can be made by assuming that the inner lining of the GroEL cavity can be modeled as a hydrophobic wall onto which the SP is adsorbed [114]. By balancing the free energy gain due to favorable hydrophobic interaction between GroEL and SP and the entropy loss due to the pinning of the SP, we estimate that the interaction energy should not exceed about $10 k_B T$ [114]. The force needed to overcome this interaction is about 200 pN. This value is large enough to unfold immunoglobin proteins with β-sandwich topology [121]. We suspect that at these values of the forces most substrate proteins can at least partially unfold. These estimates give credence to the notion that it is the generation of force in the power stroke of the chaperonin machinery that unfolds the SP [120].

The estimate of the force given above is only an average force. A given SP molecule can bind to a subset of the seven subunits [123]. Because of the heterogeneity of the conformations of the misfolded SP, we expect variations in the binding states from molecule to molecule. Thus, there should be a distribution of unfolding forces. Recent AFM experiments [124] show that this distribution is very broad, indicating that there is a large sample to sample variation in the unbinding forces. Such large variations for other SP can only be measured using single-molecule measurements. Surely, these sample-to-sample variations in lifetimes of the complexes [125] and forces imparted to SP will require revisions of the iterative annealing mechanism [103].

VIII. CONCLUSIONS

Protein folding presents significant challenges because the parameter space is extremely large. From the myriad of experimental and theoretical studies it is not clear that there are many general principles that govern the kinetics of folding. Nevertheless, using simple models, several precise predictions have been made. In this chapter we have described the utility of simple lattice models and phenomenological theories in answering very specific questions in protein folding. It is remarkable that these simple ideas have been fruitful in enabling us to formulate conceptual questions such as the physical basis for the emergence of structures and their designability. Lattice models can also be used to understand qualitatively the importance of intermediates in the folding of proteins that are controlled by the stability of disulfide bonds. Experimentally testable predictions in the field of assisted folding also have been made using caricatures of the chaperonin systems. These practical applications attest to the utility of these models in providing a conceptual understanding of the basic principles in a variety of problems.

None of the applications described here can be tackled using a "realistic" all-atom representation of proteins. The precise predictions that we and others have made using coarse-grained models of proteins are currently beyond the reach of molecular dynamics simulations. In this sense, we hope that this chapter serves as a challenge to the practitioners of all-atom simulations. Even assuming that the interaction potentials are adequate, the severe restriction on the simulation time scales acts as a major constraint. The lack of reliable potentials and the accessible computer times has prevented straightforward use of molecular dynamics calculations from being a predictive tool. There is hope that, in the next few years, unlimited computer power may be unleashed to obtain a detailed picture of how proteins fold. This potential comes from developments in distributed computing that can, in principle, be used to generate several long trajectories. In the applications that we have carried out, a rather detailed picture of β-hairpin assembly for several sequences has been obtained (D. K. Klimov, D. Newfield, and D. Thirumalai, unpublished results). A different, but related, approach has also been undertaken by Pande and co-workers (V. Pande, private communication). The development of a high-performance computer by IBM also has raised the specter of hope that computational bottlenecks may be overcome in the next few years so that challenging problems such as biomolecular folding can be undertaken. Even if these tolls are routinely available, simple concepts will play a pivotal role in formulating the issues in the study of biomolecules, because in the ultimate analysis protein folding (or any other problem in molecular biology) is not *merely* a computational problem.

Acknowledgments

We are grateful to Carlos Camacho, Zhuyan Guo, and George H. Lorimer for illuminating discussions and George Rose for providing us with the Fig. 1. We thank David Goldenberg for a critical reading of the article. This work was supported in part by the National Science Foundation grant CHE99-75150.

References

1. T. E. Creighton, *Proteins: Structures and Molecular Principles*, W. H. Freeman, New York, 1993.

2. T. Schindler, M. Herrier, M. A. Marahiel, and F. X. Schmid, Extremely rapid protein folding in the absence of intermediates. *Natur. Struct. Biol.* **2**, 663–673 (1995).

3. C. B. Anfinsen, Principles that govern the folding of protein chains. *Science* **181**, 223–230 (1973).

4. R. L. Baldwin, Why is protein folding so fast. *Proc. Natl. Acad. Sci. USA* **93**, 2627–2628 (1996).

5. C. Levinthal, in *Mossbauer Spectroscopy in Biological Systems*, P. Debrunner, J. C. M. Tsibris, and E. Munck, eds., University of Illinois Press, Urbana, IL, 1968.

6. K. A. Dill and H. S. Chan, From Levinthal to pathways to funnels. *Natur. Struct. Biol.* **4**, 10–19 (1997).

7. A. R. Fersht, Nucleation mechanisms in protein folding. *Curr. Opin. Struct. Biol.* **7**, 3–9 (1997).

8. W. A. Eaton, V. Munoz, P. A. Thompson, C. K. Chan, and J. Hofrichter, Submillisecond kinetics of protein folding. *Curr. Opin. Struct. Biol.* **7**, 10–14 (1997).

9. V. S. Pande, A. Y. Grosberg, T. Tanaka, and D. S. Rokhsar, Pathways for protein folding: Is a new view needed? *Curr. Opin. Struct. Biol.* **8**, 68–79 (1998).

10. D. Thirumalai and D. K. Klimov, Deciphering the timescales and mechanisms of protein folding using minimal off-lattice models. *Curr. Opin. Struct. Biol.* **9**, 197–207 (1999).

11. R. L. Baldwin and G. D. Rose, Is protein folding hierarchic? I. Local structure and peptide folding. *Trends Biochem. Sci.* **24**, 26–33 (1999).

12. R. L. Baldwin and G. D. Rose, Is protein folding hierarchic? II. Folding intermediates and transition states. *Trends Biochem. Sci.* **24**, 77–83 (1999).

13. J. Rumbley, L. Hoang, L. Mayne, and S. W. Englander, An amino acid code for protein folding. *Proc. Natl. Acad. Sci. USA* **98**, 105–112 (2001).

14. P. T. Lansbury, Evolution of amyloids: What normal protein folding can tell us about fibrillogenesis and disease. *Proc. Natl. Acad. Sci. USA* **96**, 3342–3344 (1999).

15. S. Williams, T. P. Cosgrove, R. Gillmanshin, K. S. Fang, R. H. Callender, W. H. Woodruff, and R. B. Dyer, Fast events in protein folding: Helix melting and formation in a small peptide. *Biochemistry* **35**, 691–697 (1996).

16. G. S. Huang and T. G. Oas, Submillisecond folding of monomeric λ repressor. *Proc. Natl. Acad. Sci. USA* **92**, 6878–6882 (1995).

17. V. Munoz, P. A. Thompson, J. Hofrichter, and W. A. Eaton, Folding dynamics and mechanism β-hairpin formation. *Nature* **390**, 196–199 (1997).

18. A. G. Ladurner and A. R. Fersht, Upper limit of the time scale for diffusion and chain collapse in chymotrypsin inhibitor 2. *Nature Struct. Biol.* **6**, 28–31 (1999).

19. K. Kuwata, R. Shastry, H. Cheng, M. Hashino, C. A. Batt, Y. Goto, and H. Roder, Structural and kinetic characterization of early folding events in β-lactoglobulin. *Natur. Struct. Biol.* **8**, 151–155 (2001).

20. M. Rief, M. Gautel, F. Oesterhelt, J. M. Fernandez, and H. E. Gaub, Reversible unfolding of individual titin immunoglobulin domains by AFM. *Science* **276**, 1109–1112 (1997).

21. M. S. F. Kellermayer, S. B. Smith, H. L. Granzier, and C. Bustamante, Folding–unfolding transitions in single titin molecules characterized with laser tweezers. *Science* **276**, 1112–1116 (1997).

22. A. F. Oberhauser, P. E. Marszalek, M. Carrion-Vazquez, and J. M. Fernandez, Single protein misfolding events captured by atomic force microscopy. *Natur. Struct. Biol.* **6**, 1025–1028 (1999).

23. H. Li, A. F. Oberhouser, S. B. Fowler, J. Clarke, and J. M. Fernandez, Atomic force microscopy reveals the mechanical design of a modular protein. *Proc. Natl. Acad. Sci. USA* **97**, 6527–6531 (2000).

24. T. E. Fisher, A. F. Oberhauser, M. Carrion-Vazquez, P. E. Marszalek, and J. M. Fernandez, The study of protein mechanics with the atomic force microscope. *Trends Biochem. Sci.* **24**, 379–384 (1999).

25. A. R. Fersht, Characterizing transition states in protein folding. *Curr. Opin. Struct. Biol.* **5**, 79–84 (1995).

26. L. S. Itzhaki, D. E. Otzen, and A. R. Fersht, The nature of the transition state of chymotrypsin inhibitor 2 analyzed by protein engineering methods: Evidence for a nucleation–condensation mechanism for protein folding. *J. Mol. Biol.* **254**, 260–288 (1995).

27. E. L. McCallister, E. Alm, and D. Baker, Critical role of β-hairpin in protein G folding. *Natur. Struct. Biol.* **7**, 669–673 (2000).

28. S. E. Jackson, How do small single domain proteins fold. *Fold. Des.* **3**, R81–R91 (1998).

29. R. A. Goldstein, Z. A. Luthey-Schulten, and P. G. Wolynes, Optimal protein-folding codes from spin-glass theory. *Proc. Natl. Acad. Sci. USA* **89**, 4918–4922 (1992).

30. A. Sali, E. Shakhnovich, and M. Karplus, Kinetics of protein folding: A lattice model study of the requirements for folding to the native state. *J. Mol. Biol.* **235**, 1614–1636 (1994).

31. J. D. Bryngelson, J. N. Onuchic, N. D. Socci, and P. G. Wolynes, Funnels, pathways and energy landscape of protein folding, *Proteins: Struct. Funct. Genet.* **21**, 167–195 (1995).

32. D. K. Klimov and D. Thirumalai, Factors governing the foldability of proteins. *Proteins: Struct. Funct. Genet.* **26**, 411–441 (1996).

33. J. N. Onuchic, Z. A. Luthey-Schulten, and P. G. Wolynes, Theory of protein folding: An energy landscape perspective. *Annu. Rev. Phys. Chem.* **48**, 545–600 (1997).

34. D. K. Klimov and D. Thirumalai, Linking rates of folding in lattice models of proteins with underlying thermodynamic characteristics. *J. Chem. Phys.* **109**, 4119–4125 (1998).

35. D. S. Riddle, V. P. Grantcharova, J. V. Santiago, E. Alm, I. Ruczinski, and D. Baker, Experiment and theory highlight role of native topology in SH3 folding. *Natur. Struct. Biol.* **6**, 1016–1024 (1999).

36. D. Baker, A surprising simplicity of protein folding. *Nature* **405**, 39–42 (2000).

37. K. Plaxco, K. T. Simons, and D. Baker, Contact order, transition state placement and the refolding rates of single domain proteins. *J. Mol. Biol.* **277**, 985–994 (1998).

38. J. Clarke, E. Cota, S. B. Fowler, and S. J. Hamill, Folding studies of immunoglobulin-like β-sandwich proteins suggest that they share a common folding pathway. *Struct. Fold. Des.* **7**, 1145–1153 (1999).

39. K. A. Dill, S. Bromberg, K. Yue, K. M. Fiebig, D. P. Yee, P. D. Thomas, and H. S. Chan, Principles of protein folding—a perspective from simple exact models. *Protein Sci.* **1995**, 561–602 (1995).

40. D. Thirumalai, From minimal models to real proteins: Time scales for protein folding. *J. Phys. I* **5**, 1457–1467, 1995.

41. A. R. Dinner, A. Sali, L. J. Smith, C. M. Dobson, and M. Karplus, Understanding protein folding via free-energy surfaces from theory and experiment. *Trends Biochem. Sci.* **25**, 331–339 (2000).

42. Y. Duan and P. A. Kollman, Pathways to a protein folding intermediate observed in a 1-microsecond simulation in aqueous solution. *Science* **282**, 740–744 (1998).

43. D. Thirumalai, D. K. Klimov, and M. R. Betancourt, Exploring the folding mechanisms using lattice models, in *Monte Carlo Approach to Biopolymers and Protein Folding*, P. Grassberger, G. T. Barkema, and W. Nadler, eds., Singapore, World Scientific, 1998, pp. 19–28.

44. D. Thirumalai and D. K. Klimov, Emergence of stable and fast folding protein structures, in S. Kim, K. J. Lee, and W. Sung, eds., *Stochastic Dynamics and Pattern Formation in Biological Systems*, American Institute of Physics, New York, 2000, pp. 95–111.

45. C. Camacho and D. Thirumalai, Theoretical predictions of folding pathways using the proximity rule with applications to BPTI. *Proc. Natl. Acad. Sci. USA* **92**, 1277–1281 (1995).

46. C. Camacho and D. Thirumalai, Modeling the role of disulfide bonds in protein folding: Entropic barriers and pathways. *Proteins Struct. Funct. Gen.* **22**, 27–40 (1995).

47. M. R. Betancourt and D. Thirumalai, Exploring the kinetic requirements for enhancement of protein folding rates in the GroEL cavity. *J. Mol. Biol.* **287**, 627–644 (1999).

48. W. J. C. Orr, Statistical treatment of polymer solutions at infinite dilution. *Trans. Faraday Soc.* **43**, 12–27 (1947).

49. H. Taketomi, Y. Ueda, and N. Gō, Studies on protein folding, unfolding, and fluctuations by computer simulation. *Int. J. Pept. Protein Res.* **7**, 445–459 (1975).

50. H. S. Chan and K. A. Dill, Intrachain loops in polymers: Effects of excluded volume. *J. Chem. Phys.* **90**, 493–509 (1989).

51. N. D. Socci and J. N. Onuchic, Kinetic and thermodynamic analysis of protein-like hetero-polymers: Monte carlo histogram technique. *J. Chem. Phys.* **103**, 4732–4744 (1995).

52. V. I. Abkevich, A. M. Gutin, and E. I. Shakhnovich, Specific nucleus as the transition state for protein folding: Evidence from the lattice model. *Biochemistry* **33**, 10026–10036 (1994).

53. S. Miyazawa and R. L. Jernigan, Estimation of effective inter-residue contact energies from protein crystal structures: Quasi-chemical approximation. *Macromolecules* **18**, 534–552 (1985).

54. A. Kolinski, A. Godzik, and J. Skolnick, A general method for the prediction of the three dimensional structure and folding pathway of globular proteins: Application to designed helical proteins. *J. Chem. Phys.* **98**, 7420–7433 (1993).

55. M. R. Betancourt and D. Thirumalai, Pair potentials for protein folding: Choice of reference states and sensitivity of predicted native states to variations in the interaction schemes. *Prot. Sci.* **8**, 1–8 (1999).

56. L. A. Mirny and E. I. Shakhnovich, How to derive a protein folding potential? A new approach to an old problem. *J. Mol. Biol.* **264**, 1164–1179 (1996).

57. D. Tobi and R. Elber, Distance-dependent, pair potential for protein folding: Results from linear optimization. *Proteins Struct. Funct. Gen.* **41**, 40–46 (2000).

58. N. Gō, Theoretical studies of protein folding. *Annu. Rev. Biophys. Bioeng.* **12**, 183–210 (1983).

59. C. Clementi, P. A. Jennings, and J. N. Onuchic, How native-state topology affects the folding of dihydrofolate reductase and interleukin-1. *Proc. Natl. Acad. Sci. USA* **97**, 5871–5876 (2000).

60. D. K. Klimov and D. Thirumalai, Cooperativity in protein folding: From lattice models with side chains to real proteins. *Fold. Des.* **3**, 127–139 (1998).

61. J. L. Martin, Computer enumerations, in *Phase Transitions and Critical Phenomena*, C. Domb and M.S. Green, eds., Academic Press, New York, 1974, pp. 102–110.

62. N. Metropolis, A. W. Rosenbluth, M. N. Rosenbluth, A. H. Teller, and E. Teller, Equation of state calculations by fast computing machines. *J. Chem. Phys.* **21**, 1087–1092 (1953).

63. A. M. Ferrenberg and R. H. Swendsen, Optimized monte carlo data analysis. *Phys. Rev. Lett.* **63**, 1195–1198 (1989).

64. R. V. Pappu, R. Srinivasan, and G. D. Rose, The flory isolated-pair hypothesis is not valid for polypeptide chains: Implications for protein folding. *Proc. Natl. Acad. Sci. USA* **97**, 12565–12570 (2000).

65. Z. Guo and D. Thirumalai, Kinetics and thermodynamics of folding of a de novo designed four-helix bundle. *J. Mol. Biol.* **263**, 323–343 (1996).

66. S. Govindarajan, R. Recabarren, and R. Goldstein, Simulating the total number of protein folds. *Proteins Struct. Funct. Gen.* **35**, 408–414 (1999).

67. H. Li, N. Wingreen, and C. Tang, Emergence of preferred structures in a simple model of protein folding. *Science* **273**, 666–669 (1996).

68. P. G. Wolynes, Symmetry and the energy landscape of biomolecules. *Proc. Natl. Acad. Sci. USA* **93**, 14249–14255 (1996).

69. Per-Anker Lindgard and H. Bohr, Magic numbers in protein structures. *Phys. Rev. Lett.* **77**, 779–782 (1996).

70. M. E. Holtzer, E. G. Lovett, D. A. d'Avignon, and A. Holtzer, Thermal unfolding in a gcn4-like leucine zipper: $^{13}C^{\alpha}$ nmr chemical shifts and local unfolding curves. *Biophys. J.* **73**, 1031–1041 (1997).

71. A. Bachman and T. Kiefhaber, Apparent two-state tendamistat folding is a sequential process along a defined pathway. *J. Mol. Biol.* **306**, 375–386 (2001).

72. V. S. Pande and D. S. Rokhsar, Folding pathway of a lattice model for proteins. *Proc. Natl. Acad. Sci. USA* **96**, 1273–1278 (1999).

73. Z. Guo and D. Thirumalai, Kinetics of protein folding: Nucleation mechanism, time scales and pathways. *Biopolymers*, **36**, 83–103 (1995).

74. T. Kiefhaber, Kinetic traps in lysozyme folding. *Proc. Natl. Acad. Sci. USA* **92**, 9029–9033 (1995).

75. T. Pan and T. R. Sosnick, Intermediates and kinetic traps in the folding of large ribozyme revealed by UV and CD spectroscopies and catalytic activity. *Nature Struct. Biol.* **14**, 931–938 (1997).

76. A. Matagne, S. E. Radford, and C. M. Dobson, Fast and slow tracks in lysozyme folding: Insight into the role of domains in the folding process. *J. Mol. Biol.* **267**, 1068–1074 (1997).

77. X. Zhuang, L. E. Bartley, A. P. Babcock, R. Russell, T. Ha, D. Herschlag, and S. Chu, A single-molecule study of RNA catalysis and folding. *Science* **288**, 2048–2051 (2000).

78. D. Thirumalai, D. K. Klimov, and S. A. Woodson, Kinetic partitioning as a unifying theme in the folding of biomolecules. *Theor. Chem. Acc.* **1**, 149–156 (1997).

79. T. E. Creighton, Renaturation of the reduced bovine pancreatic trypsin inhibitor. *J. Mol. Biol.* **87**, 563–577 (1974).

80. T. E. Creighton, Conformational restrictions on the pathway of folding and unfolding of BPTI. *J. Mol. Biol.* **113**, 275–293 (1977).

81. T. E. Creighton, Energetics of folding and unfolding of pancreatic trypsin inhibitor. *J. Mol. Biol.* **113**, 295–312 (1977).

82. T. E. Creighton, Effects of urea and guanidine. Hcl on the folding and unfolding of pancreatic trypsin inhibitor. *J. Mol. Biol.* **113**, 313–328 (1977).

83. T. E. Creighton and D. P. Goldenberg, Kinetic role of meta-stable native like two disulphide species in the folding transition of bovine pancreatic trypsin inhibitor. *J. Mol. Biol.* **179**, 497–526 (1984).

84. J. S. Weissman and P. S. Kim, Reexamination of the folding of BPTI: Predominance of native intermediates. *Science* **253**, 1386–1393 (1991).

85. J. S. Weissman and P. S. Kim, Kinetic role of non-native species in the folding of bovine pancreatic trypsin inhibitor. *Proc. Nat. Acad. Sci. USA* **89**, 9900–9904 (1992).

86. J. S. Weissman and P. S. Kim, The pro region of BPTI facilitates folding. *Cell* **71**, 841–851 (1992).

87. J. S.Weissman and P. S. Kim, Efficient catalysis of disulphide bond rearrangements by protein disulphide isomerase. *Nature* **365**, 185–188 (1993).

88. C. B. Anfinsen, Principles that govern the folding of protein chains. *Science* **181**, 223–230 (1973).

89. D. Thirumalai, Time scales for the formation of the most probable tertiary contacts in proteins with applications to cytochrome C. *J. Phys. Chem.* **103**, 608–610 (1999).

90. D. K. Klimov and D. Thirumalai, Mechanisms and kinetics of β-hairpin formation. *Proc. Natl. Acad. Sci. USA* **97**, 2544–2549 (2000).

91. N. J. Darby and T. E. Creighton, Dissecting the disulphide-coupled folding pathway of bovine pancreatic trypsin inhibitor—forming 1st disulphide bonds in analogues of the reduced protein. *J. Mol. Biol.* **232**, 873–886 (1993).

92. M. Dadlez and P. S. Kim, A third native one-disulphide intermediate in the folding of bovine pancreatic trypsin inhibitor. *Nature Struct. Biol.* **2**, 674–679 (1995).

93. M. Ferrer, G. Barany, and C. Woodward, Partially folded molten globule and molten coil states of bovine pancreatic trypsin inhibitor. *Nature Struct. Biol.* **2**, 211–217 (1995).

94. J. X. Zhang and D. P. Goldenberg, Amino acid replacement that eliminates kinetic traps in the folding pathway of pancreatic trypsin inhibitor. *Biochemistry* **32**, 14075–14081 (1993).

95a. J. Clarke and A. R. Fersht, Engineered disulfide bonds as probes of the folding pathway of barnase—increasing the stability of proteins against the rate of denaturation. *Biochemistry* **32**, 4322–4329 (1993).

95b. V. I. Abkevich and E. I. Shakhnovich. What can disulfide bonds tell us about protein energetics, function and folding: Simulations and bioinformatic analysis. *J. Mol. Biol.* **300**, 975–985 (2000).

96. G. H. Lorimer, A quantitative assessment of the role of the chaperonin proteins in protein folding *in vivo*. *FASEB J.* **10**, 5–9 (1996).

97. A. Richardson, S. J. Landry, and C. Georgopolulos, The ins and outs of a molecular chaperone machine. *Trends Biochem. Sci.* **23**, 138–143 (1998).

98. W. A. Fenton and A. L. Horwich, GroEL-mediated protein folding. *Prot. Sci.* **6**, 743–760 (1997).

99. G. H. Lorimer, Folding with a two-stroke motor. *Nature* **388**, 720–723 (1997).

100. Z. Xu and P. B. Sigler, GroEL/GroES: Structure and function of a two-stroke folding machine. *J. Struct. Biol.* **124**, 129–141 (1999).

101. C. D. Sfatos, A. M. Gutin, V. Abkevich, and E. I. Shakhnovich, Simulations of chaperone-assisted folding. *Biochemistry* **35**, 334–339 (1996).

102. H. S. Chan and K. A. Dill, A simple model of chaperonin-mediated protein folding. *Prot. Struct. Funct. Genet.* **24**, 345–351 (1996).

103. M. J. Todd, G. H. Lorimer, and D. Thirumalai, Chaperonin-facilitated protein folding: Optimization of rate and yield by an iterative annealing mechanism. *Proc. Natl. Acad. Sci. USA* **93**, 4030–4035 (1996).

104. F. J. Corrales and A. R. Fersht, Toward a mechanism of GroEL–GroES chaperone activity: An ATPase-gated and pulsed folding and annealing cage. *Proc. Natl. Acad. Sci. USA* **93**, 4509–4512 (1996).

105. K. Braig, Z. Otwinowski, R. Hegde, D. C. Boisvert, A. Joachimiak, A. L. Horwich, and P. B. Sigler, The crystal structure of the bacterial chaperonin at 2.8 Å. *Nature* **371**, 578–586 (1994).

106. J. F. Hunt, A. J. Weaver, S. J. Landry, L. Gierasch, and J. Deisenhofer, The crystal structure of the GroES co-chaperonin at 2.8 Å resolution. *Nature* **379**, 37–49 (1996).

107. Z. Xu, A. Horwich, and P. B. Sigler, The crystal structure of the asymmetric GroEL–GroES–(ADP)₇ chaperonin complex. *Nature* **388**, 741–750 (1997).

108. J. Ma, P. B. Sigler, Z. H. Xu, and M. Karplus, A dynamic model for allosteric mechanism for GroEL. *J. Mol. Biol.* **302**, 303–313 (2000).

109. O. Yifrach and A. Horovitz, Nested cooperativity in the ATPase activity of the oligomeric chaperonin GroEL. *Biochemistry* **34**, 5303–5308 (1995).

110. D. Thirumalai and G. H. Lorimer, Chaperonin-mediated protein folding. *Annu. Rev. Biophys. Biomol. Struct.* **30**, 245–269 (2001).

111. P. V. Viitanen, A. A. Gatenby, and G. H. Lorimer, Purified chaperonin 60 (GroEL) interacts with the non-native states of a multitude of *Escherichia coli* proteins. *Prot. Sci.* **1**, 363–369 (1992).

112. L. Chen and P. B. Sigler, The crystal structure of a GroEL/peptide complex: Plasticity as a basis for substrate diversity. *Cell* **99**, 757–768 (1999).

113. J. Chatellier, A. M. Buckle, and A. R. Fersht, GroEL recognizes sequential and nonsequential linear structural motifs compatible with extended β-strands and α-helices. *J. Mol. Biol.* **292**, 163–172 (1999).

114. D. Thirumalai, Theoretical perspectives on *in vitro* and *in vivo* folding, in S. Doniach, editor, *Statistical Mechanics, Protein Structure, and Protein–Substrate Interactions*, Plenum, New York, 1994, pp. 115–134.

115. K. Gulukota and P. G. Wolynes, Statistical mechanics of kinetic proof reading in protein folding *in vivo. Proc. Natl. Acad. Sci. USA* **91**, 9292–9296 (1994).

116. O. Yifrach and A. Horovitz, Coupling between protein folding and allostery in the GroE chaperonin system. *Proc. Natl. Acad. Sci.* **97**, 1521–1524 (2000).

117. J. Monod, J. Wyman, and J. P. Changeaux, On the nature of allosteric interactions: A plausible model. *J. Mol. Biol.* **12**, 88–118 (1965).

118. R. Zahn, S. Perrett, G. Stenberg, and A. R. Fersht, Catalysis of amide proton exchange by the molecular chaperones GroEL and SecB. *Science* **271**, 642–645 (1996).

119. S. E. Nieba-Axmann, M. Ottinger, K. Wuthrich, and A. Pluckthun, Multiple cycles of global unfolding of GroEL-bound cyclophilin A evidenced by NMR. *J. Mol. Biol.* **271**, 803–818 (1997).

120. M. Shtilerman, G. H. Lorimer, and S. W. Englander, Chaperonin function: Folding by forced unfolding. *Science* **284**, 822–825 (1999).

121. T. E. Fisher, P. E. Marszalek, and J. M. Fernandez, Stretching single molecules into novel conformations using the atomic force microscope. *Nat. Struct. Biol.* **9**, 719–724 (2000).

122. D. K. Klimov and D. Thirumalai, Native topology determines force-induced unfolding pathways in globular proteins. *Proc. Natl. Acad. Sci. USA* **97**, 7254–7259 (2000).

123. G. W. Farr, K. Furtak, M. B. Rowland, N. A. Ranson, H. R. Saibil, T. Kirchhausen, and A. L. Horwich, Multivalent binding of non-native substrate proteins by the chaperonin GroEL. *Cell* **100**, 561–573 (2000).

124. A. Vinckier, P. Gervasoni, F. Zaugg, U. Ziegler, P. Lidner, P. Groscurth, A. Pluckthun, and G. Semenza, Atomic force microscopy detects changes in the interaction forces between GroEL and substrate proteins. *Biophys. J.* **74**, 3256–3263 (1998).

125. M. B. Viani, L. I. Pietrasanta, J. B. Thompson, A. Chand, I. C. Gebeshuber, J. H. Kindt, M. Richter, H. G. Hansma, and P. K. Hansma, Probing protein–protein interactions in real time. *Nature Struct. Biol.* **7**, 644–647 (2000).

126. R. Sayle and E. J. Milner-White, Rasmol: Biomolecular graphics for all. *Trends Biochem. Sci.* **20**, 374–376 (1995).

127. W. A. Fenton, Y. Kashi, K. Furtak, and A. L. Horwich, Residues in chaperonin GroEL required for polypeptide binding and release. *Nature* **371**, 614–619 (1994).

PROTEIN RECOGNITION BY SEQUENCE-TO-STRUCTURE FITNESS: BRIDGING EFFICIENCY AND CAPACITY OF THREADING MODELS

JAROSLAW MELLER

*Department of Computer Science, Cornell University, Ithaca, NY, U.S.A.;
and Department of Computer Methods, Nicholas Copernicus
University, Torun, Poland*

RON ELBER

Department of Computer Science, Cornell University, Ithaca, NY, U.S.A.

CONTENTS

*Computational Methods for Protein Folding: A Special Volume of Advances in Chemical Physics,
Volume 120*, Edited by Richard A. Friesner. Series Editors I. Prigogine and Stuart A. Rice.
ISBN 0-471-20955-4. © 2002 John Wiley & Sons, Inc.

I. INTRODUCTION

The threading approach [1–8] to protein recognition is a generalization of the sequence-to-sequence alignment. Rather than matching the unknown sequence S_i to another sequence S_j (one-dimensional matching), we match the sequence S_i to a shape $\mathbf{X}_j$ (three-dimensional matching). Experiments found a limited set of folds compared to a large diversity of sequences. A shape has (in principle) more detectable "family members" compared to a sequence, suggesting the use of structures to find remote similarities between proteins. Hence, the determination of overall folds is reduced to tests of sequence fitness into known and limited number of shapes.

The sequence–structure compatibility is commonly evaluated using reduced representations of protein structures. Assuming that each amino acid residue is represented by a point in three-dimensional space, one may define an effective energy of a protein as a sum of inter-residue interactions. The effective pair energies can be derived from the analysis of contacts in known structures. Knowledge-based pairwise potentials proved to be very successful in fold recognition [2,3,6,9–11], *ab initio* folding [11–13], and sequence design [14–15].

Alternatively, one may define the so-called "profile" energy [1,5] taking the form of a sum of individual site contributions, depending on the structural environment (e.g., the solvation/burial state or the secondary structure) of a site. The above distinction is motivated by computational difficulties of finding optimal alignments with gaps when employing pairwise models.

Consider the alignment of a sequence $S = a_1 a_2 \ldots a_n$ of length, n, where a_i is one of the 20 amino acids, into a structure $\mathbf{X} = (x_1, x_2, \ldots, x_m)$ with m sites, where x_j is an approximate spatial location of an amino acid (taken here to be the geometric center of the side chain). We wish to place each of the amino acids in a corresponding structural site $\{a_i \rightarrow x_j\}$. No permutations are allowed. In order to identify homologous proteins of different length, we need to consider deletions and insertions into the aligned sequence. For that purpose we introduce an "extended" sequence, $\bar{S}$ which may include gap "residues" (spaces, or empty structural sites) and deletions (removal of an amino acid, or an amino acid corresponding to a virtual structural site).

Our goal is to identify the matching structure $\mathbf{X}_j$ with the extended sequence $\bar{S}_i$. The process of aligning a sequence S into a structure $\mathbf{X}$ provides an optimal score and the extended sequence $\bar{S}$. This double achievement can be obtained using dynamic programming (DP) algorithm [16–19]. In DP the computational effort to find the optimal alignment (with gaps and deletions) is proportional to $n \times m$, as compared to exponential number $(\approx 2^{n+m})$ of all possible alignments.

In contrast to profile models, the potentials based on pair interactions do not lead to optimal alignments with dynamic programming. A number of heuristic algorithms that provide approximate alignments have been proposed [20]. These algorithms cannot guarantee an optimal solution with less than exponential number of operations [21]. Another common approach is to approximate the energy by a profile model (the so-called frozen environment approximation) and to perform the alignment using DP [22]. In this work, we are aiming at deriving systematic approximations to pair energies that would preserve the computational simplicity of profile models.

Threading protocols that are based exclusively on pairwise models were shown to be too sensitive to variations in shapes [23]. Therefore, pairwise potentials are often employed in conjunction with various complementary "signals," such as sequence similarity, secondary structures, or family profiles [9–11,24–28]. Such additional signals enhance the recognition when the tertiary contacts are significantly altered. In GenTHREADER [9], for example, sequence alignment methods are employed as the primary detection tools. A pairwise threading potential is then used to evaluate the consistency of the sequence alignments with the underlying structures. Bryant and co-workers use, in turn, an energy function which is a weighted sum of a pairwise threading potential and a sequence substitution matrix [10].

Distant-dependent pair energies are expected to be less sensitive to variations in shapes than simple contact models, in which inter-residues interactions are assumed to be constant up to a certain cutoff distance and are set to zero at larger distances. A number of distance-dependent pairwise potentials have been proposed in the past [29,30]. We consider both simple contact models and distance-dependent power law potentials and compare their performance with that of novel profile models.

We compute the energy parameters by linear programming (LP) [31–33]. There are a number of alternative approaches to derive the energy parameters. For example, statistical analysis of known protein structures makes it possible to extract "mean-force" potentials [34–38]. Another approach is the optimization of a single target function that depends on the vector of parameters such as T_f/T_g [39], the Z score [1], or the σ parameter [40]. We note also that optimization of the gap energies has been attempted in the past [22,41]. The statistical analysis is the least expensive computationally. The optimization approaches have the advantage that misfolded structures can be made part of the optimization, providing a more complete training. The LP approach is computationally more demanding compared to other protocols. However, it has important advantages, as discussed below.

In LP training we impose a set of linear constraints (for energy models linear in their parameters) of the general form

$$\Delta E_{\text{dec, nat}} \equiv E_{\text{decoy}} - E_{\text{native}} > 0 \qquad (1)$$

where E_{native} is the energy of the native alignment (of a sequence into its native structure) and E_{decoy} represents the energies of the alignments into non-native (decoy) structures. In other words, we require that the energies of native alignments be lower than the energies of alignments into misfolded (decoy) structures.

While optimization of the Z, T_f/T_g, and σ scores led to remarkably successful potentials [1,39,40], it focuses at the center of the distribution of the $\Delta E_{\text{dec, nat}}$'s and does not solve exactly the conditions of Eq. (1). For example, the tail of the distribution of the $\Delta E_{\text{dec, nat}}$ may be slightly wrong, and a fraction f of the $\Delta E_{\text{dec, nat}}$'s may "leak" to negative values. If f is small, it may not leave a significant impression on the first and second moments of the distribution; that is, the value of the Z score remains essentially unchanged. "Tail misses" is not a serious problem if we select a native shape from a small set of structures. However, when examining a large number of constraints, even if f is small, the number of inequalities that are not satisfied can be very large, making the selection of the native structure difficult if not impossible.

In contrast to the optimization of average quantities, the LP approach guarantees that all the inequalities in Eq. (1) are satisfied. If the LP cannot find a solution, we get an indication that it is impossible to find a set of parameters that solve all the inequalities in Eq. (1). For example, we may obtain the impossible condition that the contact energy between two ALA residues must be smaller than 5 and at the same time must be larger than 7. Such an infeasible solution is an indicator that the current model is not satisfactory, and more parameters or changes in the functional form are required [31–33]. Hence, the LP approach, which focuses on the tail of the distribution near the native shape, allows us to learn continuously from new constraints and improve further the energy functions, guiding the choice of their functional form.

In the present chapter we evaluate several different scoring functions for sequence-to-structure alignments, with parameters optimized by LP. Based on a novel profile model, designed to mimic pair energies, we propose an efficient threading protocol of accuracy comparable to that of other contact models. The new protocol is complementary to sequence alignments and can be made a part of more complex fold recognition algorithms that use family profiles, secondary structures, and other patterns relevant for protein recognition.

The first half of the chapter is devoted to the design of scoring functions. Two topics are discussed: the choice of the functional form (Section II) and the choice of the parameters (Section III). The capacity of the energies is explored and optimal parameters are determined (Section IV). High capacity indicates that a large number of protein shapes are recognized with a small number of parameters.

The second part of the manuscript deals with optimal alignments. We design gap energies (Section V) and introduce a double Z-score measure (from global and local alignments) to assess the results (Section VI). Presentation of extensive tests of the algorithm (Section VII) is followed by the conclusions and closing remarks.

II. FUNCTIONAL FORM OF THE ENERGY

In a nutshell there are two "families" of energy functions that are used in threading computations, namely the pairwise models (with "identifiable" pair interactions) and the profile models. In this section we formally define both families and we also introduce a novel THreading Onion Model (THOM), which is investigated in the subsequent sections of the chapter.

A. Pairwise Models

The first family of energy functions is of pairwise interactions. The score of the alignment of a sequence S into a structure $\mathbf{X}$ is a sum of all pairs of interacting amino acids,

$$E_{\text{pairs}} = \sum_{i<j} \phi_{ij}(\alpha_i, \beta_j, r_{ij}) \tag{2}$$

The pair interaction model, ϕ_{ij}, depends on the distance between sites i and j and also depends on the types of amino acids, α_i and β_j. The latter are defined by the alignment, because certain amino acid residues $a_k, a_l \in S$ are placed in sites i and j, respectively.

We consider two types of pairwise interaction energies. The first is the widely used contact potential. If the geometric centers of the side chains are closer than 6.4 Å, then the two amino acids are considered in contact. The total energy is a sum of the individual contact energies:

$$\phi_{ij}(\alpha_i, \beta_j, r_{ij}) = \left\{ \begin{array}{ll} \varepsilon_{\alpha\beta}, & 1.0 < r_{ij} < 6.4 \,\text{Å} \\ 0, & \text{otherwise} \end{array} \right\} \tag{3}$$

where i, j are the structure site indices (contacts due to sites in sequential vicinity are excluded, $i = 3 < j$), α, β are indices of the amino acid types (we drop the subscripts i and j for convenience), and $\varepsilon_{\alpha\beta}$ is a matrix of all the possible contact types. For example, it can be a 20×20 matrix for the twenty amino acids.

TABLE I
The Definitions of Different Groups of Amino Acids That Are Used in the Present Study[a]

Hydrophobic (HYD)	ALA CYS HIS ILE LEU MET PHE PRO TRP TYR VAL
Polar (POL)	ARG ASN ASP GLN GLY LYS SER THR
Charged (CHG)	ARG ASP GLU LYS
Negatively charged (CHN)	ASP GLU

[a]Note that 10 types of amino acids are found to be sufficient to solve the Hinds–Levitt set either by pairwise interaction models or by THOM2 (in the case of continuous LJ models, HIS was replaced by CYS). The amino acid types are HYD, POL, CHG, CHN, GLY, ALA, PRO, TYR, TRP, and HIS. The list implies that when an amino acid appears explicitly, it is excluded from other groups that may contain it. For example, HYD includes in this case CYS, ILE, LEU, MET, and VAL, while CHG includes ARG and LYS only, since the negatively charged residues form a separate group.

Alternatively, it can be a smaller matrix if the amino acids are grouped together to fewer classes. Different groups that are used in the present study are summarized in Table I. The entries of $\varepsilon_{\alpha\beta}$ are the target of parameter optimization.

The advantage of the single-step potential is its simplicity. This is also its weakness. From a chemical physics perspective the interaction model is over-simplified and does not include the (expected) distance-dependent interaction between pairs of amino acids. To investigate a potential with more "realistic" shape we also consider a "distance power" potential:

$$\phi_{ij}(\alpha_i, \beta_j, r_{ij}) = \frac{A_{\alpha\beta}}{r_{ij}^m} + \frac{B_{\alpha\beta}}{r_{ij}^n} \tag{4}$$

Here two matrices of parameters are determined: one for the m power, $A_{\alpha\beta}$, and one for the n power, $B_{\alpha\beta}$ ($m > n$). The signs of the matrix elements are determined by the optimization. In "physical" potentials like the Lennard-Jones model we expect $A_{\alpha\beta}$ to be positive (repulsive) and $B_{\alpha\beta}$ to be negative (attractive). The indices m and n cannot be determined by LP techniques and have to be decided on in advance. A suggestive choice is the widely used Lennard-Jones [LJ(12,6)] model ($m = 12$, $n = 6$). In contrast to the square well, the LJ(12,6) form does not require a prespecification of the arbitrary cutoff distance, which is determined by the optimization. It also presents a continuous and differentiable function that is more realistic than the square well model.

We show in Section IV that the LJ(12,6), commonly employed in atomistic simulations, performs poorly when applied to inter-residue interactions. Therefore other continuous potentials of the type described in Eq. (5) were investigated. We propose a shifted LJ potential (SLJ) that has significantly higher capacity compared to LJ and is closer in performance to that of the square well potential.

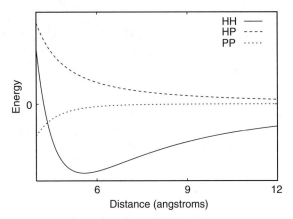

Figure 1. A sample plot of the Lennard-Jones-like potential that we developed. The functional form is $A_{\alpha\beta}/r_{ij}^6 + B_{\alpha\beta}/r_{ij}^2$ (LJ(6,2)), where the indices α and β denote the amino acid types and the indices i and j are the positions along the chain. $A_{\alpha\beta}$ and $B_{\alpha\beta}$ are optimized using the LP approach. The plot includes interactions of the types HH, HP, and PP, where H stands for hydrophobic and P stands for polar residues, respectively. The coefficients A and B are given in Table 7a. Note that the usual Lennard-Jones potential (LJ(12,6)) has a poor recognition capacity.

The SLJ is based on the replacement of $A_{\alpha\beta}/r_{ij}^{12}$ by $A_{\alpha\beta}/(r_{ij} + a)^{12}$, where a is a constant that we set to 1 Å.

The SLJ is a smoother potential with a broader minimum. An alternative potential that also creates a smoother and wider minimum is obtained by changing the distance powers. We also optimized a potential with the (unusual) ($m = 6$, $n = 2$) pair. This choice was proven most effective and with the largest capacity of all the continuous potentials that we tried (Fig. 1).

B. Profile Models

The second type of energy function assigns "environment" or a profile to each of the structural sites [1]. The total energy $E_{profile}$ is written as a sum of the energies of the sites:

$$E_{profile} = \sum_i \phi_i(\alpha_i, \mathbf{X}) \tag{5}$$

As previously, α_i denotes the type of an amino acid a_k of S that was placed at site i of $\mathbf{X}$. For example, if a_k is a hydrophobic residue and x_i is characterized as a hydrophobic site, the energy $\phi_i(\alpha_i, \mathbf{X})$ will be low (score will be high). If a_k is charged, then the energy will be high (low score). The total score is given by a sum of the individual site contributions.

We consider two profile models. The first, which is very simple, was used in the past as an effective solvation potential [1,2,42]. We call it THOM1 (THreading Onion Model 1), and it suggests a clear path to an extension (which is our prime model), namely, THOM2. The "onion" level denotes the number of contact shells used to describe the environment of the amino acid. The THOM1 model uses one "contact" shell of amino acids. The more detailed THOM2 energy model (to be discussed below) is based on two layers of contacts.

In the "profile" potential THOM1, the total energy of the protein is a direct sum of the contributions from m structural sites and can be written as

$$E_{\text{THOM1}} = \sum_i \varepsilon_{\alpha_i}(n_i) \tag{6}$$

The energy of a site depends on two indices: (a) the number of neighbors to the site, n_i [a neighbor is defined as for pairwise interaction—Eq. (2)], and (b) the type of the amino acid at site i, α_i. For 20 amino acids and a maximum of 10 neighbors we have 200 parameters to optimize, a number that is comparable to the detailed pairwise model.

THOM1 provides a nonspecific interaction energy, which, as we show in Section IV, has relatively low prediction ability when compared to pairwise interaction models. THOM2 is an attempt to improve the accuracy of the environment model, making it more similar to pairwise interactions. In order to mimic pair energies, we first define the energy $\varepsilon_{\alpha_i}(n_i, n_j)$ of a contact between structural sites i and j, where n_i is the number of neighbors to site i and n_j is the number of contacts to site j (see Fig. 2). The type of amino acid at site i is α_i. Only one of the amino acids in contact is "identifiable." The total contribution due to a site i is then defined as a sum over all contacts to this site $\phi_{i,\text{THOM2}}$ $(\alpha_i, \mathbf{X}) = \sum_j' \varepsilon_{\alpha_i}(n_i, n_j)$, with the prime indicating that we sum only over sites j that are in contact with i (i.e., over sites j satisfying the condition $1.0 < r_{ij} < 6.4$ Å and $|i - j| \geq 4$). The total energy is finally given by a double sum over i and j:

$$E_{\text{THOM2}} = \sum_i \sum_j{}' \varepsilon_{\alpha_i}(n_i, n_j) \tag{7}$$

Consider a pair of sites (i, j) which are in contact and occupied by amino acids of types α_i and α_j. Let the number of neighbors of site i be n_i, and let for site j be n_j. The effective energy contribution of the (i, j) contact is

$$V_{ij}^{\text{eff}} = \varepsilon_{\alpha_i}(n_i, n_j) + \varepsilon_{\alpha_j}(n_j, n_i) \tag{8}$$

Hence, we can formally express the THOM2 energy as a sum of approximate pair energies $E_{\text{THOM2}} = \sum_{i<j} V_{ij}^{\text{eff}}$.

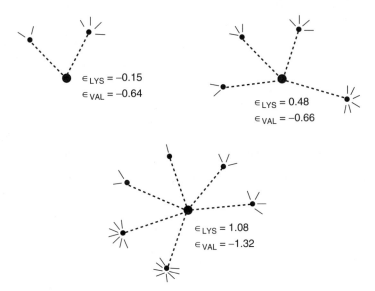

Figure 2. A schematic representation of the interactions with the THOM2 potential. THOM2 assigns scores according to two contact shells. As an example we show a sample of contacts to a site and the associated energies for valine and lysine. As expected, the hydrophobic residue (valine) strongly prefers to be at a site with a large number of neighbors in the first and second shells. Lysine is the extreme case on the polar side.

The effective energy mimics the formalism of pairwise interactions. However, in contrast to the usual pair potential the alignments with THOM2 can be done efficiently. Structural features alone (the number of the contacts) determine the "identity" of the neighbor. The structural features are fixed during the computations, making it possible to use dynamic programming. This is in contrast to pairwise interactions for which the identity of the neighbor may vary during the alignment. For 20 amino acids, the number of parameters for this model can be quite large. Assuming a maximum of 10 neighbors, we have $20 \times 10 \times 10 = 2000$ entries to the parameter array. In practice we use a coarse-grained model leading to a reduced set of structural environments (types of contacts) as outlined in Table II.

The use of a reduced set makes the number of parameters (300 when all 20 types of amino acids are considered) comparable to that of the contact potential. Further analysis of the new model is included in Section IV.

III. OPTIMIZATION OF THE ENERGY PARAMETERS

Here we consider the amino acid interactions (the gap energies are discussed in Section V). In order to optimize the energy parameters, we employ the so-called

TABLE II
Definitions of Contact Types for the THOM2 Energy Model[a]

Type of Site[b]	$n=1,2;\ \bar{1}$	$n=3,4,5,6;\ \bar{5}$	$n \geq 7;\ \bar{9}$
$n=1,2;\ \bar{1}$	$(\bar{1},\bar{1})$	$(\bar{1},\bar{5})$	$(\bar{1},\bar{9})$
$n=3,4;\ \bar{3}$	$(\bar{3},\bar{1})$	$(\bar{3},\bar{5})$	$(\bar{3},\bar{9})$
$n=5,6;\ \bar{5}$	$(\bar{5},\bar{1})$	$(\bar{5},\bar{5})$	$(\bar{5},\bar{9})$
$n=7,8;\ \bar{7}$	$(\bar{7},\bar{1})$	$(\bar{7},\bar{5})$	$(\bar{7},\bar{9})$
$n \geq 9;\ \bar{9};$	$(\bar{9},\bar{1})$	$(\bar{9},\bar{5})$	$(\bar{9},\bar{9})$

[a]The THOM2 model defines an energy of a site as a sum of contributions due to contacts to this site. A contact between two amino acids is "on" if their distance is smaller than 6.4 Å. Different types of contacts are defined by the number of neighbors to the two sites involved in contact i.e., the information about the first and second contact layer of a site is used (see Fig. 2). We consider five types of sites in the first layer (primary site i occupied by an amino acid of known type) and three types of sites in the second layer (secondary site j with no amino acid type assigned). Therefore, there are $5 \times 3 = 15$ types of contacts. The primary site i may be occupied by any of the 20 amino acids, leading to $20 \times 15 = 300$ different energy terms. A reduced set of amino acids is associated with a smaller number of parameters to optimize (for 10 types of amino acids, the number of parameters is $10 \times 15 = 150$). The notation we used for each type of site is based on a representative number of neighbors. The number of neighbors n in a given class and its representative are given in the first column (for different classes of sites in the first layer) and in the first row (for different classes of sites in the second layer). The intersections between columns and rows correspond to contacts of different types: a contact between two sites of medium number of neighbors is denoted by $(\bar{5},\bar{5})$, for example.

gapless threading in which the sequence S_i is fitted into the structure $\mathbf{X}_j$ with no deletions or insertions. Hence, the length of the sequence (n) must be shorter or equal to the length of the protein chain (m). If n is shorter than m, we may try $m - n + 1$ possible alignments varying the structural site of the first residue $\{a_1 \rightarrow x_1, x_2, \ldots, x_{m-n+1}\}$.

The energy (score) of the alignment of S into $\mathbf{X}$ is denoted by $E(S,\mathbf{X},\mathbf{p})$, where $\mathbf{X}$ stands (depending on the context) either for the whole structure or only for a substructure of length n, relevant for a given gapless alignment. The energy function, $E(S,\mathbf{X},\mathbf{p})$, depends on a vector $\mathbf{p}$ of q parameters (so far undetermined). A proper choice of the parameters will get the most from a specific functional form, where we restrict the discussion below to knowledge-based potentials.

Consider the sets of structures $\{\mathbf{X}_i\}$ and sequences $\{S_j\}$. There is a corresponding energy value for each of the alignments of the sequences $\{S_j\}$ into the structures $\{\mathbf{X}_i\}$. A good potential will make the alignment of the "native" sequence into its "native" structure the lowest in energy. If the exact structure is not in the set, alignments into homologous proteins are also considered "native." Let $\mathbf{X}_n$ be the native structure. A condition for an exact recognition potential is

$$E(S_n, \mathbf{X}_j, \mathbf{p}) - E(S_n, \mathbf{X}_n, \mathbf{p}) > 0, \qquad \forall\, j \neq n \tag{9}$$

In the set of inequalities (9) the coordinates and sequences are given and the unknowns are the parameters that we need to determine. We first describe the sets used to train the potential and then describe the technique to solve the above inequalities.

A. Learning and Control Sets

Two sets of protein structures and sequences are used for the training of parameters in the present study. Hinds and Levitt developed the first set [43] that we call the HL set. It consists of 246 protein structures and sequences. Gapless threading of all sequences into all structures generated the 4,003,727 constraints [i.e., the inequalities of Eq. (8)]. The gapless constraints were used to determine the potential parameters for the 20 amino acids. Because the number of parameters does not exceed a few hundred, the number of inequalities is larger than the number of unknowns by many orders of magnitude.

The second set of structures consists of 594 proteins and was developed by Tobi et al. [32]. It is called the TE set and is considerably more demanding. It includes some highly homologous proteins (up to 60% sequence identity) and poses a significant challenge to the energy function. For example, the set is infeasible for the THOM1 model, even when using 20 types of amino acids (see Section IV). The total number of inequalities that were obtained from the TE set using gapless threading was 30,211,442. The TE set includes 206 proteins from the HL set.

We developed two other sets that are used as control sets to evaluate the new potentials in terms of both gapless and optimal alignments. These control sets contain proteins that are structurally dissimilar to the proteins included in the training sets. The degree of dissimilarity is specified in terms of the RMS distance between the structures. The structure-to-structure alignments (necessary for RMS calculations) were computed according to a novel algorithm [45].

The new structural alignment is based on dynamic programming and provides for closely related structures results that are comparable to the DALI program [44]. Contrary to DALI, we employ (consistently with our threading potentials) the side-chain coordinates, and not the backbone (C_α) atoms, while overlapping two structures (in fact, in analogy with THOM2, we overlap the contact shells, disregarding however the identities of amino acids). Thus, the results of our structure-to-structure alignments refer to superimposed side-chain centers. Our cutoff for structural dissimilarity is 12 Å RMSD.

The first control set, which is referred to as S47, consists of 47 proteins representing families not included in the training. This includes 25 structures used in the CASP3 competition [46] and 22 related structures chosen randomly from the list of VAST [47] and DALI [44] relatives of CASP3 targets. None of the 47 structures has homologous counterparts in the HL set, and only three have counterparts in the TE set. As measured by our novel (both global and local) structure-to-structure alignments, the remaining proteins differ from those in the training sets by at least 12 Å with respect to HL set and 9.3 Å with respect to TE set (the RMS distance is larger than 12 Å for all but seven shorter proteins), respectively.

The second control set, referred to as S1082, consists of 1082 proteins that were not included in the TE set and which are different by at least 3 Å RMSD (measured, as previously, between the superimposed side chain centers) with respect to any protein from the TE set and with respect to each other. Thus, the S1082 set is a relatively dense (but nonredundant up to 3 Å RMSD) sample of protein families. The training and control sets are available from the web [48].

B. Linear Programming Protocol

The "profile" energies and the pairwise interaction models that were discussed in Section II can be written as a scalar product:

$$E = \sum_{\gamma} n_{\gamma} p_{\gamma} \equiv \mathbf{n} \cdot \mathbf{p} \tag{10}$$

where $\mathbf{p}$ is the vector of parameters that we wish to determine. The index of the vector, γ, is running over the types of contacts or sites. For example, in the pairwise interaction model the index γ is running over the identities of the amino acid pairs (e.g., a contact between alanine and arginine). In the THOM1 model it is running over the types of sites characterized by the identity of the amino acid at the site and the number of its neighbors. n_{γ} is the number of contacts, or sites of a specific type found in a fold. The "number" may include additional weight. For example, the number of alanine–alanine contacts in a protein is (of course) an integer. However, in the Lennard-Jones model, the contact type $A_{\alpha,\beta} \equiv p_{\gamma}$ is associated with additional geometric weight hidden in a continuous "number" function, $n_{\gamma} \propto 1/r^m$.

In the pairwise contact model, there are 210 types of contacts for the 20 amino acids. We have experimented with different representations and different numbers of amino acid types. While the Hinds–Levitt set can be solved with a reduced number of parameters, the more demanding requirements of the larger set necessitates (for all models presented here) the use of at least 210 parameters.

We wish to emphasize that the linear dependence of the potential energies on their parameters is not a major formal restriction. Any potential energy $E(\mathbf{X})$ can be expanded in terms of a basis set (say $\{n_{\gamma}(\mathbf{X})\}_{\gamma=1}^{\infty}$) in which the coefficients are unknown parameters:

$$E(S, \mathbf{X}, \mathbf{p}) = \sum_{\gamma=1}^{\infty} p_{\gamma} n_{\gamma}(\mathbf{X}) \tag{11}$$

Note that we deliberately used a similar notation to Eq. (11) and that the information on $\mathbf{X}$ and S is "buried" in $n_\gamma(\mathbf{X})$. A good choice of the basis set will converge the sum to the right solution with only a few terms. Of course, such a choice is not trivial to find, and one of the goals of the present chapter is to explore different possibilities.

The linear representation of the energy simplifies Eq. (9) as follows:

$$E(S_n, \mathbf{X}_j, \mathbf{p}) - E(S_n, \mathbf{X}_n, \mathbf{p}) = \sum_\gamma p_\gamma(n_\gamma(\mathbf{X}_j) - n_\gamma(\mathbf{X}_n))$$

$$= \mathbf{p} \cdot \Delta\mathbf{n}_j > 0 \qquad \forall j \neq n \qquad (12)$$

Hence, the problem is reduced to the condition that a set of inner vector products will be positive. Standard linear programming tools can solve Eq. (12). We use the BPMPD program of C. S. Meszaros [49], which is based on the interior point algorithm. We seek a point in parameter space that satisfies the constraints, and we do not optimize a function in that space. In this case, the interior point algorithm places the solution at the "maximally feasible" point, which is at the center of the accessible volume of parameters [50].

The set of inequalities that we wish to solve includes tens of millions of constraints that could not be loaded into the computer memory directly (we have access to machines with two to four gigabytes of memory). Therefore, the following heuristic approach was used. Only a subset of the constraints is considered, namely, $\{\mathbf{p} \cdot \Delta\mathbf{n} < C\}_{j=1}^J$, with a threshold C chosen to restrict the number of inequalities to a manageable size (which is about 500,000 inequalities for 200 parameters). Hence, during a single iteration, we considered only the inequalities that are more likely to be significant for further improvement by being smaller than the cutoff C.

The subset $\{\mathbf{p} \cdot \Delta\mathbf{n} < C\}_{j=1}^J$ is sent to the LP solver "as is." If proven infeasible, the calculation stops (no solution possible). Otherwise, the result is used to test the remaining inequalities for violations of the constraints [Eq. (12)]. If no violations are detected, the process was stopped (a solution was found). If negative inner products were found in the remaining set, a new subset of inequalities below C was collected and sent to the LP solver. The process was repeated, until it converged. Sometimes convergence was difficult to achieve, and human intervention in the choices of the inequalities was necessary. Nevertheless, all the results reported in the present chapter were iterated to a final conclusion. Either a solution was found or infeasibility was detected.

IV. EVALUATION OF PAIR AND PROFILE ENERGIES

In this section we analyze and compare several pairwise and profile potentials, optimized using the LP protocol. As described in the previous section, given the training set (HL or TE) and the sampling of misfolded (decoy) structures generated by gapless threading, either we obtain a solution (perfect recognition on the training set) or the LP problem proves infeasible.

We use the infeasibility of a set to test the capacity of an energy model. We compare the capacity of alternative energy models by inquiring how many native folds they can recognize (before hitting an infeasible solution). Next, using the control sets, we further test the capacity of the models in terms of generalization and the number of inequalities in Eq. (9) that can be still satisfied, although they were not included in the training. We use the same sets of proteins and about the same number of optimal parameters. The larger the number of proteins that are recognized with the same number of parameters, the better the energy model. We focus on the capacity of four models: the square well and the distance power-law pairwise potentials, as well as THOM1 and THOM2 models. We find that the "profile" potentials have in general lower capacity than the pairwise interaction models.

A. Parameter-Free Models

Perhaps the simplest comparison that we can make is for zero-parameter models, and this is where we start. Zero-parameter models have nothing to optimize. They suggest an immediate and convenient framework for comparison, independent of successful (or unsuccessful) optimization of parameters.

An example of pair interaction energy with no parameters is the famous H/P model [51]. In H/P the interactions of pairs of amino acids of the type HP and PP are set to zero and the HH interaction is $-\lambda$. The total energy of a structure is the number of HH contacts (n_i) of structure i times $-\lambda$; that is, $E_i = -n_i\lambda$. The positive parameter λ determines the scale of the energy, however, it does not affect the ordering of the energies of different structures. The difference $E_i - E_n = -\lambda(n_i - n_n)$ is positive or negative, regardless of the magnitude of $|\lambda|$. The existence of a solution of the inequalities in (9) is therefore independent of λ.

For the HL protein set with 246 structures, the HP model predicts the correct fold of 200 proteins. For the larger TE set, the HP recognizes correctly 456 of the 594 proteins. This result is quite remarkable considering the simplicity of the model used, and it raises hopes for even more remarkable performance of the pairwise interaction model once more types of pair interactions are introduced. It is therefore disappointing that the addition of many more parameters to the pairwise interaction model did not increase its capacity as significantly as one may hope, though gradual increase is still observed.

A simple, parameter-free THOM1 model can be defined as follows. As in the pairwise interaction, we consider two types of amino acids: H and P. The energy of a hydrophobic site is defined as $\varepsilon_H(n) = -\lambda n$. For a polar site it is $\varepsilon_P = 0$. It is evident from the above definitions that the parameter-free THOM1 cannot possibly do better than the HP model, because neighbors of the type HH and HP are counted on equal footing. Indeed the parameter-free THOM1 is doing poorly in both HL and TE sets (only 118 of 246 proteins were solved for HL and 211 of 594 for TE).

B. "Minimal" Models

The parameter-free models are insufficient to solve exactly even the HL set. By "exact" we mean that each of the sequences picks the native fold as the lowest in energy using a gapless threading procedure. Hence, all the inequalities in Eq. (12), for all sequences S_n and structures $\mathbf{X}_j$, are satisfied and the LP problem of Eq. (12) is feasible. This section addresses the question; What is the minimal number of parameters that is required to obtain an exact solution for the HL and for the TE sets? The feasibility of the corresponding sets of inequalities [Eq. (12)] is correlated with the number of model parameters, as listed in Table III.

Consider first the training on the HL set (the solution of the TE set will be discussed in Section IV.D). For the square well potential we require the smallest number of parameters (i.e., 55) to solve the HL set exactly. Only 10 types of

TABLE III
Comparing the Capacity of Different Threading Potentials[a]

Potential	Hinds–Levitt Set	Tobi–Elber Set
SWP, HP model, par-free	200	456
SWP, 10 aa, 55 par	246*	504
SWP, 20 aa, 210 par	246*	530
SWP, 20 aa, 210 par	237	594*
LJ 12-6, 10 aa, 110 par	246*	125
SLJ 12-6, 10 aa, 110 par	246*	488
LJ 6-2, 10 aa, 110 par	246*	530
THOM1, HP model, par-free	118	221
THOM1, 20 aa, 200 par	246*	474
THOM2, 10 aa, 150 par	246*	478
THOM2, 20 aa, 300 par	246*	428
THOM2, 20 aa, 300 par	236	594*

[a]Capacity for recognition of pairwise and profile threading potentials is measured by gapless threading on Hinds–Levitt and Tobi–Elber representative sets of proteins. We compare the capacity of "parameter-free" models (such as the HP and the HP variant of THOM1), demonstrating the superiority of pair potential on profile model in the simplest possible case. We also show that the square well potential and the LJ(6,2) potential are significantly better than THOM1. THOM2, however, is showing comparable performance and is able to learn the TE set (see also Table IV). SWP stands for square well pairwise potential, and SLJ stands for shifted Lennard-Jones potential. For each potential the number of amino acids types used and the resulting number of parameters are reported. The training set used (either HL or TE) is indicated by an asterisk in the second or third column, respectively. The number of correct predictions for structures in HL and TE sets is given in the second and third columns as well.

amino acids were required: HYD, POL, CHG, CHN, GLY, ALA, PRO, TYR, TRP, HIS (see also Table I). The above notation implies that an explicit mentioning of an amino acid excludes it from other, broader subsets. For example, HYD includes now only CYS ILE LEU MET PHE and VAL, whereas CHG includes ARG and LYS only because the negatively charged residues form a separate group, CHN. The LJ, THOM1, and THOM2 models require 110, 200, and 150 parameters, respectively, to provide an exact solution of the same (HL) set (see table IV). It is impossible to find an exact potential for the HL set without (at least) 10 types of amino acids.

Smaller number of parameters led to infeasibility. The optimized models are then used "as is" to predict the folds of the proteins at the TE set. Again, we find that the pairwise interaction model is doing the best and is followed by THOM2 and THOM1, with LJ(12,6) closing.

The above test of the models optimized on the HL set gives an "unfair" advantage to the THOM models that are using more parameters. Nevertheless, even this head start did not change the conclusion that the pairwise square well model better captures the characteristics of sequence fitness into structures. Without the need for efficient treatments of gaps (see Section V), the pairwise interaction model should have been our best choice. Moreover, so far THOM2 is not significantly better than THOM1.

C. Evaluation of the Distance Power-Law Potentials

The LJ(12,6) model, which is a continuous representation of the pairwise interaction, performs poorly. The model trained exactly on the HL set predicts correctly only 125 structures from the 594 structures of the TE set. This result is surprising because the LJ is continuous and differentiable (and more realistic), and has more parameters.

A possible explanation for the failure of LJ(12,6) is the following. The LJ(12,6) is describing successfully atomic interactions. The shape of atoms is much better defined than the shape of amino acid side chains. Amino acids may have flexible side chains and alternative conformations, making the range of acceptable distances significantly larger. To represent alternative configurations of the same type of side chains, potentials with wide minima are required.

To test the above explanation and in a search for a better model, we also tried a shifted LJ function (SLJ) as well as an LJ-like potential with different powers ($m = 6$, $n = 2$, LJ(6,2); see also Fig. 1). As can be seen from Table IV, the "softer" potentials are performing better than the steep LJ(12,6) potential. For example, a LJ(6,2) potential trained on the HL set with 110 parameters (only 10 types of amino acids were used) recognizes correctly 530 proteins of the TE set. Thus, LJ(6,2) has a similar capacity to a square well potential, trained on the same set with 210 parameters.

TABLE IV
Comparison of Performance of THOM2 and Knowledge-Based Pairwise
Potentials Using Gapless Threading[a]

Potential	Recognized Structures	Nonsatisfied Inequalities [mln]
BT	1447 (87.3%)	0.28
HL	1412 (85.2%)	3.53
MJ	1410 (85.1%)	0.48
THOM2	1396 (84.3%)	0.38
TE	1353 (81.7%)	0.33
SK	1293 (78.0%)	0.16

[a]The results of gapless threading on the TE set with 20 redundant structures excluded and extended by the S1082 set (see text for details) are reported. The resulting set of 1656 proteins generates about 226 million inequalities. The results of THOM2 potential are compared to five other knowledge-based pairwise potentials by Betancourt and Thirumalai (BT) [37], Hinds and Levitt (HL) [36], Myazawa and Jerningan (MJ) [34], Godzik, Kolinski, and Skolnick (GKS) [38] and Tobi and Elber (TE) [32]. The latter potential was trained using LP protocol and the same (TE) training set. Potentials are ordered according to the number of proteins recognized exactly (out of 1656), given in the second column (values in parentheses indicate the percentage of proteins recognized exactly). The third column contains the number of inequalities (out of 226 mln) that are not satisfied. Note lack of correlation between the number of proteins that are missed and the number of inequalities that are not satisfied.

This suggests that in *ab initio* off-lattice simulations of protein folding, which employ "residue"-based potentials, LJ(6,2) may be more successful than commonly used LJ(12,6) [12]. Finally, we comment that the training of the LJ type potential was numerically more difficult than the training of the square well potential.

D.   Capacity of the New Profile Models

We turn our attention below to further analysis of the new profile models. An indication that THOM2 is a better choice than THOM1 is included in the next comparison: the number of parameters that is required to solve exactly the TE set (see Table III). It is impossible to find parameters that will solve exactly the TE set using THOM1 (the inequalities form an infeasible set). The infeasibility is obtained even if 20 types of amino acids are considered. In contrast, both THOM2 and the pairwise interaction model led to feasible inequalities if the number of parameters is 300 for THOM2 and 210 for the square well potential (SWP). Note that the set of parameters that solved exactly the TE set does not solve exactly the HL set because the latter set includes proteins not included in the TE set.

We have also attempted to solve the TE set using SWP and THOM2 with a smaller number of parameters. For square well potential the problem was proven infeasible even for 17 different types of amino acids and only very similar amino acids grouped together (Leu and Ile, Arg and Lys, Glu and Asp).

Similarly, we failed to reduce the number of parameters by grouping together structurally determined types of contacts in THOM2. Enhancing the range of a "dense" site to be a site of seven neighbors or more also results in infeasibility.

Although the rare "crowded" sites need to be considered explicitly to solve the TE set with THOM2, a reduced form of the full THOM2 potential trained on the TE set is doing quite well. Consider the contacts $(\bar{9}, \bar{1})$, $(\bar{9}, \bar{5})$, and $(\bar{9}, \bar{9})$. These contacts are very rare and are therefore merged with the contact types $(\bar{7}, \bar{1})$, $(\bar{7}, \bar{5})$, and $(\bar{7}, \bar{9})$. After the merging the number of parameters drops to 200 (instead of 300). The "new" potential recognizes 540 proteins out of 594 of the TE set. Only 324 inequalities are not satisfied. Hence, adding 100 parameters increases the capacity of the potential only by a minute amount.

To make a comparison to potentials not designed by the LP approach and to test at the same time the generalization capacity of THOM2, we consider the set of 1656 proteins obtained by adding the S1082 set to the TE set (with 20 redundant structures i.e., structures differing by less than 3 Å with respect to other structures in the TE set removed. This is a demanding test because it contains many homologous pairs and many short proteins that may be similar to fragments of larger proteins. Using the gapless threading protocol, we evaluate the performance of five knowledge-based pairwise potentials. As can be seen from Table IV, the Betancourt–Thirumalai (BT) potential [37] recognized exactly the largest number of proteins, followed by the Hinds–Levitt (HL) [36], Miyazawa–Jernigan (MJ) [34], THOM2, Tobi–Elber (TE) [32], and Godzik–Skolnick–Kolinski (GSK) [38] potentials. However, in terms of the number of inequalities that are not satisfied, the GSK potential is the best, followed by BT, TE, THOM2, MJ, and HL potentials.

The performance of THOM2 potential (84.3% accuracy) is comparable to the performance of other square well potentials (including the TE potential trained on the same set). Because most of the proteins used in this test were not included in the training, we conclude that the perfect learning on the training set avoids overfitting the data.

E. Dissecting the New Profile Models

The THOM1 potential is the easiest to understand and we therefore start with it. In Fig. 3 we examined the statistics of THOM1 contacts from the HL learning set. The number of contacts to a given residue is accumulated over the whole set and is presented by a continuous line. We expect that polar residues have a smaller number of neighbors compared to hydrophobic residues, which is indeed the case. The distributions for hydrophobic and polar residues are shown in Figs. 3a and 3b, respectively. The distributions make the essence of statistical potentials that are defined by the logs of the distribution (appropriately normalized).

The statistical analysis employs only native structures, whereas our LP protocol is using sequences threaded through wrong structures (misthreaded) during the process of learning. As a result, the LP has the potential for accumulating more information, attempting to put the energies of the mis-threaded sequence as far as possible from the correct thread. In Fig. 4 we show the results of the LP training for valine, alanine, and leucine that are in general agreement with the statistical data above. Nevertheless, some interesting and

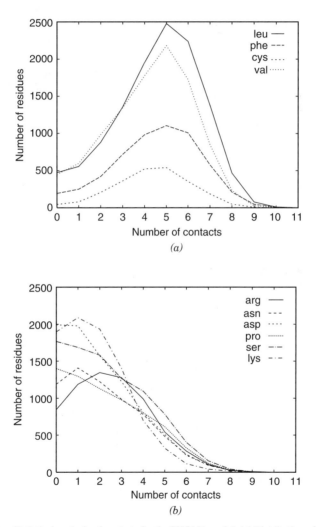

Figure 3. Statistical analysis of contacts for the THOM1 model. (a) Distribution of the number of contacts for hydrophobic residues. (b) Distribution of the number of contacts for (c) Data for alanine and glycine.

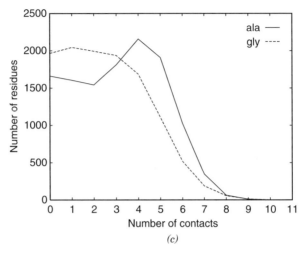

(c)

Figure 3 *(Continued)*

significant differences remain. For example, very rare valine residues with 10 neighbors obtain positive energies.

A plausible interpretation of this result is that these rare sites are used to enhance recognition in some cases, due to specific "homologous features." In Table Va we examined the type of contacts (in terms of the number of neighbors) for native and decoy structures.

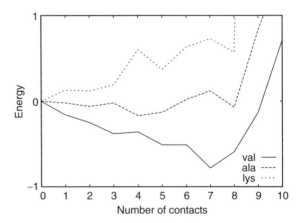

Figure 4. Potentials for THOM1 energy as extracted from LP training. Three residues are shown: alanine, lysine, and valine. Note that the minimum of the potential for valine is at seven neighbors. Note also that lysine has a minimum at zero neighbors.

TABLE V
Characterization of Native and Decoy Structures[a]

(a)

Type of Site[b]	Native (HYD/POL)	Decoys (HYD/POL)
(1)	16.97 (4.89/12.09)	24.20 (11.72/12.48)
(2)	17.30 (6.06/11.24)	21.72 (10.52/11.20)
(3)	17.72 (8.29/9.43)	18.70 (9.06/9.64)
(4)	16.60 (9.68/6.92)	15.00 (7.28/7.73)
(5)	14.62 (10.16/4.47)	10.79 (5.24/5.55)
(6)	9.96 (7.66/2.30)	6.04 (2.94/3.10)
(7)	4.95 (4.02/0.92)	2.63 (1.28/1.35)
(8)	1.57 (1.32/0.25)	0.77 (0.38/0.40)
(9)	0.26 (0.21/0.05)	0.12 (0.06/0.06)
(10)	0.04 (0.04/0.00)	0.02 (0.01/0.01)

(b)

Type of Contact	Native (HYD/POL)	Decoys (HYD/POL)
$(\bar{1},\bar{1})$	5.09 (1.59/3.50)	11.34 (5.48/5.85)
$(\bar{1},\bar{5})$	9.02 (2.99/6.04)	12.69 (6.15/6.54)
$(\bar{1},\bar{9})$	0.41 (0.15/0.26)	0.35 (0.17/0.18)
$(\bar{3},\bar{1})$	6.25 (2.88/3.37)	9.51 (4.60/4.91)
$(\bar{3},\bar{5})$	24.09 (13.01/11.08)	26.59 (12.91/13.68)
$(\bar{3},\bar{9})$	3.23 (1.88/1.35)	2.29 (1.12/1.18)
$(\bar{5},\bar{1})$	2.77 (1.81/0.96)	3.18 (1.54/1.64)
$(\bar{5},\bar{5})$	28.36 (20.96/7.40)	22.09 (10.75/11.34)
$(\bar{5},\bar{9})$	6.85 (5.11/1.74)	3.84 (1.87/1.96)
$(\bar{7},\bar{1})$	0.40 (0.31/0.09)	0.34 (0.16/0.17)
$(\bar{7},\bar{5})$	9.56 (8.00/1.56)	5.84 (2.85/3.00)
$(\bar{7},\bar{9})$	3.21 (2.60/0.61)	1.54 (0.75/0.79)
$(\bar{9},\bar{1})$	0.01 (0.01/0.00)	0.01 (0.01/0.01)
$(\bar{9},\bar{5})$	0.52 (0.44/0.08)	0.29 (0.15/0.14)
$(\bar{9},\bar{9})$	0.23 (0.19/0.04)	0.09 (0.05/0.05)

[a]Frequencies of different types of sites (relevant for the training of THOM1) found in the native structures of HL set as opposed to decoy structures generated using the HL set are presented in part a. In THOM1 the type of site is defined by number of its neighbors (n). Frequencies are defined by the percentage from the total number of 53,012 native sites in HL set and 556.14 millions of decoy sites generated using HL set, respectively. Frequencies of different types of contacts (appropriate for the training of THOM2) found in the native structures of TE set as opposed to decoy structures generated using TE are given in Table Vb. Different classes of contacts are specified in Table II. Frequencies are defined by the percentage from the total number of 439,364 native contacts in TE set and 10,089.19 millions of decoy contacts generated using TE set, respectively. The comparable site and contact distributions separated for hydrophobic and polar residues (as defined in Table I) are given in parentheses.

It is evident that native structures tend to have more contacts but that the difference is not profound. The deviations are the result of threading short sequences through longer structures (we have more threading of this kind). Such threading suggests a small number of contacts for the set of decoy structures. A sharper difference between native and decoy structures is observed when the contacts are separated to hydrophobic and polar (Table Vb). The difference in hydrophobic and polar contacts is very small at the decoy structures and much more significant for the native shapes.

Another reflection of the same phenomenon is the statistics of pair contacts. For the native structures we find that 42.6% of the contacts are of HH type, 38.2% are HP, and 19.3% are PP. This statistics is of the HL set that has a total of 93,823 contacts. For the decoy structures the statistics of pair contacts is vastly different. Only 23.5% of the contacts are HH, HP contacts are 50% of the total, and 26.5% are PP. The number of contacts that were used is 833.79 million. More details can be found in Tables Va and Vb.

THOM2 has significantly higher capacity, however the double layer of neighbors makes the results more difficult to understand. In Fig. 2 we showed the energy contributions of a few typical structural sites as defined by the THOM2 model. For example, the "lowest" picture in Fig. 2 is a site with six neighbors in the first contact shell and a wide range of neighbors in the second shell. The second shell includes a site with just two neighbors as well as a site with nine neighbors. The overall large number of neighbors suggests that this site is hydrophobic, and the corresponding energies of lysine and valine indeed support this expectation.

In Fig. 5 we present a contour plot of the total contributions to the energies of the native alignments in the TE set, as a function of the number of contacts in the first shell, n, and the number of secondary contacts to a primary contact, n', respectively. The results for two types of residues, lysine and valine, are presented. The contribution of a type of site to the native alignment is twofold: its energy $\varepsilon_\alpha(n, n')$ and the frequency of that site f. It is possible to find a very attractive (or repulsive) site that makes only negligible contribution to the native energies because it is extremely rare (i.e., f is small). For specific examples see Table VI. By plotting $f \cdot \varepsilon_\alpha(n, n')$ we emphasize the important contributions. Hydrophobic residues with a large number of contacts stabilize the native alignment, as opposed to polar residues that stabilize the native state only with a small number of neighbors.

It has been suggested that pairwise interactions are insufficient to fold proteins and higher-order terms are necessary [30]. It is of interest to check if the environment models that we use catch cooperative, many-body effects. As an example we consider the cases of valine–valine and lysine–lysine interactions. We use Eq. (8) to define the energy of a contact. In the usual pairwise

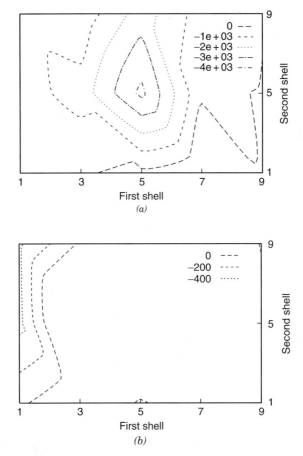

Figure 5. Contour plots of the total energy contributions to the native alignments in the TE set for valine and lysine residues as a function of the number of neighbors in the first and second shells. Part a shows that contacts involving valine residues with five to six neighbors with other residues of medium number of neighbors stabilize most the native alignments. On the other hand, as can be seen from part b, only contacts involving lysine residues with a small number of neighbors stabilize native alignments.

interaction the energy of a valine–valine contact is a constant and independent of other contacts that the valine may have.

In Table VI we list the effective energies of contacts between valine residues as a function of the number of neighbors in the primary and secondary sites. The energies differ widely from -1.46 to $+3.01$. The positive contributions refer,

TABLE VI
Cooperativity in Effective Pairwise Interactions of the THOM2 Potential[a]

(a)

	$V(\bar{1})$	$V(\bar{3})$	$V(\bar{5})$	$V(\bar{7})$	$V(\bar{9})$
$V(\bar{1})$	− 0.56	− 0.41	− 0.17	− 1.46	3.01
$V(\bar{3})$	− 0.41	− 0.34	− 0.44	− 0.30	− 0.07
$V(\bar{5})$	− 0.17	− 0.44	− 0.54	− 0.61	− 0.38
$V(\bar{7})$	− 1.46	− 0.30	− 0.61	− 0.49	− 0.76
$V(\bar{9})$	3.01	− 0.07	− 0.38	− 0.76	− 1.03

(b)

	$K(\bar{1})$	$K(\bar{3})$	$K(\bar{5})$	$K(\bar{7})$	$K(\bar{9})$
$K(\bar{1})$	− 0.03	− 0.03	− 0.19	1.18	0.69
$K(\bar{3})$	− 0.03	0.28	0.40	0.58	0.61
$K(\bar{5})$	− 0.19	0.40	0.52	0.83	0.86
$K(\bar{7})$	1.18	0.58	0.83	1.34	0.38
$K(\bar{9})$	0.69	0.61	0.86	0.38	− 0.59

[a]For a pair of two amino acids α and β in contact, we have 25 different possible types of contacts (and consequently 25 different effective energy contributions) because α and β may occupy sites that belong to one of the five different types characterized by the increasing number of contacts in the first contact shell (see Table II). Moreover, the 5×5 interaction matrix will, in general, be asymmetric. The effective energies of contact between two VAL residues with a different number of neighbors are given in part a, whereas the energies of contacts between two LYS residues are given in part b.

however, to very rare types of contacts, and the energies of the probable contacts are negative as expected. Hence, the THOM2 model is compensating for missing information on neighbor identities by taking into account significant cooperativity effects.

To summarize the study of the potentials we provide, in Table VII, the optimal parameters for LJ(6,2), THOM1, and THOM2 potentials.

V. THE ENERGIES OF GAPS AND DELETIONS

In the present section we discuss the derivation of the energy for gaps (insertions in the sequence) and deletions. A gap residue is denoted by a $-$, and a deletion is denoted by a v. For example, the extended sequence $\bar{S} = a_1 - va_3 \ldots a_n$ has a gap at the second structural position (x_2) and a deletion at the second amino (a_2).

A. Protocol for Optimization of Gap Energies

The gap (an unoccupied structural site) is considered to be an (almost) normal amino acid. We assigned to it a score (or energy) according to its environment, like any other amino acid. Here we describe how the energy function of the gap was determined. The parameters were optimized for THOM1 and THOM2, because these are the models accessible to efficient alignment with gaps.

Gap training is similar to the training of other amino acid residues. Only the database of "native" and decoy structures is different. To optimize the gap parameters we need "pseudo-native" structures that include gaps. We construct such "pseudo-native" conformations by removing the true native shape X_n of the sequence S_n from the coordinate training set and by putting instead a homologous structure, X_h. The best alignment of the native sequence into the homologous structure is $\bar{S}_n$ into X_h, and it includes gaps. We require that the

TABLE VII
Parameters of Some of the Threading Potentials Trained Using the LP Protocol[a]

(a)

	HYD	POL	CHG	CHN	GLY	ALA	PRO	TYR	TRP	CYS
HYD	9.32	1.45	−0.44	−0.4	7.35	−1.09	2.17	−0.54	2.29	9.93
POL	1.45	−1.19	−1.07	−0.95	−1.55	−0.75	−1.12	1.41	2.7	0.49
CHG	−0.44	−1.07	2.62	−0.44	−0.35	−1.23	−0.67	0.21	−2.47	−2.51
CHN	−0.4	−0.95	−0.44	1.89	−0.01	3.58	1.32	6.73	8.92	−1.61
GLY	7.35	−1.55	−0.35	−0.01	−1.15	−1.11	2.23	−1.39	−1.17	−1.52
ALA	−1.09	−0.75	−1.23	3.58	−1.11	2.9	−1.53	5.64	−2.43	3.59
PRO	2.17	−1.12	−0.67	1.32	2.23	−1.53	6.51	8.86	8.64	−2.68
TYR	−0.54	1.41	0.21	6.73	−1.39	5.64	8.86	4.98	7.19	−2.55
TRP	2.29	2.7	−2.47	8.92	−1.17	−2.43	8.64	7.19	9.95	−3.74
CYS	9.93	0.49	−2.51	−1.61	−1.52	3.59	−2.68	−2.55	−3.74	−0.12

	HYD	POL	CHG	CHN	GLY	ALA	PRO	TYR	TRP	CYS
HYD	−2.34	0.47	1.71	1.11	−0.21	−0.35	1.22	−1.33	−0.98	−5.11
POL	0.47	0.01	−0.02	0.48	−0.07	−0.7	2.38	−0.81	−0.87	0.57
CHG	1.71	−0.02	0.23	−1.65	−0.51	1.13	0.05	−1.93	1.29	3.73
CHN	1.11	0.48	−1.65	0.12	0	1.58	−2.26	0.33	4.91	3.35
GLY	−0.21	−0.07	0.51	0	1.35	0.41	−0.82	0.47	−1.93	−3.59
ALA	−0.35	−0.7	1.13	1.58	0.41	−1.59	1.3	−2.38	2.12	1.19
PRO	1.22	2.38	0.05	−2.26	−0.82	1.3	−4.08	−3.2	−7.25	−1.37
TYR	−1.33	−0.81	−1.93	0.33	0.47	−2.38	−3.2	−2.9	−5.13	1.67
TRP	−0.98	−0.87	1.29	4.91	−1.93	2.12	−7.25	−5.13	−2.73	−0.2
CYS	−5.11	0.57	3.73	3.35	−3.59	1.19	−1.37	1.67	−0.2	−7.87

TABLE VII (Continued)

(b)

	ALA	ARG	ASN	ASP	CYS	GLN	GLU	GLY	HIS	ILE	LEU	LYS	MET	PHE	PRO	SER	THR	TRP	TYR	VAL
(1)	−0.02	0.10	−0.22	0.02	−0.13	0.02	0.05	−0.05	−0.15	−0.17	−0.04	0.13	−0.40	−0.52	0.29	−0.02	0.02	−0.20	−0.23	−0.16
(2)	−0.06	−0.23	−0.07	0.20	−0.37	0.21	−0.03	0.06	−0.05	−0.30	−0.22	0.12	−0.20	−0.25	0.24	−0.01	−0.10	−0.57	−0.27	−0.25
(3)	−0.02	−0.01	−0.01	−0.43	−0.72	0.09	0.10	0.05	−0.25	−0.48	−0.37	0.19	−0.66	−0.58	0.06	0.05	−0.12	−0.77	−0.37	−0.38
(4)	−0.17	0.12	0.29	0.37	−0.70	0.22	0.40	0.14	−0.31	−0.64	−0.41	0.60	−0.50	−0.68	0.22	0.00	0.21	−0.36	−0.39	−0.36
(5)	−0.13	0.22	0.20	0.68	−1.13	0.33	0.45	0.38	0.24	−0.53	−0.50	0.37	−0.39	−0.65	0.31	0.31	0.02	−0.65	−0.78	−0.51
(6)	0.02	0.32	0.17	0.43	−1.16	0.02	0.70	0.42	0.36	−0.57	−0.58	0.63	−0.80	−0.82	0.75	0.27	0.24	−0.46	−0.72	−0.51
(7)	0.12	−0.10	0.30	0.43	−1.27	0.46	0.20	0.20	0.27	−0.76	−0.54	0.73	−0.44	−0.40	0.42	0.09	0.36	0.12	−0.39	−0.78
(8)	−0.07	0.91	−0.12	−0.01	−1.60	0.51	0.83	0.29	−0.71	−1.37	−0.72	0.57	−0.66	0.25	0.02	0.36	0.15	−0.26	−0.74	−0.59
(9)	0.83	1.36	0.11	0.35	−1.71	0.82	10.00	2.12	0.38	−0.33	1.03	10.00	1.66	−1.03	1.13	2.23	−0.57	10.00	−0.38	−0.13
(10)	1.57	10.00	10.00	10.00	10.00	10.00	10.00	0.83	10.00	−0.93	−0.47	10.00	10.00	0.40	10.00	0.00	10.00	−0.78	10.00	0.71

(c)

	ALA	ARG	ASN	ASP	CYS	GLN	GLU	GLY	HIS	ILE	LEU	LYS	MET	PHE	PRO	SER	THR	TRP	TYR	VAL
(1,1)	0.23	−0.03	−0.03	−0.08	−0.82	−0.26	0.09	0.29	0.07	−0.12	−0.16	−0.02	0.21	−0.20	0.03	0.05	−0.07	−0.50	−0.64	−0.28
(1,5)	−0.21	−0.26	−0.10	0.20	−1.11	0.00	−0.08	0.00	0.03	0.31	−0.23	−0.13	−0.15	−0.29	−0.23	0.07	−0.09	−0.60	−0.40	−0.36
(1,9)	−6.01	−4.09	−5.42	−6.14	−7.27	−5.88	−5.80	−5.81	−4.75	−5.46	−5.85	−4.91	−4.97	−5.83	−6.17	−5.89	−5.89	−5.25	−6.79	−6.99
(3,1)	−0.01	−0.10	−0.17	0.02	−0.50	−0.09	0.11	0.31	0.04	−0.10	−0.10	0.11	−0.20	−0.17	−0.02	0.40	0.06	−0.31	−0.29	−0.05
(3,5)	−0.08	0.18	0.15	0.13	−0.69	0.12	0.24	0.04	−0.03	−0.29	−0.21	0.14	0.08	−0.32	−0.05	0.06	0.08	−0.36	−0.28	−0.17
(3,9)	−0.29	0.06	−0.33	0.08	−0.78	0.18	0.02	−0.13	−0.47	−0.60	−0.49	0.09	−0.85	−0.07	0.19	0.23	−0.15	−0.15	0.03	−0.27
(5,1)	0.13	−0.21	0.04	0.22	−0.15	−0.11	0.08	0.48	0.19	−0.15	−0.32	−0.06	−0.15	−0.27	0.17	0.19	0.34	−0.07	0.02	0.19
(5,5)	0.06	0.16	0.20	0.17	−0.60	0.04	0.13	0.18	−0.04	−0.25	−0.19	0.26	−0.26	−0.28	0.09	0.11	0.02	−0.36	−0.30	−0.27
(5,9)	−0.65	0.68	−0.26	−0.19	−0.82	−0.09	0.43	−0.36	−0.19	−0.47	−0.42	0.34	0.32	0.07	0.55	0.22	0.01	0.04	−0.46	−0.58
(7,1)	6.29	5.50	5.56	6.02	5.09	5.55	5.68	6.10	5.70	5.59	5.26	6.08	5.64	5.80	5.82	5.23	5.48	6.42	5.17	5.53
(7,5)	0.17	0.29	0.36	0.39	−0.28	0.28	0.45	0.33	0.28	−0.08	−0.01	0.50	0.24	−0.16	0.42	0.13	0.34	0.04	−0.08	−0.03
(7,9)	0.08	0.41	0.00	−0.15	−0.30	0.04	−0.27	0.05	0.69	0.04	−0.17	0.67	0.06	0.03	−0.71	0.82	0.24	−0.36	0.14	−0.25
(9,1)	10.00	4.50	6.05	5.21	4.00	5.94	10.00	10.00	10.00	10.00	6.22	5.59	4.91	6.02	9.61	10.00	10.00	5.88	10.00	10.00
(9,5)	0.26	0.30	0.26	0.71	0.41	−0.02	0.32	0.83	−0.09	1.26	−0.15	0.52	−0.19	0.43	3.07	0.43	0.52	−0.08	0.08	0.21
(9,9)	0.20	0.04	−0.37	−1.34	−1.19	0.47	1.37	−1.36	1.06	−1.99	−0.25	−0.29	1.41	−1.33	6.94	3.22	−0.54	0.81	−0.53	−0.52

[a]Numerical values of the energy parameters for three potentials are given: LJ(6,2) (part a—note that the "repulsive" coefficients A are given first, followed by the "attractive" coefficients B; the unit distance is 3 Å), THOM1 trained on an HL set of proteins (part b), and THOM2 trained on TE set of proteins (part c; see text for details). The rows in the tables correspond to either different types of amino acids (LJ) or to different types of sites (THOM1) or contacts (THOM2). The columns correspond to different types of amino acids.

alignment $\bar{S}_n$ into the homologous protein will yield the lowest energy compared to all other alignments of the set. Hence, our constraints are

$$E(\bar{S}_n, \mathbf{X}_j, \mathbf{p}) - E(\bar{S}_n, \mathbf{X}_h, \mathbf{p}) = \sum_{\gamma} p_{\gamma}(n_{\gamma}(\mathbf{X}_j) - n_{\gamma}(\mathbf{X}_h)) > 0 \qquad \forall j \neq h, n$$

(13)

Equation (13) is different from Eq. (12) in two ways. First, we consider the "extended" set of "amino acids"—$\bar{S}$ instead of S. Second, the native-like structure is $\mathbf{X}_h$—a coordinate set of a homologous protein and not $\mathbf{X}_n$.

The number of inequalities that we may generate (alignments with gaps inserted into a structure and deletions of amino acids) is exponentially large in the length of the sequence, making the exact training more difficult. Some compromises on the size of samples for inequalities with gaps have to be made. To limit the scope of the computations, we optimize here the scores of the gaps only. Thus, we do not allow the amino acid energies (computed previously by gapless threading; see Section III) to change while optimizing parameters for gaps. Moreover, the sequence $\bar{S}$ (obtained by prior alignment of the native sequence against a homologous structure) is held fixed, and gapless threading against all other structures in the set is used to generate a corresponding set of inequalities [Eq. (13)]. By performing gapless threading of $\bar{S}_n$ into different structures, we consider only a small subset of all possible alignments of $\bar{S}_n$, because we fixed the number and the position of the gaps that we added to the native sequence S_n.

Pairs of homologous proteins from the following families were considered in the training of the gaps: globins, trypsins, cytochromes and lysozymes (see Table VIII). The families were selected to represent vastly different folds with a

TABLE VIII
Pairs of Homologous Structures Used for the Training of Gap Penalties[a]

Native	Homologous	Similarity
1mba (myoglobin, 146)	1lh2 (leghemoglobin, 153)	20%, 2.8 Å, 140 res
1mba (myoglobin, 146)	1babB (hemoglobin, chain B, 146)	17%, 2.3 Å, 138 res
1ntp (β-trypsin, 223)	2gch (γ-chymotrypsin, 245)	45%, 1.2 Å, 216 res
1ccr (cytochrome c, 111)	1yea (cytochrome c, 112)	53%, 1.2 Å, 110 res
1lz1 (lysozyme, 130)	1lz5 (1lz1 + 4 res insert, 134)	99%, 0.5 Å, 130 res
1lz1 (lysozyme, 130)	1lz6 (1lz1 + 8 res insert, 138)	99%, 0.3 Å, 129 res

[a]For each pair the native and the homologous structures are specified by their PDB codes, names, and lengths in the first and second column, respectively. In the third column the similarity between the native and the homologous proteins is defined in terms of sequence identity (%), RMS distance (angstroms), and length (number of residues) of the FSSP structure-to-structure alignment, obtained by submitting the corresponding pairs to the DALI server [44].

significant number of homologous proteins in the database. The globins are helical, trypsins are mostly β-sheets, and lysozymes are α/β proteins. Note also that the number of gaps differs appreciably from a protein to a protein. For example, $\bar{S}_n$ includes only one gap for the alignment of 1ccr (sequence) versus 1yea (structure), and 22 gaps for 1ntp versus 2gch.

The energy functional form that we used for the gaps is the same as for other amino acids. The "pseudo-native" structures with extended sequences are added to the HL set (while removing the original native structures). Gapless threading into other structures of the HL set results in about 200,000 constraints for the gap energies. Because we did not consider all the permutations of the gaps within a given sequence and our sampling of protein families is limited, our training for the gaps is incomplete. Nevertheless, even with this limited set we obtain satisfactory results. A representative set of homologous pairs that we used allows us to arrive at scores that can detect very similar proteins (e.g., the cytochromes 1ccr and 1yea) and also related proteins that are quite different (e.g., the globins 1lh2 and 1mba); see Table VIII.

The process of generating pseudo-native is as follows: For each pair of native and homologous proteins the alignment of the native sequence $\bar{S}_n$ into the homologous structure $\mathbf{X}_h$ is constructed. This alignment uses an initial guess for the gap energy, which is based on the THOM1 potential and was based on the following observations.

- The gap penalty should increase with the number of neighbors. For example, we require that $\varepsilon_-(n + 1) > \varepsilon_-(n)$ for the THOM1 gap energy.
- The energy of a gap with contacts must be larger than the energy of an amino acid with the same number of contacts. The gap energy must be higher; otherwise, gaps will be preferred to real amino acids. For example, the THOM1 energy of the proline residue with one neighbor is 0.29. Therefore the gap energy must be larger than 0.29; or in general, $\varepsilon_-(n) > \varepsilon_k(n)$, where $k = 1, \ldots, 20$ (types of amino acids) and $n = 1, \ldots, 10$ (number of neighbors).
- The energy of amino acids without contacts is set to zero. The gap energy is therefore greater than zero.

In Table IX we provide the initial guess for the gaps (used to determine pseudo-native states) and the final optimal gap values for THOM1 and THOM2. The value of 10 is the maximal penalty allowed by the optimization protocol that we used. However, this value is not a significant restriction. A solution vector $\mathbf{p}$ can be used to generate another scaled solution $\lambda\mathbf{p}$, where λ is a positive constant.

Nevertheless, note that the maximal value is reached rather quickly. This may indicate that our sampling of inequalities is still insufficient from the perspective of native alignment. The values of gaps that are found only in decoy states are increasing without limit in the LP protocol. For example, it is so rare

TABLE IX
The Gap Penalties for THOM1 and THOM2 Models as
Trained by the LP Protocol with the Limited Set of
Homologous Structures from Table VIII[a]

(a)

Type of Site	Initial Penalty	Optimized Penalty
(0)	0.1	2.7
(1)	0.3	3.9
(2)	0.6	9.0
(3)	0.9	10.0
(4)	2.0	10.0
(5)	4.0	10.0
(6)	6.0	10.0
(7)	8.0	10.0
(8)	9.0	10.0
(9)	10.0	10.0

(b)

Type of Contact	Penalty
(0)	1.0
$(\bar{1},\bar{1})$	8.9
$(\bar{1},\bar{5})$	5.7
$(\bar{1},\bar{9})$	10.0

[a]Initial and optimized gap penalties for different types of
sites in the THOM1 model are given in part a. Optimized gap
penalties for different types of contacts in the THOM2 model
are given in part b. Penalties that are not specified explicitly
are equal to the maximum value of 10.0.

to find a gap at the hydrophobic core of a protein that our protocol assigns to it
the maximal penalty.

The gaps are favored in sites with a small number of contacts. This observation is expected, because gaps are usually found in loops with significant solvent
exposure. Note that THOM2 is penalized for a gap for each individual contact.

In Table X we show the results of optimal threading with gaps (using
dynamic programming) for myoglobin (1mba) against leghemoglobin (1lh2)
structure. We show the initial alignment (with the *adhoc* gap parameters from
Table IXa) defining the pseudo-native state, and we also show the results for
optimized gap penalties for THOM1 and THOM2. These alignments are largely
consistent with the DALI [44] structure–structure alignment (see Table X). Note
that the gaps appear (as expected) in loop domains (e.g., the CD, EF, and GH
loops). The only "surprising" gap is at position 9. Further tests of alignments
with gaps for proteins that we did not learn are given in Section VI.

TABLE X
An Example of Output from the Program LOOPP for Sequence-to-Structure Alignments [48][a]

(a)

```
........1.........2.........3.........4.........5.........          1–59
SLSAAEADLAGKSWAPVFANKNANGLDFLVALFEKFPDSANFFADFKGKSVADIKASPK        1mba
GALTESQAALVKSSWEEFNANIPKHTHRFFILVLEIAPAAKDLFSFLKGTSEVPQNNPE        1lh2
........1.........2.........3.........4.........5.........          1–59
```

```
6.........7.........8.....ii...9.........0.........1......          60–116
LRDVSSRIFTRLNEFVNNAANAGKMSA–MLSQFAKEHVGFGVGSAQFENVRSMFPGFV         1mba
LQAHAGKVFKLVYEAAIQLEVTGVVVTDATLKNLGSVHVSKGVADAHFPVVKEAILKTI        1lh2
6.........7.........8.........9.........0.........1........         60–118
```

```
...2..i..i.....3.........4..i...i.i     117–146
ASVAAP-PA-GADAAWTKLFGLIIDALK-AAG-A-      1mba
KEVVGAKWSEELNSAWTIAYDELAIVIKKEMDDAA      1lh2
.2.........3.........4.........5...      119–153
```

(b)

```
........i.1........i.2.i........3.....i....4.........5.....         1–55
SLSAAEAD-LAGKSWAPVF-ANK-NANGLDFLVALFEK-FPDSANFFADFKGKSVADIK         1mba
GALTESQAALVKSSWEEFNANIPKHTHRFFILVLEIAPAAKDLFSFLKGTSEVPQNNPE        1lh2
........1.........2.........3.........4.........5.........          1–59
```

```
....6.........7.....i....8..i.......9.........0.........1..        56–112
ASPKLRDVSSRIFTRLNEFV-NNAANAG-KMSAMLSQFAKEHVGFGVGSAQFENVRSMF         1mba
LQAHAGKVFKLVYEAAIQLEVTGVVVTDATLKNLGSVHVSKGVADAHFPVVKEAILKTI        1lh2
6.........7.........8.........9.........0.........1........         60–118
```

```
....i...2.........3.........4......     113–146
PGFV-ASVAAPPAGADAAWTKLFGLIIDALKAAGA     1mba
KEVVGAKWSEELNSAWTIAYDELAIVIKKEMDDAA     1lh2
.2.........3.........4.........5...     119–153
```

(c)

```
........i.1.........2.........3.........4.........i....i.i.        1–55
SLSAAEAD-LAGKSWAPVFANKNANGLDFLVALFEKFPDSANFFADFKGK-SVAD-I-K        1mba
GALTESQAALVKSSWEEFNANIPKHTHRFFILVLEIAPAAKDLFSFLKGTSEVPQNNPE        1lh2
........1.........2.........3.........4.........5.........          1–59
```

```
....6.........7.......i.8.i........9.........0.........1..         56–112
ASPKLRDVSSRIFTRLNEFVNNA-ANA-GKMSAMLSQFAKEHVGFGVGSAQFENVRSMF         1mba
LQAHAGKVFKLVYEAAIQLEVTGVVVTDATLKNLGSVHVSKGVADAHFPVVKEAILKTI        1lh2
6.........7.........8.........9.........0.........1........         60–118
```

TABLE X   (*Continued*)

. . . . 2. i. . . . . 3. . . . . 4.	113–146
PGFVASVAA-PPAGADAAWTKLFGLIIDALKAAGA	1mba
KEVVGAKWSEELNSAWTIAYDELAIVIKKEMDDAA	1lh2
. 2. . . . . 3. . . . . 4. . . . . 5. . .	119–153

[a]We compare alignments of myoglobin (1 mba) sequence into leghemoglobin (1lh2) structure using the intial (part a) and trained gap penalties (part b for THOM1 and part c for THOM2). Note that the location of insertions in the initial alignment (which is used for training of gap energies) is to a large extent consistent with the DALI structure to structure alignment [44], which aligns: residues 2–50 of 1 mba to 3–51 of 1lh2 (helices A, B, and C), residues 53–56 of 1 mba to 52–55 of 1lh2 (implying deletions at positions 51 and 52 in 1 mba), residues 59–80 of 1 mba to 56–77 of 1lh2 (E helices), residues 81–86 of 1 mba to 82–87 of 1lh2, residues 87–121 of 1 mba to 89–123 (with the implied insertion at position 88 in 1 lh2), residues 122–139 of 1 mba to 126–143 of 1lh2 (implying two insertions at positions 124 and 125 in 1lh2) and residues 140–145 of 1 mba to 145–150 of 1lh2 (with an insertion at position 144 in 1lh2), respectively. Note also that F and G helices are shifted considerably in the DALI alignment (there is no counterpart of the D helix in 1lh2). The initial THOM1 alignment (part a) is in perfect agreement with the DALI superposition between residues 88 and 150 of 1lh2, except for two insertions at positions 128 and 147 (shifted by three residues with respect to the DALI alignment). The insertions at positions 88, 125, 151, and 153 coincide with the DALI alignment. The THOM2 alignment, with trained gap penalties (part c), is in perfect agreement with the DALI superposition for residues 10 to 50 of 1lh2 and then departs from the DALI alignment, overlapping E, F, and G helices with a smaller shift.

B.   Deletions

Yet another technical comment is concerned with "deletions" that were mentioned above. A single deletion makes the native sequence shorter by one amino acid, leaving the structure unchanged. In sequence–sequence alignment, deletions can be made equivalent to insertion of gaps. In threading, however, the sequence and the structure are asymmetric. Deleting of residues (amino acids with no corresponding structural sites) or the insertion of gap residues (empty structural sites) is not the same operation.

Nevertheless, in the present chapter we exploit an assumed symmetry between insertion of a gap residue to a sequence and the placement of a "delete" residue in a "virtual" structural site. The deletions are assigned an environment dependent value that is equal to the averaged gap insertion penalty for the mirror image problem (shorter sequence instead of longer). The deletion penalty is set equal to the cost of insertion averaged over two nearest structural sites. No explicit dependence on the amino acid type is assumed.

While optimization for deletions is not performed in the present chapter, such an optimization is similar to the optimization of gaps. Consider a partial alignment of the sequence $\bar{S}_n = \ldots a_{j'-1}v_{j'}a_{j'+1} \ldots$ into a homologous structure, $\mathbf{X}_h = (\ldots, x_j, x_{j+1}, \ldots)$, in which $a_{j'-1}$ is placed into x_j, $a_{j'+1}$ is placed into x_{j+1}, and $v_{j'}$ is a deletion. What is the energetic cost associated with deleting $v_{j'}$? An estimate would be based on an analogous formulation to the gap residue:

$$\varepsilon_{v_j}(\bar{S}_n, \mathbf{X}_h) = \varepsilon_v(x_j, x_{j+1}) \tag{14}$$

We denoted the "deletion" residue by "v" because it corresponds to a virtual site inserted into the structure. The deletion is designed as a special energy term that depends on the nearest structural sites: x_j and x_{j+1}. The optimization of the new energy function is the target of a future work.

VI. TESTING STATISTICAL SIGNIFICANCE OF THE RESULTS

In the following we will consider optimal alignments of an extended sequence $\bar{S}$ with gaps into the library structures $\mathbf{X}_j$. We focus on the alignments of complete sequences to complete structures (global alignments [16]) and alignments of continuous fragments of sequences into continuous fragments of structures (local alignment [17]). In global alignments, opening and closing gaps (gaps before the first residue and after the last amino acid) reduce the score. In local alignments, gaps or deletions at the C and N terminals of the highest scoring segment are ignored. Only one local segment, with the highest score, is considered.

Threading experiments that are based on a single criterion (the energy) are usually unsatisfactory. While we do hope that the (free) energy function that we design is sufficiently accurate so that the native state (the native sequence threaded through the native structure) is the lowest in energy, this is not always the case. Our perfect training is for the training set and for gapless threading only. The results were not extended to include (a) perfect learning with gaps or (b) perfect recognition of shapes of related proteins that are not the native.

Despite significant efforts to eliminate all "false-positive" signals, the present authors are not aware of any energy function that can achieve this goal. Tobi and Elber [30] conjectured, based on significant numerical evidence, that it is impossible to use a general pair interaction model and to make the native structure the lowest in energy from a set of protein-like structures. The evidence was given for the (simpler) problem of gapless threading. In the present chapter we discuss the more complex problem of threading with gaps that makes the robust detection of the native state even more difficult.

Other investigators use the Z score as an additional filter or as the primary filter [18,52,4,6], and we follow their steps. The novelty in the present protocol is the combined use of global and local Z scores to assess the accuracy of the prediction. This filtering mechanism was found to provide improved discrimination as compared with a single Z score test.

A. The Z-Score Filter

The Z score, which may be regarded as a dimensionless, "normalized" score, is defined as

$$Z = \frac{\langle E \rangle - E_p}{\sqrt{\langle E^2 \rangle - \langle E \rangle^2}} \tag{15}$$

The energy of the current "probe"—that is, the energy of the optimal alignment of a query sequence into a target structure—is denoted by E_p. The averages, $\langle \ldots \rangle$, are over "random" alignments (that still need to be defined). The Z score is designed as measure of the deviation of our "hits" from random alignments. The larger the value of Z, the more significant the alignment. This is because the score is far from the "random" average value.

A nontrivial question is how we define a random alignment. The randomness can come from two sources: random structure or random sequence. It is common in *ab initio* folding to assess the correctness of a given structure by comparing its energy to the energies of other structures assumed random. This approach is useful if the number of structures is much larger than the number of sequences (typical of *ab initio* computations). However, in threading protocols the number of structures is relatively small and the number of sequences (with gaps) is significantly larger.

It is therefore suggestive to use a measure, which is based on random sequences instead of random structures. Following the common practice [52–54] we generate this distribution numerically, employing sequence shuffling of the probe sequence. Let $S_p = a_1 a_2 \ldots a_n$ be the probe sequence. We consider the family of sequences that is obtained by permutations of the original sequence.

The set of shuffled sequences has the same amino acid composition and length as the native sequence. This leads to a deviation from "true" randomness (no constraints) that is used in analytical models. Nevertheless, the constraints are convenient to "solve" the problem of the energy of the unfolded state. In the unfolded state all amino acids are assumed to have no contacts with other amino acids. Therefore all the shuffled sequences have the same energy in the unfolded state.

We address the convergence of the Z score in Fig. 6. How many shuffled sequences do we need before we get a reliable estimate? For example, after 100 shuffles the Z score of the global alignment of 1pbxA into 2lig (two different families) suggests that the result is significant. However, enlarging the sample to include 1000 random probes significantly reduces the Z score below the "cutoff" of 3. Hence, especially when the signal is not very strong, it is important to fully converge the value of the Z score. The large number of alignments that are performed for the shuffled sequences (between 50 and 1000) makes the process computationally demanding and underlines the need of an efficient algorithm for genomics scale threading experiments.

An essential decision needed is what is a "good" score and what is a "bad" score. Intuitively, negative energies are assumed "good." Negative energies are lower than the state with no contacts—that is, contacts with water molecules as in the unfolded state. However, no such intuition is obvious for the Z score. To establish a cutoff for the Z score that eliminates false positives, we consider the probability $P(Z_p)$ of observing a Z score larger than Z_p by chance. Clearly our

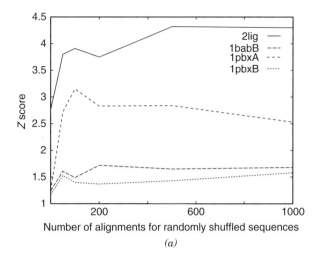

(a)

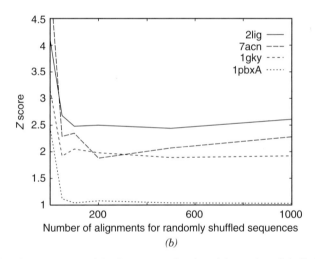

(b)

Figure 6. The convergence of the Z scores as a function of the number of shuffled sequences. The results for global and local alignments are presented in the parts a and b, respectively. The sequence of the aspartate receptor protein 2lig (not included in the training set) is aligned to all the structures of the HL set, and the best matches are shown. Note that hemoglobin 1pbxA is found among the good matches (false positive) with a global Z score of about 3 when using only 100 shuffled sequences to estimate the distribution for random sequences. Converging the Z scores makes it possible to better separate the native alignment with respect to incorrect alternatives. The Z score for local alignment of 2lig into 1pbxA is small (about 1) and suggests that this match is indeed a false positive. The initial values in the figure correspond to scaled energies of the alignments.

results will be statistically significant only if $P(Z_p)$ is very small. The expectation value of the number of occurrences of false positives in N alignments with a Z score larger than Z_p is $N \cdot P(Z_p)$.

To estimate $P(Z_p)$, we thread sequences of the S47 set through structures included in the Hinds–Levitt set. The probe sequences of known structures were selected to ensure no structural similarity between the HL set and the structures of the probe sequences (see Section III.A). Therefore any significant hit in this set may be regarded as a false positive.

Z scores of local alignments are employed to estimate $P(Z_p)$. In local alignments the number of "good" energies (significantly lower than zero) is large, underlining the need for an additional selection mechanism to eliminate false positives. It also makes it possible for us to estimate $P(Z_p)$ for a population of alignments with "good" scores. For each probe sequence, Z scores are calculated for 200 structures with the best energies. Only alignments with matching segments of at least 60% of the total sequence length are considered. One hundred shuffled sequences are used to compute the averages required for a single Z-score evaluation. A histogram of the resulting 6813 pairwise alignments is presented in Fig. 7.

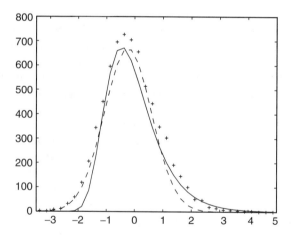

Figure 7. The probability distribution function of the Z scores computed for the population of false positives. A set of 47 sequences from the 547 set of proteins with known structures without homologs in the HL set is used to sample the distribution of Z scores for false positives. Each of the sequences is aligned to all the structures included in HL set. The Z scores are calculated for the 200 best matches (according to energy) using 100 shuffled sequences. The observed distribution of Z scores is represented by $+$. The dashed line shows the attempted analytical fit to a Gaussian distribution, whereas the solid line the analytical fit to the expected extreme value (double exponential) distribution. Note the significant tail to the right, which is the probability of obtaining a relatively large Z score by chance. See text for more details.

Let us denote by $\hat{p}(Z)$ the probability density of finding a Z-score value between Z and $Z + dZ$. Hence, $P(Z_p)$ is given by $P(Z_p) = \int_{-\infty}^{Z_p} \hat{p}(Z)\, dZ$. We approximate the observed distribution ('+') by an analytical fit to the extreme value distribution (represented by a continuous line in Fig. 6), which is defined by [55]

$$\hat{p}(Z) = 1/\sigma \cdot \exp[-(Z - a)/\sigma - e^{(Z-a)/\sigma}] \qquad (16)$$

In the realm of sequence comparison, the extreme value distribution has been used to model scores of random sequence alignments for local, ungapped alignments [56] as well as for local alignments with gaps [57].

The observed distribution is asymmetric and has a long tail toward high Z-score values (which is the tail that we are mostly interested in). Note, however, that there are significant differences between the numerical data and the analytical fit (and of course from the symmetric Gaussian distribution; dotted line in Fig. 7). Some deviations are expected because the distribution we extracted numerically differs from a random distribution. As discussed above, we use, for example, only alignments with negative energies. Hence, the energy filter was already employed.

Using analytical fit, we find that $P(Z_p) = 1 - \exp[-\exp(-1.313 \cdot (Z_p + 0.466))]$ with the 98% confidence intervals 1.313 ± 0.112 and 0.466 ± 0.079. For example, we estimate that the probability of observing a random Z score that is larger than 4 is 0.003. We emphasize, however, that the analytical fit is an upper bound as is shown in Fig. 6. For example, the observed number of Z scores larger than 4.0 is equal to 3—as opposed to the expected number of finding a Z score larger than 4.0, which is equal to (according to the analytical fit) $6813 \cdot 0.003 = 20.4$.

We observe similar discrepancy for global threading alignments of all the sequences from the HL set into all the structures in the HL set. For each probe sequence we select the 10 best matches (with lowest energies) that are subsequently subject to the statistical significance test, resulting in a sample of 2460 Z scores. Only five of the calculated Z scores, which are larger than 3.0, correspond to false positives. Using the analytical fit from Fig. 7 the expected number of observing by chance Z scores larger than 3.0 is equal to 24.6. Thus, it seems that the conservative estimate of the tail of the extreme value distribution indeed provides an upper bound for the probability of observing a false positive with a low energy and a high Z score.

B. Double Z-Score Filter

When searching large databases, the probability of observing false positives is growing, because the expected number of false positives is $N \cdot P(Z_p)$, where N is the number of structures in the database. Therefore, only relatively high Z

scores may result in significant predictions. Unfortunately, there are many correct predictions with low Z scores that overlap with the population of false positives. A high cutoff will therefore miss many true positives. Restricting the Z score test to only best matches (according to energy) is still insufficient. Therefore we propose an additional filtering mechanism, based on a combination of Z scores for global and local alignments. The double Z-score filter eliminates false positives, missing much smaller number of correct predictions.

Global alignments (in contrast to local alignments) are influenced significantly by a difference in the lengths of the structure and the threaded sequence. The matching of lengths was considered too restricted in previous studies [58]. However, at our hands and using environment-dependent gap penalty, the Z

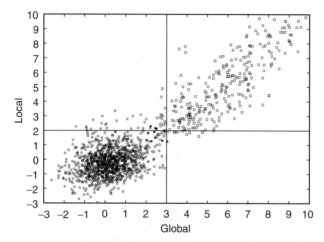

Figure 8. The joint probability distribution for the Z scores of global and local alignments. The distribution at the lower left corner (circles) is the result of the alignments of the 547 set sequences against all structures in HL set. The Z scores for the false positives are computed using 1000 shuffled sequences for both global and local alignments to ensure convergence. Only weak energy constraint are used; that is, 100 best global and 200 best local matches are subject to a Z score test, and then a given pair (global Z score, local Z score) is included if the energy of the global alignment is negative. The resulting 1081 pairs are included in the figure. The best pair in this population is slightly below the threshold (3.0,2.0). The population in the right upper corner represents (square boxes) 331 pairs of HL sequences aligned to HL structures with global Z scores larger than 2.5 and local Z scores larger than 1 [some of the Z scores fall beyond the (10,10) range). This set includes 236 native alignments and 95 non-native alignments. There are 10 matches that are false positives (filled squares), and they are all below the threshold (3, 2). Four of them are marginally so. The Z scores of this distribution were generated using 1000 shuffled sequences for global alignments, but only 50 for local alignments. Stiffer energy constraints were employed in which only the 10 best matches (according to energy) for global alignments and with 200 best matches for local alignments were considered. Of course, there is still a population of matches below (2.5,1.0) threshold (including 10 native alignments). However, the number of false positives below this threshold grows quickly, making predictions with Z scores in this range difficult.

score of the global alignment was proven a useful independent filter. This filter is an addition to the use of energy (of local and global alignments) and of the Z score of local alignments.

In Fig. 8 we present the joint probability distribution for global and local Z scores for a population of false positives versus a population of correct predictions. The squares at the upper right corner represent correct predictions, resulting from 331 native alignments (of a sequence into its native structure) and homologous alignments (of a sequence into a homologous structure) of the HL set proteins. The circles at the left lower corner are false positives obtained from the alignments of the sequences of the S47 set against all structures in the HL. The procedure is the same as the one used previously to generate the probability density function for the Z scores of local alignments (see Fig. 7). However, the Z scores are computed using 1000 shuffled sequences for both global and local alignments, which is sufficient to converge the values of the Z scores. The converged results reduce somewhat the tails of the distribution. For example, the number of false positives with a global Z score larger than 2.5 and a local Z score larger than 1.0 is equal to 3, as compared to 7 with only 100 shuffled sequences.

Figure 8 shows that the thresholds of 3.0 for global Z scores and of 2.0 for local Z scores are sufficient to eliminate all the false predictions. These cutoffs result in a number of misses, for example, 23 native alignments are dismissed as insignificant (see also the next section). However, this is a price we have to pay for high confidence levels in our predictions. The total number of pairwise alignments for which we compute the global and the local Z scores, and subsequently test for the presence of false positives, is about 10,000. Hence, we estimate that the probability of observing a single false positive with a global and a local Z score larger than the 3.0 and 2.0 thresholds is smaller than 0.0001.

VII. TESTS OF THE MODEL

There are three tests that we perform in this section on the THOM2 potential. We use optimal alignments and the double Z-score test proposed in Section VI. First, we analyze the results of threading the sequences of the HL set into all the structures of the HL set. Self-recognition and family recognition are discussed. Next, threading of the CASP3 sequences into an extended TE set is used to test the performance of the new threading protocol on the set of folds that were not included in the training. Finally, further tests of family recognition are presented, including the comparison of THOM2 results with those of a pairwise model using the frozen environment approximation.

A. The HL Test

The HL set was partially learned (using gapless threading). The first test verifies that the additional flexibility of gaps and deletion maintain good prediction

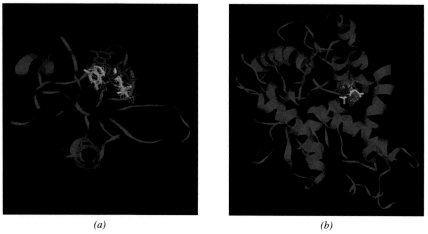

(a)　　　　　　　　　　　　　(b)

Figure 4. (See Chapter 4.) For the predicted protein structure of 2sarA (2cmd_) generated by GeneComp using a template provided by the Fischer Database [34], the red-colored ligand represents the superposition of the ligand bound to the native receptor. The highest-scored match is colored in yellow.

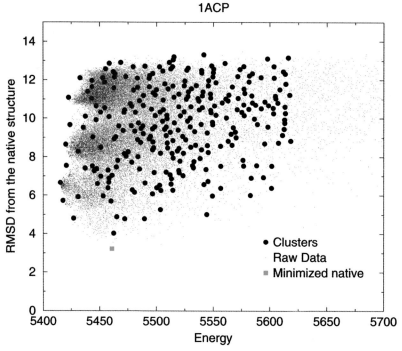

Figure 7. (See Chapter 6.) Comparison of raw data and clustered results (red dots: raw simulation data, black circles: cluster representatives, green square: locally minimized native structure).

Figure 7. (See Chapter 2.) The native-state conformation of the bovine pancreatic trypsin inhibitor (BPTI). The figure was produced with the program RasMol 2.7.1 [126] from the PDB entry 1bpi. There are three disulfide bonds in this protein: Cys5–Cys55 shown in red, Cys14–Cys38 shown in black, and Cys30–Cys51 shown in blue. The corresponding Cys residues are in the ball-and-stick representation and are labeled. The two helices (residues 2–7 and 47–56) are shown in green.

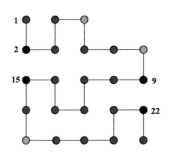

Figure 8. (See Chapter 2.) (a) The ground-state conformation of the two-dimensional model sequence with $M = 23$ beads and four covalent (S) sites. The red, green, and black circles represent, respectively, the hydrophobic (H), polar (P), and S sites.

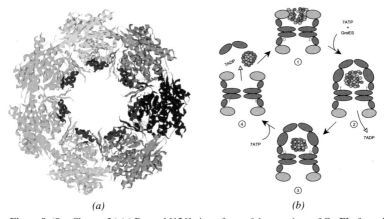

(a)   (b)

Figure 9. (See Chapter 2.) (a) Rasmol [126] view of one of the two rings of GroEL, from the PDB file 1oel. The seven chains are indicated by different colors. The amino acid residues forming the binding site of the apical domain of each chain (199-204, helix H: 229-244 and helix I: 256-268) are shown in red. The most exposed hydrophobic amino acids that are facing the cavity and are implicated in the binding of the substrate as indicated by mutagenesis experiments [112, 127] are : Tyr199, Tyr203, Phe204, Leu234, Leu237, Leu259, Val263, and Val264. (b) A schematic sketch of the hemicycle in the GroEL–GroES-mediated folding of proteins. In step 1 the substrate protein is captured into the GroEL cavity. The ATPs and GroES are added in step 2, which results in doubling the volume, in which the substrate protein is confined. The hydrolysis of ATP in the *cis*-ring occurs in a quantified fashion (step 3). After binding ATP to the *trans*-ring, GroES and the substrate protein are released that completes the cycle (step 4).

ability such as self- and family recognition. We note that our training did not include the Z score, so successful predictions based on only the Z score are useful tests even if performed on the training set of structures. The second test is a prediction experiment on proteins not included in the learning set. There are 40 new proteins that are included in Table XIa.

TABLE XI

A Summary of the THOM2 Threading Alignments of All the Sequences of the HL Set Into All the Structures of the HL Set[a]

(a)

1bbt1, 1gp1A, 1grcA, 1ipd, 1lap, 1lpe, 1phd, 1prcL, 1prcM, 1rbp, 1rhd, 1rnh, 1stp, 1wsyB, 2cna, 2cts, 2gbp, 2snv, 2wrpR, 3sicE, 4dfrA, 4gcr, 4rcrH, 4rcrL, 4rcrM, 7acn, 8adh, 4cms, 4i1b, 5fd1, 1atnA, 1tfd, 2aaiA, 2aaiB, 2bbkA, 2bbkB, 2lig, 2mnr, 2plv1, 2sas

(b)

Energy	Z Score	N
First	First	234
First	Second	4
First	Fourth	1
Second	Second	3
Weak	Weak	4

(c)

Z Score	N
First	177
Second or Third	35
Fourth and lower	14
Weak	11
Very Weak	9

[a]A list of proteins of the HL set that were not included in the training (TE) set is given in part a. A summary of the native global alignments is included in part b. Part c contains a summary for the native local alignments. The number of native alignments N, with ranks specified in terms of energies (first column in part b) and Z scores (second column in part b and the first column in part c), is given in the last column. For global alignments, "weak" is used to mark alignments with a weak energy or Z-score signals. There are four weak alignments corresponding to the photosynthetic centers membrane domains that were not included in the training set. Only five out of the remaining 242 native alignments obtain Z scores smaller than 3.0 (four alignments with Z scores larger than 2.5 and one alignment with a Z score smaller than 2.5). For local alignments, "very weak" denotes native alignments with Z scores smaller than 1.0, whereas "weak" marks alignments having Z scores larger than 1.0 and smaller than 2.0. There are 226 local native alignments with Z scores larger than 2.0. Note also that energy is not used to filter local alignments (beyond the initial restriction to 200 best candidates).

The self-recognition of the HL set proteins in terms of optimal alignments and Z-score filters is summarized in Tables XIb and XIc (see also Fig. 8). In Table XIb we provide the data for the global alignment. Energy and Z-score filters are considered. Of the total of 246 proteins, 234 are clear-cut cases (the energy and the Z scores of the native alignment are at the top). The four failures are membrane proteins (photosynthetic reaction centers) that were not included in the training set. In Table XIc the data for the local alignments are provided. We use only the Z score as a filter because there are many incorrect alignments with good (negative) energies. Among nine native alignments that are clear failures ($Z < 1.0$), six refer to structures that were included in the training set.

As examples of protein families, represented in the HL set, we discuss cytochromes, dehydrogenases, and acid proteases. Cytochromes were included in the training of the gaps, so we might expect that identification of cytochromes will be easy. Yet, this is not the case and we report a "bad" case scenario for some of the members of the family in Table XIIa. The Z-score values are below what we usually consider as a significant hit. Even though the correct proteins make it to the top, the global Z scores are too low (1.3–1.4) to confirm the prediction. The successful recognition of dehydrogenases and acid proteases families is shown in Tables XIIb and XIIc. We comment that most of the family members of the HL set are recognized irrespective of the choice of the probe sequence, as long as it belongs to a given family. More extensive tests of family recognition are discussed in Section VII.C.

Global Z scores reported in Tables XI and XII are converged using 1000 shuffled sequences. Local Z scores are, however, computed using only 50 shuffled sequences. The constraint here is of computational resources. Global Z scores are computed only for 10 energy-best structures and can be done

TABLE XII
Examples of Predictions for Families of Homologous Proteins[a]

(a)				
Query srequence: 5cytR	Structure	Energy	Z score	RMS
Global alignments	5cytR	−22.1	4.1	0.0
	1ccr	−10.4	1.4	6.9
	3c2c	−10.4	1.4	4.9
	1rro	−11.2	1.3	—
	256bA	−12.0	1.0	—
Local alignments	5cytR	−31.0	3.9	0.0
	1ccr	−35.6	3.2	1.9
	1yea	−23.9	3.2	1.9
	2ccyA	−22.8	3.0	—
	2fox	−27.6	2.3	—

TABLE XII   (*Continued*)

(b)

Query Sequence: 1llc	Structure	Energy	Z Score	RMS
Global alignments	1llc	− 80.0	7.0	0.0
	1lldA	− 60.7	4.4	5.3
	1ldnA	− 52.9	4.2	4.6
	4mdhA	− 47.4	2.1	6.7
	6ldh	− 45.8	1.6	4.6
Local alignments	1ldnA	− 73.4	5.2	4.1
	1llc	− 89.8	5.2	0.0
	1lldA	− 74.1	4.4	5.0
	6ldh	− 73.4	4.3	4.4
	1ipd	− 82.7	2.8	—

(c)

Query Sequence: 1pplE	Structure	Energy	Z Score	RMS
Global alignments	1pplE	− 77.3	9.5	0.0
	2er7E	− 61.4	7.3	2.9
	3aprE	− 51.9	4.3	3.9
	4cms	− 45.0	4.2	5.4
	4pep	− 43.1	3.6	5.7
Local alignments	1pplE	− 79.2	12.9	0.0
	2er7E	− 68.6	8.3	2.9
	3aprE	− 59.6	4.5	5.2
	4pep	− 55.4	3.3	5.7
	1prcH	− 46.6	2.2	—

[a]The results of global and local threading alignments for representatives of three families in the HL set are reported. The families are cytochromes (part a), lactate and malate dehydrogenases (part b), and pepsin-like acid proteases (part c). Five best alignments, ordered according to their Z scores (fourth column), are reported. The names of the query sequences are specified in the first column, target structures in the second, and the energy of the alignment in the fourth column, respectively. In the last column the RMS distance between the (known) structure of the probe (query) and the target structure, according to a novel structure-to-structure alignment (Meller and Elber [45]), is provided. RMS distances larger than 12 Å are indicated by a dash. Note that in a "bad" case scenario a distance of about 5 Å between the superimposed side-chain centers of 5cytR and 3c2c is sufficient to make threading identification virtually impossible because the Z score is too low (see part a). The local alignment provides a significantly improved Z score in this case. On the other hand, there are homologous structures that are not detected by the local alignments, although their global Z scores are high. Examples are malate dehydrogenase 4mdh (see part b) and acid protease 4cms (see part c). The structures with the PDB codes 1rro and 2fox (part a), 1ipd (part b) and 1prcH (part c) do not belong to the families of interest.

accurately. Local Z scores are computed for 200 alignments. The number of alignments with negative energies, which needs to be probed by an additional filter, is much larger for local alignments. With limited computational resources and/or a large-scale alignment project, it may be necessary to use Z scores that are not fully converged. For example, when aligning a 1pplE sequence into a 1prcH structure, a Z score of 1.8 with 1000 shuffled sequences is obtained, as opposed to 2.2 with only 100 shuffles sequences.

Finally, we remark that we were able to find alignments (with gaps) that have energies lower that the energy of the native state. Moreover, even aligning a sequence into its own structure may result in lower energy than the native if the addition of gaps and deletions is favorable. One such example is the alignment (with gaps) of 1llc onto its native shape.

B.   Recognition of Folds Not Included in the Training

In order to assess the generalization capacity of THOM2 in terms of optimal alignments, we use the S47 set again. Let us recall that the S47 set is composed of CASP3 [46] targets and their relatives. Using CASP3-related structures is a convenient way of finding protein shapes that are not sampled in the training. The experiment we perform is for self-recognition and is not aimed at finding remote relatives (as in CASP). The results are summarized in Table XIII. The native and

TABLE XIII
Self-Recognition for Folds That Were Not Learned[a]

PDB Code (len)	FSSP Z-score (RMS)	THOM2 Global Z score	THOM2 Local Z score
1HKA (158)	33.0 (0.0)	**7.1**	**7.1**
1VHI (139)	4.3 (5.2)	0.2	0.3
2A2U (158)	33.8 (0.0)	**2.5**	**4.0**
1BBP (173)	11.6 (3.3)	**3.5**	**3.0**
2EZM (101)	55.3 (0.0)	**3.7**	**3.2**
1QGO (257)	46.0 (0.0)	**5.6**	**7.6**
1ABE (305)	6.4 (3.4)	0.5	0.4
1BYF (123)	29.5 (0.0)	1.8	2.8
1YTT (115)	16.4 (2.2)	-0.1	1.4
1JWE (114)	26.9 (0.0)	2.6	2.3
1B79 (102)	18.7 (1.3)	0.3	1.3
1B7G (340)	61.5 (0.0)	**8.7**	**8.8**
1A7K (358)	25.1 (2.9)	-0.4	-0.9
1EUG (225)	43.0 (0.0)	**3.4**	**3.0**
1UDH (244)	30.8 (1.7)	-1.0	2.9
1D3B (72)	18.4 (0.0)	**3.5**	**2.8**
1B34 (118)	13.4 (1.1)	1.9	2.0
1DPT (114)	24.8 (0.0)	**6.2**	**6.0**
1CA7 (114)	18.7 (1.2)	**4.0**	**2.5**
1BG8 (76)	19.1 (0.0)	**3.4**	**3.5**

TABLE XIII   (*Continued*)

PDB Code (len)	FSSP Z-score (RMS)	THOM2 Global Z score	THOM2 Local Z score
1DJ8 (79)	16.2 (0.7)	**5.1**	**3.9**
1QFJ (226)	42.7 (0.0)	**8.1**	**8.4**
1VID (214)	7.1 (3.1)	− 2.0	0.5
1BKB (132)	25.1 (0.0)	2.7	1.5
1EIF (130)	17.4 (1.6)	**3.5**	**2.0**
1B0N (103)	19.5 (0.0)	**4.7**	**5.0**
1LMB (87)	8.0 (5.3)	0.3	0.1
1BD9 (180)	38.8 (0.0)	**4.5**	**5.8**
1BEH (180)	36.0 (0.3)	**7.4**	**5.8**
1BHE (376)	70.2 (0.0)	6.7	0.6
1RMG (422)	36.9 (2.2)	0.9	—
1B9K (237)	39.7 (0.0)	**8.1**	**8.2**
1QTS (247)	36.1 (0.7)	**3.5**	**6.4**
1EH2 (95)	24.3 (0.0)	**6.0**	**6.5**
1QJT (99)	7.6 (2.5)	**3.6**	**3.7**
1BQV (110)	20.9 (0.0)	**3.5**	**2.3**
1B4F (82)	3.2 (3.3)	0.0	1.7
1CK2 (104)	26.0 (0.0)	**5.2**	**4.3**
1CN8 (104)	14.3 (2.2)	**5.3**	**2.0**
1BL0 (116)	24.9 (0.0)	0.5	0.5
1JHG (101)	3.4 (6.6)	1.1	1.0
1BNK (100)	24.9 (0.0)	**5.4**	**6.3**
1B93 (148)	31.4 (0.0)	**4.0**	**3.2**
1MJH (143)	6.1 (3.4)	0.3	1.3
1BK7 (190)	37.2 (0.0)	**7.7**	**9.0**
1BOL (222)	19.7 (2.3)	0.1	− 1.0
1BVB (211)	37.3 (0.0)	**5.3**	**4.3**

[a]Twenty-two pairs of CASP3 targets and their structural relatives, as well as an additional three singleton targets, are added to the TE set. Their PDB codes are given in the first column (with lengths in parentheses). The actual CASP3 targets are given as the first structure of each pair (e.g., 1HKA from the pair 1HKA, 1VHI). If the domain is not specified and one refers to a multidomain protein, then the A (or first) domain is used. The results of global and local THOM2 threading of the 25 CASP3 sequences into an extended TE set (594 + 47 structures) are reported in the third and fourth column, respectively. Two of 25 native alignments (594 + 47) gave weak signals (DNA-binding protein 1BLO and glycosidase 1BHE). Four other native alignments (2A2U, 1BYF, 1JWE, and 1BKB) provide global Z scores somewhat smaller than 3. The DALI Z scores and RMS deviations for structure-to-structure alignments into native and homologous structures are reported in the second column (the native structures have RMS distances of zero). Note that low Z scores indicate that only short fragments of the respective structures are aligned and the resulting RMS deviation may not be representative. Nine related structures, among the 14 pairs with the DALI Z score larger than 10, obtain Z scores larger than 3.0 and 2.0 for the global and local THOM2 threading alignments, respectively. The alignment of 2A2U sequence into the 1BBP structure was the only significant hit of any of the target sequences into the structures included in the training (TE) set. Thus, no false positives with scores above our confidence cutoffs were observed. All the predictions that can be made with a high degree of confidence are indicated by Z scores printed using boldface type.

homologous shapes were embedded in the structures of the TE set, and the sequences of CASP targets were aligned into all the structures of such extended set. We provide in the table the results of the native alignments and the alignments into related homologous structure, irrespective of their rank.

One encouraging observation is that the native structures are found with high probability. Twenty of 25 structures would have been found if the native structure was included in the set. A less encouraging observation is the sensitivity of the results to structural fluctuations. The THOM2 model can identify related structures only if their distance is not too large. Nine out of 14 homologous structures with the DALI [44] Z score for structure-to-structure alignment larger than 10 are detected with high confidence. Only one homologous structure with the DALI Z score lower than 10 is detected.

Only three among the 25 structures of the CASP3 targets included in Table XIII had homologous counterparts in the training set. These are 2a2u, 1byf, and 1eug with their respective homologous proteins 1bbp, 2msb, and 1akz. It is therefore reassuring that most of the native structure and a significant fraction of relatives are recognized in terms of both their energies and the Z scores. Also, there are no further significant hits into other structures from the TE set. Hence, no false positives above our confidence thresholds are observed in this test. We conclude that our nearly perfect learning (on a training set) preserves significant capacity for identification of new folds using optimal alignments with gaps.

Note also that good scores with the global alignment are obtained for length differences (between sequence and structure) that are on the order of 10%. This was made possible by using environment-dependent gaps. When the differences in length are profound (e.g., 1bqv versus 1b4f), it is obvious and expected that the global alignment will fail. Large differences are clearly focused on identification of domains and not a whole protein. This is a different problem, which the present chapter does not address.

C. Recognition of Protein Families: THOM2 Versus Pair Energies

Three families are considered here: globins (92 proteins), immunoglobins (Fv fragments, 137 proteins), and the DNA-binding, POU-like domains (26 proteins). Sequences of all family members are aligned optimally to all the structures in the family. Both the local and global alignments are generated for each sequence–structure pair, and the results are compared in terms of the sum of Z scores for global and local alignments. Thus we employ here a simplified version of the double Z-score filter discussed before. The THOM2 results are compared to the results of the TE pairwise potential, which was trained on the same set of 594 proteins using the LP protocol. The difference in the LP protocol was that an objective function was optimized.

The alignments due to the pairwise potential are computed using the first iteration of the frozen environment approximation (FEA) [22]. That is, when evaluating fitness of a query sequence into a structure, we assume that types of contacts are fixed according to the native identities of sites making contacts to a primary site occupied by a query residue. Such an approach is in fact a different profile approximation to the "true" pair energies. In THOM2, the number of neighbors to a secondary site approximates its identity, whereas in FEA it is approximated by the identity of the native residue at that site. In principle, the FEA should be iterated until self-consistency is achieved [22]. Purely structural characterization of contact types in THOM2 avoids this problem.

In order to compute optimal alignments with the FEA, we need to set the gap penalties for the TE potential. After some experimentation the insertion penalties are chosen to be proportional to the number of neighbors to a site, $\varepsilon_-(n) = 0.2 \cdot (n+1)$. This choice is consistent with the THOM2 gap energies, which also penalize sites of no neighbors. The proportionality coefficient was gauged using the same families that were used to train THOM2 gap energies. However, no LP training was attempted. The deletion penalties are also consistent with the THOM2 model, and they are defined in the way described in Section V.

Figures 9a to 9f show the joint histograms of the sum of Z scores for local and global threading alignments versus the RMS deviations between superimposed (according to our novel structure-to-structure alignments; see Section III.A) side-chain centers. Figures 9a, 9c, and 9e show the results for THOM2 (for globins, immunoglobins, and POU-like domains, respectively), whereas Figs. 9b, 9d, and 9f show the corresponding results for TE potential with FEA. The vertical lines in the figures correspond to the sum of global and local Z scores equal to 5, which roughly discriminates the high confidence matches (with higher Z scores) and lower confidence matches that might be obscured by the false positives.

The population of matches that are difficult to identify by pairwise sequence-to-sequence alignments is represented by the filled squares. Sequence alignments are generated using Smith–Waterman algorithm with the BLOSUM50 substitution matrix (with the signs inverted) and structurally biased gap penalties [$\varepsilon_-(n) = 8 + (n-5)$, where n is the number of neighbors to a site]. Confidence of matches is estimated using Z scores defined, analogously to threading alignments, by the distribution of scores for shuffled sequences. We find that structurally biased gap penalties improve the recognition in case of weak sequence similarity. We do not observe false positives with more than 50% of the query sequence aligned and with a Z score larger than 8 (the distribution of Z scores for sequence substitution matrices is vastly different from that of threading potentials, with very high Z score for homologous sequences). All the matches represented by circles can be identified with high confidence by pairwise sequence-to-sequence alignments.

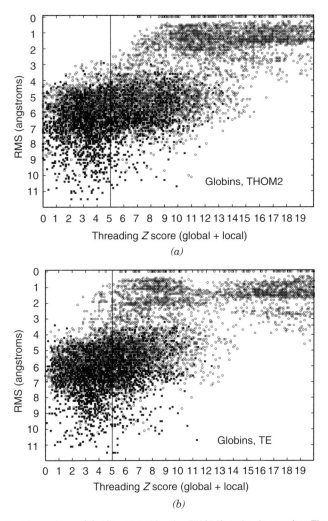

(a)

(b)

Figure 9. Comparison of family recognition by THOM2 and pair energies. The results of THOM2 for families of globins, immunoglobins (Fv fragments), and POU-like domains are compared to the results of Tobi–Elber (TE) pairwise potential. TE potential was optimized using LP protocol (with different target function) and the same training set. The first iteration of the so-called frozen environment approximation is performed to obtain approximate alignments for the TE potential. Parts a–f show the joint histograms of the sum of Z scores for local and global threading alignments versus the RMS deviations between superimposed (according to structure-to-structure alignments; see text for details) side-chain centers. Parts a, c, and e show the results for THOM2 (for globins, immunoglobins, and POU-like domains, respectively), whereas parts b, d, and f show the corresponding results for TE potential and the frozen environment approximation. The population of matches that are difficult to identify by pairwise sequence-to-sequence alignments is represented by the filled squares (see text for details). Note that the number of low THOM2 Z scores (for example, smaller than 5) is, on the average, smaller for families of globins and POU-like proteins. This is further highlighted in parts g and h, which show one-dimensional histograms of the sum of Z scores for local and global threading alignments for globins and POU-like domains. On the other hand, the TE potential and FEA perform better for immunoglobins family, which is also easier for sequence alignment methods (see text for details).

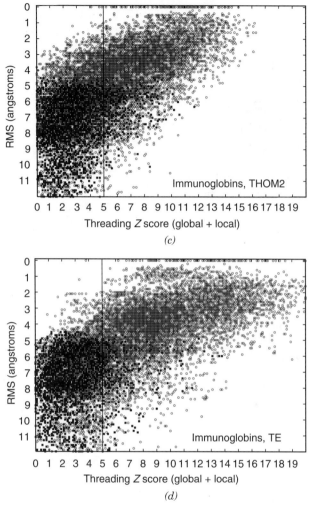

Figure 9   (*Continued*)

Nearly all pairs differing by less than 3 Å RMSD can be identified by THOM2 threading alignments. Most of the matches in the range between 3 and 5 Å can be still identified with high confidence. However, the number of confident matches (to the right with respect to vertical lines representing our cutoff of 5 in terms of sum of local and global Z scores) quickly decreases with the growing RMS distance. Essentially all the pairs with RMSD smaller than 3 Å can be also identified by pairwise sequence alignments. Below this threshold,

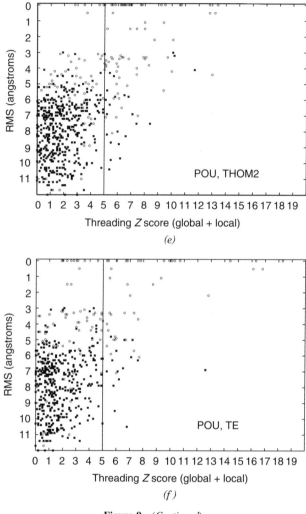

Figure 9 (*Continued*)

however, we observe many matches that can be still identified by threading but not by sequence alignment (filled rectangles corresponding to threading Z score higher than 5).

On the other hand, there are many matches due to the sequence alignment that are not detected by threading. Because we do not incorporate family profiles in our threading protocol, we do not include here a systematic comparison with the results of PsiBLAST [59]. However, we found examples

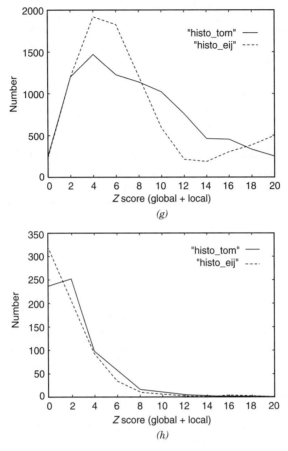

Figure 9   (*Continued*)

of matches detected with high confidence by threading and not detected by PsiBLAST in each of the families considered here (e.g, globins 1flp and 1ash or POU-like proteins 1akh and 1mbg).

Note that for the families of globins and POU-like domains the number of low THOM2 Z scores (for example smaller than 5) is, on the average, smaller than the number of low Z scores obtained with the TE potential and FEA. This is further highlighted in Figs. 9g and 9h showing one-dimensional histograms for the sum of Z scores for local and global threading alignments for globins and POU-like domains. For example, the number of low confidence matches ($Z < 5$) for globins increases from 2851 in case of THOM2 to 3265 in case of the TE potential. One can also notice that the distribution of Z

scores is different, with many very high Z scores for alignments into very close homologs as opposed to lower scores for more divergent pairs, in case of the TE potential.

Interestingly, FEA with the TE potential fails also for a larger number of native alignments. This is especially clear for the family of DNA binding proteins (see Figs. 9e and 9f). The number of native alignments with very low Z scores (smaller than 4) is equal to 7 in case of pairwise model and only 2 in case of THOM2. Because DNA binding proteins may be stabilized by contacts that are not included in our model, the energies of native alignments are quite poor. One striking example is the 1hdp. According to TE potential, 1hdp has the native energy equal to -0.42. An alternative alignment into its native structure with one insertion and one deletion in the sequence improves the energy to -0.63 despite the cost of gaps. On the other hand, THOM2 model seems to be capable of compensating for that using the information about the shape of the protein as encoded in the contact (solvation) shell characterization of each contact. The THOM2 native alignment for 1hdp is the lowest in energy and leads to higher Z scores.

The relatively worse performance of the pairwise model may result from the suboptimality of alignments that we generate using FEA, especially that our gap penalties for the TE potential were not optimized by LP protocol and we did not attempt to converge the FEA until self-consistency is achieved. However, as discussed above, in many instances it is clear that even better gap penalties will not be able to improve the observed scores. The specific functional form of our new profile model contributes to the relatively better performance too.

On the other hand, there are families for which the pairwise model works better. As can be seen from Figs. 9c and 9d, one such example is the family of immunoglobins. The FEA is expected to perform well when the sequence similarity is sufficiently high, because the information about the native sequences is used to generate optimal alignments. The divergence in terms of what can be detected by sequence similarity is larger for globins and POU-like proteins than for immunoglobins. For example, contrary to other families considered here, all the immunoglobins with RMSD smaller than 4 Å can be detected by sequence alignments. Therefore, good performance of the FEA with the TE potential is expected in this case.

VIII.   CONCLUSIONS AND FINAL REMARKS

In the present chapter we proposed and applied an automated procedure for the design of threading potentials. The strength of the procedure, which is based on linear programming tools, is the automation and the ability of continuous exact learning. The LP protocol was used to evaluate different energy functions for accuracy and recognition capacity. Keeping in mind the necessity for efficient threading algorithms with gaps, we selected the THOM2 model as our best choice.

Statistical filters based on local and global Z scores were outlined. We observe that, while using very conservative Z scores that essentially exclude false positives, the new protocol recognizes correctly (without any information about sequences) most of the family members with the RMS distance between the superimposed side chain centers of up to 4 Å. We also observe many instances of successful recognition of family members that cannot be confidently recognized by pair energies with the so-called frozen environment approximation.

The present approach is based on fitness of sequences into structures. Nevertheless, it is easily extendable to include also sequence similarity, family profiles, secondary structures, and other relevant signals. Because the THOM2 model provides an effective and comparable in performance alternative to pairwise potentials, it can be used as a fast component of fold recognition methods employing pair energies. It is the target of a future work.

The algorithms and threading potentials presented in this chapter are available in the program LOOPP (Learning, Observing, and Outputting Protein Patterns). The program (including the source code and sets of proteins for training and recognition) is available from the web [48]. It is also possible to submit sequences directly to our server.

Acknowledgments

This research was supported by an NIH NCRR grant to the Cornell Theory Center (acting director Ron Elber) for the developments of Computational Biology Tools. It was further supported by a seed grant from DARPA to Ron Elber. Jaroslaw Meller acknowledges also partial support from the Polish State Committee for Scientific Research (Grant 6 P04A 066 14).

References

1. J. U. Bowie, R. Luthy, and D. Eisenberg, A method to identify protein sequences that fold into a known three-dimensional structure. *Science* **253**, 164–170 (1991).

2. D. T. Jones, W. R. Taylor, and J. M. Thornton, A new approach to protein fold recognition. *Nature* **358**, 86–89 (1992).

3. M. J. Sippl and S. Weitckus, Detection of native-like models for amino acid sequences of unknown three-dimensional structure in a database of known protein conformations. *Proteins* **13**, 258–271 (1992).

4. A. Godzik, A. Kolinski, and J. Skolnick, Topology fingerprint approach to the inverse folding problem. *J. Mol. Biol.* **227**, 227–238 (1992).

5. C. Ouzounis, C. Sander, M. Scharf, and R. Schneider, Prediction of protein structure by evaluation of sequence–structure fitness. Aligning sequences to contact profiles derived from 3D structures. *J. Mol. Biol.* **232**, 805–825 (1993).

6. S. H. Bryant and C. E. Lawrence, An empirical energy function for threading protein sequence through folding motif. *Proteins* **16**, 92–112 (1993).

7. Y. Matsuo and K. Nishikawa, Protein structural similarities predicted by a sequence–structure compatibility method. *Protein Sci.* **3**, 2055–2063 (1994).

8. L. A. Mirny and E. I. Shakhnovich, Protein structure prediction by threading. Why it works and why it does not. *J. Mol. Biol.* **283**, 507–526 (1998).

9. D. T. Jones, GenTHREADER An efficient and reliable protein fold recognition method for genomic sequenecs. *J. Mol. Biol.* **287**, 797–815 (1999).

10. A. R. Panchenko, A. Marchler-Bauer, and S. H. Bryant, Combination of threading potentials and sequence profiles improves fold recognition. *J. Mol. Biol.* **296**, 1319–1331 (2000).

11. M. J. E. Sternberg, P. A. Bates, L. A. Kelley, and R. M. MacCallum, Progress in protein structure prediction: Assessment of CASP3. *Curr. Opin. Struct. Biol.* **9**, 368–373 (1999).

12. A. Liwo, S. Oldziej, M. R. Pincus, R. J. Wawak, S. Rackovsky, and H. A. Scheraga, A united-residue force field for off-lattice protein structure simulations: Functional forms and parameters of long range side chain interaction potentials from protein crystal data. *J. Comp. Chem.* **18**, 849–873 (1997).

13. Y. Xia, E. S. Huang, M. Levitt, and R. Samudrala, *Ab initio* construction of protein tertiary structures using a hierarchical approach. *J. Mol. Biol.* **300**, 171–185 (2000).

14. A. Babajide, I. L. Hofacker, M. J. Sippl, and P. F. Stadler, Neural networks in protein space: a computational study based on knowledge-based potentials of mean force. *Fold. Des.* **2**, 261–269 (1997).

15. A. Babajide, R. Farber, I. L. Hofacker, J. Inman, A. S. Lapedes, and P. F. Stadler, Exploring protein sequence space using knowledge based potentials. *J. Comp. Biol.* submitted, Santa Fe Institute preprint 98-11-103 (1999).

16. S. B. Needleman and C. D. Wunsch, A general method applicable to the search for similarities in the amino acid sequences of two proteins. *J. Mol. Biol.* **48**, 443–453 (1970).

17. T. F. Smith and M. S. Waterman, Identification of common molecular subsequences. *J. Mol. Biol.* **147**, 195–197 (1981).

18. M. S. Johnson, J. P. Overington, and T. L. Blundell, Alignment and searching for common protein folds using a data bank of structural templates. *J. Mol. Biol.* **231**, 735–752 (1993).

19. H. T. Croman, C. E. Leiserson, and R. L. Rivest, *Introduction to Algorithms*, MIT Press, Cambridge, MA, 1985, Chapter 16.

20. R. H. Lathrop and T. F. Smith, Global optimum protein threading with gapped alignment and empirical pair score functions. *J. Mol. Biol.* **255**, 641–665 (1996).

21. R. H. Lathrop, The protein threading problem with sequence amino-acid interaction preferences is NP-complete. *Protein Eng.* **7**, 1059–1068 (1994).

22. R. A. Goldstein, Z. A. Luthey-Schulten, and P. G. Wolynes, The statistical mechanical basis of sequence alignment algorithms for protein structure prediction, in *Recent Developments in Theoretical Studies of Proteins*, Ron Elber, ed., World Scientific, Singapore, 1996, Chapter 6.

23. S. H. Bryant, Evaluation of threading specificity and accuracy. *Proteins* **26**, 172–185 (1996).

24. A. Elofsson, D. Fischer, D. W. Rice, S. Le Grand, and D. Eisenberg, A study of combined structure–sequence profiles. *Fold. Des.* **1**, 451–461 (1998).

25. B. Rost, R. Schneider, and C. Sander, Protein fold recognition by prediction-based threading. *J. Mol. Biol.* **270**, 471–480 (1997).

26. K. T. Simons, I. Ruczinski, E. Huang, and D. Baker, Improved recognition of native-like protein structures using a combination of sequence-dependent and sequence-independent features of proteins. *Proteins* **34**, 82–95 (1999).

27. A. R. Ortiz, A. Kolinski, and J. Skolnick, Fold assembly of small proteins using Monte Carlo simulations driven by restraints derived from multiple sequence alignments. *J. Mol. Biol.* **277**, 419–448 (1998).

28. J. Park, K. Karplus, C. Barret, R. Hughey, D. Haussler, T. Hubbard, and C. Chothia, Sequence comparisons using multiple sequences detect three times as many remote homologues as pairwise methods. *J. Mol. Biol.* **284**, 1201–1210 (1998).

29. I. Bahar and R. L. Jerningan, Inter-residue potentials in globular proteins and the dominance of highly specific hydrophilic interactions at close separation. *J. Mol. Biol.* **266**, 195–214 (1997).

30. D. Tobi and R. Elber, Distance-dependent pair potential for protein folding: Results from linear optimization. *Proteins: Struct. Funct. Genet.* **41**, 40–46 (2000).

31. V. N. Maiorov and G. M. Crippen, Contact potential that recognizes the correct folding of globular proteins. *J. Mol. Biol.* **227**, 876–888 (1992).

32. D. Tobi, G. Shafran, N. Linial, and R. Elber, On the design and analysis of protein folding potentials. *Proteins: Struct. Funct. Genet.* **40**, 71–85 (2000).

33. M. Vendruscolo and E. Domany, Pairwise contact potentials are unsuitable for protein folding. *J. Chem. Phys.* **109**, 11101–11108 (1998).

34. S. Miyazawa and R. L. Jernigan, Residue-residue potentials with a favorable contact pair term and an unfavorable high packing density term for simulation and threading. *J. Mol. Biol.* **256**, 623–644 (1996).

35. A. Godzik, A. Kolinski, and J. Skolnick, Knowledge-based potentials for protein folding: What can we learn from protein structures? *Proteins: Struct Funct Genet* **4**, 363–366 (1996).

36. D. A. Hinds and M. Levitt, A lattice model for protein structure prediction at low resolution. *Proc Natl. Acad. Sci. USA* **89**, 2536–2540 (1992).

37. M. R. Betancourt and D. Thirumalai, Pair potentials for protein folding: choice of reference states and sensitivity of predicted native states to variations in the interaction schemes. *Protein Sci.* **2**, 361–369 (1999).

38. A. Godzik, A. Kolinski, and J. Skolnick, Are proteins ideal mixtures of amino acids? Analysis of energy parameter sets. *Protein Sci.* **4**, 2107–2117 (1995).

39. J. D. Bryngelson and P. G. Wolynes, Intermediates and barrier crossing in a random energy-model (with applications to protein folding). *J. Phys. Chem.* **93**, 2902–6915 (1989).

40. D. K. Klimov and D. Thirumalai, Linking rates of folding in lattice models of proteins with underlining thermodynamic characteristics. *J. Chem. Phys.* **109**, 4119–4125 (1998).

41. W. R. Taylor and R. E. Munro, Multiple sequence threading: conditional gap placement. *Fold. Des.* **2**, S33–S39 (1997).

42. G. N'emethy, K. D. Gibson, K. A. Palmer, C. N. Yoon, G., Paterlini, A. Zagari, S. Rumsey, and H. A. Scheraga, Energy parameters in polypeptides. *J. Phys. Chem.* **96**, 6472–6484 (1992).

43. D. A. Hinds and M. Levitt, Exploring conformational space with a simple lattice model for protein structure. *J. Mol. Biol.* **243**, 668–682 (1994).

44. L. Holm and C. Sander, The FSSP database of structurally aligned protein fold families. *Nucleic Acids Res.* **22**, 3600–3609 (1994).

45. J. Meller and R. Elber, to be published.

46. CASP3. Third community wide experiment on the critical assessment of techniques for protein structure prediction. Asilomar, USA, 1998, http://Predictioncenter.lnl.gov/casp3

47. H. M. Berman, J. Westbrook, Z. Feng, G. Gilliland, T. N. Bhat, H. Weissig, I. N. Shindyalov, and P. E. Bourne: The Protein Data Bank. *Nucleic Acids Research* **28**, 235–242 (2000).

48. J. Meller and R. Elber, Learning, Observing and Outputting Protein Patterns (LOOPP)—a program for protein recognition and design of folding potentials, http://www.tc.cornell.edu/CBIO/loopp

49. C. S. Meszaros, Fast Cholesky factorization for interior point methods for linear programming. *Comput. Math. Appl.* **31**, 49–51 (1996).

50. I. Adler and R. D. C. Monteiro, Limiting behavior of the affine scaling continuous trajectories for linear programming problems. *Math. Program.* **50**, 29–51 (1991).

51. H. S. Chan and K. A. Dill, Protein folding in the landscape perspective: Chevron plots and non-Arrhenius kinetics. *Proteins: Struct. Funct. Genet.* **30**, 2–33 (1998).

52. S. H. Bryant and S. F. Altschul, Statistics of sequence–structure threading. *Curr. Opin. Struct. Biol.* **5**, 236–244 (1995).

53. W. M. Fitch, Random sequences. *J. Mol. Biol.* **163**, 171–176 (1983).

54. S. F. Altschul and B. W. Erickson, Significance of nucleotide sequence alignments: A method for random sequence permutation that preserves dinucleotide and codon usage. *Mol. Biol. Evol.* **2**, 526–538 (1985).

55. E. J. Gambel, *Statistics of Extremes*, Columbia University Press, New York, 1958.

56. S. Karlin and S. F. Altschul, Methods for assessing the statistical significance of molecular sequence features by using general scoring schemes. *Proc. Natl. Acad. Sci. USA* **87**, 2264–2268 (1990).

57. W. R. Pearson and D. J. Lipman, Improved tools for biological sequence comparison. *Proc. Natl. Acad. Sci. USA* **85**, 2444–2448; W. R. Pearson, Empirical statistical estimates for sequence similarity searches. *J. Mol. Biol.* **276**, 71–84 (1998).

58. D. Fischer, A. Elofsson, D. Rice, and D. Eisenberg, Assesing the performance of fold recognition methods by means of a comprehensive benchmark, in *Pacific Symposium on Biocomputing, Hawaii*, 1996, pp. 300–318.

59. S. F. Altschul, T. L. Madden, A. A. Schaffer, J. Zhang, Z. Zhang, W. Miller, and D. J. Lipman, Gapped BLAST and PSI-BLAST: A new generation of protein database search programs. *Nucleic Acid Res.* **25**, 3389–3402 (1997).

60. J. F. Gibrat, T. Madej, and S. H. Bryant, Surprising similarities in structure comparison. *Curr. Opin. Struct. Biol.* **6**, 377–385 (1996).

61. A. G. Murzin, S. E. Brenner, T. Hubbard, and C. Chothia, SCOP: A structural classification of proteins database for the investigation of sequences and structures. *J. Mol. Biol.* **247**, 536–540 (1995).

A UNIFIED APPROACH TO THE PREDICTION OF PROTEIN STRUCTURE AND FUNCTION

JEFFREY SKOLNICK

Laboratory of Computational Genomics, Danforth Plant Science Center, Creve Coeur, MO, U.S.A.

ANDRZEJ KOLINSKI

Laboratory of Computational Genomics, Danforth Plant Science Center, Creve Coeur, MO, U.S.A.; and Department of Chemistry, University of Warsaw, Warsaw, Poland

CONTENTS

Computational Methods for Protein Folding: A Special Volume of Advances in Chemical Physics, Volume 120, Edited by Richard A. Friesner. Series Editors I. Prigogine and Stuart A. Rice. ISBN 0-471-20955-4. © 2002 John Wiley & Sons, Inc.

I. INTRODUCTION

In this postgenomic era, a key challenge is to interpret the information provided by the knowledge of the proteome, the set of protein sequences found in a given organism. Unfortunately, having a list of protein sequences in and of itself provides little insight; the key question is, What is the function of all of the proteins? Function covers many levels, ranging from molecular to cellular or physiological to phenotypical. By employing sequence-based methods that exploit evolutionary information, between 40% and 60% of the open reading frames (ORFs) in a given genome can be assigned some aspect of function ranging from physiological to biochemical function. Indeed, because of their considerable success, sequence alignment methods such as PSI-BLAST [1,2] and sequence motif (that is, local sequence descriptors) methods such as Prosite [3], Blocks [4], Prints [5,6], and Emotif [7] set the standard against which all alternative approaches must be measured. However, sequence-based approaches increasingly fail as the protein families become more diverse [8]. The remaining unassigned ORFs, termed ORFans, represent an important challenge and represent an area where structure-based approaches to function prediction can play a significant role. One structure-based method combines one-dimensional information about sequence and structure and has had some success [9]. An alternative structure-based approach to function prediction that employs the sequence–structure–function paradigm has recently been developed [8,10–15]. Here, low-resolution models predicted by threading or *ab initio* folding are screened for matches to known active sites; if a match is found, then a functional assignment is made. However, this method requires a predicted structure of appropriate resolution. Structure prediction techniques will also play an important role in probe selection in structural genomics, where the ultimate goal is to experimentally determine the structure of all possible protein folds such that any newly found sequence is within modeling distance of an already solved structure. Thus, in this review, we examine the status of contemporary structure prediction approaches and demonstrate that the resulting (quite often low-resolution) models can be used both to identify the biochemical function of the protein and to dock known ligands to the correct binding sites.

Presently, there exist three approaches to protein structure prediction: homology modeling, threading, and *ab initio* folding. In homology modeling, the probe and template sequences are clearly evolutionarily related, and the structures of the probe and template are quite close to each other. The second structure prediction method is threading, where one attempts to find the closest matching structure in a library of already solved structures but where the structures can be analogous; that is, the two proteins are not necessarily evolutionarily related, but they adopt very similar structures. Ideally, threading should extend sequence-based approaches. Threading and homology modeling suffer from the fundamental disadvantage that an example of the fold of the sequence of interest must already have been solved in order for the method to be successful. Finally, there is *ab initio* folding where one attempts to fold a protein from a random conformation; obviously this is the hardest of the three methods of structure prediction, but it has the advantage that an example of the fold need not have been seen before. As detailed in what follows, a number of variants of *ab initio* folding use extensive information from threading. Such information might include local secondary structure information, supersecondary structure information, and/or predicted tertiary contacts. Indeed, the major focus of this review is to describe a unified approach to protein structure prediction that reduces to threading plus structure refinement when an example of the probe sequence is found; but if not, it incorporates information from weakly significant probe sequence–template structure matches and then does *ab initio* folding with the structural information gleaned from such matches. It has the advantage that it can predict a novel fold even though some of the information comes from threading on already solved structures.

II.   OVERVIEW AND HISTORICAL PERSPECTIVE

A.   Comparative Modeling Methods

Comparative modeling can be used to build the structure of those proteins whose sequence identity is above 30% or so with a protein template structure [16]. This usually consists of three steps: (1) Search for sequence similarity to a member of a set of carefully selected sequences with known three-dimensional structure; (2) use the detected structural template to build a molecular model; and (3) carefully validate the resulting models. In the recent CASP3 prediction experiment [17], encouraging results were reported by Bates and Sternberg [18], Blundell and co-workers [19], Yang and Honig [20], Dunbrack [21], and Fischer [22]. While the automated approach of Sali's MODELLER [23,24] did not do as well as others, it is nevertheless a widely used comparative modeling package. The results of CASP3 suggest that the key to a good model is to generate the best possible initial sequence alignment and to modify it as little as possible [25,26]. Thus, as the sequence identity of the probe and template moves into the twilight zone, sequence alignments degrade with a comparable degradation in the quality of the model structures.

As an example of genome-scale comparative modeling using standard sequence alignment algorithms and MODELLER, Sanchez and Sali [27] recently scanned a portion of the yeast genome, *S. cerevisiae* [28]. They found homologous proteins of known structure for about 17% of the proteins (1071 sequences), and they built three-dimensional models for these yeast proteins. Only 40 of these modeled proteins had a previously determined experimental structure, and 236 proteins were related to a protein of known structure for the first time.

An obvious limitation of the above approach is that it requires a homologous protein whose structure is known. Depending on the genome, 15–25% of all sequences now have a homologous protein of known structure [29]. This percentage is slowly increasing as new structures are being solved at an increasing rate. Interestingly, the majority of newly solved structures exhibit an already known fold. At this point, it is still uncertain whether this indicates that proteins can adopt a limited number of folds or if it simply indicates a bias toward certain types of protein folds that crystallize relatively readily.

B.   Threading

Threading is another means of predicting the tertiary structure of proteins. Here, for the sequence of interest, one attempts to find the closest matching structure in a library of known folds [30,31]. The paradigm of homology modeling is still followed with its three steps: (1) identifying the structural template, (2) creating the alignment, and (3) building the model. Thus, threading has limitations that are similar to classical homology modeling. First and foremost, an example of the correct structure must exist in the structural database that is being screened. If not, the method will fail. Second, the quality of the model is limited by the extent of actual structural similarity between the template and the probe structure. Until recently [32], one could not readjust the template structure to more correctly accommodate the probe sequence. While the quality of alignments generated by threading algorithms improved from CASP1 to CASP3 [17], it nevertheless remains problematic. Another question is whether threading recognizes distant homologies (i.e., a protein that is evolutionarily distant but still related to the template protein) as opposed to pure fold recognition targets (where the two proteins are evolutionarily unrelated, but have converged to the same fold). We note that for sequences that are evolutionarily very distant, convergent versus divergent evolution is very difficult to prove. Nevertheless, we still have the problem of identifying two proteins as having the same fold, when only about 65% of their sequences share a common core, with the possibility that the remainder of the fold differs significantly.

Next, we describe the features of existing threading algorithms that performed well in CASP3 as well as in the intervening period prior to CASP4. In the construction of a threading algorithm, one is faced with three choices: the type of energy used to assess the probe sequence–template structure suitability, the degree of detail used to describe interaction centers if multibody interactions are included, and the conformational search scheme employed to find the optimal sequence-structure alignment. In what follows, we address each of these three features in turn.

The first step in constructing a threading algorithm involves the choice of the potential used to describe the sequence-structure fitness and the potential for scoring functions containing more than one term; weights must be established. Among the kinds of energy terms that have been previously considered are the burial status of residues, secondary structure propensities and/or predicted secondary structure, additional penalty terms [33,34] (for example, those that compensate for different protein lengths), and the inclusion of pair or higher-order interactions between side chains. Contemporary algorithms often include an evolutionary component related to the sequence similarity between the template and the probe sequence [35]. Inclusion of such sequence-based terms improves the ability of the algorithm to recognize the correct structural template as well as the quality of the predicted alignment in the structural template [34, 36–39]. While such terms should not be needed in a structure-based approach, in practice they are found to be quite important.

If pair interactions are included, then the interaction centers must be selected, with common choices being the Cαs [40,41], the Cβs [42,43], the side-chain centers of mass, specially defined interaction centers [30,44], or any side-chain atom [45]. This defines the protein representation. Then, one must again choose the form of the interaction. Contact potentials [45,46], continuous distance-dependent potentials [42,47], and interaction environments [48] are the choices that have been made for the functional form of the pair energy.

Third, given an energy function, the optimal alignment between the probe sequence and each structural template must be found. Dynamic programming [49] is the best choice when local interaction schemes are used (e.g., when the energy consists of mutation matrices and secondary structure propensities). The situation when a nonlocal scoring function is used (e.g., pair interactions) is not as straightforward. Here, the problem is to update the interactions in the template structure to include the actual partners present in the probe sequence. To retain speed (a crucial feature if entire genomes are to be scanned), some workers employ dynamic programming with the "frozen" approximation (where the interaction partners or a set of local environmental preferences are taken from the template protein in the first threading pass) [45,50]. Iterative updating might follow this [45,48,51]. Still others employ double dynamic programming, which updates a subset of interactions recognized as being the most important in the first pass of the dynamic programming algorithm [42]. Other, more computationally intensive approaches evaluate the nonlocal scoring function directly and search for the optimal probe–template alignment by Monte Carlo [44] or branch-and-bound search strategies [30]. These have the advantage that the correct energy is evaluated, but unfortunately they are very CPU-intensive.

A problem with almost all threading search protocols is that they do not allow the actual template structure to adjust to reflect the actual structural modifications relative to the template structure that are actually present in the native conformation of the probe. For example, Monte Carlo and branch-and-bound strategies allow the partner from the probe sequence provided by the current probe–template alignment to be used, but they do not allow the template's backbone structure to readjust to accommodate the probe sequence. Such structural modifications should be quite important when the probe and template structure are analogous. As a simple example, when the probe's TYR replaces a GLY in the template protein, then the contacts associated with the amino acid at that position in the structure would be radically different. Yet, this effect is not accounted for at all in threading. However, the potential ability to recognize analogous structures is precisely the realm where threading should be the most valuable as compared to pure sequence-based methods.

As indicated above, because threading uses structure, it should be superior to sequence-based approaches that are one-dimensional and that assess the evolutionary relationship between sequences and thereby, by inference, their structural relationship. In practice, however, many of the most successful fold-recognition approaches in CASP3 were pseudo one-dimensional and used evolutionary information that contributed a significant fraction of the selectivity [52] (typically implemented in the form of sequence profiles) plus predicted secondary structure. In particular, the Jones [53] and the Koretke groups [39] employed this type of approach, where secondary structure played an ancillary role. The Nishikawa group [54] also employed a hierarchy of local scoring functions to describe hydration, secondary structure, hydrogen bonding, and side-chain packing.

There were other successful approaches in CASP3 where structure played a more prominent role. For example, the Sippl group [55] employed burial energy and the frozen approximation to evaluate pair interactions, but unlike many others, they used a single sequence rather than sequence profiles or other implementations of multiple sequence information. While the Sippl approach is more structure-based, in order for dynamic programming to be used all interactions were made pseudo one-dimensional. The Bryant group [56] was unique in that they explicitly treated pair interactions within a structural core identified from the evolutionary conservation of structure across each protein family. In order for the core to be identified, a number of structures in the protein family must be solved. While this approach embodies the original idea of threading, they too employ a PSI-BLAST sequence-profile component. Indeed, they conclude that the combination of both sequence profiles and contact potentials improves the success rate relative to that when either of the terms is used alone. Because the Bryant group employs a nonlocal scoring function that *a priori* precludes dynamic programming, a Monte Carlo search procedure was used to find the best sequence–structure fitness. Unfortunately, these calculations are very CPU-intensive, thereby precluding the application of this approach on a genomic scale unless there are very substantial computer resources.

The general consensus was that CASP3 saw some progress in threading, with alignment quality improving from CASP2 [17,26,52], but, as pointed out by Murzin [52], threading "performs better on distant homology recognition targets than on 'pure' folding recognition targets. This bias probably resulted from the implementation of 'distant homology' filters." Thus, techniques that extend the ability of threading techniques to address "pure" fold recognition situations are still required. But, as Bryant and co-workers [35] have pointed out, the best results are found when a sequence–profile term is combined with threading potentials. These observations motivated the development of a new threading algorithm, PROSPECTOR (*PRO*tein *S*tructure *P*redictor *E*mploying *C*ombined *Th*reading to *O*ptimize *R*esults) [57], where it was demonstrated that pair interactions could significantly improve the sequence–structure specificity over that when only sequence–profile terms are used. However, when multiple scoring functions are combined, the resulting recognition ability is even larger. In Section IV, we discuss the results of this new approach in some detail, because it is a key component of a recently developed unified approach to protein structure prediction. But here we note that while considerable progress has been made in threading by a number of workers, we will have to await the results of CASP4 to assess the full extent of this progress as well as the limitations of such approaches.

C. *Ab Initio* Protein Structure Prediction

Due to the time scale of the protein folding process, which takes from milliseconds to minutes, at present, it is rather impractical to attempt protein structure assembly using all-atom detailed models. Indeed, contemporary computers allow classical molecular dynamics simulations of a protein surrounded by an appropriate number of water molecules over a much shorter period of time, corresponding to tens or hundreds of nanoseconds (depending on protein size). This inability to routinely access longer time scales stimulated numerous attempts to simplify the problem by reducing the number of explicitly treated degrees of freedom of the polypeptide chain and by simplifying the model of intra and intermolecular interactions. Such a reduction of the number of degrees of freedom could be achieved by assuming a united-atom representation of entire amino acid residues, by assuming a single-atom representation of the main chain and a similar representation of the side groups. The internal degrees of freedom of the side groups were frequently ignored in such models or were treated in an approximate fashion. Such a simplified protein representation also led to simplifications in the interaction scheme; for example, all reduced models either ignored the effect of water or implicitly treated it.

The first attempts at the reduced modeling of protein folding were undertaken about 25 years ago. In their classical work, Levitt and Warshel [58] proposed a model that later inspired other analogous simplifications of protein representation. They assumed two centers of interaction per residue, one associated with the alpha carbon and the second with the center of mass of the side group. There was a single degree of freedom per amino acid—the rotation around the Cα–Cα virtual bond—while the planar angle for the Cα trace was assumed to be constant [59]. A knowledge-based potential controlled the short-range interactions, while the interactions between the side groups were in the form of a Lennard-Jones potential (partially corrected for the hydrophobic effect). The sampling was done by means of classical molecular dynamics. Simulations of a small protein bovine pancreatic trypsin inhibitor sometimes produced structures resembling the native fold. The best structures had a root-mean-square-deviation (RMSD), from native in the range of 6.5 Å. Later, Kuntz et al. [60,61], Hagler and Honig [62], and Wilson and Doniach [63] studied somewhat similar continuous models. The results were of comparable quality; some qualitative features of small protein folds were sometimes recovered in their simulations.

More recently, continuous-space models with more structural details were proposed and investigated with respect to their ability to predict the native conformation of a protein. Sun [64] examined models with an all-atom representation of the main chain and a single united atom representation of the side groups. Knowledge-based statistical potentials described the interactions between the side groups. Interestingly, his study demonstrated that a genetic algorithm could quite efficiently sample the conformational space of the chain. For small peptides (mellitin, pancreatic polypeptide inhibitor, and apamin), proper structures were predicted whose accuracy ranged from 1.66 Å to 4.5 Å, depending on peptide size. A similar model, but with two united atoms per side chain (for the larger amino acids), was studied by Wallqvist and Ullner [65]. Results for pancreatic polypeptide inhibitor were slightly more accurate, probably due to the better packing of the model side chains. Such reduced continuous models were explored not only as a means of protein structure prediction but also as a tool for investigating the general aspects of protein folding dynamics and thermodynamics [66,67].

Pedersen and Moult [68] proposed a very interesting approach to protein structure prediction. They assumed an all-heavy atom representation of the protein with knowledge-based potentials describing intraprotein interactions. As a sampling method, they used a combination of Monte Carlo (MC) and genetic algorithms. The MC runs produced a set of structures for the starting population of the genetic algorithm (GA). The crossover points were selected in the regions of the largest structural flexibility, as detected during the MC runs. MC simulations were also performed between crossover events in the GA scheme. Low- to moderate-resolution protein fragments and the approximate folds of small proteins have been successfully predicted by this method. Unfortunately, it appears that the applicability of this method is limited to rather small proteins.

Even reduced models of proteins have a large number of conformational degrees of freedom, and an effective sampling of the long-time processes for larger proteins in a continuous space could be very difficult if not impossible. To further simplify the problem, discrete or lattice models were proposed and examined. Early studies of the lattice proteins focused not on structure prediction but rather on understanding the fundamentals of protein folding thermodynamics and some aspects of the folding dynamics. These works were pioneered by Gō et al. [69], and then followed by Krigbaum and Lin [70,71], Skolnick and Kolinski [72–84], Sikorski and Skolnick [85–88], Chan and Dill [89–92], Dill et al. [93–96], Sali et al. [97,98], Shakhnovich et al. [99–105], and others [106–111]. Since the subject of this chapter is protein structure prediction and due to the existence of excellent reviews on the subject, we refrain from a more detailed review of these works.

Probably the first attempt to predict the native structure of a protein in an *ab initio* fashion within the framework of a lattice representation is due to

Dashevskii [112]. A diamond lattice chain was used to approximate the poly-peptide conformations. A chain growth algorithm executed the sampling of conformational space. Compact structures resembling native folds of small polypeptides were generated and identified by a simple force field. Next, Covell investigated a simple cubic lattice model of real proteins [113]. The behavior was controlled by the force field that consisted entirely of long-range interactions that included a pairwise, knowledge-based potential, a surface term, and a potential that corrects the local packing of the model chain. The quality of crude folds generated by this method were not worse than the quality of folds obtained using early continuous models. Covell and Jernigan [114] studied five small globular proteins by the enumeration of all possible compact conformations of a body-centered cubic lattice chain. They found that the closest to native conformation could always be found within the top 2% of the lowest-energy structures, as assessed by a knowledge-based interaction scheme.

Hinds and Levitt [115] proposed an interesting lattice model of proteins. In a diamond lattice chain, a single lattice vertex represents several residues of a real protein. An elaborate statistical potential was employed to mimic the mean interactions between such defined protein segments. Frequently, correct folds of low resolution were generated among the compact structures enforced by the sampling scheme.

Kolinski and Skolnick [75–84,116–120] developed a series of high-coordination lattice models of globular proteins. Lattices of various resolution were employed to mimic the conformation of the $C\alpha$ trace of real proteins, from three-dimensional "chess-knight"-type lattices to a high coordination lattice with 90 lattice vectors to represent possible orientations of the $C\alpha$–$C\alpha$ virtual bonds. The models employed in the test structure predictions [118,121–123] had additional interaction centers to represent the side groups. For each side chain, a single-sphere, multiple rotamer representation was assumed. The force field of each of these models contained several terms mimicking the short-range interactions, explicitly cooperative hydrogen bonds, one body, and pairwise and multibody long-range interactions with an implicit averaged effect of the water molecules. It has been shown for several cases of small globular proteins [118] and simple multimeric molecular assemblies [124–126] that such models can generate correct low- to moderate-resolution (high-resolution in the case of leucine zippers) folds during Monte Carlo simulated annealing computer experiments.

Various recently developed methods for *ab initio* protein structure predictions were tested during the CASP3 (Critical Assessment of Techniques for Protein Structure Prediction) exercises, concluded in December 1998 in Asilomar, California [127]. A number of new techniques have been developed before that time, and a number of them constitute qualitative progress in *ab initio* prediction with respect to the previous CASPs (held every two years).

The ROSETTA method proposed by Baker and co-workers [128] is very innovative. The method consists of several steps. First, a multiple sequence alignment for a sequence of interest was prepared, and the secondary structure prediction is made using the PHD server based on Rost and Sander's [129–131] secondary prediction technique. Secondary structure predictions and sequence alignments were then used to extract the most plausible 3- to 9-residue structural fragments (25 fragments for each segment of the query sequence) from the structural database (according to the secondary structure prediction and the sequence similarity). Then a Monte Carlo algorithm employing a random insertion of fragments into the structure was used to build the three-dimensional structure. The scoring function contained a hydrophobic burial term, elements of electrostatics, a disulfide bond bias, and a sequence-independent term that evaluates the packing of secondary structure elements. The top 25 (of 1200 generated) structures frequently contained the proper fold. The best five structures exhibiting a single hydrophobic core were selected by "visual inspection." This could be considered to be a flaw of the method (at this stage of development). It would be difficult to do a manual evaluation of the predictions on a massive scale. Nevertheless, for 18 targets, four predictions were globally correct (with an RMSD range of 4–6 Å for the native structure), and the majority of their predictions contained significant fragments of structure that were correct. It should be noted that a somewhat similar idea of protein structure assembly using predefined fragments and the Monte Carlo method was also pursued in the method developed by Jones [132] and tested during the CASP2 exercise.

A number of other groups made good predictions on a fraction of difficult *ab initio* target proteins. Ortiz et al. [133] applied a high coordination lattice model developed by Kolinski and Skolnick [122,123] to a number of small target proteins. Monte Carlo simulated annealing calculations started from random expanded conformations of the target proteins. The model assumed a 90-basis vector representation of the alpha carbon trace that has a 1.2 Å resolution due to the spacing of the underlying cubic lattice grid. Off-lattice single-sphere side chains could assume multiple orientations with respect to the backbone, thereby mimicking the distribution of rotamers for particular amino acids. The generic force field of the model consisted of knowledge-based potentials (derived from the statistics of the regularities seen in known protein structures) for short-range interactions, one body burial, pairwise and multibody surface long-range interactions, and terms simulating the regularity and cooperativity of the main-chain hydrogen bond network. Additionally, a weak bias toward predicted secondary structure (obtained from multiple sequence alignments + secondary structure prediction from PHD [129–131]) and weak theoretically predicted long-range contact restraints from correlated mutation analysis were implemented in the interaction scheme [134–138]. Contact prediction was based on the analysis of correlated mutations in sequences detected by multiple sequence alignments. For some targets, the globally correct fold or large fragments of the structure were correctly predicted. The method was capable of assembling low-resolution novel folds. The level of success during the CASP3 exercise was on the same level as reported for test predictions made for a series of small globular proteins prior to CASP3 [137].

A similar methodology, but one based on a completely different protein representation [139,140] (that are discussed in Sections V and VI), was employed by Kolinski and co-workers with a similar fraction of correctly predicted structures [133]. An important advantage of this method was its computational speed and nicer scaling of computational cost against protein chain length. Thus, the prediction of structures of larger proteins via *ab initio* folding became possible.

Osguthorpe [141] employed a continuous model and molecular dynamics simulated annealing. In spite of the use of a quite detailed model (main chain united atoms and up to three united atoms per residue), its very flexible chain geometry enabled efficient sampling. The potentials were derived from the statistics of known protein structures. The method enabled us to obtain correct predictions of substantial fractions of the structure of the attempted targets, and for one of the difficult targets, the prediction resulting from this method was the most accurate.

A very interesting hierarchical procedure has been used by Samudrala et al. [142]. First, as previously proposed by Hinds and Levitt [143], all compact conformations of test proteins were enumerated using the diamond lattice model with multiple residues per chain unit. The best (according to the force field of the lattice model) structures were then selected for further consideration. Subsequently, the all-atom structures were reconstructed by fitting the predicted secondary structure fragments to the lattice models. These structures were subject to energy minimization using an all-atom force field and spatial restraints of the lattice models. The optimized structures were scored by a combination of all-atom and residue-based knowledge-based potentials [144]. Then, distance geometry [145] was used to generate a number of possible "consensus" models. The local geometry of predicted secondary structure was again fitted to the resulting models. Finally, the resulting all-atom models were optimized and rank-ordered according to energy. A number of qualitatively correct protein fragments of significant size were correctly predicted. The method appears to be very robust and (as pointed out by the authors) it was likely that it could be further improved. Probably the major weakness of the method in its present form is in the small fraction of good structures in the initial pool of lattice models.

The method developed by Scheraga and co-workers [146] and used in CASP3 is based on the global optimization of the potential energy of a united atom model [147]. Due to the force-field design of the model, which is based on basic physical principles, this method is very close to a purely thermodynamic approach. In this respect, it qualitatively differs from the previously outlined methods. This off-lattice protein model has a united atom representation of the alpha carbons, side groups, and peptide bond group, with fixed bond lengths and variable bond angles. The interaction potentials between united atoms describe the mean free energy of interactions and account in an implicit way for the average solvent effect and cooperativity of the hydrogen bonds [148]. The optimization is performed by means of the Conformational Space Annealing technique [147], which subsequently narrows the search regions and finally finds distinct families of low-energy conformations. The lowest-energy, reduced model conformations are subsequently converted into the all-atom models and optimized by electrostatically driven Monte Carlo simulations [149]. For a fraction of CASP3 targets, this method produced exceptionally good predictions. The method seems to perform much better on helical proteins than on β or α/β proteins.

D.   Choice of Sampling Scheme

In the past, different methods of sampling of protein model conformational space have been employed with various degrees of success. Traditional molecular dynamics can be used only in the case of continuous models. Other sampling schemes, including a variety of Monte Carlo methods, genetic algorithms, and combinations of these methods, could be applied to continuous as well as to the discrete (including lattice representation) models.

In general, the choice of the simulation/optimization algorithm depends on the aim of the studies. Different procedures are needed for the study of protein dynamics and folding pathways from those procedures that are just targeted to find the lowest-energy conformations of model polypeptides.

Monte Carlo procedures for chain molecules [150] use a wide spectrum of strategies for conformational updating. In some algorithms, the updates are global, as in the chain growth algorithms, whereas other algorithms employ pivot moves of a large part of the model chain. In other algorithms, the trial modifications are local, involving only a small portion of the chain or a small distance displacement of a larger part of the chain. Sometimes, the local and global modifications were combined in the same algorithm.

What is the relationship between the molecular dynamics simulations of a continuous model and an isothermal Monte Carlo trajectory of an otherwise similar discretized (or lattice) model? When only local (and small distance) moves are applied in a properly controlled random (or rather pseudorandom) scheme, the discrete models mimic the coarse-grained Brownian dynamics of the chain. The Monte Carlo trajectory could be then interpreted as the numerical solution to a stochastic equation of motion. Of course, the short-time dynamics (the time scale of a single elementary move in the Monte Carlo scheme) of the discrete model has no physical meaning. However, the long-time dynamics should be qualitatively correct, albeit with possible distortions of the time scale of various dynamic events. Such an equivalence of the molecular dynamics and stochastic dynamics of equivalent off-lattice and lattice-simplified protein models has been demonstrated in the past by Rey and Skolnick [151], and by Skolnick and Kolinski [152]. Recent studies have shown that Monte Carlo folding pathways observed for high-coordination lattice models reproduce the qualitative picture of folding dynamics seen in experiments [153]. Thus, it could be rather safely assumed that Monte Carlo lattice dynamics can be used in meaningful studies of protein dynamics, folding pathways, the mechanism of multimeric protein assembly and other aspects of biopolymer dynamics. The validity of protein dynamics studies using discrete models depends more on the assumed accuracy of the protein representation and its force field than on the particular sampling scheme. However, some oversimplified discrete models may face serious ergodicity problems. This aspect of Monte Carlo simulations always needs to be carefully examined.

Isothermal simulations (molecular dynamics or Monte Carlo) provide characteristics of the system's properties at a single temperature. Numerous simulations at various temperatures (above and below the folding transition temperature) are needed to gain some insight into the thermodynamics of the folding process. There is a very serious problem associated with the extremely slow relaxation of protein models in the dense globular state. The local barriers in the energy landscape near the folded state are high and the sampling becomes ineffective. Thus the computer studies employing straightforward MD or canonical MC algorithms became prohibitively expensive. Essentially, the same applies to various simulated annealing strategies. In all cases, the design of sampling details could be very important. For example, properly designed local moves can "jump over" the high local energy barriers, thereby speeding up the sampling of the entire conformational space.

Mulicanonical [154] (or entropy sampling Monte Carlo [108–110]) simulations provide more complete data on folding thermodynamics [116,155–157]. Due to their differently defined transition probabilities in the sampling scheme, energy barriers became much less important, but are substituted by entropic barriers. From a single series of simulations, it is possible to obtain an estimation of all thermodynamic functions (energy, free energy, and entropy) over a wide range of temperatures. However, the cost of such computations grows rapidly with the system size and its complexity.

A somewhat simpler, but by no means trivial, task is to find the lowest energy state of the model polypeptide. Due to the thermodynamic hypothesis [158], which postulates that native proteins are in the global minimum of the conformational energy, the minimum energy state of a properly designed protein model should closely mimic the folded conformation. A variety of strategies have been developed to solve this global minimum problem [159]. For a relatively simple system, when the total energy could be expressed in the analytical form, it is possible to solve the problem in a deterministic fashion [160]. For more complex (i.e., realistic models of proteins) systems, existing methods do not guarantee that the lowest energy conformation will be found. The number of possible conformations and the rugged energy landscape make a systematic search impractical.

Simulated annealing, ESMC [108,109,161], Monte Carlo with minimization [162], genetic algorithms [64,163–165], and the combination of genetic algorithms with Monte Carlo sampling have been successfully used in the past to find the near-native conformations of reduced models of small proteins [68].

Recently, a number of studies have focused on the comparison of various Monte Carlo strategies for finding the global minimum of a protein model [166–168]. Probably the most straightforward of these search strategies is simulated annealing, where the system temperature is gradually lowered during the simulations, starting from a relatively high temperature (above the folding transition) and ending at a low temperature below the folding temperature (usually well below due to thermal fluctuations). When on repeated runs starting from different initial states, the same conformation is recovered; one may assume that there is a good chance that the global minimum has indeed been found. However, for difficult problems, simulated annealing runs (or at least a substantial fraction of the runs) could be trapped in local energy minima. Some of the local minima could be close to the model's representation of the native state, whereas others could correspond to conformations that are far away from the properly folded state. There is no simple test of convergence in the simulated annealing method. The efficiency of the simulated annealing method could be considerably improved by a certain modification of transition acceptance criteria. For instance, one may perform local minimization before and after the transition and then apply the Metropolis criterion to the locally lowest energy pairs or conformations [16]. This way, the sampling procedure can avoid visits to a large fraction of irrelevant local energy minima.

In contrast to simulated annealing, sampling techniques within the multicanonical ensemble have some internal convergence tests. In a version of this technique, called entropy sampling Monte Carlo [108–110], the estimation of the system's entropy is built by a sampling process that is controlled by the density of states of particular discretized levels of conformational energy. When converged, all energy levels, including the lowest energy, should be sampled with the same frequency. The ESMC method is "quasi-deterministic": The data from the preceding simulations could be used to improve the accuracy in the successive runs. In principle, when converged, ESMC should find the lowest energy state. In practice, the energy spectrum near the lowest energy state could be associated with large entropy barriers, and the lowest energy state could be not detected in spite of the apparent convergence—that is a constant density of visited states in the remaining low-energy portion of the energy spectrum. The rate of convergence of the ESMC method into the low-energy portion of the energy landscape could be accelerated by the artificial deformation of the entropy curve (artificial increase of the density of states) in the less important, high-energy range [156].

The replica exchange Monte Carlo method [169] addresses the problem of local minima in a different way. A number of copies of the model system are simulated by means of a standard Metropolis scheme at various temperatures. The temperature range covers temperatures from a temperature well above the folding temperature down to a temperature below the folding transition temperature. Occasionally, the replicas are randomly swapped according to a criterion that depends on temperature difference and the energy difference. Thus, the low-energy conformations at a higher temperature have a chance to be moved to a lower temperature. As a result, the copies of the system sample not only the conformational space but also move between various temperatures. At high temperatures, the energy barriers could be surmounted easily; at low temperatures the vicinities of energy landscape "valleys" are efficiently sampled.

Comparison of the computational cost of finding the lowest energy state for a simple protein-like copolymer model [168] shows that replica exchange Monte Carlo (REMC) is much more efficient than simple Metropolis sampling with a simulated annealing protocol in spite of the fact that multiple copies of the system have to be simulated. The REMC method also finds the low-energy conformations many times faster than the ESMC method. Thus, it appears that the REMC method (or its variants) could be a method of choice for use in the *ab initio* folding of reduced protein models, where finding the lowest energy state is the main goal of computational experiment. Due to the very efficient sampling by the REMC method, the samples at various temperatures could be used for the "umbrella"-type estimation of the system entropy. That may extend the applications of the REMC method into cost-efficient studies of protein folding thermodynamics.

III.   OVERVIEW OF THE UNIFIED FOLDING METHOD

When faced with the problem of predicting the tertiary structure of an unknown sequence, one typically runs PSI-BLAST [170] over sequences from the structures in the protein data bank [171]. Then, if this does not work, one runs a threading program to see if it detects a significant probe–template match. Even if either of these two cases is successful, for nontrivial cases often the alignments of the probe sequence may be in error, and there may be gaps in the alignment of the probe sequence to the template structure and/or sometimes there are long unaligned regions. If both methods fail, then *ab initio* folding is the requisite structure prediction method. Thus, ideally one would like to have a unified approach that automatically treats these possibilities. In what follows, we describe one recently developed unified approach.

An overview of the idea is given in Fig. 1. First, one runs our threading algorithm, PROSPECTOR [57], and establishes if there is a significant probe sequence–template structure match. If so, the template is used as a soft bias in a generalized comparative modeling approach that involves *ab initio* folding in the vicinity of the template in a reduced protein model. Threading also provides predicted secondary structure and tertiary contacts that are not restricted to the template structure but can be extracted from other structures. This allows the possibility of fold prediction in those regions absent in the alignment of the probe sequence to the template structure. The advantage of this generalized comparative modeling is that it can improve the initial alignment generated by the threading algorithm and can provide a structure prediction for the unaligned

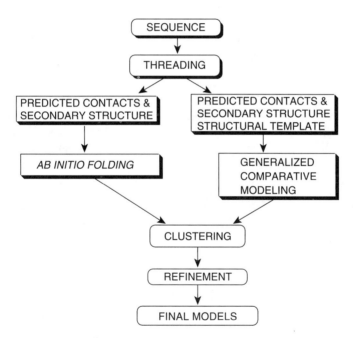

Figure 1. Flow chart describing the unified approach to protein structure prediction. First, threading is done. If a significant hit to a template is found, then generalized comparative modeling in the vicinity of the template but supplemented by predicted secondary structure and contacts possibly from other templates is done. If no significant probe sequence–template structure match is found, then consensus contacts and sets of local distances in the top 20 scoring structures are extracted and employed as restraints in an *ab initio* folding algorithm. Once a sufficient number of simulations (typically 100) are done, the structures are clustered, full atomic models are built in the refinement step; and then using a new, distant-dependent atomic pair potential [204], the top five scoring structures are selected.

regions of the probe sequence. On the other hand, if there is no significant match to a template, then the predicted secondary structure and tertiary contacts extracted from threading are passed to an *ab initio* folding algorithm that uses the same reduced protein model. Then, for both generalized comparative modeling and *ab initio* folding, the resulting structures are clustered, atomic detail is added and the results are reported.

IV. THREADING RESULTS

A. First-Pass Threading

Recently, to build on the strengths and address the weaknesses of existing threading approaches, we have developed a new threading algorithm called PROSPECTOR (*PRO*tein *S*tructure *P*redictor *E*mploying *C*ombined *T*hreading to *O*ptimize *R*esults) [57], which runs sufficiently quickly so that entire genomes can be scanned in the matter of several days on a standard workstation or PC. During the course of the development of this program, we noticed that sequence profiles generated from the BLOSUM 62 matrix [172] often generated reasonable alignments between the probe and template sequences, even when the alignment score was insignificant. This suggested that the first stage of a hierarchical approach to threading should employ a sequence-profile [170,173, 174] (using a sequence profile plus a three-state secondary structure prediction scheme gave worse results) to generate the initial probe sequence to template structure alignment. We call this the "partly thawed" approximation. Then, the resulting alignment of the probe sequence in the template structure is used to calculate the partners for the evaluation of the pair interactions. Previously, in the first iteration of the frozen approximation [45], the partners were taken from the template structure. This worked well only when the environments in the probe and template structures were similar, but more often than not the environments were quite different. On successive iterations, in the so-called defrosted approximation [45] where the partners were taken from the previous alignment, there were times when the resulting algorithm never converged. Here, after the first initial alignment, quite good results were obtained.

The database for multiple sequence alignment (MSA) generation used in the construction of the sequence profile combines Swissprot (http://www.expasy.ch/sprot/) and the genome sequence database (ftp://kegg.genome.ad.jp/genomes/genes). First, a profile for relatively closely related sequences, whose sequence identity lies between 35% and 90%, is calculated. These sequences are selected from the composite database by FASTA [175,176]. Then, pairwise sequence alignments with the probe sequence are generated using CLUSTALW [177], and a sequence profile is generated. We term this the "closely" related set of alignments. To this set, we add additional sequences whose E value in FASTA is less than 10, use CLUSTALW to generate pairwise alignments, and then generate a profile for distantly related sequences; these are termed the "distantly" related set of alignments. The goal here is to have two sequence profiles: one that is more sensitive to more closely related sequences and another that can sometimes detect more distantly related sequences.

The first step of the threading protocol is to independently scan the structural database of interest using each of the sequence-profiles with a Needleman–Wunsch type of global alignment program [49]. Each of the two sequence profiles generates an alignment of the sequence in each of the template structures. Each alignment is used to identify the partners in the probe sequence to be used in the calculation of the pair interactions. Here we use our previously developed side-chain contact potential averaged over all homologs which includes a contribution from contacting fragments that have weak sequence similarity to each member of the close set of probe sequences [178]. Furthermore, we also use a pseudo energy term that describes the preferences for consecutive types of amino acids to adopt a given type of secondary structure. This secondary structure propensity term is also averaged over homologs, and thus it results in a secondary structure propensity profile. For each scoring function, close (distant) sequence profile, and close (distant) sequence plus pair interactions plus the secondary structure propensity profile, we scan the structural database and output the top five scoring structures. Thus, a total of 20 possible structures are output, along with their alignments.

B. Application to the Fischer Database

As a test case, we have focused on the Fischer database [179] that is comprised of 301 template structures and 68 probe sequences. We tried a variety of approaches on this database before deciding on the aforementioned combination of parameters. We just summarize the results of these studies here. For a given scoring function, the Needleman–Wunsch global alignment algorithm recognized more correct probe–template pairs than did the Smith–Waterman [180] local alignment algorithm. We also tried using the secondary structure profiles as the initial step in generating the probe–template alignment for pair evaluation. Secondary structure profiles alone only correctly recognize 18 cases in the first position, whereas secondary structure profiles plus pair profiles correctly assign 29 cases. This clear improvement shows the utility of pair potentials in this approach; nevertheless, even 29 recognized pairs is rather poor performance. The major improvement in fold recognition comes, as others have observed, when sequence profiles are used. Even if the sequence profile is turned off completely but is used to generate the alignment, the number of correctly recognized pairs increases to 35 correct probe–template pairs in the top position. In all cases, inclusion of pair interactions improves the yield of correct probe–template matches.

We summarize our results using PROSPECTOR1 in Table I (the first pass of PROSPECTOR). One of the best alternative methods is that of Gonnet, which

TABLE I
Summary of Threading Results on the Fischer Database for Different Scoring Functions[a]

Method	Number of Fischer Pairs in the First Position	Number of Fischer Pairs in the Top 5(4) Positions	Number of Fischer Pairs in the Top 10(8) Positions
PROSPECTOR1			
"Close" sequence profile	44	46(46)	49(47)
"Close" sequence profile plus secondary structure plus pair profile	45	55(53)	56(55)
"Distant" sequence profile	46	53(51)	53(53)
"Distant" sequence profile plus secondary structure plus pair profile	52	56(56)	59(57)
Hierarchy of four scoring methods	59	63(62)	65(63)
Hierarchy of three scoring functions (as above but without the "distant" sequence-profiles)	58	62	64
PROSPECTOR2			
"Close" PROSPECTOR2. sequence profile plus protein specific pair and secondary structure potentials profile	48	51(51)	58(58)
"Distant" sequence profile plus protein specific pair and secondary structure potentials	51	59(59)	59(59)
Hierarchy of four scoring methods	61	64(64)	65(65)
Hierarchy of three scoring functions (as above but without the "Distant" sequence profiles)	60	64	65
Other Methods			
Simple Blast[1]	27	—	—
PSI-BLAST restricted to the Fischer database [170,182]	24	37(36)	40(39)
PSI-BLAST using extensive sequence database and PSSM constructed using IMPALA [247]	41	46(46)_	47(46)
Original GKS threading program [45]	22	30	34
Hybrid threading [181]	52	57	60
Best UCLA benchmark results as of 2/4/00 which is prediction of secondary structure plus mult-gonnet [34]	52	(56)	(58)

[a]Results are reported in both the top 5(4) and top 10(8) positions [181], with the number in parenthesis given by the UCLA benchmark website (http://www.doembi.ucla.edu/people/fischer/BENCH/table1.html).

recognizes 52 proteins in the top position, the same number as the distant profile plus pair interactions recognizes, but if a hierarchical method is used, then ours is clearly the best, because 59 proteins are recognized in the top position. It is clearly superior to all our early efforts as well as to the alternative hybrid method [181], BLAST [1], and PSI-BLAST [170,182]. It might be argued that because we use four scoring functions while the hybrid method uses only three, this is not a strictly fair comparison. If we eliminate those results obtained from the "distant" sequence profiles, then we obtain 58, 62, and 64 cases in the top 1, 5, and 10 position as compared to 52, 57, and 60, respectively, of Gonnet.

We then applied the method to a second Fischer benchmark comprised of 29 probe–template pairs and scanned each probe sequence against the original Fischer structural database plus an additional 19 template structures (http:// www.doembi.ucla.edu/people/fischer/BENCH/tablepairs2.html). We have only been able to find 27 of the 29 probe sequences and have reported our results accordingly. PROSPECTOR1 places 17 correct pairs in the top position, and it also places 21 and 22 in the top four and eight positions, respectively. This is the same as the best reported results of 17 correctly identified pairs. However, in our case one probe, "stel," which is supposed to be matched to 2azaA, selects 2pcy in the top position, which has the same core as 2azaA. Then, we have 18, 19 (19), and 20 (20) correct matches in the top position and top five (four) and ten (eight) positions, respectively. Thus, we have somewhat better results than previous workers.

C. Iterative Threading

1. General Idea

Just as PSI-BLAST [170] can increase its specificity by iteration, so can threading. In fact, the set of structures selected by PROSPECTOR contains additional information even beyond providing for a structural match. If we look at the set of 20 structures that are selected as being the best scoring sequence– template structure pairs, it is possible to extract additional information by looking for consensus predictions. By way of illustration, we consider the prediction of tertiary contacts. We focus on all contacts between residues that are at least five residues apart, and we count the predicted contacts generated by the aligned regions of structure. If there is a consensus (i.e., at least three contacts are consistently predicted), then we employ this information in two ways: (1) to enhance the specificity of threading by constructing a protein-specific, threading-based pair potential and (2) as described in Section IV.F, to predict tertiary contacts.

Using a previously derived formalism to convert contacts into a pair potential [178], we derive a set of protein-specific potentials, where the contacts are not only extracted from fragments with weak sequence similarity, but rather are generated by consensus contacts in the threaded structures. We use the arithmetic average of this potential and the previous iteration's pair potential in the next iteration of threading. This case is termed the "close" and "distant" protein-specific potentials, and we call the threading method that employs these terms PROSPECTOR2.

2. Application of PROSPECTOR2 to the Fischer Database

The results from PROSPECTOR2 are also reported in Table I. The "close" case now recognizes 48 proteins as compared to 45 in the top position. The "distant" case recognizes 51 as compared to 52 previously, but the composite of the four scoring functions now recognizes 61, 64, and 65 proteins in the top position as compared to 59, 63, and 65 in the top, top five, and top ten positions, respectively, for PROSPECTOR2. In all cases, the method improves when pair potentials are used as compared to that when the corresponding sequence profile alone is used. Similarly for in the second Fischer database, a total of 17, 20, and 20 proteins are recognized in the top, top five, and top ten positions, respectively.

D. Genome-Scale Iterative Threading

In tests on genome scale threading, we found that the optimum number of correctly recognized folds was found on the third iteration, PROSPECTOR3. However, because of the computational cost of constructing pair potentials that used local sequence fragment similarity, in our preliminary study and in the interest of computational tractability we employed the best quasi-chemical pair scale [183]. We term this PROSPECTORQUASI1-3. Furthermore, to deal with the problem of very large proteins that may contain more than one domain, in addition to threading the entire sequence, we also threaded 150 residue fragments, starting at the first residue and then shifting by 25 residues until the final fragment of possibly shorter length is scanned. This allows for the detection of domains. For genome-scale threading, our structure library consists of 2466 sequences constructed so that no pair of proteins has greater than 35% sequence identity between them.

1. M. genitalium

This genome consists of 480 ORFs [184]. The first pass of PROSPECTOR, PROSPECTORQUASI1 assigns 153 proteins to a structure in the protein data-bank. The second pass, PROSPECTORQUASI2, assigns 182, and the third pass, PROSPECTORQUASI3, assigns 194. This constitutes an assignment of 40% of the genome. All assignments are made using an automated protocol based on the score significance. Of these 194 structural predictions, all but three are correct. In contrast, several years ago Fischer and Eisenberg [185] assigned the folds of 103 out of a total of 468 proteins by their threading algorithm. Gerstein has reported identification of 211 proteins using PSI-BLAST [186,187]. Genethreader assigns

200 proteins, but for 15 of them the assignment appears to be incorrect [188] as assessed by a consensus of Gerstein's results (http://bioinfo.mbb.yale.edu/genome/MG/) and PROSPECTORQUASI3.

2. E. coli

The *E. coli* genome contains 4289 ORFs [189], for which PROSPECTOR-QUASI3 assigns 1716 ORFs to structures in the Protein Data Bank. This constitutes about 40% of the genome. Interestingly, this is the same percentage of structures as was assigned in *M. genitalium*. In contrast, without the use of active site filters, a total of 1250 confident structure predictions have been made, using a sequence profile-based method [190].

E.   Extension of PROSPECTOR to Include an Orientation-Dependent Pair Potential

To enhance specificity, we next replaced the pair potential by one that is orientation dependent and again perform three iterations of modified PRO-SPECTOR, PROSPECTORIEN1-3. In applications to the Fischer database, we found that, on average, PROSPECTORIEN3 generates the most accurate probe–template alignments. The resulting set of structures constitutes the initial model that will be subjected to the generalized comparative modeling described in Section V.

F.   Threading-Based Prediction of Tertiary Contacts

For a given iteration, the set of 20 top-scoring structures can also be used to predict the tertiary contacts in the probe protein. Again we demand that a given pair of contacts occurs in at least 25% of the top-scoring structures. For each interaction of PROSPECTOR1-3 and PROSPECTORIEN1-3, we collect the predicted contacts. The sets of contacts are then pooled.

Next we report our results for the set of 18 small proteins that constituted part of the validation set for the MONSSTER *ab initio* folding algorithm [191]. Of course, in this 18-protein test set, care is taken to remove all homologous proteins to the probe sequence from our structural database, and all proteins whose global root-mean-square deviations (RMSD) from native that are less than 8.5 Å are also excluded. On average, 28% of the contacts are correct, and 69% are correct within two residues. The correlated mutation analysis gives, on average, 34% correct and 82% correct within ±2 residues [191–193]. While the threading-based method has somewhat lower accuracy, in contrast to the correlated mutation analysis, it can be readily automated. Note that a contact-prediction accuracy of about 70% correct within ±2 residues is sufficient for the successful assembly of the global fold using the MONSSTER *ab initio* structure prediction program [191,193].

TABLE II
Comparison of Contact Prediction Accuracy for CASP3 Targets for Threading and Correlated Mutation Based Approaches[a]

Name of Protein	Number of Contacts Predicted	$\delta = 0$ From Threading	$\delta = 0$ Mutation Analysis	$\delta = 2$ From Threading	$\delta = 2$ Mutation Analysis	$\delta = 3$ From Threading	$\delta = 3$ Mutation Analysis
1jwe_	16	0.19	0.14	0.5	0.44	0.5	0.65
1eh2_	22	0.68	0.14	0.91	0.73	0.91	0.98
1bqv_	19	0.05	0	0.53	0.13	0.53	0.5
1ck5B	22	0.14	0.02	0.59	0.4	0.55	0.51
Average		0.265	0.075	0.63	0.43	0.62	0.66

[a] % of contacts correct with $\delta = 1m1$ residues of a correctly predicted contact.

Turning to the results of CASP3, the correlated mutation analysis performed considerably poorer, whereas threading-based contact prediction was better [133]. In Table II, for four of these proteins, we show the predicted contact results and compare them to correlated mutation analysis. Now, within ± 2 residues, 63% of the contacts are correct as predicted by the threading-based method as compared to 43% from the correlated mutation analysis; this is a qualitatively significant improvement. Within ± 3 residues, correlated mutation analysis is slightly more accurate at 66% versus 62% from the threading-based contact predictions. Here again, we excluded all analogous and homologous proteins in the prediction of contacts from the analysis of consensus contacts in the alignments generated by PROSPECTOR1-3 and PROSPECTORORIEN1-3.

In Table III we present the set of predicted contact results for 28 proteins that that will be subject to *ab initio* folding in Section VI. Again the requisite contact prediction accuracy is achieved, with 31% of the contacts exactly predicted on average and 73% correctly predicted on average within ± 2 residues. If we use the threshold of 70% prediction accuracy as indicative that the folding simulation will be successful, then, as shown in Section IV, 20 of these 28 proteins should be foldable. The asterisk indicates those proteins that are foldable, as assessed by the presence of a cluster of structures whose RMSD from native is less than 6.5 Å. In practice, of the 28 proteins, 13 are foldable. In addition, another two whose contact prediction accuracy is less than 70% correct within ± 2 residues are also foldable. Of course, the presence of reasonably accurate contacts in and of themselves do not guarantee that the native topology will be found; but in all cases of accurate contacts, if there are a sufficient number of such contacts, then rather low RMSD structures are found in the pool; see Table VI. Thus, this is a reasonably effective method of predicting acceptably accurate tertiary contacts.

TABLE III
Predicted Contact Accuracy from Threading for 28 Proteins Used in an *Ab Initio* Folding Test[a]

Name of Protein	Number of Contacts Predicted	$\delta = 0$	$\delta = 1$	$\delta = 2$[b]	$\delta = 3$
1stfI	25	0.28	0.48	0.8*	0.88
1poh_	37	0.3	0.54	0.7*	0.7
1pou_	30	0.33	0.47	0.73*	0.9
1ife_	56	0.18	0.39	0.54	0.79
2azaA	47	0.38	0.53	0.79*	0.85
256bA	1	0	0	1*	1
1tlk_	53	0.81	0.94	1*	1
2pcy_	45	0.4	0.51	0.91*	0.91
1tfi_	52	0.19	0.35	0.60	0.79
2sarA	29	0.21	0.55	0.76	0.86
5fd1_	23	0	0.17	0.30	0.52
1cewI	7	0.57	0.86	0.86	0.86
1ctf_	46	0.11	0.3	0.50	0.7
1mba_	12	0.58	0.67	0.67	0.75
1shaA	41	0.34	0.66	0.85*	0.88
1thx_	53	0.23	0.55	0.72*	0.83
1shg_	42	0.19	0.57	0.76	0.86
1ubi_	23	0.61	0.65	0.78	0.83
6pti_	54	0.26	0.56	0.61	0.8
1cis_	19	0.21	0.58	0.95	0.95
1fas_	22	0.27	0.59	0.77	0.86
1ftz_	18	0.5	0.72	0.78*	0.89
1c5a_	20	0.1	0.3	0.4*	0.5
1fc2C	18	0.44	0.78	0.83*	1
1gpt_	19	0.37	0.53	0.79	0.89
1hmdA	33	0.18	0.36	0.52*	0.73
1ixa_	14	0.43	0.64	0.79*	0.86
1lea_	23	0.3	0.52	0.74*	0.96
Average		0.31	0.53	0.73	0.83

[a]$\delta = m$ is the number of contacts predicted within $\pm m$ residues of a correctly predicted contact. Correlated mutation analysis is from the CASP3 predictions of Ortiz et. al. [133].
[b]An asterisk indicates that this protein is foldable by *ab initio* (see Section VI).

V. GENERALIZED COMPARATIVE MODELING

Quality sequence-to-structure alignments generated by the threading procedure depend on the level of sequence identity of the target and the template proteins. In the cases of high sequence similarity, the protein folds are very similar, and classical methods of comparative modeling [194,195] led to good-quality models, frequently to models of similar quality to those obtained from the refinement of the X-ray data or good NMR data. When the sequence similarity becomes low or nondetectable by sequence comparison methods, the template proteins could be weakly homologous or just analogous—that is having similar folds without any obvious evolutionary relations. As a consequence, the resulting alignments are usually incomplete, with a substantial number of gaps and insertions. A fraction of residues of the probe protein, which is sometimes substantial, are not aligned to the template. Moreover, in the aligned parts of the structure, the true structure of the probe protein may differ in many important details from the structure resulting from the alignment to the template. Also, an optimal structural alignment of the two structures could be quite far from the threading-based alignment. Due to low sequence similarity, the threading alignment might not be the optimal one.

Is it possible to build a good-quality model based on poor alignments? Usually, it is not possible by means of contemporary procedures for comparative modeling. When the template structure differs substantially from the probe structure, the resulting models are typically much closer to the template structure than to the true structure of the probe protein [196]. The models do not move (in conformational space) in the direction of the probe structure, but instead wander around the template structure. Moreover, in the cases of large gaps in the alignment, the filled-in pieces of structure are sometimes completely nonphysical (non-protein-like).

A recently proposed method is described in the next sections that attempt to address this problem. The idea is to perform a kind of *ab initio* folding in the vicinity of the template structure, with the model force field controlling details of the folding. The template is used only to reduce the searchable portion of conformational space and loosely defines the general topology of the probe protein fold. The lattice model employed in these procedures has a limited resolution and accuracy. Consequently, the obtained models, in general, cannot achieve the accuracy of the experimental structures. As a result, it is rather pointless to apply the proposed methodology to those cases when the alignments are very good and complete. In such cases, the obtained structures would be slightly worse than structures built by classical comparative modeling tools. Such situations could be easily detected. In the remaining cases of low homology (or just analogy of the folds), the method is robust in the sense that it does not do any "harm" to the initial threading-based models and, for a substantial fraction of cases, leads to a qualitative improvement of the models. The resulting structures move toward the true probe structure. Because this approach bears some similarity to the comparative modeling, we call this method of homology/analogy-based structure prediction generalized comparative modeling (GeneComp, GC). The applied methodology is essentially the same for the template-restrained folding as for purely *ab initio* folding, the crossover is smooth, and there is no sharp boundary between threading-based and *ab initio* approaches.

A. Description of the Method

The method of generalized comparative modeling consists of several steps, which sequentially transform the threading alignment into a full-atom model of the probe protein. They are the following:

1. Build the threading alignment by a method described in the previous sections.
2. Construct the starting lattice model using the partial template from the threading as a structural scaffold.
3. Fold/optimize the lattice model using the threading alignment as a loosely defined structural template.
4. Cluster the lattice folding results [197] and/or calculate a mean structure by means of distance geometry (DG).
5. Refine the averaged model by Monte Carlo simulated annealing of an intermediate resolution off-lattice continuous model.
6. Reconstruct atomic details.

B. The Lattice Model and Its Force Field

Before describing the particular steps of the comparative modeling methodology, we outline the lattice model employed in all coarse-grained simulations (restrained or *ab initio*). Due to assumed reduced representation, we have named this protein model the side-chain-only (SICHO) model [139,198]. Technical details of the model design and its force field could be found elsewhere [199]. Here, an outline is provided for the reader's convenience. Most of the reduced models of proteins assume a more or less explicitly reduced (all-complete) representation of the main-chain backbone [200]. Frequently the alpha-carbon trace is used to represent the main-chain conformations, and the side chains are neglected or represented on various levels of simplification. When designing the present model, two partially contradictory goals were taken into consideration. First, for computational simplicity, there should just be a single degree of conformational freedom per residue. Second, the model should enable straightforward implementation of as accurate and selective a force field as possible. Thus, we assumed a single center of interactions that corresponds to the center of mass of the side group and the alpha carbon atoms.

This side-chain representation has several advantages over the alpha-carbon reduced representation. It is known that the sequence-specific interactions in proteins are due to different character of the side chains. The interactions of the main chain are rather generic. Then, having the coordinates of the side chains, it is very easy to reconstruct the main chain-coordinates [200]. In contrast, the reconstruction of the side-chain positions from the positions of the main chain is not trivial [201] and requires extensive optimization. Additionally, the side chains are bigger and their size varies between amino acids. Thus, this side-chain representation provides for better and more protein-like packing, with a well-defined first coordination shell.

The model chain is restricted to an underlying simple cubic lattice with the lattice spacing 1.45 Å. The set of possible virtual bonds between consecutive side chains is defined by a set of 646 lattice vectors. The shortest are of the vector type $|\pm3,0,0|$ and $|\pm2,\pm2,\pm1|$ while the longest are of the type $|\pm5,\pm2,\pm1|$, expressed in lattice units. The distribution of the length of the chain bond covers the majority (except for the wings) of the distribution seen in proteins. The main excluded volume is simulated by a cluster of the 19 closest (to the center of the model side chain) points on the underlying cubic lattice. This hard core of the chain is supplemented by soft-core repulsion spheres for the larger amino acids. The size of these spheres is adjusted in such a way that the folded model chains mimic average packing density of globular proteins.

The force field of the model consists of three types of potentials. First are the generic contributions that are independent of sequence and enforce the protein-like chain stiffness and internal packing. Potentials of the second type are amino acid-dependent and are used to reproduce the short-range interactions describing secondary structure propensities and orientation-dependent pair interactions. The potentials of the third type (short-range potentials identical in form to that described above and pairwise potentials [202]) are protein-dependent. Their derivation involves multiple sequence alignments of the sequence of interest, and the strength of interactions depends on the sequence similarity of protein fragments.

C. Construction of the Starting Lattice Chain

The threading alignment was used as a template to construct the initial lattice models. First, the aligned parts of the probe sequence were fitted to the template, and pieces of the lattice chain were built by taking into consideration the excluded volume of the model chain and the necessity of "stretching" the chain between the gaps in the template. Then, starting from the shortest loop, the loops and nonaligned chain ends were randomly inserted, again taking into account the excluded volume. The proper geometry of the model chain (avoiding nonphysical distances between side groups close along the chain) was preserved during the chain-building procedure. For good alignments, this procedure produces good models that need very little refinement. For extremely bad alignments, it may fail; in these (very rare) cases a less restrictive algorithm that allows for a larger deviation from the template could be used.

D. Restrained Lattice Folding: Optimization of the Initial Model

As discussed in Section II. D, the replica exchange Monte Carlo method appears to be an efficient tool for searching the conformational space of reduced protein models. This technique was therefore used for the restrained folding (or refinement) of the probe proteins using the threading alignments as loosely defined structural templates. In the beginning of the procedure, a number of copies of the initial model are created and placed at various temperatures, according to the REMC scheme. Two subsequent runs were performed. In the first run, the range of temperatures is wider and shifted toward higher temperatures to allow for the fast equilibration of all replicas. In the subsequent longer run, the temperature range was smaller so that approximately half of the replicas run below the folding temperature and half above. About 20 replicas were usually simulated. This number of copies guarantees very fast and efficient swapping of conformations between the various temperature levels (the temperature increment between replicas has been assumed to be temperature-independent—a linear temperature set). A somewhat larger number of replicas may be required for fast convergence of larger proteins—250 residues or more. The conformations seen at the lowest temperature of the REMC scheme rapidly find the global energy minimum.

Three types of restraints are used to keep the sampling process in a broad conformational neighborhood of the template conformation.

The first is the most straightforward. The aligned portion of the template structure is placed at the center of the Monte Carlo working box. Then, at the beginning of the simulation, the starting chains are superimposed on the template. During the simulations, there are weak and somewhat ambiguous attractions (linear with distance) between aligned (according to the threading results) residues of the template and the moving probe chain. Thus during the simulation, the initial alignments have the chance to be corrected or even overridden by the model force field.

The set of tertiary contacts predicted by threading comprise the second set of restraints. Because only about one-third are correct and a much larger fraction are "almost" correct (i.e., they are shifted by ± 1 or ± 2 residues), the energy of attraction between the two residues of the probe predicted to be in contact grows linearly with the closest distance between the ± 2 segments of the model chain. For very good alignments, the predicted contacts are, to a large extent, consistent with the template structure, and this set of restraints is essentially redundant to the restraints of the first type. For poorer alignments, a number of other locally similar proteins may contribute to the contact prediction. Consequently, the predicted contacts may significantly modify the resulting structures of the probe with respect to the template; that is, an averaged effect of other weak "templates" is introduced.

The third set of restraints contains the probe distances predicted from the fragment threading procedure. The distance restraints are limited to the pairs of residues that are no farther away than the length of the largest secondary structure element in the protein, which is equivalent to the estimated diameter (from the number of residues) of the probe protein.

E. Building the Average Models

For each probe protein, several independent simulations (10–20) were executed. From each simulation in the second pass, 200 conformations were stored in a constant interval of simulation time. The collected structures were averaged using a two-step distance geometry (DG), procedure. After the first pass, those structures far away from the average were rejected, and the final DG conformation was constructed from the remaining set of structures. Interestingly, DG averaging always led to a lower RMSD from the native than the average RMSD for the original set of conformations from the lattice simulations. Sometimes the structures from DG were close to the best structures seen in the folding simulations. Alternatively, our recently developed clustering procedure [197] could be used to identify clusters of the lowest energy conformations. The centroid of this cluster can then be treated as an averaged model. In the case of generalized comparative modeling, the two approaches are essentially equivalent. However, for *ab initio* folding, the clustering procedure is more powerful in identifying the most plausible fold from the sometimes-diverse results of *ab initio* lattice-folding simulations.

F. Reconstruction of Detailed Atomic Models

A very fast procedure was designed for reconstruction of the atomic details from the known positions of the alpha carbons and the side chains. The only constraints are the positions of the side-chain centers of mass. The initial local alpha-carbon trace geometry that is approximately reconstructed from the SICHO center-of-mass positions is not perfect. Therefore, the positions of alpha carbons are optimized in the first step. This is done by a gradient-optimization procedure using a very simple force field to improve the local geometry. At the next stage, positions of backbone atoms are reconstructed according to the local $C\alpha$ trace conformation. In this step, the vector normal to the plane defined by three consecutive alpha carbons is calculated. This vector is almost parallel to a peptide bond plane. Thus, the remaining atoms of the peptide bond can be positioned quite accurately. Next, positions of side chain atoms are rebuilt. The conformations of the side chains are chosen from a representative database of rotamers. For rigid amino acids (e.g., phenylalanine), there is a single conformation in the database. There are up to 20 conformations for large, flexible side chains (e.g., lysine). The conformation of the rotamer depends on (a) the distance between the $C\alpha$ atom and the center of mass of the side chain and (b) local chain conformation (i.e., $C\alpha$–$C\alpha$–$C\alpha$ angle). Next, as a final stage of the reconstruction procedure, the side chains are rotated around a virtual $C\alpha$—center-of-mass bond—to avoid excluded volume conflicts. This procedure produces reasonable structures; however, the packing of side chains after all-atom reconstruction is not optimized. This can be done by one of the standard procedures of molecular mechanics. For the data reported in this work, this step was omitted.

G. Summary of Results on Fischer Database and Comparison with an Earlier Version of Generalized Comparative Modeling

Fischer's database of protein sequences and structures [34] is a standard benchmark set for validation of threading approaches. As mentioned previously, PROSPECTOR recognizes a majority of the related sequences correctly. Here, we would like to test our generalized comparative modeling approach on the same test set. Probably, Fischer's database [34] provides a very good test for the method. It contains closely related pairs of proteins (typical of homology modeling cases), pairs of weakly related proteins, and some pairs of very weakly similar ones. As suggested above, one may expect that for very closely homologous pairs of proteins, our method is not recommended. Indeed, the geometrical fidelity of the lattice model is in the range of 1 Å, and the model accuracy (due to deficiencies of the force field and to other factors associated with the reduced character of the model) is probably significantly lower and could be estimated to be about 2–3 Å. Also, for very weakly analogous proteins, where the template structure is far away from the probe structure and when the alignment is sparse or when alignment covers only a small fraction of the probe sequence, the method applied here will not provide good models: The restraints from the template prohibit the requisite large-scale rearrangements of the modeled structure. In most intermediate cases, one may expect a qualitative improvement of the model with respect to the quality of the initial threading-based models.

The above expectations are based on an earlier version of the generalized homology modeling with lattice folding in the neighborhood of the template structure [199]. The test results of the earlier approach are summarized in Table IV where an automated modeling by Modeller [203] (using the threading templates as starting points) is compared with lattice modeling refined by Modeller. While the number of cases given in this table is small, one may conclude that in a fraction of cases the improvement of the threading models is of a qualitative nature. Also, as expected, already-good models (see the example of 1aba_) do not improve. The threading procedure [181] used to generate the initial alignments for these 12 pairs produced worse alignments on average than the PROSPECTOR threading algorithm employed for the more massive test involving Fischer's database. To make the comparison more complete, for the few pairs that were not properly detected by PROSPECTOR, the match (and resulting alignments) was enforced, that is, the highest-scoring structural match was not taken as a template, but rather the correct structural template was used. The results for the proteins from Fischer's database are compiled in Table V.

TABLE IV
α-Carbon RMSD from Native for Models Built from the Initial Threading Alignments and Refined by Lattice Simulations[a]

Probe/Template Proteins	Threading + Modeller	SICHO + Modeller
1aba_/1ego_	4.43	4.86
1bbhA/2ccy_	6.77	6.82
1cewI/1molA	14.96	14.38
1hom_/1lfb_	7.82	3.70
1stfI/1molA	6.40	5.95
1tlk_/2rhe_	7.23	4.17
256bA/1bbh_	6.09	4.36
2azaA/1paz_	21.95	10.77
2pcy_/2azaA	6.56	4.41
2sarA/9rnt_	10.28	7.83
3cd4_/2rhe_	6.74	6.39
5fd1_/2fxd_	25.67	12.40

[a]The first column gives the PDB codes of the probe and template proteins detected by the threading algorithm. The second column gives the results of automated comparative modeling using the threading alignments as a template definition. The RMSD is given for the alpha-carbon trace. The right column contains the results of SICHO modeling followed by a refinement using the Modeller program. In the refinement stage the lattice models were used as a "template" for Modeller. Original alignments are the same for both approaches compared in the table.

Similar to the earlier version [199] of the comparative homology modeling, there are essentially three possibilities. First, when the threading model is very good the lattice modeling does not improve the overall quality of the molecular model; however, "no harm" to the quality of the model by application of the entire methodology could be assumed. Then, there are cases of topologically correct templates with moderate overall distance from the true probe structure. Here, in most cases a qualitative improvement of the model quality could be observed. Finally, for very bad initial models the final models are still not satisfactory; the accuracy is too low to be sure that the overall fold has been properly recovered. Some of these models can even contain topological errors.

A number of very interesting observations can be extracted from analysis of the data compiled in Table V. The first is that the lowest energy criterion for selection of the final model is not the best one. On the contrary, the distance geometry averaging or clustering procedures almost always provide models of better accuracy. The two methods (DG and clustering) lead to essentially the same (on average) quality of molecular models and are quite consistent. At the same time, it should be pointed out that the structure selection is not perfect. Usually the structures generated by clustering or DG are worse than the best structures observed in simulations. Definitely, better methods of selection (for example, based on all-atom structures) of the best structures from the lattice folding trajectories need to be developed.

TABLE V

Compliation of Results of Generalized Comparative Modeling on Proteins from the Fischer Database[a]

Target	Template	Alignment Coverage	Aligned Part	Best RMSD	Lowest Energy	DG	First Cluster
1aaj_	1paz_	82.86	6.74	6.15	9.26	9.37	9.00
1aba_	1ego_	90.81	6.52	3.55	5.90	4.75	3.95
1aep_	256bA	64.05	18.36	18.31	18.36	21.45	22.38
1arb_	4ptp_	80.99	16.32	15.78	17.47	17.46	17.69
1atnA	1atr_	75.27	12.42	12.00	13.25	13.16	13.04
1bbhA	2ccyA	93.89	2.74	2.71	3.65	3.07	2.99
1bbt1	2plv1	93.59	12.55	9.57	10.81	10.70	10.80
1bgeB	1gmfA	66.67	7.89	4.93	6.27	5.45	5.71
1c2rA	1ycc_	85.35	4.35	4.31	5.75	5.34	5.30
1cauB	1cauA	89.63	5.18	4.04	5.69	5.45	5.41
1cewI	1molA	70.37	4.85	4.10	8.00	7.79	7.83
1chrA	2mnr_	92.97	3.50	3.77	5.35	4.90	4.78
1cid_	2rhe_	55.93	19.76	14.05	18.88	18.44	16.97
1cpcL	1colA	81.40	15.71	12.30	13.43	13.58	13.17
1crl_	1ede_	47.75	20.01	21.35	24.21	24.09	24.93
1dsbA	2trxA	51.65	12.46	11.58	15.94	16.47	15.30
1dxtB	1hbg_	92.52	2.74	2.91	3.54	3.01	3.08
1eaf_	4cla_	78.13	13.25	9.27	10.09	10.32	10.10
1fc1A	2fb4H	96.62	12.99	2.63	3.21	13.12	2.74
1fxiA	1ubq_	61.46	10.94	8.53	10.28	10.18	10.14
1gal_	3cox_	74.01	15.03	14.03	17.74	17.80	17.38
1gky_	3adk_	85.48	6.68	6.13	8.75	6.36	8.87
1gp1A	2trxA	54.89	11.48	9.08	14.75	13.74	15.06
1hip_	2hipA	80.00	3.55	3.92	4.86	4.26	4.13
1hom_	1lfb_	97.73	1.62	1.50	2.30	1.57	1.70
1hrhA	1rnh_	91.30	7.15	4.90	5.50	5.07	5.07
1isuA	2hipA	95.16	6.06	3.20	4.35	5.07	4.08
1lgaA	2cyp_	77.60	12.45	12.44	17.14	15.59	16.53
1ltsD	1bovA	59.00	9.99	8.11	12.16	10.21	9.47
1mdc_	1ifc_	96.97	2.62	2.55	3.12	2.66	2.65
1mioC	1minB	88.38	14.48	14.05	15.19	14.71	14.94
1mup_	1rbp_	93.63	5.56	4.14	4.89	4.38	4.51
1npx_	3grs_	92.17	14.56	13.61	14.15	14.12	14.09
1onc_	7rsa_	98.08	3.81	3.08	3.53	3.51	3.29
1osa_	4cpv_	70.27	16.84	16.56	18.02	17.90	17.81
1pfc_	3hlaB	89.22	3.84	3.81	4.69	4.28	4.46
1rcb_	1gmfA	71.32	6.28	3.91	5.51	6.09	4.25
1sacA	1ayh_	76.47	18.13	16.89	18.52	18.81	18.93
1stfI	1molA	69.47	8.46	4.97	7.38	7.07	8.11
1tahA	1tca_	56.92	19.00	18.90	21.60	21.51	20.96
1ten_	3hhrB	93.33	5.60	3.14	3.98	3.62	3.45
1tie_	4fgf_	66.87	7.88	7.88	8.80	8.60	8.94
1tlk_	2rhe_	95.83	4.61	2.35	3.49	3.42	3.03
2afnA	1aozA	95.83	25.27	22.60	23.68	25.05	23.50

TABLE V   (*Continued*)

Target	Template	Alignment Coverage	Aligned Part	Best RMSD	Lowest Energy	DG	First Cluster
2ak3A	1gky_	78.26	15.63	14.65	15.51	15.46	15.27
2azaA	1paz_	62.79	7.60	6.33	8.40	7.87	7.30
2cmd_	6ldh_	95.83	5.02	4.22	4.74	4.44	4.49
2fbjL	8fabB	94.37	10.30	7.04	7.72	8.78	8.37
2gbp_	2liv_	80.94	10.72	9.50	10.66	10.07	10.35
2hhmA	1fbpA	71.69	15.26	15.99	18.30	17.57	17.83
2hpdA	2cpp_	85.33	6.44	5.41	6.75	5.83	5.81
2mnr_	4enl_	95.52	14.92	13.55	14.07	14.28	14.27
2mtaC	1ycc_	65.31	14.35	14.04	16.01	16.49	16.51
2omf_	2por_	82.06	23.61	21.82	23.51	23.45	24.17
2pia_	1fnr_	79.44	15.72	15.64	17.29	16.77	18.24
2pna_	1shaA	46.55	10.69	7.27	11.31	8.92	10.89
2sarA	9rnt_	91.67	6.36	4.88	6.11	5.76	5.84
2sas_	2scpA	86.49	6.45	5.51	6.42	6.11	5.95
2sga_	4ptp_	98.82	17.74	9.78	11.87	10.49	11.94
2sim_	1nsbA	66.14	14.34	16.52	19.79	18.57	17.47
2snv_	4ptp_	84.11	14.28	12.78	14.07	13.84	13.31
3cd4_	2rhe_	92.78	7.02	5.98	7.40	7.15	7.05
3chy_	4fxn_	86.72	6.07	3.58	4.91	4.36	4.59
3hlaB	2rhe_	83.15	10.30	4.72	9.76	8.63	8.62
3rubL	6xia_	74.13	20.91	22.26	24.19	24.15	23.71
4sbvA	2tbvA	97.49	18.68	17.73	18.47	18.53	18.97
5fd1_	2fxb_	55.66	10.95	10.70	12.13	11.99	11.61
8i1b_	4fgf_	73.97	11.31	10.77	12.58	12.88	12.65

[a]The first two columns contain the PDB codes of the target and template proteins, respectively. The percentage of a target sequence aligned to a template is given in column 3. The fourth column provides RMSD (all values for alpha-carbon traces) for the aligned part of the template from "true" structure of the target—a measure of the alignment quality. The fifth column gives the best RMSD for the model chains observed in a set of sparely written trajectories (a few hundred photographs). The sixth column gives the RMSD for the lowest energy (according to the SICHO force field) conformation observed in the trajectories. The RMSD values in the two last columns correspond to the average structures obtained via distance geometry and clustering algorithm. The two methods of averaging are almost equivalent, with slightly better performance of the DG approach. In number of cases, the final models for the entire structure are better (as measured by RMSD from the crystallographic structure) than the initial threading models—that is the aligned part.

H.   Comparison to Modeller

Recently, several tools were developed for the fast building of all-atom models of proteins by various means of comparative modeling. Probably, the most efficient is Modeller, developed by Sali and Blundel [195]. Modeller allows for the high-throughput modeling of protein structures on a genomic scale. The method proposed here is more complex and more computationally demanding; however, it is still feasible in large-scale applications. The key question is, Are the results worth the increased computational cost? To answer this question, we compared various models for the Fischer database proteins [34] in Table VI, where the results of generalized comparative modeling described in this contribution are compared with models generated by Modeller. Both procedures started from exactly the same templates and the same alignments generated by PROSPEC-TOR. If we consider all models, then GeneComp performs better than Modeller in 53 cases, worse in 13, and the same in two cases. If only templates whose RMSD is less than 10 Å are considered, then GeneComp performs better in 29 cases, Modeller performs better in five cases, and they perform the same in one case. However, in the latter, the two structures differ by a small amount. In many cases of very good (or good) templates, the two methods generate models of similar quality. The situation changes when the homology becomes weaker and when, consequently, the threading models become more distant from the probe structure. In these cases, the models generated by GeneComp are almost always of noticeably better accuracy. We can most likely ignore the cases when both methods lead to very bad models. It is safe to say that there is usually no difference between models 12 and 14 Å from the true probe structure. The utility of such models for structural genomics is at least problematic (of course, it depends somewhat on protein size—a very large protein may still be of a correct overall topology with this high RMSD). However, there is quite a difference between a model that is 4 Å from the true structure and a 6 Å model (or even more between a 6 Å model and 10 Å model). As can easily be seen from the data compiled in Table VI, in the range of 4–8 Å, the GeneComp models are in most cases significantly more accurate than the models generated by Modeller. The typical difference is 1–2 Å; however, in a few cases it is as much as 4–5 Å. Interestingly, the models generated by GeneComp frequently have a lower RMSD for the entire structure than the RMSD of the original aligned fragments. These are the cases when a qualitative improvement with respect to simple comparative modeling was observed. The lattice simulations improve entire structures. Thus, on average the proposed method leads to qualitatively better molecular models with pronounced consequences for structure-based protein function prediction and other aspects of proteomics.

VI. *AB INITIO* FOLDING

A. Description of the Method

The method for *ab initio* folding of small globular proteins employs the same modeling tools as in generalized comparative modeling. There are, however, some differences. Of course, now there is no template to restrict the

TABLE VI
Comparison of Generalized Comparative Modeling with Automated Modeling via Modeller[a]

Target	GeneComp + DG	Modeller	GeneComp + DG + Modeller
1aaj_	9.37	10.13	9.30
1aba_	4.75	6.66	4.73
1aep_	21.45	21.56	21.32
1arb_	17.46	18.56	17.35
1atnA	13.16	15.61	13.15
1bbhA	3.07	3.02	3.03
1bbt1	10.70	10.21	10.68
1bgeB	5.45	10.34	5.42
1c2rA	5.34	5.84	5.30
1cauB	5.45	5.93	5.93
1cewI	7.79	8.47	7.76
1chrA	4.90	4.57	4.91
1cid_	18.44	20.19	18.44
1cpcL	13.58	15.62	13.52
1crl_	24.09	25.89	23.98
1dsbA	16.47	16.37	16.45
1dxtB	3.01	3.05	3.00
1eaf_	10.32	10.82	10.18
1fc1A	13.12	15.02	12.48
1fxiA	10.18	11.27	10.11
1gal_	17.80	18.86	17.66
1gky_	6.36	11.82	6.45
1gp1A	13.74	15.22	13.66
1hip_	4.26	4.06	4.09
1hom_	1.57	1.73	1.57
1hrhA	5.07	6.95	5.05
1isuA	5.07	5.84	5.20
1lgaA	15.59	14.72	15.68
1ltsD	10.21	10.88	10.22
1mdc_	2.66	2.66	2.71
1mioC	14.71	16.78	14.68
1mup_	4.38	4.93	4.40
1npx_	14.12	14.48	14.05
1onc_	3.51	5.14	3.50
1osa_	17.90	16.89	17.91
1pfc_	4.28	4.39	4.49

[a]The same alignments (see Table V) were used as starting templates for GeneComp (RMSD for the DG averaged models) and Modeller. The last column provides RMSD for the models generated by Modeller starting from the complete models obtained by GeneComp. In almost all cases the models generated by GeneComp are more accurate than the models generated by Modeller, and in 15–20 cases the improvement is of a qualitative nature (see the text for explanation). Refinement of the GeneComp models by Modeller (compare columns 2 and 4) leads to marginal changes of the molecular models, indicating the consistency of the GeneComp models, with local atomic details of the PDB structures.

conformational search. The generic and protein-independent components of the force field for the lattice models are the same, and the protein-specific potentials have a similar form [202]. The difference is that in *ab initio* folding they are less specific. For the test purposes, all homologous (and analogous) proteins have been excised from the structural database used to derive the potentials. As a result, the number and accuracy of the predicted contacts are lower, as is the accuracy of the short-range terms. As before, a conservative prediction of the regular elements of secondary structure was used to bias the short-range interactions. Thus the requirements for the folding simulations are much higher. A much larger number of independent simulations were executed to check the reproducibility of the results and to provide a representative sample for the clustering procedure and final fold selection.

The selection of the initial conformations for the REMC simulations requires some comment. In principle, random expanded conformations could be used. However, this slows down the convergence of the process. For this reason, a different strategy was adopted. Having a prediction of secondary structure, gapless threading of structures of comparable size is performed using the matching fractions of the predicted secondary structure to the actual secondary structure of the templates as a scoring function. Of course, all homologous and analogous proteins were removed from the pool. Fifty lattice chains were built using the 50 best scoring structures as templates. While these starting structures are different from the probe fold, they may have the proper element(s) of secondary structure that may serve as a fast nucleation site for the folding process. In the preliminary simulation runs, 50 replicas were used. The second iterations used the top 20 (20 lowest-energy replicas) as the input pool. The simulation results from the last iteration of the lattice-folding algorithm were subject to a clustering procedure [197] that was also used to make the final fold selection.

B.   Results of *Ab Initio* Folding on 28 Test Proteins

Sequences of 28 globular proteins were selected as the test set for the *ab initio* folding protocol. The set is representative of single-domain small proteins. It contains alpha proteins with α/β-, $\alpha + \beta$-, and β-type folds. In about 50% of the cases, low-resolution folds of correct topology were obtained as one of a number of clusters. The results are compiled in Table VII that also contains the RMSD for the best structures observed during simulations at the lowest temperature replica of the system as well as the RMSD of all structures that cluster [197]. It is clear that simulations generate a small subset of very good structures for the majority (22 of 28) of the tested proteins. Unfortunately, the fold selection procedure rarely selects structures close to the very best ones. The discrepancy is more drastic than in the case of template-restricted folding. It could be proven rigorously that to obtain a 3 Å structure by random in a set of trajectories

TABLE VII
Summary of *Ab Initio* Folding Results

Protein Name[a]	Best RMSD	Lowest-Energy RMSD	RMSD of Centroid of Each Cluster
1c5a_	4.86	10.87	11.20 11.63 **5.70** 8.75
1cewI	6.71	10.08	8.77 13.84 15.29 12.00 11.66
1cis_	4.98	11.52	10.41 10.34 9.36 9.67 10.43 6.81 7.25
1ctf_	7.10	11.06	10.72 11.40 11.54
1fas_	5.30	8.55	9.30 7.47 11.68 10.15 11.89 **6.36** 12.87
1fc2C	2.91	7.34	7.21 7.61 **3.35**
1ftz_	2.65	8.79	8.78 6.52 **3.05** 7.11 6.50 8.18
1gpt_	4.92	7.45	7.58 8.66 9.70 9.59
1hmdA	5.02	10.57	10.36 12.95 14.20 12.52 **5.51**
1ife_	6.53	9.23	11.57 9.24 13.64 11.71 12.12 11.41
1ixa_	4.02	6.62	**6.36** 6.92 9.28 10.65 10.53
1lea_	3.23	11.85	10.93 9.95 8.32 8.44 **5.82**
1mba_	9.61	12.72	12.63 15.28 12.01 15.44 13.51
1poh_	2.90	12.63	12.76 11.91 **3.87**
1pou_	2.70	4.98	**3.95** 9.88 9.93 10.93 11.61
1shaA	3.94	13.07	13.82 12.08 12.75 9.00 10.49 6.00
1shg_	4.40	9.00	8.99 9.06
1stfI	5.47	10.19	8.06 12.86 11.17 13.68 11.99 16.74
1tfi_	7.62	9.48	10.15 8.88 10.56 10.20
1thx_	2.97	12.72	12.83 11.27 **3.89** 13.04 14.40
1tlk_	3.13	7.38	11.02 **6.35**
1ubi_	3.05	10.98	10.71 10.51 11.57 12.07 8.13 10.54
256bA	3.09	3.73	**3.52** 8.38 14.88 10.01 14.91 12.13
2azaA	3.83	7.20	**5.75 12.86 13.01 14.00 13.30 13.30**
2pcy_	3.72	7.75	**5.56** 7.12 11.39 13.46 13.19
2sarA	8.45	13.11	10.71 11.92 12.18 12.71 14.10 13.93 14.10 13.79
5fd1_	8.67	12.53	12.20 10.84 12.48 10.94 14.35 14.26
6pti_	5.36	7.36	6.68 10.81 10.99 10.14 9.14

[a]Bold indicates that this protein is foldable; that is, one of the clusters has an average RMSD from native less than 6.5 Å.

containing a few thousand photographs is practically impossible. Thus, the model force field and the sampling scheme do a reasonably good job in sampling protein-like regions of conformational space, including the neighborhood of the native state. At the same time, the force field lacks a sufficient discriminatory ability to select the closest-to-native fold generated from a large number of competing protein-like structures. These competing structures have elements of native topology with misfolded fragments of structure; sometimes they are mirror images of native-like folds.

Overall, though, if one defines a successful simulation as one with a native topology whose backbone RMSD is less than 6.5 Å, then in 15/28 cases (i.e., about 54% of the cases) the simulations are successful. Again, a different, more efficient fold selection method needs to be developed; such efforts are currently underway. An alternative recently being explored is the method of inserting atomic detail and then scoring the structures using a recently developed distance-dependent potential of mean force [204]. If this is done, then 1stfI is not foldable, but 1fas_,1gpt_,1mba_ are foldable, giving a total of 17 (i.e., 61%) of the test set proteins successfully folded.

VII. COMPATIBILITY OF REDUCED AND ATOMIC MODELS

A. Reproducibility of Structural Details

Reduced models have a long history. Some reproduce just the overall fold of globular proteins, whereas other (more complex) models maintain some details of protein structure. The SICHO model, based on just a single center of interaction per residue, appears at first glance to be a drastic simplification. However, due to its flexibility, the model is more accurate than it may appear at first. First of all, the mesh size of the underlying cubic lattice is equal to 1.45 Å, which means that a simple fit of the lattice model to a detailed PDB [171] structure has an average accuracy of 0.7–0.8 Å with respect to the side-chain centers of mass. Due to the coarse-grained character of the potentials, correctly folded (say, by a pure *ab initio* approach) structures are of somewhat lower accuracy. Very small proteins or peptides could be folded to 1.5 Å to 2.0 Å from the native structure. The accuracy of larger proteins decreases due to an accumulation of errors across the structure. For 100-residue proteins, properly folded structures have an RMSD in the range of 3.5–6.5 Å from native. When looking for elements of secondary structure as helices and β-hairpins, the accuracy is of the same range as for very small proteins or slightly better and ranges between 1.0 and 2.0 Å. The above numbers are given for the side-chain centers of mass. Our model employs a very crude and simple reconstruction of the α-carbon coordinates as a simple combination (with the coefficients extracted from a statistical analysis of the structural database) of the positions of three consecutive side-chain centers. This estimation is contaminated by a small systematic error (there is no correction from deviation of the α-carbon from the plane defined by three corresponding side-chain united atoms) and by some statistical error related to errors in the side-chain positions. Compensating for this is a statistical reduction of the absolute error of Cαs because the main-chain units are "inside" the secondary structure elements defined by the side-chain centers of mass. Consequently, errors in the side-chain positions translate into a slightly smaller error in the positions of the α-carbons. As a result, the accuracy of the crude α-carbon trace is the same or slightly better than the accuracy of the explicit virtual chain of the side groups.

The level of local (and global) accuracy of the model is sufficient to allow for quite accurate reproductions of the most important structural details. First, the contact maps of the side chains extracted from the model are very similar to the contact maps calculated from the crystallographic structures, assuming a 4.5 Å cutoff for contacts between heavy atoms of the side chains (side groups are considered to be in contact when any pair of their heavy atoms are at a distance smaller than the above cutoff). The overlap with native for properly folded structures is 85–90%. There are some excess contacts in the lattice models, and some contacts are missed due to the spherical shape of the model side chains and the statistical character of the cutoff distances for the model residues. More interestingly, the model hydrogen bond network (properly calculated from the estimated coordinates of alpha carbons) of the main chain coincides with similar (85–90%) accuracy with the main-chain hydrogen bonds assigned by the DSSP procedure [205] to the corresponding native structures. Bifurcated hydrogen bonds (the weaker ones) are ignored in this comparison, because the model does not allow for H-bond bifurcation. As in real proteins, the model structures have very regular networks of hydrogen bonds. Helices, except for their ends, exhibit a regular pattern of two hydrogen bonds per residue. The same is observed for internal β-strands in β-sheets. The edge strands usually have a single model H-bond per residue. Sometimes, even patterns characteristic of β-bulges are reproduced with high fidelity. The model network of H-bonds is explicitly cooperative. This leads to protein-like cooperative folding. Interestingly, misfolded structures also look very protein-like unless they violate some "rules" of protein folding—for example, the handedness of the β–α–β connections [206].

The protein-like geometry of such a simple model is enforced by the proper design of the force field that has two distinct types of components: sequence-dependent (or even protein-specific), which drive folding toward a specific fold, and generic, which strongly bias the model chain toward the average protein-like local conformational stiffness. The force field also has packing preferences. This way a vast majority of the irrelevant portion of the conformational space of the high coordination lattice (containing 646 possible side-chain–side-chain virtual bonds) model is efficiently avoided during the sampling process.

B. Reconstruction of Atomic Details

The lattice SICHO model exhibits good compatibility with detailed all-atom models. Projection of the all-atom structures onto the lattice model is trivial, and the accuracy of the projection is about 0.8 Å RMSD for the side-chain centers of mass or for the coarse reconstruction of all the α-carbon positions. More interesting, and certainly more challenging, is the reconstruction of the atomic details from the lattice models. A couple of similar procedures have recently been developed for this purpose [200]. In one, the crude estimated coordinates of the α-carbons are refined using the distance restraints typical for proteins and simple potentials for optimization of the backbone geometry. In the next stage, the remaining atoms of the main chain are reconstructed using a library of backbone fragments. Finally, a library of side-chain rotamers is employed to build the side-group conformations that are the most consistent with the lattice model. The side-group geometry and packing can be optimized relatively easily because the gross overlaps are by definition excluded by placing the rotamers as close as possible to the lattice chain (which itself exhibits a reasonable approximation of the packing in a protein). When starting from the lattice fit to the crystallographic structure, this reconstruction process returns a full atom structure that differs on average by about 1 Å RMSD from the original one. Further minimization by the CHARMM force field [207] leads to a small improvement of the model. The same accuracy of all-atom reconstruction is expected for all conformations generated during the lattice simulations.

A somewhat different procedure that has an advantage of computational speed leads to structures that are about 1.5 Å from the original all-atom model. Thus, there is the possibility of multiscale simulations of protein systems. The computational speed of the SICHO model enables simulations that correspond to the time-scales characteristic of real protein folding. At specific interesting points of MC trajectory, one can perform all-atom reconstruction, followed by detailed MD simulations. Another possibility that is now being explored is to use the all-atom models (derived from lattice structures) as a means of selecting the "best," possibly closest to native, structures generated in lattice folding simulations by the SICHO model.

C. Feasibility of Structural Refinement

As discussed in other parts of this chapter (see Sections VIII and IX), low-resolution models could be successfully employed in the functional annotation of new proteins and even for docking ligands. Of course, the more accurate the model, the wider its applications. The SICHO model is of limited resolution. Typical, well-folded structures have an RMSD that is 2 to 6.5 Å from native. Is it possible to improve such models using more a detailed representation and a more exact force field? Is it possible to include the solvent successfully in an explicit way at this stage? It appears that at least for moderately small proteins with a reasonable starting lattice structure, sometimes the models can be refined to a resolution close to that of experimental structures. Successful refinement of a small protein, CMTI, from a low-resolution MONSSTER folding algorithm [137] to a structure close to the experimental one was recently done by Simmerling et al. [208]. Earlier, for similar low-resolution lattice models, several structures of leucine zippers were also successfully refined to experimental resolution [124,125]. These studies were subsequently extended using ESMC to provide a treatment of the GCN4 leucine zipper folding thermodynamics as well as the prediction of the native state [209], and it was subsequently shown that the CHARMM force field, when supplemented by a generalized Born/surface area treatment, is highly correlated with the lattice-based force field [210]. These studies are extremely encouraging, although it is now unclear how soon the gap between low-resolution lattice folds and high-resolution all-atom structures for larger proteins will be closed.

VIII.   FROM STRUCTURE TO BIOCHEMICAL FUNCTION

A.   Does Knowledge of Protein Structure Alone Imply Protein Function?

Because proteins can have similar folds but different functions [211,212], determining the structure of a protein does not necessarily reveal its function. The most well-studied example is the $(\alpha/\beta)_8$ barrel enzymes, of which triose phosphate isomerase (TIM) is the archetypal representative. Members of this family have similar overall structures but different functions, including differing active sites, substrate specificities, and cofactor requirements [213,214]. An analysis of the 1997 SCOP database [211] shows that the five largest fold families are the ferredoxin-like, the (α/β) barrels, the knottins, the immunoglobulin-like, and the flavodoxin-like fold families with 22, 18, 13, 9, and 9 subfamilies, respectively. In fact, 57 of the SCOP fold families consist of multiple superfamilies [15]. These data only show the tip of the iceberg: Each superfamily is further composed of protein families, and each individual family can have radically different functions. For example, the ferredoxin-like superfamily contains families identified as Fe–S ferredoxins, ribosomal proteins, DNA-binding proteins, and phosphatases, among others. More recently, a much more detailed analysis of the SCOP database has been published [215], which finds broad function–structure correlation for some structural classes, but also finds a number of ubiquitous functions and structures that occur across a number of families. The article provides a useful analysis of the confidence with which structure and function can be correlated [215]. For a number of functional classes, knowledge of protein structure alone is insufficient information to assign the specific details of protein function.

B.   Active Site Identification

It has been suggested that the active sites in proteins are better conserved than the overall fold [27]. If so, then one should be able to identify not only distant ancestors with the same global fold and same biochemical activity, but also proteins with similar functions but different global folds. Nussinov and co-workers empirically demonstrated that the active sites of eukaryotic serine proteases, subtilisins, and sulfhydryl proteases exhibit similar structural motifs [216]. Furthermore, in a recent modeling study of *S. cerevisiae* proteins, active sites were found to be more conserved than other regions [27]; this was also seen in the study of the catalytic triad of the α/β hydrolases [11]. Kasuya and Thornton [217] have created structural analogs of a number of Prosite sequence motifs and showed, for the 20 most frequent Prosite patterns, that the associated local structure is rather distinct [3]. These results provide clear evidence that enzyme active sites are structurally more highly conserved than other regions of a protein.

C. Identification of Active Sites in Experimental Structures

Several groups have identified functional sites in proteins with the goal of engineering or inserting functional sites into new locations, and success has been achieved for several metal-binding sites [218–226]. However, because highly accurate site descriptors of backbone and side-chain atoms were used, this fueled the idea that significant atomic detail is required if protein structure is to be used to identify protein function. Similarly, detailed side-chain active site descriptors of serine proteases and related proteins were employed to identify functional sites [227], while more automated methods for finding spatial motifs in protein structures have been developed [37,216,228–233].

Unfortunately, such methods require the exact placement of atoms within protein side chains and are inapplicable to the inexact, low-resolution predicted structures generated by the state-of-the-art *ab initio* folding and threading algorithms (see Sections IV–VI). These methods are required when the sequence identity of the sequence of interest to solved structures is too low to use comparative modeling. To address this need, Skolnick and Fetrow have recently developed "fuzzy," inexact descriptors of protein functional sites [8]. They are applicable to both high-resolution, experimental structures and low-resolution (backbone RMSD 4–6 Å from native) structures. These descriptors are α-carbon-based, "fuzzy functional forms" (FFFs). Initially, they created FFFs for the disulfide oxidoreductase [8,10] and α/β-hydrolase catalytic active sites [11] (an additional 198 have now been built, with comparable results [234]).

The disulfide oxidoreductase FFF was originally applied to screen 364 high-resolution structures from the Brookhaven protein database [235]. For the true positives, the proteins used to create the FFF have different structures and low sequence identity to those proteins used to build the FFF, but the active sites are quite similar [8]. Here, the FFF accurately identified all disulfide oxidoreductases [8]. In a larger dataset of 1501 proteins, the FFF again accurately identified all of the disulfide oxidoreductases, but it also selected another protein, 1fjm, a serine-threonine phosphatase. Initially this was a discouraging result, but subsequent examination of the sequence alignments combined with an analysis of the subfamily clustering strongly suggested that this putative active site might indeed be a site of redox regulation in the serine-threonine phosphatase-1 family [12]. If experimentally verified, this would highlight the advantages of using structural descriptors to analyze multiple functional sites in proteins. In particular, function prediction would not be restricted to the "primordial" function that characterizes the sequence family, but could also include additional functions gained during the course of evolution.

D.   Requirements of Sequence–Structure–Function Prediction Methods

Any sequence–structure–function method that does function prediction by analogy relies on three key features. First, the function of the template protein must be known. Second, the active site residues must be identified and associated with the function of the protein. Third, a crystal structure of a protein that contains the active site must be solved so one can excise the active site for constructing the corresponding three-dimensional active site motif. Evolutionary approaches to function prediction often just require that the first criterion be satisfied, but for more distant homologs the second should be checked as well, because functions can be modified during evolution. The third requirement is unique to structure-based approaches to function prediction. Based on studies to date [8,10–12,14,15], identification of an enzyme's active site requires a model whose backbone RMSD from native near the active sites is about 4–6 Å for structures generated by *ab initio* folding. This predicted structure quality is due to the fact that the errors in the active site geometry found in the predicted structure tend to be systematic rather than random. However, threading does not suffer from this problem because, in the predicted structure, if the alignment does not include the active site residues, no functional prediction is made. If it does, the local geometry is the same as in the template's native structure. Threading can have alignment problems, but locally—at least in the vicinity of the active site—these can often be overcome if the threading score includes a sequence similarity component or if Generalized Comparative Modeling is done. Nevertheless, in practice, for both *ab initio* and threading models, the quality of the predicted structures is better in the core of the molecule than in the loops, so prediction of the function of a protein whose active site is in loops may be problematic. Currently, the method has only been applied to identify enzyme active sites. Recent work described in Section VIII suggests that at least in some situations, low-resolution structures can also be used to at least partially address the problem of substrate and ligand binding. But in general, techniques that will further refine inexact protein models will be necessary to extend the approach.

E.   Use of Predicted Structures from *Ab Initio* Folding

As noted above, the recent CASP3 results suggest that for small proteins, current tertiary structure prediction schemes can often (but far from always) create inexact protein models of the global fold. Are these structures useful for identifying functional sites in proteins? To explore this issue, using the *ab initio* structure prediction program MONSSTER [191,193], the tertiary structure of the glutaredoxin, lego, was predicted whose backbone RMSD from the crystal structure was 5.7 Å. To determine if this inexact model could be used for function identification, the set of correctly folded structures and a set of 55 incorrectly folded structures were screened with the FFF for disulfide oxidoreductase activity [8,10]. The FFF uniquely identified the active site in the correctly folded structure but not in a library of incorrectly folded ones [15]. This is a proof-of-principle demonstration that inexact models produced by the *ab initio* prediction of structure from sequence can be used for the prediction of biochemical function.

F.   Use of Threaded Structures to Predict Biochemical Function

In a very important paper, Lathrop demonstrated that use of functionally conserved residues could filter threading predictions to correctly identify globins even when the threading score was insignificant [30]. While suggestive, the key question was whether or not this result could be generalized on a genomic scale. Over the past few years, we have been exploring this issue in great detail [8,10–15], and, as discussed below, we demonstrate that the use of the sequence–structure–function paradigm, when appropriately employed, allows one to predict biochemical function with a much smaller false-positive rate than BLOCKS [236,237], the best competing sequence-based approach. Indeed, we have developed a very promising approach to the problem of genome-scale function annotation.

The methodology is as follows: We use PROSPECTOR1 [57] (although, any threading algorithm could, in principle, be used) to identify the set of 20 structures that are the best scoring matches between the probe sequence and the template structure (four scoring functions times five best scoring structures for each function). Then, each structure was searched for matches to the active site residues and geometry of the FFF. If a match to the FFF is found, then for those sequences for which homologous sequences are available, a sequence-conservation profile was constructed [11]. If the putative active site residues are not conserved in the sequence subfamily to which the protein belongs, that sequence is eliminated as having the predicted function; otherwise the sequence is predicted to have the function. Using this sequence–structure–function method, 99% of the proteins in the eight genomes that have known disulfide oxidoreductase activity were found [15]; 10% to 30% more correct functional predictions are made than in alternative sequence-based approaches [15]; similar results are seen for the α/β-hydrolases [11].

In Fig. 2, we show the distribution of scores (blue) for the *E. coli* genome [238] when any of the 11 disulfide oxidoreductases in our structural database was selected as being in the top five scoring structures using the "close" sequence plus secondary structure plus pair profile scoring function. Similarly, those proteins identified on application of the disulfide oxidoreductase FFF to

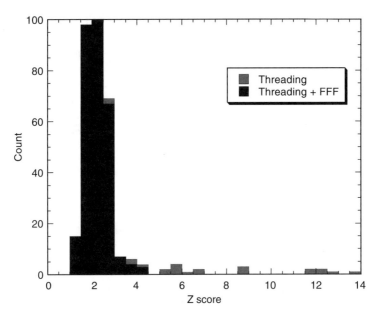

Figure 2. For the *E. coli* genome, the distribution of threading scores for the "close" sequence plus secondary structure/pair profile scoring function is shown in dark gray and those proteins identified by use of the disulfide oxidoreductase FFF are shown in light gray.

these threading models (all are known true positives) are indicated in red. Clearly, the use of the FFF allows one to extract proteins (e.g., those to the immediate right of the maximum) when their raw threading score would require one to also include a significant (in this case overwhelming) number of false positives. We note that full use of PROSPECTOR1-3 identifies all the known disulfide oxidoreductases in the *E. coli* and *M. genitalium* genomes. Note that, in general, structures whose Z-score is greater than 1 can be successfully searched for a match to a known active site.

Importantly, using structural information, the false-positive rate is much less than that found using sequence-based approaches. This conclusion arises from a detailed comparison of the FFF structural approach and the Blocks sequence-motif approach [15]. Here, the sequences in eight genomes, including *B. subtilis* [239], were analyzed for disulfide oxidoreductase function using the disulfide oxidoreductase FFF, the blocks thioredoxin block 00194 [236], and the blocks glutaredoxin block 00195 [236]. In Fig. 3 we plot the distribution of scores when the *B. subtilis* genome is threading through these two blocks. By way of example, if we assume that those sequences identified by both the FFF and Blocks [236] are "true positives," we find 13 such sequences in the *B. subtilis*

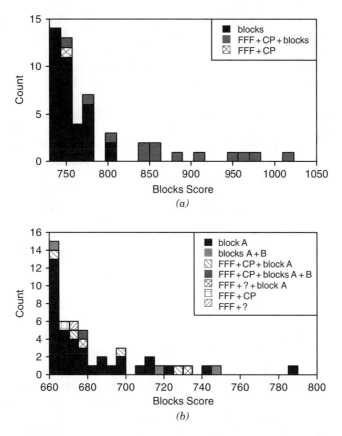

Figure 3. For the *B. subtilis* genome, the distribution of Blocks scores [236, 237] for the thioredoxin block and glutaredoxin blocks are presented. FFF indicates that the threaded structure satisfies the disulfide oxidoreductase active site descriptor, CP indicates that the sequence identified by threading and FFF satisfies the conservation profile, and ? indicates that there is just one sequence so that a CP analysis cannot be done.

genome. (Recognize that the experimental evidence validating all of these "true positives" is lacking; thus, they are more accurately termed "consensus positives.") To find these 13 "consensus positive" sequences, the FFF hits 7 false positives. In contrast, Blocks hits 23 false positives. It was previously suggested that the use of a functional requirement adds information to threading and reduces the number of false positives [30]. These data validate this claim on a genome-wide basis. Similarly, using active site descriptors as a filter, one can identify the true positives even when the threading score is barely significant (as in Fig. 2) and where selection of the structure based on the threading score alone would yield a significant number of false positives. Thus, what we require is a method that places such structures where their score is sufficiently significant that on subsequent filtration by a functional descriptor, they can be reliably identified. This is the origin of use of multiple scoring functions in PROSPECTOR1, which, in combination, selects 59 of 68 Fischer pairs in the top scoring position.

Surprisingly, despite the fact that threading algorithms have problems generating good sequence–structure alignments, we have found that active sites are often accurately aligned, even for very distant matches. This observation would agree with the above-mentioned experimental results that active sites are well-conserved in protein structures. Of course, because no genome has the function of all its proteins experimentally annotated, it is impossible to know how many proteins with the specified biochemical function are missed, nor is there yet experimental characterization of most of these predictions.

IX. USE OF LOW-RESOLUTION STRUCTURES FOR LIGAND IDENTIFICATION

One of the important elements of protein function is the ability of a protein to interact with and bind various ligands. This ability is closely related to the three-dimensional structure of the protein. Because the quality of theoretical structure prediction methods has recently improved considerably, we are developing a docking procedure that will utilize these relatively low-quality models of proteins for the prediction of plausible conformations of receptor-small ligand complexes as well as for the prediction of interactions between particular subunits of a protein in the quaternary structures.

Our approach to the problem of low-resolution docking focuses on the steric and quasi-chemical complementarity between the ligand and the receptor molecules. Because the predicted structures that result from theoretical predictions usually resemble very low-resolution experimental structures, in our method we use only approximate models of both the ligand and its receptor. Vakser et al. [240] have demonstrated that by averaging the structural details of interacting molecules it is possible to drive the docking procedure toward the real binding site, thus avoiding, in many cases, the local minima problem. It also turns out in our case that this averaging procedure allows for the compensation of the numerous structural inaccuracies that result from the theoretical predictions of the receptor structure.

In the first stage of our docking procedure, structures of both molecules, the receptor and the ligand, are projected onto a uniform cubic lattice, thus giving two clusters of adjacent cubes. These two clusters approximate the shapes of both molecules with the accuracy of the grid size. Some of the receptor cubes ("surface" cubes) can be penetrated by the ligand, leading to favorable interactions when overlapped with the ligand, whereas others (interior cubes) contribute to the repulsive contacts. As elegantly demonstrated by Vakser et al. [240], when such a procedure is correctly implemented, this simple steric matching protocol is often quite successful in rebuilding correctly docked complexes.

While the steric method described above is very efficient, in many cases, geometric criteria alone are insufficient to correctly dock the two molecules. This is especially true when the structure of the receptor is of poor quality or a ligand molecule is relatively small so that shape complementarity is insufficient to specify the correct conformation. To overcome this problem, we decided to build a statistical potential that could be used for additional evaluation of the quality of the match. In order to build the potential, we defined 20 general atom types and built the contact statistics on the basis of the structures of known complexes available in the PDB [171]. After projection of the two molecules onto the grid, every cube is additionally labeled with the properties defined by the atom types that were projected onto it. Once the approximate representation of the system is ready, the best match of these two cube-clusters is determined by exhaustive scanning over the six-dimensional conformational space of the three relative translations and the three rotations. Calculating the value of the correlation function between these two sets of cubes and the value of the potential function, the quality of the particular ligand-receptor orientation is scored.

We applied this algorithm to predict (actually postdict) the structures of several complexes available in the protein data bank. These complexes include members of the Fischer database that had co-crystallized ligands that were generated by the procedure that was described in Section V. In most cases, not only is the location of the binding site on the receptor surface correctly identified, but the proper orientation of the bound ligand was reasonably well recovered as well, within the level of accuracy of the modeled receptor itself. In many cases, even structures of receptors as far as 5–6 Å away from native turned out to be accurate enough for the docking procedure to succeed.

Table VIII below shows five examples of the homology-modeled structures that were used in our docking calculations. The quality of the modeled receptor

TABLE VIII
Results of Docking Ligands to Low-resolution Predicted Structures[a]

Structure Name	RMSD of the Receptor from Native	Relative Shift of the Ligand from Native
2sarA	5.99	3.1
2cmd_	5.57	1.3
1bbhA	3.16	1.6
1mdc_	4.92	2.6
1c2rA	4.94	3.3

[a]All dimensions are in angstroms.

(in RMSD) and shift of the docked ligand relative to its position in the superimposed native complex are also shown.

Two examples of docked ligands to the generalized homology modeled receptors are shown in Fig. 4. The red is the native orientation of the ligand, and the yellow is the best scoring match. As is immediately evident, the algorithm does a reasonably good job in docking the ligand to the correct binding site in the correct orientation. While our method is still under active development, it has already revealed its usefulness in the successful docking calculations of even small ligands to the theoretically modeled receptors. When complete, this methodology could hopefully be used for the large-scale screening of the potential ligands for the receptors predicted from genomic sequences.

X. OUTLOOK FOR THE FUTURE

A.   Possible Improvements of the Structure Prediction Methodology

The methodology for protein structure prediction outlined in this contribution, while partially successful, needs further improvement. First of all, some elements of the force field of the lattice model are not yet satisfactory. The threading algorithm PROSPECTOR, which forms the core of this approach, needs improvement. For example, it currently uses a very simple sequence profile, and more powerful techniques for generating more sensitive sequence profiles [241] need to be exploited. PROSPECTOR also generates high-scoring local sequence fragments that are often, but not always, quite accurate. This information needs to be incorporated into subsequent threading iterations as well as into partial seed structures in *ab initio* folding, akin to ROSETTA [242,243]. Better means of assessing the quality of the alignments also need to be developed.

The most promising way to improve generalized homology modeling is to couple the strength of template restraints to the quality of the template. Now, for all tested cases, the template-related restraints are of the same strength. Much better results may be possible if, for the templates that are close to the probe's structure, the restraints were very strong. For templates that are far from the probe's structure, the restraints should be very weak. The template should be used only for a loose definition of the fold topology. This requires an up-front estimation of the template quality in a semiquantitative fashion. Better scoring of the threading results and comparison with related cases (size of protein, percentage of alignment, comparison of the template alignments to other related proteins, etc.) might provide necessary data for the case-dependent scaling of the template-related restraints in the generalized homology modeling procedures.

Turning to issues associated with *ab initio folding* and, to a lesser extent, generalized comparative modeling, some elements of the force field of the lattice model are not yet satisfactory. The scaling of various contributions to the

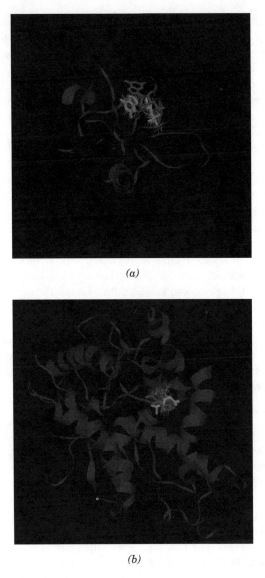

(a)

(b)

Figure 4. (See also color insert.) For the predicted protein structure of 2sarA (2cmd_) generated by GeneComp using a template provided by the Fischer Database [34], the red-colored ligand represents the superposition of the ligand bound to the native receptor. The highest-scored match is colored in yellow.

interaction scheme is now to a large extent arbitrary and adjusted essentially by a trial-and-error method. A more precise scaling will be attempted by an automated procedure targeted to generating strong (as strong as possible) correlations between RMSD from correct folds and energy. A large set of decoys (lattice structures at various distances from native) will be used for this purpose. The weakest elements of the force field will be reexamined. Probably the largest improvement of the model could be achieved via introduction of approximate electrostatics into the interaction scheme. This should include more implicit treatment of the solvent and other than intra-main-chain hydrogen bonds.

For *ab initio* folding, a better means of the fold selection is needed. As mentioned above, for the majority of small proteins, the SICHO simulations produce a fraction of very good low-to-moderate resolution structures. Unfortunately, the model force field is capable of selecting these good folds in only a fraction of cases. Perhaps the folding simulations and the fold selection procedures should be separated in a more radical way. It appears to make sense that different force fields may be more efficient for folding simulations than those used for the fold selection. Indeed, folding requires an interaction scheme that discriminates not only against the wrong folds but also against a huge part of model-chain conformational space that does not correspond to any protein structures. The fold selection stage needs potentials that essentially discriminate between various protein-like conformations. Fortunately, fold selection involves a few hundred structures. Thus, more detailed, including all-atom, interaction schemes could be employed.

B. In Combination with Experiment

A variety of fragmentary experimental data could be used to increase the accuracy and to extend the range of applicability of the described methodology for protein structure prediction. The *ab initio* folding procedure employs predicted secondary structure (in a three-letter code) and predicted contacts between side groups. None of these predictions are exact; this has a consequence for the overall performance of the method. Knowledge of the exact protein secondary structure or some elements of secondary structure significantly increases the precision and accuracy of the three-dimensional structure predictions. Also, the exact knowledge of a few side-chain contacts increases the applicability of the method. As demonstrated recently [139] for an older version of the SICHO model, knowledge of secondary structure and as few as $N/7$ to $N/5$ side-chain contacts (where N is the number of residues in the protein) enable reproducible structure assembly for proteins up to 240 residues. The larger the number of known contacts, the better the accuracy of the predicted structures. Such fragmentary structural data could be extracted from NMR experiments. When more extensive data are difficult (or impossible) to obtain, the lattice folding provides a low-to-moderate resolution molecular model of the protein of interest. In those cases where a lot of NMR-based restraints are collected, the possibility of obtaining of an approximate model from just a few identified long-range contacts may aid with assignment processes for the other signals. Such a procedure can be iterated. Alternately, such constraints could be implemented in PROSPECTOR as a potential to help improve the quality of fold selection as well as the quality of alignments. Structural restraints for the *ab initio* folding can originate not only from NMR data but also from electron microscopy. Fluorescence data or crosslinking experiments could also provide some information about the side-chain contacts. Sometimes, mutation experiments can identify residues that are involved with ligand binding. Information about the spatial arrangement of these residues could be easily incorporated into the folding algorithm. Another type of possible connection with experiment is probably worth mentioning. Sometimes, as a result of *ab initio* folding simulations, not one but a few plausible folds are generated. When compared with experiments required for structure determination from scratch, a much simpler experiment could be designed and executed for the selection between a few possible structures.

C. Improvement of Structure-Based Biochemical Function Prediction

A key component of the ability to predict the biochemical function of a protein using a structure-based approach is the availability of an extensive active site library. Once this is available, then the assignment of biochemical function can be done with a far smaller false-positive rate than alternative sequence-based approaches [15,244]. While active site FFFs can be built by hand, such a process is very time consuming, and automated approaches to active site identification must be developed. One such approach used PDB descriptors to assign active site residues [14], but more recent work using conservation profile analysis of these site descriptors indicates a significant false-positive rate [245]. However, if the identified active site residues are conserved, then one can tentatively build a functional descriptor on this basis. Alternatively, one could use BLOCKS [236] to identify conserved positions and attempt to build a three-dimensional descriptor on a unique subset of highly conserved residues [246]. We are currently undertaking such an approach.

To date, no large-scale refinement of the alignments generated by threading has been undertaken. If the alignment is in error and active site residues are not correctly aligned, then a false negative will result. Thus, we plan to apply GeneComp to demonstrate the stability of correct alignments (i.e., to show that true positives do not become false negatives). Next we plan to test the method on the weakly significant alignments (Z score > 1) first for *M. genitalium* and then for *E. coli.* If our results on the Fischer database are a guide, not only will this provide a set of better models for a significant fraction of both genomes, but perhaps, using a more complete active site library, additional ORFans can be assigned.

D. Improvement of Low-to-Moderate Resolution Docking of Ligands

Thus far we have demonstrated that in roughly 50% of the cases, the binding conformation of a known ligand can be identified using a low-resolution (backbone RMSD from native up to about 6 Å) predicted structure. While these results are encouraging, much more must be done. The energetic description describing the interaction of ligand and receptor must be improved so that the accuracy of the method is enhanced, and systematic clustering of the results using our clustering algorithm [197] must be done. Moreover, it remains to be demonstrated that unknown ligands can be identified using such an approach. Even if it turns out that in a library of several hundred thousand to millions of compounds, one could only place true ligands in the 500th position or so (a realistic goal for a low-resolution model), this would be quite valuable. Future work is proceeding along these lines.

The low-resolution description could also be used to dock macromolecular complexes. We have had very encouraging preliminary results on correctly docking the dimer in the tobacco mosaic virus, but clearly much more thorough benchmarking is required. One might imagine predicting the tertiary structure of two molecules and then docking them, but such studies are in the very preliminary stage.

E. Summary

In this review, we have described a number of approaches to the prediction of protein structure and biochemical function. A key theme of this review is that low-to-moderate resolution structures by state-of-the-art techniques are quite valuable. If the structure has a backbone RMSD from native in the range of 4–6 Å, it can be used to identify the biochemical function of a protein, and known ligands can be docked to identify the binding site as well as a low-resolution prediction of the location of the ligand in the receptor. The question then is, What are contemporary techniques for low-resolution protein structure prediction? After having reviewed the state of the field, which includes a number of promising *ab initio* studies [128,133,141,142,146] and threading algorithms [39, 53–56], we then introduced a unified approach to protein structure prediction. This methodology involves the use of a newly developed, iterative threading algorithm, PROSPECTOR [57], where one threads first (see Fig. 1). If there is no significant match to a template structure, the consensus contacts and secondary structure in the top 20 scoring structures are used as restraints in an *ab initio* folding algorithm. On average, this contact prediction predicts about one-third of the contacts correctly and predicts above 70% correctly within two residues. Application of this methodology to a representative test set of 28 structures results in the native state (of low-resolution structures up to 6.5 Å) being in one of the well-defined clusters in 15 cases. If fold selection is done not in the reduced model but in an atomic model, then 17 cases are foldable. Conversely, if PROSPECTOR identifies a global template, then we perform generalized comparative modeling, GeneComp, to refine the structures. This procedure uses the template alignment, as well as predicted contacts and secondary structure (not necessarily from the template structure), as restraints. In practice, when applied to representative probe proteins in the Fischer database [34,179], GeneComp tends to perform better on average than Modeller [23,27]. Moreover, it does no harm, that is, the quality of the model is either left the same or improves. Thus, it can be used with impunity. As in *ab initio* folding, the resulting structures are clustered and representative folds selected.

PROSPECTOR itself has been used to predict the tertiary structures of the proteins in two genomes, *M. genitalium* and *E. coli*, and successfully matches about 40% of the sequences to a known fold. Application of the three-dimensional active site descriptors designed for low-resolution structures, FFFs [8,10], allows one to select all known true positives, even when the Z score is close to 1. Furthermore, threading followed by application of the FFF has a far smaller false-positive rate than alternative sequence-based approaches such as BLOCKS [236,246]. Such approaches need to be generalized from treating enzymes to more generalized binding and macromolecular recognition.

This review describes one such way to use low-resolution structures to identify the binding site and conformation when one has a known ligand. The methodology was applied to those probe structures in the Fischer database that co-crystallized with ligands. As shown in Table VIII, it is possible to identify the binding conformation with moderate accuracy, even when the backbone RMSD from native is 6 Å. This opens up the possibility of genome scale screening of low resolution predicted structures for ligand binding.

While considerable progress has been made, there are significant challenges remaining. The generalized comparative modeling approach, GeneComp, needs to be extended so that it can treat highly homologous as well as analogous structures. Furthermore, given that *ab initio* folding algorithms quite often generate native-like structures, as also seen in generalized comparative modeling, development of better protein representations and energy functions that can select native folds from misfolded states is more crucial than ever. Clustering helps to reduce the problem by selecting representative folds, but routine unequivocal selection of native-like structures is not yet possible. It seems that the most promising approach is to convert the reduced models to full-atom models and then use either physics or knowledge-based energy functions to select the native structure. Use of active site descriptors can also help in this regard, because they act like a filter. Because of their utility in biochemical function assignment, better techniques for the construction of functionally relevant active sites is a must. Finally, while considerable progress has been made in the docking of known small-molecule ligands to low-resolution structures, methods must be developed that can identify such ligands, at the least by enriching the yield of true positives. Work in this direction is underway.

In conclusion, while techniques for the prediction of low-resolution structures have improved, they still have a way to go before structure prediction becomes routine. Nevertheless, this is a very laudable goal because low-resolution structures are of considerable utility both in the identification of biochemical function and in ligand docking. Such efforts will have to be applied on a genomic scale if structure-based approaches to function prediction are to play a role in the post genomic era. A number of such efforts are underway, and doubtless there will be more in the future.

Acknowledgment

This research was supported in part by NIH grants Nos. GM37408, GM48835, and RR-12255 and NSF grant No. 9986019. The contributions of Drs. Betancourt, Fetrow, Kihara, Ilkowski, Rotkiewicz, Wojciechowski, and N. Siew to some of the research described in this proposal are greatly appreciated. Andrzej Kolinski also acknowledges support by the University of Warsaw, grant No. 120-501/68-BW-1483/18/2000.

References

1. S. F. Altschul, W. Gish, W. Miller, et al., *J. Mol. Biol.* **215**, 403 (1990).

2. W. R. Pearson, *Methods Enzymol.* **266**, 227 (1996).

3. A. Bairoch, P. Bucher, and K. Hofmann, *Nucleic Acids Res.* **24**, 189 (1995).

4. S. Henikoff and J. G. Henikoff, *Genomics* **19**, 97 (1994).

5. T. K. Attwood, M. E. Beck, A. J. Bleasby, et al., *Nucleic Acids Res.* **22**, 3590 (1994).

6. T. K. Attwood, M. E. Beck, A. J. Bleasby, et al., *Nucleic Acids Res.* **25**, 212 (1997).

7. C. G. Nevill-Manning, T. D. Wu, and D. L. Brutlag, *Proc. Natl. Acad. Sci. USA* **95**, 5865 (1998).

8. J. S. Fetrow and J. Skolnick, *J. Mol. Biol.* **281**, 949 (1998).

9. L. Yu, J. V. White, and T. F. Smith, *Protein Sci.* **7**, 2499 (1998).

10. J. S. Fetrow, A. Godzik, and J. Skolnick, *J. Mol. Biol.* **282**, 703 (1998).

11. L. Zhang, A. Godzik, and J. Skolnick, et al., *Fold. Des.* **3**, 535 (1998).

12. J. S. Fetrow, N. Siew, and J. Skolnick, *FASEB J.* **13**, 1866 (1999).

13. N. Siew, J. Skolnick, and J. Fetrow, in preparation (2000).

14. B. Zhang, L. Rychlewski, and K. Pawlowski, et al., *Protein Sci.* **8**, 1104 (1999).

15. J. Skolnick and J. Fetrow, *TIBTECH* **18**, 34 (2000).

16. Z. Li and H. A. Scheraga, *Proc. Natl. Acad. Sci. USA* **84**, 6611 (1987).

17. J. Moult, T. Hubbard, K. Fidelis, et al., *Proteins Suppl.* **2** (1999).

18. P. A. Bates and M. J. Sternberg, *Proteins Suppl.* **47** (1999).

19. D. F. Burke, C. M. Deane, H. A. Nagarajaram, et al., *Proteins Suppl.* **55** (1999).

20. A. S. Yang and B. Honig, *Proteins Suppl.* **66** (1999).

21. R. L. Dunbrack, Jr., *Proteins Suppl.* **81** (1999).

22. D. Fischer, *Proteins Suppl.* **61** (1999).

23. R. Sanchez and A. Sali, *Proteins Suppl.* **50** (1997).

24. A. Sali, L. Potterton, F. Yuan, et al., *Proteins* **23**, 318 (1995).

25. T. Alwyn Jones and G. J. Kleywegt, *Proteins Suppl.* **30** (1999).

26. A. Zemla, C. Venclovas, and J. Moult, et al., *Proteins Suppl.* **22** (1999).

27. R. Sanchez and A. Sali, *Proc. Natl. Acad. Sci. USA* **95**, 13597 (1998).

28. A. Goffeau, B. G. Barrell, H. Bussey, et al., *Science* **274**, 546 (1996).

29. A. Elofsson and E. L. Sonnhammer, *Bioinformatics* **15**, 480 (1999).

30. R. Lathrop and T. F. Smith, *J. Mol. Biol.* **255**, 641 (1996).

31. R. T. Miller, D. T. Jones, and J. M. Thornton, *FASEB* **10**, 171 (1996).

32. A. Kolinski, P. Rotkiewicz, B. Ilkowski, et al., *Proteins* **37**, 592 (1999).

33. M. Wilmanns and D. Eisenberg, *Proc. Natl. Acad. Sci. USA* **90**, 1379 (1993).

34. D. Fischer, A. Elofsson, D. Rice, et al., *Pac. Symp. Biocomput.* **300** (1996).

35. A. R. Panchenko, A. Marchler-Bauer, and S. H. Bryant, *J. Mol. Biol.* **296**, 1319 (2000).

36. T.-M. Yi and E. S. Lander, *Protein Sci.* **3**, 1315 (1994).

37. Y. Matsuo and K. Nishikawa, *Protein Sci.* **3**, 2055 (1994).

38. D. T. Jones, *J. Mol. Biol.* **292**, 195 (1999).

39. K. K. Koretke, R. B. Russell, R. R. Copley, et al., *Proteins Suppl.* **141** (1999).

40. V. N. Maiorov and G. M. Crippen, *J. Mol. Biol.* **277**, 876 (1992).

41. A. Tropsha, R. K. Singh, I. I. Vaisman, et al., in *Pacific Symposium on Biocomputing '96*, L. Hunter and T. E. Klein, eds. World Scientific, Singapore, 1996, p. 614.

42. D. T. Jones, W. R. Taylor, and J. M. Thornton, *Nature* **358**, 86 (1992).

43. K. K. Koretke, Z. Luthey-Schulten, and P. G. Wolynes, *Protein Sci.* **5**, 1043 (1996).

44. S. H. Bryant and C. E. Lawrence, *Proteins* **16**, 92 (1993).

45. A. Godzik, J. Skolnick, and A. Kolinski, *J. Mol. Biol.* **227**, 227 (1992).

46. J. Selbig, *Protein Eng.* **8**, 339 (1995).

47. M. J. Sippl and S. Weitckus, *Proteins* **13**, 258 (1992).

48. M. Wilmanns and D. Eisenberg, *Protein Eng.* **8**, 626 (1995).

49. S. B. Needleman and C. D. Wunsch, *J. Mol. Biol.* **48**, 443 (1970).

50. J. U. Bowie, R. Luthy, and D. Eisenberg, *Science* **253**, 164 (1991).

51. R. Thiele, R. Zimmer, and T. Lengauer, *ISMB* **3**, 384 (1995).

52. A. G. Murzin, *Proteins* **37**, 88 (1999).

53. D. T. Jones, M. Tress, K. Bryson, et al., *Proteins Suppl.* **104** (1999).

54. M. Ota, T. Kawabata, A. R. Kinjo, et al., *Proteins Suppl.* **126** (1999).

55. F. S. Domingues, W. A. Koppensteiner, M. Jaritz, et al., *Proteins Suppl.* **112** (1999).

56. A. Panchenko, A. Marchler-Bauer, and S. H. Bryant, *Proteins Suppl.* **133** (1999).

57. J. Skolnick and D. Kihara, *Proteins*, in press (2000).

58. M. Levitt and A. Warshel, *Nature* **253**, 694 (1975).

59. M. Levitt, *J. Mol. Biol.* **104**, 59 (1976).

60. I. D. Kuntz, *J. Am. Chem. Soc.* **97**, 4362 (1975).

61. I. D. Kuntz, G. M. Crippen, P. A. Kollman, et al., *J. Mol. Biol.* **106**, 983 (1976).

62. A. T. Hagler and B. Honig, *Proc. Natl. Acad. Sci. USA* **75**, 554 (1978).

63. C. Wilson and S. Doniach, *Proteins* **6**, 193 (1989).

64. S. Sun, *Protein Sci.* **2**, 762 (1993).

65. A. Wallqvist and M. Ullner, *Proteins* **18**, 267 (1994).

66. D. Hoffmann and E. W. Knapp, *Eur. Biophys. J.* **24**, 387 (1996).

67. D. Hoffmann and E. W. Knapp, *Phys. Rev. E* **53**, 4221 (1996).

68. J. T. Pedersen and J. Moult, *Proteins Suppl.* **179**, 1 (1997).

69. N. Gō and H. Taketomi, *Proc. Natl. Acad. Sci. USA* **75**, 559 (1978).

70. W. R. Krigbaum and A. Komoriya, *Biochim. Biophys. Acta* **576**, 204 (1979).

71. W. R. Krigbaum and S. F. Lin, *Macromolecules* **15**, 1135 (1982).

72. A. Kolinski and J. Skolnick, *Proc. Natl. Acad. Sci. USA* **83**, 7267 (1986).

73. A. Kolinski, J. Skolnick, and R. Yaris, *J. Chem. Phys.* **85**, 3585 (1986).

74. A. Kolinski, J. Skolnick, and R. Yaris, *Biopolymers* **26**, 937 (1987).

75. J. Skolnick, A. Kolinski, and R. Yaris, *Proc. Natl. Acad. Sci. USA* **85**, 5057 (1988).

76. J. Skolnick and A. Kolinski, *Annu. Rev. Phys. Chem.* **40**, 207 (1989).

77. J. Skolnick, A. Kolinski, and R. Yaris, *Biopolymers* **28**, 1059 (1989).

78. J. Skolnick, A. Kolinski, and R. Yaris, *Proc. Natl. Acad. Sci. USA* **86**, 1229 (1989).

79. J. Skolnick and A. Kolinski, *J. Mol. Biol.* **212**, 787 (1990).

80. J. Skolnick and A. Kolinski, *Science* **250**, 1121 (1990).

81. A. Kolinski, M. Milik, and J. Skolnick, *J. Chem. Phys.* **94**, 3978 (1991).

82. A. Kolinski and J. Skolnick, *J. Phys. Chem.* **97**, 9412 (1992).

83. A. Kolinski and J. Skolnick, *Proteins* **18**, 338 (1994).

84. A. Kolinski, M. Milik, J. Rycombel, et al., *J. Chem. Phys.* **103**, 4312 (1995).

85. A. Sikorski and J. Skolnick, *Biopolymers* **28**, 1097 (1989).

86. A. Sikorski and J. Skolnick, *Proc. Natl. Acad. Sci. USA* **86**, 2668 (1989).

87. A. Sikorski and J. Skolnick, *J. Mol. Biol.* **215**, 183 (1990).

88. A. Sikorski and J. Skolnick, *J. Mol. Biol.* **212**, 819 (1990).

89. H. S. Chan and K. A. Dill, *J. Chem. Phys.* **92**, 492 (1989).

90. H. S. Chan and K. A. Dill, *Macromolecules* **22**, 4559 (1989).

91. H. S. Chan and K. A. Dill, *Proc. Natl. Acad. Sci. USA* **87**, 6388 (1990).

92. H. S. Chan and K. A. Dill, *Annu. Rev. Biophys. Biophys. Chem.* **20**, 447 (1991).

93. K. A. Dill, *Biochemistry* **24**, 1501 (1985).

94. K. A. Dill, D. O. V. Alonso, and K. Hutchinson, *Biochemistry* **28**, 5439 (1989).

95. K. A. Dill, *Curr. Biol.* **3**, 99 (1993).

96. K. A. Dill, S. Bromberg, K. Yue, et al., *Protein Sci.* **4**, 561 (1995).

97. A. Sali, E. Shakhnovich, and M. Karplus, *J. Mol. Biol.* **235**, 1614 (1994).

98. A. Sali, E. Shakhnovich, and M. Karplus, *Nature* **369**, 248 (1994).

99. E. I. Shakhnovich and A. V. Finkelstein, *Biopolymers* **28**, 1667 (1989).

100. E. I. Shakhnovich and A. M. Gutin, *Biophys. Chem.* **34**, 187 (1989).

101. E. I. Shakhnovich, G. Farztdinov, and A. M. Gutin, *Phys. Rev. Lett.* **67**, 1665 (1991).

102. E. I. Shakhnovich and A. M. Gutin, *Proc. Natl. Acad. Sci. USA* **90**, 7195 (1993).

103. E. I. Shakhnovich and A. M. Gutin, *Protein Eng.* **6**, 793 (1993).

104. E. I. Shakhnovich, *Phys. Rev. Lett.* **72**, 3907 (1994).

105. E. I. Shakhnovich, *Fold. Des.* **1**, R50 (1996).

106. A. R. Dinner, A. Sali, M. Karplus, et al., *J. Chem. Phys.* **101**, 1444 (1994).

107. A. R. Dinner, A. Sali, and M. Karplus, *Proc. Natl. Acad. Sci. USA* **93**, 8356 (1996).

108. M.-H. Hao and H. A. Scheraga, *J. Phys. Chem.* **98**, 4940 (1994).

109. M.-H. Hao and H. A. Scheraga, *J. Phys. Chem.* **98**, 9882 (1994).

110. M.-H. Hao and H. A. Scheraga, *J. Chem. Phys.* **102**, 1334 (1995).

111. A. Kolinski and P. Madziar, *Biopolymers* **42**, 537 (1997).

112. V. G. Dashevskii, *Mol. Biol.* (translation from) **14**, 105 (1980).

113. D. G. Covell, *Proteins* **14**, 409 (1992).

114. D. G. Covell and R. L. Jernigan, *Biochemistry* **29**, 3287 (1990).

115. D. A. Hinds and M. Levitt, *Proc. Natl. Acad. Sci. USA* **89**, 2536 (1992).

116. A. Kolinski and J. Skolnick, *Lattice Models of Protein Folding, Dynamics and Thermodynamics*, R. G. Landes, Austin, TX, 1996.

117. J. Skolnick and A. Kolinski, *J. Mol. Biol.* **221**, 499 (1991).

118. J. Skolnick, A. Kolinski, C. Brooks III, et al., *Curr. Biol.* **3**, 414 (1993).

119. A. Kolinski and J. Skolnick, *Acta Biochim. Polon.* **44**, 389 (1998).

120. A. Kolinski, P. Rotkiewicz, and J. Skolnick, in *Monte Carlo Approaches to Biopolymers and Protein Folding*, P. Grassberger, G. T. Barkema, and W. Nadler, eds., World Scientific, Singapore, 1998, p. 110.

121. A. Kolinski, A. Godzik, and J. Skolnick, *J. Chem. Phys.* **98**, 7420 (1993).

122. A. Kolinski and J. Skolnick, *Proteins* **18**, 353 (1994).

123. A. Kolinski and J. Skolnick, *Proteins* **18**, 338 (1994).

124. M. Vieth, A. Kolinski, C. L. Brooks III, et al., *J. Mol. Biol.* **1994**, 361 (1994).

125. M. Vieth, A. Kolinski, I. Brooks, C. L., et al., *J. Mol. Biol.* **251**, 448 (1995).

126. M. Vieth, A. Kolinski, and J. Skolnick, *Biochemistry* **35**, 955 (1996).

127. J. Moult, T. Hubbard, K. Fidelis, et al., *Proteins Suppl.* **3**, 2 (1999).

128. K. T. Simons, R. Bonneau, I. Ruczinski, et al., *Proteins Suppl.* **3**, 171 (1999).

129. B. Rost and C. Sander, *J. Mol. Biol.* **232**, 584 (1993).

130. B. Rost and C. Sander, *Proteins* **19**, 55 (1994).

131. B. Rost and C. Sander, *Proteins* **23**, 295 (1996).

132. D. T. Jones, *Proteins Suppl.* **185** (1997).

133. A. R. Ortiz, A. Kolinski, P. Rotkiewicz, et al., *Proteins Suppl.* **3**, 177 (1999).

134. D. Kihara, H. Lui, A. Kolinski, and J. Skolnick, *Proc. Natl. Acad. Sci. USA* **14**, 14 (2001).

135. A. R. Ortiz, W.-P. Hu, A. Kolinski, et al., in *Pacific Symposium on Biocomputing '97*, R. B. Altman, A. K. Dunker, L. Hunter, and T. E. Klein, eds., World Scientific, Singapore, 1997, p. 316.

136. A. R. Ortiz, A. Kolinski, and J. Skolnick, *Proc. Natl. Acad. Sci. USA* **95**, 1020 (1998).

137. A. R. Ortiz, A. Kolinski, and J. Skolnick, *J. Mol. Biol.* **277**, 419 (1998).

138. A. R. Ortiz, A. Kolinski, and J. Skolnick, *Proteins* **30**, 287 (1998).

139. A. Kolinski and S. A., *Proteins* **32**, 475 (1998).

140. A. Kolinski, P. Rotkiewicz, B. Ilkowski, et al., *Progress of Theoretical Physics (Kyoto) Suppl.* **138**, 292 (2000).

141. D. J. Osguthorpe, *Proteins Suppl.* **3**, 186 (1999).

142. R. Samudrala, H. Xia, E. Huang, et al., *Proteins Suppl.* **3**, 194 (1999).

143. D. Hinds and M. Levitt, *J. Mol. Biol.* **243**, 668 (1994).

144. B. Park, E. Huang, and M. Levitt, *J. Mol. Biol.* **266**, 831 (1997).

145. E. Huang, R. Samudrala, and J. Ponder, *Protein Sci.* **7**, 1998 (1998).

146. J. Lee, A. Liwo, D. R. Ripoll, et al., *Proteins Suppl.* **3**, 204 (1999).

147. J. Lee, A. Liwo, and H. A. Scheraga, *Proc. Natl. Acad. Sci. USA* **96**, 2025 (1999).

148. A. Liwo, R. Kazimierkiewicz, C. Czaplewski, et al., *J. Comput. Chem.* **19**, 259 (1988).

149. D. R. Ripoll, A. Liwo, and H. A. Scheraga, *Biopolymers* **46**, 117 (1988).

150. A. Baumgaertner, in *The Monte Carlo Method in Condensed Matter Physics*, K. Binder, ed., Springer, Heidelberg, 1995.

151. A. Rey and J. Skolnick, *Chem. Phys.* **158**, 199 (1991).

152. J. Skolnick and A. Kolinski, in *Computer Simulations of Biomolecular Systems. Theoretical and Experimental Studies*, W. F. van Gunsteren, P. K. Weiner, and A. J. Wilkinson, eds., ESCOM Science Publishers, 1996.

153. A. Kolinski, B. Ilkowski, and J. Skolnick, *Biophys. J.* **77**, 2942 (1999).

154. U. H. E. Hansmann and Y. Okamoto, *J. Comput. Chem.* **14**, 1333 (1993).

155. U. H. E. Hansmann and Y. Okamoto, *J. Chem. Phys.* **110**, 1267 (1999).

156. A. Kolinski, W. Galazka, and J. Skolnick, *Proteins* **26**, 271 (1996).

157. A. Kolinski, W. Galazka, and J. Skolnick, *J. Chem. Phys.* **108**, 2608 (1998).

158. C. B. Anfinsen, *Science* **181**, 223 (1973).

159. H. A. Scheraga, *Biophys. Chem.* **59**, 329 (1996).

160. L. Piela, J. Kostrowicki, and H. A. Scheraga, *J. Phys. Chem.* **93**, 3339 (1989).

161. H. A. Scheraga and M.-H. Hao, *Adv. Chem. Phys.* **105**, 243 (1999).

162. Z. Li and H. A. Scheraga, *Proc. Natl. Acad. Sci. USA* **84**, 6611 (1987).

163. A. A. Rabow and H. A. Scheraga, *Protein Sci.* **5**, 1800 (1996).

164. T. Dandekar and P. Argos, *J. Mol. Biol.* **256**, 645 (1996).

165. Z. Sun, X. Xia, Q. Guo, et al., *J. Protein Chem.* **18**, 39 (1999).

166. U. H. E. Hansmann and Y. Okamoto, *J. Comput. Chem.* **18**, 920 (1997).

167. U. H. E. Hansmann and Y. Okamoto, *Curr. Opin. Struct. Biol.* **9**, 177 (1999).

168. D. Gront, A. Kolinski, and J. Skolnick, *J. Chem. Phys.* **113**, 5065 (2000).

169. R. H. Swedensen and J. S. Wang, *Phys. Rev. Lett.* **57**, 2607 (1986).

170. S. F. Altschul and E. V. Koonin, *Trends Biochem Sci.* **23**, 444 (1998).

171. F. C. Bernstein, T. F. Koetzle, G. J. B. Williams, et al., *J. Mol. Biol.* **112**, 535 (1977).

172. J. G. Henikoff and S. Henikoff, *Methods Enzymol.* **266**, 88 (1996).

173. A. Ogiwara, I. Uchiyama, T. Takagi, et al., *Protein Sci.* **5**, 1991 (1996).

174. C. Ouzounis, C. Sander, M. Scharf, et al., *J. Mol. Biol.* **232**, 805 (1993).

175. W. R. Pearson, *Methods Mol. Biol.* **24**, 307 (1994).

176. W. R. Pearson, *J. Mol. Biol.* **276**, 71 (1998).

177. J. D. Thompson, D. G. Higgins, and T. J. Gibson, *Nucleic Acids Res.* **22**, 4673 (1994).

178. J. Skolnick, A. Kolinski, and A. Ortiz, *Proteins* **38**, 3 (2000).

179. UCLA, http://www.doembi.ucla.edu/people/fischer/BENCH/table1.html, Los Angeles, 1996.

180. M. S. Waterman and M. Eggert, *J. Mol. Biol.* **197**, 723 (1987).

181. L. Jaroszewski, L. Rychlewski, B. Zhang, et al., *Protein Sci.* **7**, 1431 (1998).
182. S. F. Altschul, T. L. Madden, A. A. Schaffer, et al., *Nucleic Acids Res.* **25**, 3389 (1997).
183. J. Skolnick, L. Jaroszewski, A. Kolinski, et al., *Protein Sci.* **6**, 676 (1997).
184. C. M. Fraser, J. D. Gocayne, O. White, et al., *Science* **270**, 397 (1995).
185. D. Fischer and D. Eisenberg, *Proc. Natl. Acad. Sci. USA* **94**, 11929 (1997).
186. S. A. Teichmann, C. Chothia, and M. Gerstein, *Curr. Opin. Struct. Biol.* **9**, 390 (1999).
187. M. Gerstein, *Proteins* **33**, 518 (1998).
188. D. T. Jones, *J. Mol. Biol.* **287**, 797 (1999).
189. F. R. Blattner, G. Plunkett, 3rd, C. A. Bloch, et al., *Science* **277**, 1453 (1997).
190. L. Rychlewski, B. Zhang, and A. Godzik, *Protein Sci.* **8**, 614 (1999).
191. A. R. Ortiz, A. Kolinski, and J. Skolnick, *J. Mol. Biol.* **277**, 419 (1998).
192. A. R. Ortiz and Skolnick, *Biophys. J.* **79**, 1787 (2000).
193. A. Ortiz, A. Kolinski, P. Rotkiewicz, et al., *Proteins Suppl.* **3**, 177 (1999).
194. A. Aszodi and W. R. Tylor, *Fold. Des.* **1**, 325 (1996).
195. A. Sali and T. L. Blundel, *J. Mol. Biol.* **234**, 779 (1993).
196. L. Jaroszewski, K. Pawlowski, and A. Godzik, *J. Mol. Modelling* **00**, 000 (1998).
197. M. Betancourt and J. Skolnick, *J. Comput. Chem.* **22**, 339 (2001).
198. A. Kolinski, L. Jaroszewski, P. Rotkiewicz, et al., *J. Phys. Chem.* **102**, 4628 (1998).
199. A. Kolinski, P. Rotkiewicz, B. Ilkowski, et al., *Proteins* **37**, 592 (1999).
200. M. Feig, P. Rotkiewicz, A. Kolinski, et al., *Proteins* **41**, 86 (2000).
201. E. S. Huang, P. Koehl, M. Levitt, et al., *Proteins* **33**, 204 (1998).
202. J. Skolnick, A. Kolinski, and A. R. Ortiz, *Proteins* **38**, 3 (2000).
203. A. Sali, L. Potterton, F. Yuan, H. van Vlijmen, and M. Karplus, Evaluation of comparative protein modeling by MODELLER, *Proteins* **23,** 318–326 (1995).
204. H. Lu and J. Skolnick, *Proteins* **44**, 223 (2001).
205. W. Kabsch and C. Sander, *Biopolymers* **22**, 2577 (1983).
206. C. Chothia and A. V. Finkelstein, *Annu. Rev. Biochem.* **59**, 1007 (1990).
207. B. R. Brooks, R. Bruccoleri, B. Olafson, et al., *J. Comput. Chem.* **4**, 187 (1983).
208. C. Simmerling, M. Lee, A. R. Ortiz, et al., *J. Am. Chem. Soc.* **122**, 8392 (2000).
209. D. Mohanty, A. Kolinski, and J. Skolnick, *Biophys. J.* **77**, 54 (1999).
210. D. Mohanty, B. N. Dominy, A. Kolinski, et al., *Proteins* **35**, 447 (1999).
211. A. G. Murzin, S. E. Brenner, T. Hubbard, et al., *J. Mol. Biol.* **247**, 536 (1995).
212. C. A. Orengo, A. D. Michie, S. Jones, et al., *Structure* **5**, 1093 (1997).
213. A. M. Lesk, C.-I. Branden, and C. Chothia, *Proteins* **5**, 139 (1989).
214. G. K. Farber and G. A. Petsko, *Trends Biochem. Sci.* **15**, 228 (1990).
215. H. Hegyi and M. Gerstein, *J. Mol. Biol.* **288**, 147 (1999).
216. D. Fischer, H. Wolfson, S. L. Lin, et al., *Protein Sci.* **3**, 769 (1994).
217. A. Kasuya and J. M. Thornton, *J. Mol. Biol.* **286**, 1673 (1999).
218. C. D. Coldren, H. W. Hellinga, and J. P. Caradonna, *Proc. Natl. Acad. Sci. USA* **94**, 6635 (1997).
219. A. L. Pinto, H. W. Hellinga, and J. P. Caradonna, *Proc. Natl. Acad. Sci. USA* **94**, 5562 (1997).
220. H. W. Hellinga and F. M. Richards, *J. Mol. Biol.* **222**, 763 (1991).
221. H. W. Hellinga, J. P. Caradonna, and F. M. Richards, *J. Mol. Biol.* **222**, 787 (1991).

222. M. Klemba and L. Regan, *Biochemistry* **34**, 10094 (1995).

223. M. Klemba, K. H. Gardner, S. Marino, et al., *Nature Struct. Biol.* **2**, 368 (1995).

224. E. Farinas and L. Regan, *Protein Sci.* **7**, 1939 (1998).

225. M. W. Crowder, J. D. Stewart, V. A. Roberts, et al., *J. Am. Chem. Soc.* **117**, 5627 (1995).

226. S. Halfon and C. S. Craik, *J. Am. Chem. Soc.* **118**, 1227 (1996).

227. A. C. Wallace, R. A. Laskowski, and J. M. Thornton, *Protein Sci.* **5**, 1001 (1996).

228. G. J. Kleywegt, *J. Mol. Biol.* **285**, 1887 (1999).

229. A. C. Wallace, N. Birkakoti, and J. M. Thornton, *Protein Sci.* **6**, 2308 (1997).

230. R. B. Russell, *J. Mol. Biol.* **279**, 1211 (1998).

231. K. F. Han, C. Bystroff, and D. Baker, *Protein Sci.* **6**, 1587 (1997).

232. P. J. Artymiuk, A. R. Poirrette, H. M. Grindley, et al., *J. Mol. Biol.* **236**, 327 (1994).

233. S. Karlin and Z. Y. Zhu, *Proc. Natl. Acad. Sci. USA* **93**, 8344 (1996).

234. J. Fetrow, personal communication (2000).

235. E. E. Abola, F. C. Bernstein, S. H. Bryant, et al., *Protein Data Bank in Crystallographic Databases—Information Content, Software Systems, Scientific Application*, Data Commission of the International Union of Crystallography, Bonn/Cambridge/Chester, 1987.

236. S. Henikoff, J. G. Henikoff, and S. Pietrokovski, *Bioinformatics* **15**, 471 (1999).

237. J. G. Henikoff, S. Henikoff, and S. Pietrokovski, *Nucleic Acids Res.* **27**, 226 (1999).

238. F. R. Blattner, G. Plunkett, 3rd, C. A. Bloch, et al., *Science* **277**, 1453 (1997).

239. F. Kunst and N. Ogasawara and I. Moszer, et al., *Nature* **390**, 249 (1997).

240. I. A. Vakser, O. G. Matar, and C. F. Lam, *Proc. Natl. Acad. Sci. USA* **96**, 8477 (1999).

241. L. Rychlewski, L. Jaroszewski, W. Li, et al., *Protein Sci.* **9**, 232 (2000).

242. C. Bystroff and D. Baker, *J. Mol. Biol.* **281**, 565 (1998).

243. K. T. Simons, R. Bonneau, I. Ruczinski, et al., *Proteins Suppl.* **171** (1999).

244. J. Skolnick, J. S. Fetrow, and A. Kolinski, *Nature Biotech.* **18**, 283 (2000).

245. T. Chiu and J. Skolnick, unpublished results (2000).

246. J. G. Henikoff, S. Pietrokovski, C. M. McCallum, et al., *Electrophoresis* **21**, 1700 (2000).

247. A. A. Schaffer, Y. I. Wolf, C. P. Ponting, et al., *Bioinformatics* **15**, 1000 (1999).

KNOWLEDGE-BASED PREDICTION OF PROTEIN TERTIARY STRUCTURE

PIERRE-JEAN L'HEUREUX, BENOIT CROMP, AND ÉRIC MARTINEAU

*Département de Chimie, Université de Montréal, Montréal, Québec, Canada;
Centre de Recherche en Calcul Appliqué, Montréal, Québec, Canada;
and Protein Engineering Network of Centers of Excellence,
Edmonton, Alberta, Canada*

JOHN R. GUNN

*Schrödinger, Inc., New York, NY, U.S.A.; Centre de Recherche en Calcul
Appliqué, Montréal, Québec, Canada; and Protein Engineering Network
of Centers of Excellence, Edmonton, Alberta, Canada*

CONTENTS

*Computational Methods for Protein Folding: A Special Volume of Advances in Chemical Physics,
Volume 120,* Edited by Richard A. Friesner. Series Editors I. Prigogine and Stuart A. Rice.
ISBN 0-471-20955-4. © 2002 John Wiley & Sons, Inc.

I. INTRODUCTION

Under the general heading of "protein folding" there is an ever-increasing body of methodology that has been rapidly evolving over the past few years. The simply stated objective of computationally determining the three-dimensional atomic coordinates of a protein starting from knowledge of the amino acid sequence remains a somewhat idealistic academic challenge, but it has led to the development of a technology base that is gaining in practical applicability. This corresponds to some extent to a shift in philosophy in which a fundamental understanding of the folding process is of less immediate interest than obtaining the best model possible with whatever means are available. Fundamental questions are of course still important and are being actively pursued [1–5], but the field is being driven more and more by the pragmatic approach [6,7]. This is highlighted by the effort being devoted to the CASP experiments, where the emphasis is placed squarely on the bottom line [8]. In this context, the methods used must be tailored to the particular problem at hand, and no available information can be left unused. Much work therefore has been devoted to making use of prior information and accumulated knowledge in the generation of computer models of proteins. This review will describe some of the ways in which such methods are being incorporated within the traditional *ab initio* framework.

A. The Knowledge-Based Approach

The label of "knowledge-based" is to some extent artificial, in that there is a spectrum of methodologies and it is not always easy to draw a clear distinction. The intended contrast is with a purist's *ab initio* approach in which one seeks a numerical solution to the fundamental laws of physics (as one would like to do in quantum chemistry) with no theoretical limit on the problems that can be addressed. A knowledge-based method, on the other hand, requires some form of *a priori* knowledge and is therefore limited in its applicability by the data that are available. If the term is used in its broadest sense, referring to methods that make explicit use of the Protein Data Bank (PDB) of known structures, this would still cover a range extending from methods which require there to be a similar structure in the PDB to those that apply observed patterns in a more general way. In principle, this includes virtually all methods because even the most determined *ab initio* practitioner still has recourse to an empirical force field that typically uses the PDB in its parameterization [9]. Even though such force fields are as general as possible, the reliance on the PDB does represent a real limitation, as anyone who has ever tried to use one to fold a membrane protein can attest.

In the context of the CASP experiments [8], the distinction is drawn between *ab initio* and "fold recognition" predictions, but there as well some overlap occurs [7]. Fold recognition often involves some refinement to model parts of the structure not found by homology, and conversely many *ab initio* methods make some use of structural fragments from the PDB. It is precisely this middle ground where the different categories are converging that is of interest and where much recent success can be found. It has become clear that there is a great deal of information to be had in the PDB and that progress is being made by extending the ways in which it can be used. The knowledge-based approach is therefore to develop methods to take advantage of what is there, even if the underlying physical principles are not fully understood.

B. Recent Trends

One of the patterns that has emerged from the CASP experiments is the relative success of the fold-recognition methods in identifying distant homologies, even in some cases where none was originally thought to exist [8]. Until recently, *ab initio* methods lagged far behind, but significant progress is now being made [7]. As mentioned above, however, this is coming from knowledge-based methods that have incorporated some of the methods that have proven successful in comparative modeling and fold recognition. It has been shown that so-called "hybrid" methods can outperform more traditional fold-recognition and *ab initio* techniques [10]. The more general methods of fold-recognition have also been shown to outperform direct homology modeling in cases of weak homology [11], suggesting that a flexible approach has the potential to cover a broad spectrum of possible targets.

Another pattern that is emerging is an increased recognition that the PDB can be used to identify structural motifs at different scales, not just individual residues (as used to derive contact potentials) or entire domains (as used in traditional fold recognition). Much recent work has gone into using the PDB to develop databases of smaller fragments which can be used to construct protein models [12], and an approach based on using local homology with a fragment library has been shown to be quite successful at generating new folds [13]. This building-block approach has also been used to generate improved sensitivity and more accurate alignments when applied to fold recognition [14].

The trend toward a more generalized approach is also reflected in recent work on scoring functions. It has been shown that traditional empirical potentials perform poorly at discriminating the correct structure [15] and that the functional form of pairwise contact energies is not even sufficient in principle [16]. The importance of evolutionary relationships has also been established, and information from multiple sequences can be used to improve recognition of misfolded structures [17]. This idea has led to the use of conformational tendencies and contact predictions from multiple sequence alignments [18] and the development of scoring functions which take into account sequence homology [19]. Scoring functions can therefore be constructed as a set of complementary components: contributions that are unique to a given sequence, those that depend on a family or class of sequences, and those that apply to all proteins.

C. Practical Considerations

The bottom line in structure prediction is to provide a useful answer to a question that is actually being posed. *Ab initio* predictions alone are rarely accurate enough to be useful; however, as NMR spectroscopy is being used to obtain structures for larger and larger proteins, there is a great practical benefit in using computational methods to aid in this process. Structure prediction methods, when coupled with experimental data, can be used to obtain higher-quality structures [20] and even to help in interpreting and assigning the spectra [21]. For this reason there is a great interest in developing methods that can make the best use of various types of experimental data (often in the form of constraints) in addition to that gleaned from the PDB.

The enormous progress that has been made in genome sequencing has also led to increased efforts in functional genomics; that is, it has enabled the use of prediction techniques to assign probable functions to newly discovered sequences [22]. In this case, the emphasis is less on obtaining accurate coordinates and more on being able to detect weak homologies in distantly related families of structures. Improved prediction methods therefore have an important role in improving the sensitivity of fold-recognition techniques, providing better alignments, and ultimately allowing weaker relationships to be detected thereby classifying more of the genome. Even though the protein folding problem may still be a long way from being fully solved, there is a great opportunity for knowledge-based methods to have a significant impact in improving structure prediction's bottom line.

II. PROTEIN MODELING

The most direct approach to modeling protein folding would be to carry out a simulation that replicates the actual folding process as it occurs in nature. Although some progress has been made in pursuing that approach [23–25], it remains impractical in most cases for two reasons: The time scale of the folding transition for moderately sized proteins exceeds that which can be attained in simulations, and the physical forces involved are not modeled with sufficient accuracy to ensure the desired outcome. Because highly simplified models are unsuitable for predicting structural details, a different point of view is needed to carry out tractable simulations of realistic models. If one is not interested in the thermodynamics of folding and wishes only to produce the folded structure, any number of nonphysical buildup or pattern-generation techniques could be imagined; however, many methods retain the basic model of a molecular simulation, albeit with a number of simplifying approximations.

A. The Computational Model

The principal simulation paradigm is based on the thermodynamic hypothesis, namely that the equilibrium structure corresponds to the global minimum of the thermodynamic free energy. Whether or not this is strictly true for a given sequence is not known; however, for the purposes of the simulation it is generally assumed that some sort of energy-like function can in principle be constructed for which the native structure is a minimum. This can be thought of as some sort of modified free energy or as a purely empirical scoring function; either way the mathematical problem is the same, namely to find the global minimum. The general problem of global minimization is nontrivially difficult, and therefore additional approximations are required in order to obtain a solution in a reasonable time. The thermodynamic analogy is often used to model this as an annealing process; however, in general any minimization method can be applied.

In its general formulation, a simulation within the framework of global function minimization consists of three basic elements. As mentioned above, the target function of the minimization must be defined so as to allow comparison of different possible structures. Secondly, there must be a procedure to search through the possible conformations in order to find the global minimum (or other acceptable solution). Finally, the conformational space—that is, the range of conformations that can be constructed and the means to transform one conformation into another—must be specified in order to constrain the search. Clearly, these elements are not independent and must fit together in order to form a coherent model. For example, an energy function need not evaluate a conformation that is not part of the allowable space. Nonetheless, each of the three components offers a different means to incorporate empirical information into the simulation.

B.   Geometrical Representations

In order to reduce the number of degrees of freedom, most simulations use a reduced model description of the protein in which only a subset of the atoms are present. There are many variations on this theme, most of which have been previously reviewed [26]. The most common approach is to represent the main-chain $N–C_\alpha–C'$ atoms explicitly, with the side chain either being represented by the C_β atom or by an extended model atom corresponding to the approximate center of mass of the side chain. The bond distances and bond angles are usually fixed to standard values, thereby leaving the backbone dihedral angles ϕ and ψ as the only degrees of freedom (with the conjugated peptide dihedral angle fixed at 180°, in some cases allowing 0° as well for proline residues). The dihedral angles can either be restricted to a limited number of allowable conformations or be allowed to continuously vary within a specified region, and both of these approaches have been explored in our group and others.

Another method we are currently developing divides the molecule into segments based on the assigned secondary structure. The relative positions of the segments and the positions of the residues within each segment are optimized in distinct steps, thereby allowing the overall topology to evolve using a long-range potential with the detailed atomic coordinates to be adapted accordingly. The protein backbone is initially not required to be continuous from one segment to the next; and each segment can be deformed as the topology changes, creating unnatural bond lengths and angles. The correct covalent connectivity, rather than being rigid from the start, is gradually annealed in using a special constraint potential during the course of the simulation.

The details of side-chain conformation are generally determined by local interactions and have relatively little influence on the overall topology of the fold. Methods have been developed to assign probable side-chain conformations based on backbone dihedral angles and observed preferences in the PDB, and this technique has been shown quite effective in correctly placing side-chain atoms on a fixed backbone [27]. The task becomes more difficult if there are significant deviations in the backbone, because the details of the side-chain contacts will no longer be the same [28]. In a recent approach, the side-chain conformations are represented by specifying a distribution of discrete rotamer states without actually including any additional coordinates. The ability of the backbone conformation to adequately accomodate the side chains can be eva-luated using a rotamer-dependent mean-field energy and a conformational entropy [29].

C. Search Algorithms

The most common minimization technique is based on the principle of simulated annealing, which involves generating an ensemble of structures which is slowly converged toward the lowest-energy region of the conformational space. This method requires that the conformational sampling be able to avoid becoming trapped in a local minimum, and a number of techniques have been developed to overcome this problem [9,30]. Other successful approaches include using a branch-and-bound algorithm to limit the scope of local searches [31], as well as combining discrete Monte Carlo trial moves with local gradient minimizations [32].

Lattice models have also been used in order to discretize the conformational space in three dimensions. A relatively fine-grained model can be searched using methods similar to those described above [33], or a coarser model can be used to generate a set of possible topologies which can then be further refined using a more detailed model [34]. Further refinement can be carried out by using consensus inter-residue contacts from simulations to generate new structures that attempt to reproduce as many as possible [35,36]. Searches can even be carried out directly in terms of inter-residue contacts and then used to generate three-dimensional coordinates [37]. Another means to simplify the conforma-tional search is to increase the range of the potential interactions during the simulation in order to build up larger-scale features of the structure [38].

Our approach is the hierarchical algorithm [39,40], in which trial moves are generated and evaluated in three different steps. At the simplest level, segments of three residues (triplets) are generated by choosing three sets of (ϕ, ψ) values at random from an allowed list. Each triplet is immediately accepted or rejected according to whether or not the orientation of its endpoints falls into an allowed region of triplet conformational space. The second level consists of complete loop segments as determined by the secondary structure. These loops are evolved from previously existing structures by using the set of triplets from the first level as trial moves and by evaluating new loops based on the difference in overall geometry from the starting loop. The final level then corresponds to the entire molecule, for which the trial moves consist of substituting entire loops with the new loops generated in the second level. It is only at this final level that the structure is evaluated by calculating the full scoring function, which is then minimized using a genetic algorithm consisting of separate mutation, hybridization, and selection steps.

D. Scoring Functions

In most current prediction methods, the objective of the scoring function is not to reproduce the physical properties of the system, but to provide the best possible recognition of the native structure. These functions can be parameterized strictly on a statistical basis to optimize their performance [41]. Although there is some correlation between statistical potentials and those developed from physical principles [42], the former generally provide better results for predictions [43]. The energetic point of view is often used to motivate the development of a scoring function, but in practice the goal is simply to evaluate the relative probability that a given structure corresponds to a real protein. A typical energy can be defined as

$$E = \sum_{ij} E_{ij}$$

where the pairwise residue–residue energy is

$$E_{ij} = -kT_0 \ln P_{ij}(r_{ij})$$

and P_{ij} is the relative probability of finding residue pair i–j at a distance r_{ij}. If one then uses the Metropolis test to accept or reject a trial move from initial energy E_i to final energy E_f according to the value of $\exp(-(E_f - E_i)/kT)$, the same algorithm could be equivalently formulated in terms of accepting moves with a probability of $(P_f/P_i)^{\alpha}$, where $\alpha = T/T_0$ and

$$P = \prod_{ij} P_{ij}(r_{ij})$$

In principle, one could try to maximize the probability, its logarithm, or for that matter any other monotonic function of it.

Empirical scoring functions generally consist of multiple components, both sequence-independent and sequence-dependent [44,45]. The former include terms to control the overall size and shape of the molecule, as well as characteristic features of local structure depending on the geometrical model being used, whereas the latter take into account the specific interactions among residues. Some scoring functions are based on physical principles, such as electrostatic interactions [38] and van der Waals forces [46], with additional parameterization based on the PDB. The most common type of scoring function, however, is based directly on observed distances between different amino acid pairs in the PDB, and it is formulated as a table of (possibly distance-dependent) pairwise contact probabilities between amino acid types [47,48]. They differ mainly in the functional form to which they are fit, as well as in the details of the normalization of the probabilities, which is a nontrivial task for a heterogeneous data set like the PDB [48,49]. The scoring functions used in our group are of this type, the details of which have been published elsewhere [32,39].

Additional specificity can be built into the scoring function in several ways. Specialized pattern-recognition and multibody terms can be included to generate more realistic secondary and supersecondary structural motifs [45,50]. The secondary structure can also be explicitly taken into account when calculating residue contact probabilities, in order to distinguish interactions between amino acids in different secondary-structure units [51]. In a more sophisticated approach, the local sequence homology is used to adjust the statistics for a particular target sequence [19]. The trend toward more explicit pattern recognition and sequence specificity in the generation of scoring functions allows more of the subtle homologies in the PDB to be exploited, although some chemical insight is still required to express it in an appropriate functional form.

III.   CONSTRAINT METHODS

Constraints provide a very direct means to add information to a simulation— simply requiring all generated structures to satisfy certain additional conditions. This approach has been used extensively to generate three-dimensional structures from NMR spectra [52], which provide data in the form of inter- atomic distances. In principle, if one had enough distance constraints, the problem would be overdetermined and could be solved mathematically with no further information required. It has been shown, however, that the use of knowledge-based simulations based on homologous structures or fragment libraries from the PDB provides more accurate models than constraint-based methods alone [20,53].

In the case where the constraints alone are insufficient to determine the structure, they can still be used to supplement energy-based simulations. The goal in this case is to make the most effective use of the constraint information and to obtain good results with a minimum of additional information required. Because the source of the constraints is typically experimental spectra that must be assigned and interpreted, or theoretical methods (such as multiple sequence alignments) that may be incorrect, it is also important to take into account errors especially in difficult cases where the input data is incomplete or uncertain. Under these conditions, the constraints can be regarded as an additional component of the scoring function, expressing the probabilities of different structures, rather than as a rigid requirement. In many implementations, these interpretations are in fact equivalent.

A.   Types of Constraints

Constraints can in principle be applied to any property of the structure where some sort of prefered value can be determined; however, the most common are those that correspond to experimental information. Some common types are outlined in the following sections.

1.   Distance Constraints

Although the use of distance constraints to determine structures from NMR spectroscopy is well-established [52], these are experimentally determined structures rather than predictions in the sense used here. Applying a limited number of distance constraints to the simulation of an unknown structure in order to determine the gross topology rather than the detailed coordinates is a more recent approach [54]. This work showed, however, that the number of distances required for this purpose was at least an order of magnitude less than that needed for a complete structure determination. The emphasis in recent years has therefore been to reduce this number even further and to increase the size of protein that can be studied, with the goal of obtaining better structural information while requiring fewer experiments. In practice, tests are usually carried out on known structures where a given number of distances can be chosen at random to simulate such data.

2.   Angle Constraints

There are currently experimental techniques to extract dihedral angles from NMR chemical shifts and coupling constants [55,56]. There is, however, a considerable margin of error on the order of $\pm 45°$ in the actual values, which varies according to secondary structure [57]. These values are therefore insufficient for purposes of constructing the backbone by a sequential buildup; however, the target values and corresponding uncertainties can be applied as constraints in a torsional scoring function. The same applies to local backbone distance constraints, which in a reduced model are more conveniently expressed as limits on the dihedral angles rather than as specific interatomic distances. Although the dihedral angles in principle determine the structure directly, it is possible to have significant local variations in ϕ and ψ without appreciably changing the overall fold. The goal is therefore to use local dihedral constraints to bias the simulation toward the native structure while maintaining sufficient flexibility to avoid propagating errors due to incorrect values. Angle constraints can also be effectively combined with distance constraints to obtain greater precision from experimental data [58].

3.   Other Types of Constraint

Data from NMR experiments which measure residual dipolar coupling [59] and paramagnetic relaxation [60] can be used to derive long-range geometrical constraints and global features of the structure. These methods allow one to determine the relative orientation of N–H bonds relative to a common (unknown) reference frame, although not directly to one another. Although it is difficult to extract detailed information from this type of data due to the inherent degeneracy of the relative orientations, it is complementary to the types of constraints mentioned above and therefore can be very useful in folding simulations to screen out incorrect structures. This type of constraint lends itself well to a scoring-function approach, because it is easier to calculate the values that would be produced by a predicted structure and compare them with the experimental data than to impose *a priori* constraints in generating the structure. Although this type of constraint shows considerable promise, its use in simulating larger proteins is still less well developed than the more traditional distance and angle constraints.

B.   Deriving Constraints from Predictions

Although the emphasis so far has been mostly on experimentally determined constraints, the same techniques that have been developed, especially in the case of uncertain or ambiguous constraints, can be just as well applied to theoretically predicted data. In cases where this is derived from sequence homology and/or multiple sequence alignments, the use of predicted constraints effectively generates a sequence-specific scoring function where any additional information is added to the generic scoring function already in place. Probable contacts can be derived from correlated mutations in a family of aligned sequences [18,61]. If a structure is known for at least one member of the family, contacts that are observed in the known structure which are likely to be conserved can be identified by looking at correlated mutations across the sequences, using the hypothesis that pairs of sites which have an increased probability of changing in concert are more likely to be in physical contact. Because there is a large number of possible pairs in a given sequence, as well as a relatively low signal-to-noise ratio in evaluating correlations, this method is less effective when based solely on sequence data without a reference to identify pairs that are likely to be in contact at all. On the other hand, extracting probable contact pairs can provide better results than direct homology modeling when the homology is weak and the structural alignment is uncertain.

Probable backbone dihedral angles can be predicted using sequence-based methods similar to those used in predicting secondary structure [62,63]. Although this could be considered a simple torsional potential, it is included in this section because it nonetheless incorporates sequence-specificity into the potential and can be implemented using the techniques of flexible angle constraints. In another method, contact distances between residues in different helices were determined by first selecting likely hydrophobic residues to form helix–helix contacts and then using a distance range typical of observed helix pairs in the PDB [36]. Distance constraints can also be generated directly from the simulation results themselves [35]. In an ensemble of predicted structures, the frequencies of inter-residue contacts can be analyzed to identify those that are observed across a range of structures. These "consensus" contacts can then be imposed as constraints and used to generate structures that are better than any of those used to derive the constraints. A similar approach has been used in our group to correctly identify inter-residue contacts using an ensemble of structures in which no structure individually had the correct topology.

C.   Constraint Implementation

Constraints are typically applied as a penalty function that is added as an extra term in the scoring function, often as some simple function (e.g., harmonic) of the difference between the actual and target values. Other strategies are possible, however, and constraints have also been used systematically in the construction of model structures. This can be applied to distance constraints, where a buildup procedure is used to generate structures that satisfy all constraints [64]. Angle constraints can also be used to systematically search the conformational space, both using a branch-and-bound procedure [65] or in a tree-search algorithm in combination with distance constraints [66].

In the case of sparse constraints, however, it has been shown that there is an advantage to using more flexible, or "floppy" constraints that allow for a more effective conformational search [67]. In our work, we apply inter-residue cons-traints to the C_β–C_β distances, regardless of the atoms involved in the original data. This is partly due to the practical problem of not representing side-chain atoms, but it also serves to simplify the calculation. The range of possible C_β–C_β distances consistent with the data is accounted for by using generous limits on the constraints. Rather than corresponding to a loss of precision, this actually improves the efficiency of the minimization.

We have studied a variety of functional forms for the constraint penalty functions and have found that a flat-bottom well with an exponential tail provides the best results. This penalty function has the form

$$U(r) = \begin{cases} -1, & r < c \\ -\exp(-r/d), & r > c \end{cases}$$

where c is the maximum constraint distance and d is the width of the tail. The best results are obtained with a square-well width of 8 Å and a tail width of 3 Å. The width is held constant independent of the actual constraint distance, because this allows greater flexibility and gives better scores to nearly correct structures. In fact, even for distances known to be less than 6 Å, setting c to 8 Å gave better results than a c of 6 Å, due to the fact that correct contacts are better recognized despite local errors in the structure. For the same reason, no inner cutoff was used other than the usual excluded volume term.

In cases where the constraints are known to be accurate, good results can also be obtained for penalty functions that become large at long distances, such as linear or quadratic tails. This gives a large energy for any structure that severely violates any constraint. This is fatal, however, in cases where some constraints are incorrect or even contradictory. It is therefore important to ensure that while there is a favorable score for satisfied constraints and an attractive force in their vicinity, in the limit of grossly violated constraints the corresponding score goes to zero and is simply ignored.

1. Ambiguous Constraints

Ambiguous constraints arise in working with NMR NOE data that haven't been completely assigned [68]. In cases where similar residues have virtually the same chemical shifts, it can be difficult to identify which sites in the sequence are responsible for an observed contact. The same principle also applies to cysteine (S–S) linkages where several different pairings of cysteine residues may be possible. In such cases, carrying out a simulation with simultaneous constraints corresponding to each possibility can be used to determine the correct pairings [69]. The results of simulations with conflicting distance constraints have even been used to eliminate incorrect assignments for subsequent simulations and eventually deduce the correct contacts [21,70]. Another approach that gives rise to ambiguous constraints is the simulation of predicted secondary structure, where the different possible assignments can be expressed as a weighted combination of short-range distance constraints [71].

In our implementation, ambiguous distance constraints are simply expressed as a linear combination of all possibilities; in other words, all constraints are treated equally. As the penalty function goes to zero for violated constraints, the score is essentially the same for a residue that satisfies any one of the possible constraints, and the structure as a whole is optimized to satisfy as many as possible. An optional weighting factor can be included to represent the relative probabilities associated with different assignments.

D. Results

In order to test some of the ideas discussed above, we have carried out a number of experiments on known structures by artifically generating constraints from the PDB coordinates. Although this is far removed from real-world applications, having precise control over the quantity and quality of the supplemental data allows the methods to be carefully evaluated and allows their limits to be better determined. In the following sections, some representative examples are presented to illustrate the progress that has been made, and comparisons are shown with similar work from other groups.

TABLE I
Results of Simulations with Constraints for 3ICB

	Constraints	Low RMS	Average RMS	Standard Deviation
Present work	0	4.6	9.8	1.9
	10	3.0	4.9	1.3
	89	3.0	3.3	0.2
Aszódi et al.	0		10.0	1.5
	10		6.3	2.0
	86		2.9	0.2

1. Distance Constraints

The implementation of distance constraints was tested using two small globular proteins that have been previously studied in the literature: calcium-binding protein (3ICB), an α protein with 72 residues, and tendamistat (3AIT), a β protein with 62 residues [72]. In each case, a total of 10 constraints were chosen at random from among the eligible pairs of residues in the crystal structure. This was repeated for 20 simulations, each using a different set of constraints, and compared with earlier literature results [73]. The results are summarized in Tables I and II. For 3ICB, 10 constraints are sufficient to find as good a structure as was found using all of the constraints. Because of the use of ideal β-strands without any sort of strand-pairing potential, 3AIT proved to be much more difficult, although the addition of 10 constraints does also lead to a significant improvement. Other published simulations [74] show better results when all of the constraints are used, but fail completely for small numbers of constraints. A test was also carried out with a larger molecule, myoglobin (1MBA), an α protein with 140 residues, the results of which are shown in Table III. Using 20 constraints in this case, a structure with an RMS deviation of 4.5 Å was obtained, comparable to 4.9 Å reported elsewhere for the same set

TABLE II
Results of Simulations with Constraints for 3AIT

	Constraints	Low RMS	Average RMS	Standard Deviation
Present work	0	8.4	9.7	0.4
	10	4.8	8.4	1.3
	116	3.6	6.8	1.6
Aszódi et al.	0		9.4	0.7
	10		5.8	0.6
	120		3.7	0.2

TABLE III
Results with Constraints for 1MBO

	Constraints	Low RMS	Average RMS	Standard Deviation
Present work	0	7.1	12.3	1.8
	20	4.5	10.3	1.8
	30	3.2	5.7	1.2
	50	3.6	5.3	1.6
	100	2.9	4.5	1.0
Skolnick et al.	20	4.9	5.6	

of constraints [75]. This result improved to 3.2 Å with a random selection of 30 constraints, which was essentially equivalent to results obtained with larger numbers of constraints.

2. Angle Constraints

Within our hierarchical model, it is more convenient to implement angle constraints in a different manner. Instead of using a scoring-function approach, we introduce the constraint information at the level of the list of allowed ϕ–ψ pairs. Because the pairs are selected randomly, the number of values in each region will determine the corresponding bias in the simulation. Test calculations were carried out for myoglobin (1MBO) in which part of the dihedral list corresponded to the usual distribution and the other part was limited to a region with a width of 30° around the target values. Clearly, if the weight of the latter region is 100%, this represents a rigid constraint, however, in order to maintain the flexibility of the simulation and allow for the possibility of incorrect data, it is useful to retain some of the original distribution. Simulation results are summarized in Table IV as a function of the relative weight of the constraint region. Good results are obtained with a 50% weighting, indicating that there is

TABLE IV
Results of Simulations of 1MBO Using Angle Constraints with Different Relative Weights

Constraint Weight (%)	Low RMS	Average RMS	Average Score
0	8.1	11.1	−172
6	7.4	11.7	−172
20	4.9	9.8	−173
30	5.1	6.5	−218
50	2.5	4.1	−226
100	1.7	2.7	−226

TABLE V
Results for 1MBO with 100 Total Constraints

Number of Good Constraints	Number of False Constraints	Low RMS	Average RMS
100	0	2.6	4.7
75	25	3.7	5.2
50	50	4.0	6.8
30	70	5.3	8.9
20	80	6.0	10.8

a strong cooperative selection. On the other hand, a control experiment was carried out also with 50% weighting, in which the target values were chosen at random, thus giving a nonsensical structure if taken together. The results in this case were essentially the same as those with no constraints at all, showing that the simulation is nonetheless able to ignore incorrect data.

3. Ambiguous Constraints

In order to test the sensitivity of the simulation with respect to incorrect data, a series of experiments was carried out in which the total number of distance constraints was held fixed, but the number of which were correct was varied. In a first trial, again with myoglobin (1MBO), 100 constraints were used. The correct constraints were derived by randomly selecting from among the possible contacts observed in the PDB structure, and the remaining number were randomly selected from pairs of residues known to be at least 20 Å apart in the correct structure. This was repeated with several different sets of constraints, to avoid any bias due to a lucky choice of correct constraints. The results are shown in Table V. Compared with the results in Table III, there is clearly a loss in performance due to the presence of incorrect constraints; however, reasonable results can still be obtained in cases where the nonsensical constraints actually outnumber the real ones. A similar experiment using flavodoxin (2FX2), a mixed α/β protein with 143 residues, is shown in Table VI. Although there is an increasing number of misfolded structures, as indicated by the average RMS

TABLE VI
Results for 2FX2 with 100 Total Constraints

Number of Good Constraints	Number of False Constraints	Low RMS	Average RMS
100	0	4.6	7.2
75	25	5.2	9.4
50	50	5.2	11.9

TABLE VII
Results for 1MBO with 150 Total Constraints

Number of Good Constraints	Number of False Constraints	Low RMS	Average RMS
100	50	4.0	5.9
50	100	3.7	7.7
30	120	6.1	10.7
20	130	(9.2)	(13.2)

deviation, the simulation is still able to find reasonable structures with only half of the constraints correct. A further experiment on myoglobin with 150 constraints, shown in Table VII, shows that the constraints remain useful with as few as 20% correct. Values in parentheses are actually higher than in a comparable simulation with no constraints at all. These results support the idea that, up to a certain limit, more data is better even if it becomes less reliable.

4. Predicting Constraints

The most promising method for predicting distance constraints is based on correlated mutations in multiply aligned sequences. This approach has been used in folding simulations with on average about 25% of tertiary contacts predicted to within ±1 residue in the sequence, and it was shown that this is sufficient to generate reasonable fold predictions [18,61]. In experiments carried out in our group, summarized in Table VIII, the predicted constraints were found to be more than sufficient to generate reasonable structures. Predictions in this case are considered correct if the two C_β atoms are in fact within the 8 Å

TABLE VIII
Contact Prediction Accuracy

Target:	1CCR	2LHB	1MIL
Sequence length:	107	134	84
Aligned sequences:	10	7	6
Maximum indentity:	62	31	29
	Low Sensitivity		
Predicted contacts:	88	81	84
Percent accurcy:	93	89	88
	High Sensitivity		
Predicted contacts:	33	47	45
Percent accuracy:	100	89	87

well used in the simulation. Results are shown for both low and high sensitivity, meaning that the criterion used to predict contacts based on the statistical significance of the sequence correlations was more strict in the latter case. Although this improves slightly the accuracy of the predictions, the larger number of total contacts provides a clear advantage for the low sensitivity predictions. In particular, in the case of 1MIL where there are relatively few aligned sequences with low homology, the selection criterion was of little use and yet the overall quality of the predictions was quite good.

IV. LIMITING THE SEARCH SPACE

Generic information about protein structure can be incorporated in a simulation by restricting *a priori* the conformations that can be generated. If the simulation is only capable of producing structures with certain realistic properties, the odds of finding the correct fold are greatly enhanced. In the extreme case, the choices would consist of a limited number of compact folded structures for the entire sequence. In such a "simulation" the global minimization problem is trivial (exhaustive enumeration becomes feasible) and the scoring function need only distinguish among topologicaly different structures without reproducing any of the interactions that stabilize such structures in the first place. Clearly, all the work is being done in the initial definition of possible trial structures, which therefore becomes the determining element of the algorithm. There is a necessary tradeoff between using the characteristics of known folds to limit the search and running the risk of incorrectly excluding a structure that had not been previously seen.

A trivial application of this principle, however, is the use in the hierarchical algorithm of a list of allowed ϕ–ψ pairs in generating new segments. This eliminates the need for a scoring function to penalize unfavorable regions of the Ramachandran map, as well as the need to sample such unlikely regions of the conformational space. Although the definition of this list is entirely empirical, based on observation of the PDB, it still represents real interactions that a new structure would be very unlikely to violate.

A. The Principle of Threading

The most obvious way to select realistic structures is to simply use those that are already known in the PDB, and this is the basis of what is commonly known as threading. Threading is normally associated with the problem of fold recognition—that is, identifying homologous structures in the PDB—rather than in the context of simulation. It is included here as the limiting case of a restricted search in order to establish a relationship between the *ab initio* and fold-recognition approaches and also to provide a framework for describing various intermediate methods that have been developed.

In its simplest incarnation, threading consists of attempting to map the target sequence onto the backbone coordinates of all structures in the PDB of equal or greater length. This can be visualized as stringing a flexible chain of amino acids along the fixed scaffold of a known structure—hence the name threading. In this simple approach, the number of possible alignments (mappings of the target sequence onto the corresponding residues of a known structure) is limited, and most empirical scoring functions are capable of recognizing truly homologous structures. The method fails, however, to identify distant structural homologs and is obviously incapable of generating any new folds. More realistic methods allow the connectivity of the template structures to be modified [76] and allow gaps and insertions to be introduced in the alignments. This, however, greatly increases the number of possible alignments and makes the problem of recognizing homologous structures that much more difficult [77].

B.   Local Threading and Fragment Lists

One way to overcome the combinatorial problem is to divide the problem into smaller local alignments. This can be done as a first step in generating a global alignment to a single known structure [14], or alternatively to identify shorter segments that align to parts of different structures. The structure of a known fold can be described by specifying the local environment of each residue: secondary structure, polarity, and solvent exposure [78]. This allows the threading to be carried out locally, aligning a linear sequence to a series of profiles by the same methods used for sequence–sequence alignments, independently of the rest of the molecule.

The resulting local alignments lead to a large number of possible combinations that must still be reassembled into a single structure. In this situation, rather than attempting to either select the best local homologs or carry out an exhaustive enumeration, it is more effective to return to a stochastic simulation where the local templates act as lists of trial structures for each segment. In this way, the principle of using a restricted set of conformations can be extended across various levels of structure: from individual amino acids (as in a typical simulation) to multiresidue fragments, loop and secondary structure elements, supersecondary motifs, and ultimately entire domains (as in a typical threading calculation).

1.   Using a Motif Library

A set of commonly occurring structural motifs, along with their associated sequence profiles (the probability for each amino acid to occupy each site in the structure), have been extracted from the PDB using local sequence and structure alignments [13]. Experimental evidence has even shown that some peptides do in fact adopt the corresponding motif structure in isolation and that strong fits to the sequence profile can possibly be used to identify sites of folding initiation

[79]. These motifs have been successfully used in a simulation algorithm to predict new folds, providing one of the more impressive achievements in the CASP-3 experiment (see Simons et al. in Ref. 8). This motif library has now been united into a global prediction scheme using a hidden Markov model to encode extended sequence profiles [63]. A similar library of loop motifs has also been used to model loop regions in homology models using the flanking secondary structure as a guide [80].

2. Mapping Conformational Space

Rather than attempt to identify common motifs, another approach is to try to identify a minimal set of building blocks that can be used to represent any known structure [12]. This essentially corresponds to a redefinition of the geometrical model in which the smallest unit of structure becomes a five- or six-residue fragment. The result of using this model is a greatly reduced number of degrees of freedom and a more efficient exploration of conformational space.

C. Fragment Screening and Enrichment

As an alternative to using preselected fragment lists to build up a model structure, a more general approach is to use the characteristics of homologous structures to screen possible conformations. The idea is still to allow arbitrary conformations, as in a traditional simulation, but to increase selectively the proportion of generated structures with the desired protein-like qualities. By using homologous motifs from the PDB to define the selection criteria, sequence-dependent conformational preferences can be introduced into the simulation without reducing the flexibility of the model.

1. The Hierarchical Approach

In the hierarchical algorithm [40], the structures of the residue triplets are generated from independent residue conformations which are determined by the three amino acid types. These triplets are then screened according to the relative orientations of the end residues, which determine the positions of the flanking segments. For a given target sequence, the distribution of triplet geometries is calculated for segments in the PDB which have a local sequence homology greater than a specified cutoff. This distribution is used to accept or reject randomly generated triplets so as to reproduce the observed probabilities of finding a triplet with a given geometry. This generates a sequence-specific list of triplet conformations which can then be used to generate larger fragments. In preliminary tests using this method on a set of test proteins, both the average energy and deviation from the native structures was found to decrease as the selectivity of the screening (the homology threshold) was increased. In these tests, any structure with significant global homology to the target sequence was excluded from the fragment database.

Loop segments of varying lengths are then built up by randomly selecting from the lists of triplet conformations. Loops are again selected by comparing the end-to-end distances and rotations with homologous loops in the PDB. Although the internal structure of the loops is free to vary, the goal is to generate structures that are more likely to produce a favorable positioning of the flanking segments. Because this type of selection is applied successively at three different levels of structure, the overall process is quite efficient and the cutoff parameters can be freely adjusted to give the desired level of structural similarity and sequence homology at various stages of the simulation.

D. Modeling Secondary Structure

Due to its well-characterized regular motifs, secondary structure is an obvious candidate for fragment-based modeling. Indeed, a common approach, and the one traditionally used in our group, is to simply hold the secondary structure fixed during the calculation, which is an extreme application of the principles described in this section. In cases where the secondary structure is predicted from the sequence, this is a crude application of fragment selection by sequence profile. This effectively removes a large number of degrees of freedom and eliminates the need to use the scoring function to stabilize α-helix and β-sheet conformations.

This approach can be generalized, and some flexibility reintroduced into the structure, by developing specific models to reproduce the observed variability within the regular structures. A list of strand or helix structures can be assembled from the PDB, with associated error tolerances on the dihedral angles to account for kinks and imperfections, and this can be used to define the possible conformations of an arbitrary helix as a single unit. Sheets are in general more complex and show more natural variability; however, the possible collective structures have been extensively studied and characterized [81,82]. Using the generic properties of β-sheets, a library of conformations with varying twist and curvature can be constructed for an arbitrary sequence.

As a preliminary test, we have carried out a series of simulations with a range of possible helix and strand geometries to determine if the tertiary contacts would be sufficient to identify the native structure. The list of trial structures consisted of a continuous deformation from an ideal geometry to the (known) native geometry, with the same deformation vector extended to also generate even more deformed structures. The results for a set of test sequences are shown in Figs. 1 and 2 for helices and sheets, respectively. The deformations are grouped into discrete bins, and in each case the corresponding native structure falls into bin number six. For helices, which have a smaller average deformation, the distribution is relatively smooth with a maximum at the native geometry. In the case of β-strands, the distribution is more-or-less flat with a

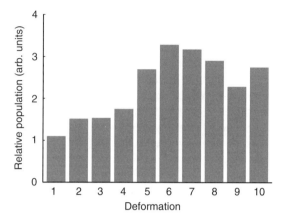

Figure 1. Helix selection frequency as a function of relative deformation for a set of test proteins. In each case, bin 1 corresponds to an ideal structure, bin 6 corresponds to the native structure, and the other bins correspond to a linear extrapolation.

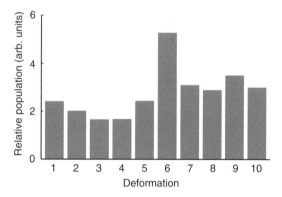

Figure 2. Strand selection frequency as a function of relative deformation for a set of test proteins. The bin deformations are as in Fig. 1.

more pronounced peak at the native geometry, suggesting that the correct deformation does in some sense better "fit together" and is energetically favored.

V. HOMOLOGY AND STRUCTURAL TEMPLATES

For homologous proteins, a threading alignment as described in the previous section can be used to provide a template for the entire structure. In the absence of global homology, however, local alignments can still be used to extract localized structural constraints. This approach, unfortunately, results in the loss of any information about the overall topology of the tertiary structure, which is the most difficult part of the folding problem. An alternative is try to identify a smaller number of aligned residues, possibly with significant gaps, in order to provide key reference points for determining the overall structure. In this case, much local information will be missing, and local structure must still be determined using standard simulation methods; however, the relative three-dimensional positions of different parts of the structure can be controlled. This is consistent with the chemical interpretation of a relatively small number of conserved residues playing an important role in both fold stability and function (although of course there are many exceptions to this picture.) When the homology is weak, it may be more effective to try to identify the most probable conserved residues than to rely on a global alignment that is likely incorrect.

A. Identifying Structural Templates

The most straightforward approach is to carry out a standard threading calculation and exclude regions with a poor alignment score. Template residues can also be excluded in regions where the target is not predicted to have a regular secondary structure, or where the template secondary structure differs from that predicted for the target. In this way, the parts of the alignment most likely to correspond to a stable core can be identified and the simulation can be used to fill in the gaps. In our implementation, the superposition of the selected residues with their corresponding coordinates in the aligned template structure is then used as an additional contribution to the scoring function. Another approach is to constrain the simulation to follow the template structure, but to allow the specific alignment to change during the simulation [83].

Positions likely to be conserved in a sequence can also be identified by searching through a database of known sequence patterns such as PROSITE [84]. In our approach, patterns identified in the target sequence were then used to search the PDB for structures containing the same patterns. The coordinates of the conserved residues were then averaged over all matching structures to generate a composite template that was then used in the simulation. An experiment was carried out for the myoglobin sequence (1MBO) using coordinates from seven structures in the PDB having less than 20% sequence identity with 1MBO to obtain the template coordinates. The results of the simulation are shown in Table IX as a function of the number of template sites used. Good structures were obtained using a template with a relatively small number of aligned residues.

B. Multiple Templates

In many cases there may be more than one possible template for a given target sequence. This can arise from different choices of reference structure, or for the

TABLE IX
Performance as a Function of Template Size for 1MBO

Size of Template	Low RMS	Average RMS
0	9.5	13.8
11	4.7	9.4
25	3.0	4.2
50	2.4	3.3
146	2.1	3.0

same reference structure different alignments and choices of predicted secondary structure. In addition, different parts of the target sequence might align well to different structures in the PDB. In such cases, the simulation can be used to choose among conflicting alignments and to combine different templates.

1. Template Competition

In our implementation, the same philosophy is used as in the case of distance constraints. The scoring function is a spline-fit switching function of the actual superposition RMS deviation with the template coordinates. This function is equal to -1 below a lower cutoff value, equal to zero above an upper cutoff value, and varies smoothly in between the two. Conflicting templates can therefore be used simultaneously, and a favorable score will be obtained for structures that superpose well on any one or more of them and no penalty is assessed for distant templates. The simulation can therefore be used to identify which of the possible templates gives the best fit consistent with the connectivity of the sequence and the generic scoring function.

2. Results

This methodology was used in the most recent CASP experiment, from which two representative examples will be described which illustrate how the methodology was applied. For sequence T0089, threading results suggested eight possible templates for the N-terminal region, four possible templates for the C-terminal region, and three or four different alignments and secondary-structure assignments in each case. None of the alignments had a sequence identity greater than 15%, and in addition there was a gap of about 120 residues between the two templates. Simulations were run using all possible combinations of two templates, and the final prediction was selected based on the fit to the templates, the overall energy, and the ability of the connecting segment to fold.

The situation was reversed in the case of sequence T0087, where instead of a gap there was an overlap of over 100 residues between the two proposed templates. In this case, 11 choices for the N-terminal region and six choices for the C-terminal region were identified, all with about 10–15% sequence identity. For each possible combination the two templates were used simultaneously, thereby generating a conflicting set of constraints for the region in which the two overlapped. The final prediction was selected as that which provided the best simultaneous fit for both templates, thus hopefully giving a relative orientation of the two domains consistent with the context of each.

Unfortunately, the preliminary results indicate that none of the proposed templates was correctly aligned to the native structure, so it is difficult to judge the performance of the simulation methodology. In each case, however, the submitted structures were correctly ranked, with the best one selected as the first choice.

C. Local Templates

The use of multiple simultaneous templates can also be extended to model generic structural motifs. In contrast to the method of segment libraries discussed earlier, these are structural relationships which are nonlocal in sequence; rather than describing the local backbone conformation, the goal is to describe the relative spatial orientations of different structural elements. The use of multiple templates allows different possibilities to be considered, thereby providing a library of three-dimensional relationships. This use of generic structural templates provides a general alternative to local multibody scoring functions that recognize specific structural motifs.

1. β-Strand Pairing

Generating realistic β-sheet structures is a notoriously difficult problem due to the specific relative orientation of noncontiguous backbone segments produced by the H-bonding pattern. The H-bonds themselves, however, are short-range interactions that are difficult to simulate and often fail to produce the desired overall structure. Specific multibody interactions that take into account strand orientation are therefore often used to overcome this problem [45,85–87].

An alternative approach for correctly aligning two β-strands is to extract a template of a similar strand pair from the PDB, which can then be used to superimpose the target strands. A library of possible pairings can be generated based on sequence homology, and the technique of multiple templates described above can be used to select a suitable candidate for each interacting strand pair. To determine whether or not templates derived from unrelated structures could provide correct strand-pairing geometries, the closest structural homologs in the PDB were identified for a number of strand pairs, along with the best superposition in a list of the top 10 sequence homologs. Shown in Table X are the results of this experiment for the mixed α/β protein ribonuclease A (2RAT). (Sequences with more than 20% overall similarity to the target were excluded

TABLE X
Strand-Pairing Templates for 2RAT

Strand Pair	Length	Best Possible	Homologous
1–4	5, 8	0.64 Å (1BIA.1)	0.99 Å (1ZXQ)
4–5	8, 8	0.82 Å (1BYT)	1.93 Å (2MEV.2)
2–3	3, 3	0.08 Å (8FABA.A)	0.16 Å (2ENG)
3–6	3, 6	0.34 Å (1EFT)	0.81 Å (1BLI.A)
6–7	6, 8	0.91 Å (1A62)	2.56 Å (1CBJ.A)

from the calculation.) Reasonable models can be obtained for each pair, despite the lack of global homology.

2. Hydrophobic Contacts

A similar approach can also be applied to helix pairs, which, despite being linked only by hydrophobic contacts, tend to pack in well-defined relative orientations. It has been shown that by identifying conserved hydrophobic contacts between different helices, a model can be found in the PDB which reproduces the correct helix–helix packing and can be used to reconstruct the tertiary structure [88]. Because the helix structure is very regular, a single contact geometry is sufficient to generate a helix template of arbitrary length using a standard backbone conformation.

VI.   NEW DIRECTIONS

The next logical step in the evolution of structure prediction is to generalize further the knowledge-based methods described so far in order to make maximum use of the motifs in the PDB, even in the absence of any detectable *a priori* homology, and to eventually replace the physically motivated idea of a universal energy function. Local structure will be modeled using fragment libraries, inter-residue interactions through generalized distance constraints, and multibody correlations through localized motif templates. The scoring function will become a moving target that adapts itself to the results of the simulation, adding a knowledge-based component to the already sophisticated search methods currently in use.

A.   Sequence-Specific Potentials

Flexible distance constraints can be used to express the probability of forming different specific contacts in the structure, based on the context of each residue. Conceptually, if contact probabilities were to be predicted solely on the basis of amino acid type (hydrophobic residues are more likely to be in contact with other hydrophobic residues), this simply reduces to a traditional generic energy function. Pair potentials have already been developed which derive contact probabilities based on local sequence [19], local secondary structure [51], and β-strand pairing [50]; any other observed correlation can be combined and expressed in the same way. Generalized sequence-based methods (such as in Ref. 63) can also be used to derive sequence-specific scoring functions for local conformation and structural context, allowing for a customized selection of fragments and templates.

B.   Constraint Refinement

The results of the simulation itself can also be used to improve the prediction of inter-residue contacts, thus allowing an iterative series of simulations to generate successively more specific scoring functions. This is analogous to the use of iterative simulations in assigning NOE signals in NMR spectroscopy [21], except with purely theoretical input. It has been shown, however, that the statistical analysis of an ensemble of predicted structures can be used to derive more accurate contact information than any of the structures individually [35]. Preliminary experiments in our group have shown that it is possible to start with a large number of possible contacts and, by successively eliminating those that are observed less frequently in the ensemble, to eventually identify the correct native contacts.

VII.   CONCLUSION

Considerable progress has been made over the past few years in developing practical tools for structure prediction. Geometrical models, empirical scoring functions, and global minimization algorithms have all evolved together to increase the efficiency and selectivity of simulation-based methods. Different techniques have advantages and disadvantages: Discretized models gain in sampling efficiency at the expense of resolution, template models carry more three-dimensional information, constraint-based methods are less sensitive to alignments, and so on. The result, however, is an increasingly complete spectrum of methods that are beginning to achieve meaningful results in a variety of real-world applications. As more and more information is being added to sequence and structure databases, there is every reason to expect this trend to continue.

Acknowledgments

The authors would like to thank Sabrina Bédard, Geneviève Dufresne, Martin Éthier, Olivier LeBel, Éric Morneau, and Marc-André Thivièrge, who contributed to some of the results discussed here.

References

1. A. R. Fersht, *Curr. Opin. Struct. Biol.* **7**, 3 (1997).
2. H. S. Chan and K. A. Dill, *Proteins* **30**, 2 (1998).

3. N. D. Socci, J. N. Onuchic, and P. G. Wolynes, *Proteins* **32**, 136 (1998).

4. B. Honig, *J. Mol. Biol.* **293**, 283 (1999).

5. D. J. Bicout and A. Szabo, *Protein Sci.* **9**, 452 (2000).

6. J. Moult, *Curr. Opin. Biotechnol.* **10**, 583 (1999).

7. D. J. Osguthorpe, *Curr. Opin. Struct. Biol.* **10**, 146 (2000).

8. Third Community Wide Experiment on the Critical Assessment of Techniques for Protein Structure Prediction, *Proteins Suppl.* **3** (1999).

9. J. Lee, A. Liwo, D. R. Ripoll, J. Pillardy, J. A. Saunders, K. D. Gibson, and H. A. Scheraga, *Int. J. Quant. Chem.* **77**, 90 (2000).

10. L. Jaroszewski, L. Rychlewski, B. Zhang, and A. Godzik, *Protein Sci.* **7**, 1431 (1998).

11. M. J. Schoonman, R. M. Knegtel, and P. D. Grootenhuis, *Comput. Chem.* **22**, 369 (1998).

12. C. Micheletti, F. Seno, and A. Maritan, *Proteins* **40**, 662 (2000).

13. C. Bystroff and D. Baker, *J. Mol. Biol.* **281**, 565 (1998).

14. R. Thiele, R. Zimmer, and T. Lengauer, *J. Mol. Biol.* **290**, 757 (1999).

15. Y. Wang, H. Zhang, W. Li, and R. A. Scott, *Proc. Natl. Acad. Sci. USA* **92**, 709 (1995).

16. M. Vendruscolo and E. Domany, *J. Chem. Phys.* **109**, 11101 (1998).

17. O. Olmea, B. Rost, and A. Valencia, *J. Mol. Biol.* **293**, 1221 (1999).

18. A. R. Ortiz, A. Kolinski, and J. Skolnick, *J. Mol. Biol.* **277**, 419 (1998).

19. J. Skolnick, A. Kolinski, and A. Ortiz, *Proteins* **38**, 3 (2000).

20. B. L. Podlogar, G. C. Leo, P. A. McDonnell, D. A. Loughney, G. W. Caldwell, and J. F. Barrett, *J. Med. Chem.* **40**, 3453 (1997).

21. M. Nilges, M. J. Macias, S. I. O'Donoghue, and H. Oschkinat, *J. Mol. Biol.* **269**, 408 (1997).

22. D. T. Jones, *Curr. Opin. Struct. Biol.* **10**, 371 (2000).

23. C. L. Brooks III, *Curr. Opin. Struct. Biol.* **8**, 222 (1998).

24. X. Daura, B. Jaun, D. Seebach, W. F. van Gunsteren, and A. E. Mark, *J. Mol. Biol.* **280**, 925 (1998).

25. Y. Duan and P. A. Kollman, *Science* **282**, 740 (1998).

26. R. A. Friesner and J. R. Gunn, *Annu. Rev. Biophys. Biomol. Struct.* **25**, 315 (1996).

27. R. L. Dunbrack, Jr., and F. E. Cohen, *Protein Sci.* **6**, 1661 (1997).

28. E. S. Huang, P. Koehl, M. Levitt, R. V. Pappu, and J. W. Ponder, *Proteins* **33**, 204 (1998).

29. A. S. Lemak and J. R. Gunn, *J. Phys. Chem. B* **104**, 1097 (2000).

30. H. A. Scheraga and M. H. Hao, *Adv. Chem. Phys.* **105**, 243 (1999).

31. D. M. Standley, V. A. Eyrich, A. K. Felts, R. A. Friesner, and A. E. McDermott, *J. Mol. Biol.* **285**, 1691 (1999).

32. V. A. Eyrich, D. M. Standley, and R. A. Friesner, *J. Mol. Biol.* **288**, 725 (1999).

33. A. Kolinski and J. Skolnick, *Proteins* **32**, 475 (1998).

34. Y. Xia, E. S. Huang, M. Levitt, and R. Samudrala, *J. Mol. Biol.* **300**, 171 (2000).

35. E. S. Huang, R. Samudrala, and J. W. Ponder, *Protein Sci.* **7**, 1998 (1998).

36. E. S. Huang, R. Samudrala, and J. W. Ponder, *J. Mol. Biol.* **290**, 267 (1999).

37. M. Vendruscolo, E. Kussell, and E. Domany, *Fold. Des.* **2**, 295 (1997).

38. F. Avbelj and L. Fele, *Proteins* **31**, 74 (1998).

39. J. R. Gunn, *J. Phys. Chem.* **100**, 3264 (1996).

40. J. R. Gunn, *J. Chem. Phys.* **106**, 4270 (1997).

41. D. J. Ayers, T. Huber, and A. E. Torda, *Proteins* **36**, 454 (1999).

42. D. Mohanty, B. N. Dominy, A. Kolinski, C. L. Brooks III, and J. Skolnick, *Proteins* **35**, 447 (1999).

43. T. Lazaridis and M. Karplus, *Curr. Opin. Struct. Biol.* **10**, 139 (2000).

44. K. T. Simons, I. Ruczinski, C. Kooperberg, B. A. Fox, C. Bystroff, and D. Baker, *Proteins* **34**, 82 (1999).

45. A. Kolinski, B. Ilkowski, and J. Skolnick, *Biophys. J.* **77**, 2942 (1999).

46. A. Liwo, R. Kazmierkiewicz, C. Czaplewski, M. Groth, S. Oldziej, R. J. Wawak, S. Rackovsky, M. R. Pincus, H. A. Scheraga, *J. Comput. Chem.* **19**, 259 (1998).

47. M. J. Sippl, *Curr. Opin. Struct. Biol.* **5**, 229 (1995).

48. R. L. Jernigan and I. Bahar, *Curr. Opin. Struct. Biol.* **6**, 195 (1996).

49. F. Seno, A. Maritan, and J. R. Banavar, *Proteins* **30**, 244 (1998).

50. H. Zhu and W. Braun, *Protein Sci.* **8**, 326 (1999).

51. C. Zhang and S.-H. Kim, *Proc. Natl. Acad. Sci. USA* **97**, 2550 (2000).

52. A. T. Brünger, G. M. Clore, A. M. Gronenborn, and M. Karplus, *Proc. Natl. Acad. Sci. USA* **83**, 3801 (1986).

53. L. Kirnarsky, O. Shats, and S. Sherman, *J. Mol. Struct. (Theochem.)* **419**, 213 (1997).

54. M. J. Smith-Brown, D. Kominos, and R. M. Levy, *Protein Eng.* **6**, 605 (1993).

55. M. Hong, J. D. Gross, W. Hu, and R. G. Griffin, *J. Magn. Res.* **135**, 169 (1998).

56. K. Kloiber and R. Konrat, *J. Biomol. NMR* **17**, 265 (2000).

57. R. D. Beger and P. H. Bolton, *J. Biomol. NMR* **10**, 129 (1997).

58. M. J. Bayley, G. Jones, P. Willett, and M. P. Williamson, *Protein Sci.* **7**, 491 (1998).

59. G. M. Clore, A. M. Gronenborn, and N. Tjandra, *J. Magn. Res.* **131**, 159 (1998).

60. J.-C. Hus, D. Marion, and M. Blackledge, *J. Mol. Biol.* **298**, 927 (2000).

61. A. R. Ortiz, A. Kolinski, and J. Skolnick, *Proteins* **30**, 287 (1998).

62. H. S. Kang, N. A. Kurochkina, and B. Lee, *J. Mol. Biol.* **229**, 448 (1993).

63. C. Bystroff, V. Thorsson, and D. Baker, *J. Mol. Biol.* **301**, 173 (2000).

64. D. A. Debe, M. J. Carlson, J. Sadanobu, S. I. Chan, and W. A. Goddard III, *J. Phys. Chem. B* **103**, 3001 (1999).

65. J. L. Klepsis, C. A. Floudas, D. Morikis, and J. D. Lambris, *J. Comput. Chem.* **20**, 1354 (1999).

66. G. P. Gippert, P. E. Wright, and D. A. Case, *J. Biomol. NMR* **11**, 241 (1998).

67. D. Bassolino-Klimas, R. Tejero, S. R. Krystek, W. J. Metzler, G. T. Montelione, and R. E. Bruccoleri, *Protein Sci.* **5**, 593 (1996).

68. M. Nilges, *J. Mol. Biol.* **245**, 645 (1995).

69. J. Boisbouvier, M. Blackledge, A. Sollier, and D. Marion, *J. Biomol. NMR* **16**, 197 (2000).

70. C. Mumenthaler, P. Güntert, W. Braun, and K. Wüthrich, *J. Biomol. NMR* **10**, 351 (1997).

71. C. C. Chen, J. P. Singh, and R. B. Altman, *Bioinformatics* **15**, 53 (1999).

72. J. R. Gunn, *Intell. Syst. Mol. Biol.* **6**, 78 (1998).

73. A. Aszódi, M. J. Gradwell, and W. R. Taylor, *J. Mol. Biol.* **251**, 308 (1995).

74. O. Lund, J. Hansen, S. Brunak, and J. Bohr, *Protein Sci.* **5**, 2217 (1996).

75. J. Skolnick, A. Kolinski, and A. R. Ortiz, *J. Mol. Biol.* **265**, 217 (1997).

76. F. A. Hamprecht, W. Scott, and W. F. van Gunsteren, *Proteins* **28**, 522 (1997).

77. X. De La Cruz and J. M. Thornton, *Protein Sci.* **8**, 750 (1999).

78. D. W. Rice and D. Eisenberg, *J. Mol. Biol.* **267**, 1026 (1997).

79. Q. Yi, C. Bystroff, P. Rajagopal, R. E. Klevit, and D. Baker, *J. Mol. Biol.* **283**, 293 (1998).

80. J. Wojcik, J.-P. Mornon, and J. Chomilier, *J. Mol. Biol.* **289**, 1469 (1999).

81. F. R. Salemme and D. W. Weatherford, *J. Mol. Biol.* **146**, 101 (1981).

82. F. R. Salemme and D. W. Weatherford, *J. Mol. Biol.* **146**, 119 (1981).

83. A. Kolinski, P. Rotkiewicz, B. Ilkowski, and J. Skolnick, *Proteins* **37**, 592 (1999).

84. K. Hofmann, P. Bucher, L. Falquet, and A. Bairoch, *Nucleic Acids Res.* **27**, 215 (1999).

85. A. Monge, E. J. P. Lathrop, J. R. Gunn, P. S. Shenkin, and R. A. Friesner, *J. Mol. Biol.* **247**, 995 (1995).

86. T. Dandekar and P. Argos, *J. Mol. Biol.* **256**, 645 (1996).

87. D. M. Standley, J. R. Gunn, R. A. Friesner, and A. E. McDermott, *Proteins* **33**, 240 (1998).

88. M. Parisien, F. Major, and M. Peitsch, *Pac. Symp. Biocomput.* **1998**, 425 (1998).

AB INITIO PROTEIN STRUCTURE PREDICTION USING A SIZE-DEPENDENT TERTIARY FOLDING POTENTIAL

VOLKER A. EYRICH AND RICHARD A. FRIESNER

Department of Chemistry and Center for Biomolecular Simulation, Columbia University, New York, NY, U.S.A.

DARON M. STANDLEY

Schrödinger Inc., New York, NY, U.S.A.

CONTENTS

Computational Methods for Protein Folding: A Special Volume of Advances in Chemical Physics, Volume 120, Edited by Richard A. Friesner. Series Editors I. Prigogine and Stuart A. Rice. ISBN 0-471-20955-4. © 2002 John Wiley & Sons, Inc.

I. INTRODUCTION

In previous work [1–4], we have investigated the ability of simple potential functions, derived from statistics in the Protein Data Bank (PDB [5,6]), to generate correct predictions of protein tertiary structure given the native secondary structure as input. Most recently [2], we studied an unbiased sample of 95 proteins in the size range of 30–160 residues, and we were able to locate native-like low energy structures in a significant number of cases. However, there were also many examples of unsatisfactory performance; furthermore, the utilization of native secondary structure derived from PDB coordinates is an obvious limitation in terms of the utility of the method for protein structure prediction. Thus, a significant improvement in the potential function, along with tests under more realistic conditions, were required before one could consider applying the methodology to problems of practical interest.

A principal reason for carrying out the studies described above was to generate a large database of plausibly misfolded structures in the hope of elucidating systematic flaws in the database potential function that we employed, a principal component of which is the pairwise potential of mean force developed by Sippl and co-workers [7]. We have recently uncovered one systematic error in the Sippl formulation of the statistical pair potential, and we remedied this deficiency in a straightforward fashion: The potential function, at least as applied to the problems discussed here, should be dependent upon the size of the protein, a feature that has also been uncovered in other, more theoretical work [8]. To this end, we developed a statistical potential that is derived from proteins that are similar in size to the protein for which a prediction is to be made. The result is a new type of statistical pair potential with qualitatively improved predictive properties in tertiary folding simulations. While the new potential function is still not rigorously predictive of the native structure in all cases, application to actual protein structure prediction problems is now a much more feasible goal.

Having achieved this advance in the potential function, we relaxed the assumption of accurate knowledge of native secondary structure and examined the capabilities of the methodology with more realistic types of input data. In the present chapter, we approach this objective in two stages. First, we carry out simulations using ideal, rather than crystallographic, representations of the secondary structure elements (while still deriving the location and length of the various elements from the PDB). For α-helices, the use of ideal helices leads in some (but not all) cases to a quantitative degradation of the quality of the results; in general, however, qualitatively similar success is achieved. For all α- and mixed α/β-proteins, there is an occasional substantial diminishment of the ranking of the lowest energy low RMSD structure, when idealized strands are used.

Second, we carry out computational experiments using secondary structure assignments derived from secondary structure prediction methods in conjunction with ideal secondary structural elements. This protocol constitutes an actual attempt at *ab initio* protein structure prediction; no experimental data other than sequence information is input into the calculations (other than, of course, the input of PDB statistics to derive the tertiary folding potential and secondary structure prediction algorithms). Because secondary structure prediction methods have not yet reached a high degree of robustness, we perform calculations using several different predictions generated by a variety of alternative secondary structure prediction methods (which are conveniently available on Web-based servers). While there are nontrivial cases where the native-like fold is uniquely determined by the algorithm, our objective at present is not to demonstrate successful *ab initio* prediction. Instead, we ask whether the protocol is capable of generating a prediction with a good RMSD that is highly ranked (e.g., within the top five predictions, a condition compatible with the rules of the CASP3 prediction contest). For a significant number of cases, this goal has been accomplished. Furthermore, in most cases where our algorithm fails to generate a native-like fold in the top five predictions, we are able to rationalize the results in terms of limitations of our model and propose straightforward extensions to generalize and improve the model. These proposed extensions are briefly discussed at the end of this chapter.

We have chosen in this chapter to focus our efforts on α-helical and mixed α/β-proteins below 100 residues in size. In previous work [2] we showed that β-strand proteins present more of a challenge to our prediction methodology than α-helical or mixed α/β-proteins [9–12]; the modified size-dependent potential function discussed above improves the results of earlier work on β-strand containing proteins, but does not change the basic conclusion. For larger systems, our results are quite promising but not yet at the stage of completeness that we have been able to achieve for the smaller proteins. Consequently, we defer discussion of these cases to a subsequent publication.

The chapter is organized as follows. Section II describes the new potential function, discussing its novel qualitative features and presenting an algorithm for optimization of parameters using a large training set derived from the PDB. Section III briefly reviews the computational methodology used to carry out the tertiary folding simulations (previously described in detail [2]) and then presents simulation results using native secondary structure and ideal secondary structure. As a test set in this section, we employ a subset of the proteins studied previously [2] so that comparisons can be made with the results reported in that publication, and improvements in the potential functions quantified. In Section IV, we utilize predicted secondary structure lengths and positions and ideal secondary structure elements to carry out *ab initio* prediction experiments; we focus in this chapter on helical proteins, and include, in addition to proteins from the test set of Section IV, two targets from CASP3 [13]. Section V, the conclusion, summarizes our efforts.

II.   DEVELOPMENT OF A SIZE-DEPENDENT POTENTIAL ENERGY FUNCTION

A.   Identification of Systematic Errors in Previous Tertiary Folding Simulations

Although the tertiary structure prediction protocol employed in our previous work [2] was more or less able to consistently generate native-like structures for α- and mixed α/β-proteins, the energetic rank of these structures was not always satisfactory. An analysis of high-RMSD, low-energy structures obtained from those simulations reveals a systematically incorrect behavior of the statistical potential function of Sippl and co-workers [7] at large separations, most prominently for pairs of hydrophilic residues. This feature of statistical potentials has been uncovered in several other computational experiments [8,14].

The hydrophobicity term developed by Sippl was originally used only for recognition (i.e., threading), so it is not surprising that some modifications would be required for the asymptotic large-distance parts of the energy surface. It remains to be seen whether or not the general type of systematic errors uncovered in our tertiary structure predictions are present in the threading studies of others using similar potentials. A complete derivation of the coefficients by Sippl and co-workers can be found in Ref. 7. The two key elements of interest in the derivation of the hydrophobicity function are the inclusion of proteins of many sizes in the definition of a statistical "potential of mean force" (PMF) and the asymptotic behavior of these potentials when they are linearly extrapolated to large distances.

In Ref. 7 an individual PMF for residues i and j, separated by a distance d, is defined as

$$E_{ij} = -kT \ln \left(\frac{p_{ij}^1(d)}{p^2(d)} \right) \tag{1}$$

where $p_{ij}^1(d)$ is the normalized distribution of d for all i,j pairs in a training set and $p^2(d)$ is the normalized distribution of d of irrespective of residue pair. The training set Sippl used consisted of 88 proteins that ranged in length from 46 to 374 residues. Note also that Eq. (1), which is sometimes known as the "quasi-chemical approximation," applies only to residues separated in sequence by more than 20 amino acids (at least in Ref. 7).

Equation (1) is only defined for distances that correspond to nonzero values of both distribution functions. For this set of distances, E_{ij} is well-approximated by a linear function

$$E_{ij}^{\text{hyd}} = (H_{ij} + H_0)d \qquad (2)$$

where H_{ij} is one of 400 "pairwise hydrophobicities" and H_0 is an adjustable "average hydrophobicity," for which Sippl suggest the value 0.36. (In our own simulations, H_0 was increased if local minimization starting from the native structure yielded noncompact structures.)

The basic idea inherent in the development of the Sippl hydrophobicity potential, that of extracting a potential of mean force using PDB statistics, is an essential component of our empirical tertiary folding potential. However, based on our analysis of the low-energy misfolded structures generated in our previous experiments [2] described above, we propose to improve upon the detailed methodology for construction of the PMF by implementing the following modifications:

1. The derivation of an individual PMF for tertiary structure prediction of protein P is to be based only on proteins of roughly the same size as P.
2. In the large and small distance limits, a functional form other than Eq. (1) is to be used. The precise representation of the potential that we use to accomplish this is described below.

The first of these objectives appears rather straightforward to implement. However, a reduction in the number of proteins used to derive the distributions means we will most likely reduce the signal to noise ratio in the PMF. We addressed this problem in the following fashion. At short range, where no systematic errors were observed, we generated the usual distance statistics for each amino acid pair, averaging over proteins of various sizes. In addition to considering amino acid type, we also took into account the secondary structure type (α-helix, β-strand, loop/coil) of the residue pair for short-range statistics. At a pair separation larger than a cutoff distance R_0 (a value of 15 Å was used in all calculations), we grouped the amino acids together according to hydrophobicity. A total of four classes are defined (Table I). The statistics of residue pair i, j were grouped together with those of pair j, i so the total number of pairs was given by $N_{\text{class}}[N_{\text{class}} - 1]/2 + N_{\text{class}}$.

The reduction in the number of pairs from 210 to only 10 offsets the reduction in the number of proteins well enough that we can obtain an adequate

TABLE I
Hydrophobicity Class[a]

Class	Amino Acids
Weakly hydrophobic	Ala, Cys, His, Leu, Met, Phe, Tyr
Strongly hydrophobic	Ile, Trp, Val
Weakly hydrophilic	Asn, Gln, Gly, Pro, Ser, Thr
Strongly hydrophilic	Arg, Asp, Glu, Lys

[a]The definitions used to bin long-range distance statistics according to hydrophobicity are listed.

signal-to-noise ratio. The justification for this approach is that at large separation the probability distribution should not be sensitive to the specifics of the amino acid pair (e.g., the size of the side chain) but only to the propensity to reside on the surface of the protein as opposed to the interior. Support for this idea comes from the work of Yue and Dill [15], who carried out tertiary folding simulations with fixed secondary structure for a series of small proteins, many of which were also studied by us using a Sippl-based potential. What is striking is that, although Yue and Dill used only a two-letter code (hydrophobic and hydrophilic), in many cases their results were qualitatively similar to the ones we obtained using a much higher level of detail in the amino acid pair functions. This suggests that the considerably less drastic simplification we are making (including the retention of a fully detailed pair distribution for short distances, allowing packing effects to be described more accurately) is plausible, although this must of course be validated by the actual results.

The proteins are binned according to radius of gyration using the following formula

$$\text{size} = \text{int}(15Rg^{1/3} - 29) \qquad (3)$$

where $\text{int}(x)$ is the largest integer that is less than or equal to the real number x. Once the long- and short-range pair statistics are accumulated, they can be spliced together to generate a complete distribution for each amino acid pair. The assumption is that in the region around R_0, the individual pair distributions have already converged toward the hydrophobicity class pair distributions. By appropriately scaling the data, a potential valid over all distance ranges is generated for each amino acid pair in each size class.

The second modification was implemented by setting the PMF to a constant at distances outside of the observable range:

$$E_{ij} = \begin{cases} -kT \ln(\varepsilon_1) & (d < d_{\min}) \\ -kT \ln\left[\frac{p_{ij}^1(d)}{p^2(d)}\right] & (d_{\min} < d < d_{\max}) \\ -kT \ln(\varepsilon_2) & (d > d_{\max}) \end{cases} \qquad (4)$$

where d_{min} and d_{max} are the lower and upper bounds, respectively, on the distance range over which we were able to collect good distance statistics (the distribution function had to be greater than or equal to 0.001). The parameters ϵ_1 and ϵ_2 were also set to 0.001. In addition to the residue pair potential above, we included a second long-range energy term that is somewhat analogous to the average hydrophobicity H_0 in the linear case, in that it ensures compactness. This term, which we will refer to as the density profile, is given by

$$E_{ij} = \begin{cases} -kT \ln(p^2(d_x)) & (d < d_x) \\ -kT \ln(p^2(d)) & (d_x < d < d_{max}) \\ -kT \ln(\epsilon_2) & (d > d_{max}) \end{cases} \tag{5}$$

where d_x is the distance at which the residue independent distribution function $p^2(d)$ is a maximum. The final long-range energy is a linear combination of Eqs. (4) and (5) (with weights 1 and 0.6, respectively). The optimization of the density profile in the scoring function is a key ingredient in properly constraining the potential in the large separation limit.

In Eqs. (1)–(5) inter-residue distances are defined in terms of a single side-chain interaction point. This point, which we will refer to for simplicity as C_β, is actually the projection of the average side-chain geometric center onto the C_α–C_β bond vector.

The only function that depends on distances other than C_β–C_β is the excluded volume potential, which depends on C_α–C_α, C_α–C_β, and C_β–C_β distances. The functional form of the excluded volume term is the same as in previous work [16]:

$$E_{ij}^{exvol} = \exp\left(-\left(\frac{d_{ij}}{d_{ij}^0}\right)^{10}\right)$$

where the width of the excluded volume region d^0 is derived from the distance of closest approach for the residue pair in question in the training set.

Equations (4) and (5) are not evaluated explicitly in the minimization program, but are fit using a combination of spline [17] methods, which provide stability, the ability to filter noise easily, and the flexibility to describe an arbitrarily shaped potential curve. Moreover, the final functional form is inexpensive to evaluate, making it amenable to global minimization. The initial step in our methodology is to fit the statistical pair data for each amino acid and for the density profile to Bezier splines [17]. In contrast to local representations such as cubic splines, the Bezier spline imposes global as well as local smoothness and hence effectively eliminates the random oscillatory behavior observed in our data.

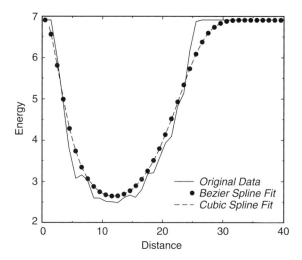

Figure 1. Data smoothing via Bezier and cubic splines. Bezier splines are shown as circular data points which approximate a typical noisy density profile (black line). Cubic splines (dashed line) are then fit to the Bezier data points (at a higher resolution than is shown here).

While Bezier splines are an optimal approach for smoothing noisy data, they cannot be rapidly evaluated using local interpolation methods. We therefore next fit a cubic spline to the Bezier spline curve. Figure 1 compares the Bezier spline and cubic spline curves for the same dataset; it can be seen that there is no meaningful difference between the two. Cubic splines can be evaluated rapidly at an arbitrary value of the residue pair separation using a standard interpolation formula (see, e.g., Ref. 17 for details). The spline coefficients needed for carrying out the interpolation are preprocessed and stored in fast memory during the simulation; the computational effort required to evaluate the spline potential is not much larger than that, for example, to determine the inter-residue distance.

B. Further Improvement of the Potential Energy Function

As Eq. (5) shows, the original form of the PMF used by Sippl and co-workers (1)] remains essentially intact in regions where good statistics are available, although more weight is given to the density distribution. The validity of treating different amino acid pairs as essentially independent, as in Eq. (1), has recently been questioned by Thomas and Dill [18]. They proposed an improved approach based on an iterative algorithm, the goal of which is to have the Boltzmann distribution of distance pairs associated with the potential energy function agree with the distribution derived from native structures. The following are the components of the iterative cycle:

1. Initialize the potential to values obtained from the quasi-chemical approximation.
2. Use this potential to generate structures; determine the relevant distribution functions (in our case, residue pair separation probabilities) from the simulated data.
3. If there are deviations between the two, the potential is corrected so as to minimize them.
4. A simulation is carried out with the new potential, and a new set of statistics is generated.
5. Steps 3 and 4 are repeated until the deviation between the statistics from the simulated data and the experimental data have been reduced to an acceptable level.

For tertiary folding, there are three major problems in implementing this strategy. First, generation of simulated data is computationally expensive if a large training set is to be used. Second, one has to define the ensemble of simulated structures from which to extract statistics. For example, does one keep, only the lowest-energy structure for each protein or keep an ensemble of low-energy structures? Third, there is the question of how to update the potential function. In what follows, we adopt a heuristic approach to these issues; the protocols presented here represent preliminary explorations of this strategy and no doubt can be improved upon. In the present work we have chosen to optimize the potential function by comparing the distribution of locally minimized native structures with that of the native structure itself. The idea is that if the minimized native structure is as close to the native structure as possible, the basin of attraction associated with the minimized native will yield acceptable low RMSD predictions. From numerous computational experiments that we have carried out, resemblance of the minimized native structure to the native structure is clearly a *necessary* condition for obtaining useful predictive results; if the minimized native structure has, for example, a high RMSD from the native, one typically will fail to locate anything reasonable in a full-scale tertiary folding simulation starting from an unfolded state. Whether this is a *sufficient* condition for robust results in such simulations is one of the principal subjects of the present chapter. We briefly summarize here the entire optimization cycle, drawing on the results of the previous sections as well as on the basic idea described above. The steps of the optimization cycle are outlined as follows:

1. Initialization:
   a. The training set of native structures, with secondary structure assigned by DSSP [19], is read into the optimization program.

    b. Proteins are sorted into size bins according to their radius of gyration using Eq. (3).

    c. The iteration counter is initialized to zero ($it = 0$).

    d. A potential energy function E_0 is computed from distance distribution functions based on the native structures.

2. All native proteins are locally minimized using E_{it}.

3. A potential energy function E_{min} is computed for each size bin based on statistics derived from the minimized structures.

4. The difference between E_{min} and E_0 is calculated:

$$E_{diff} = E_0 - E_{min}$$

5. The iteration counter is incremented and E_{it} is updated by adding a correction that is proportional to E_{diff}:

$$it = it + 1$$
$$E_{it} = E_{it-1} + k_{diff}E_{diff}$$

(A proportionality constant equal to 0.1 was chosen empirically so as to damp oscillations in the optimization procedure.)

6. Steps 2–5 are repeated until substantive improvements are no longer produced in the RMSDs of the minimized native structures.

In addition to the RMSD, the energy gap between the native and the minimized native was monitored. The smaller this energy gap, the better in general we have observed the performance of the potential to be in tertiary folding simulations. In our initial efforts we utilized a more elaborate short-range potential function that, in addition to the C_β–C_β term described in Section II A (above), included both C_β–C_α and C_α–C_α terms. The additional terms involving C_α were included in the iteration process described above. Subsequently, however, the extra terms in the short-range potential were not used in the tertiary structure predictions, because we did not see an overall improvement in the results when they were included. Another important difference between the potential energy function used in the above iterative procedure and the one used in actual tertiary structure predictions involves the density profile function. In the iterative procedure, this function was not flattened at $d < d_x$ [see Eq. (5)]. However, we found that we could improve the ranking of native-like structures with this simple modification. Thus the improvement of the potential energy function was ultimately achieved by a combination of the iterative algorithm described above and manual inspection of the individual terms after parameter optimization.

    Because it is computationally expensive to carry out global minimizations on a large test set, we are unable to objectively determine the amount of improvement with respect to the zeroth-order potential (E_0) realized by the optimization procedure outlined above. But given the fact that several proteins, which were unstable in local minimizations starting from the native using E_0, yield acceptable RMSDs using the optimized potential, we believe that parameter optimization can effectively remedy some of the deficiencies of reduced model approaches. The issue of parameter optimization along the lines of the procedure outlined above as well as other approaches in the literature (for a review see Ref. 20) will be the subject of future work.

C. Resulting Potential Energy Function

Table II lists the proteins used in the training set, a subset of the PDB Select database of nonhomologous proteins [21]. We avoided inclusion of proteins that form dimers (or other oligomers) in solution because one would expect the distributions in this case to be significantly altered due to the oligomerization process. For each protein we list the PDB code, number of residues, radius of gyration, and classification in our size bin scheme.

Figures 2–4 show the size dependence of three representative terms in Eq. (4) (after being fit to splines, as described below) for the amino acid pairs arginine–arginine, arginine–isoleucine, and isoleucine–isoleucine for the first six size bins (the bins relevant to the prediction results discussed in this chapter). Figure 5 shows the density profile [Eq. (5)] for the same size bins. Note that because the total energy is a linear combination of Eqs. (4)–(6), the oscillatory behavior at large distances (>15 Å) of the potentials in Figs. 2–4 is effectively masked by the density profile; in the short-distance limit, the excluded volume term serves a similar purpose. The energy plots in Figs. 2–4 show clearly that a linear function is a good approximation over the most populated distance ranges (10–20 Å). Moreover, the slopes in these regions can

TABLE II
Training Set[a]

Size Bin	PDB Name	N_{res}	R_g	Size Bin	PDB Name	N_{res}	R_g	Size Bin	PDB Name	N_{res}	R_g
1	1chl	36	8.8	5	1svr	94	12.1	7	1bvh	153	14.5
1	1erd	35	8.4	5	1vcc	77	12.1	7	1c25	154	14.8
1	1ret	37	8.8	5	1wkt	88	12.1	7	1cdb	101	14.0
1	2erl	35	8.2	5	2abd	86	12.6	7	1cfe	135	14.0
1	3bbg	40	8.7	5	2bby	69	12.0	7	1chd	198	15.0
2	1bor	52	9.3	5	2ezh	65	11.9	7	1cur	150	14.2
2	1dec	39	9.7	5	2fow	76	11.8	7	1def	147	14.0
2	1gps	47	9.6	5	2hgf	97	12.5	7	1eal	127	14.3
2	1sco	38	8.9	5	2hp8	68	11.7	7	1hfc	157	14.6
2	1zwa	29	9.1	5	2rgf	93	12.5	7	1ido	184	14.9
2	2bds	43	9.3	5	2sxl	88	12.6	7	1jpc	108	14.1
3	1afp	51	9.8	6	1a1x	106	13.5	7	1lcl	141	14.3

TABLE II   (*Continued*)

Size Bin	PDB Name	N_{res}	R_g	Size Bin	PDB Name	N_{res}	R_g	Size Bin	PDB Name	N_{res}	R_g
3	1afp	51	9.8	6	1a1x	106	13.5	7	1lcl	141	14.3
3	1apf	49	9.7	6	1a2p.A	108	13.6	7	1mak	113	14.0
3	1ark	56	9.9	6	1acz	108	13.8	7	1mup	157	14.7
3	1awo	57	10.4	6	1bea	116	13.6	7	1mut	129	14.6
3	1brf	53	10.1	6	1bfg	126	13.0	7	1poa	118	14.3
3	1cka.A	57	10.1	6	1bkf	107	13.3	7	1rcf	169	14.5
3	1tih	53	10.6	6	1btn	106	13.1	7	1svp.A	155	14.8
3	1zaq	44	9.9	6	1buz	116	13.2	7	1vhh	157	14.5
3	2brz	53	10.5	6	1bw3	125	13.7	7	2a0b	118	14.7
3	5pti	55	10.6	6	1c52	131	13.5	7	2ezl	99	14.7
4	1ab7	89	11.6	6	1exg	110	13.6	7	2hbg	147	14.7
4	1ah9	66	10.9	6	1fna	91	13.4	7	2hfh	93	13.9
4	1c5a	65	11.2	6	1hcd	118	13.4	7	2i1b	153	14.7
4	1ehs	48	11.6	6	1irs.A	108	13.4	7	2sns	136	14.4
4	1hoe	74	11.4	6	1jer	110	13.5	7	2vil	126	14.0
4	1kbs	60	11.3	6	1krt	110	13.6	7	3cyr	102	14.2
4	1leb	72	11.3	6	1ksr	100	13.8	7	5p21	166	14.8
4	1msi	66	10.7	6	1kte	105	13.2	8	1amx	150	15.4
4	1nkl	78	11.3	6	1kuh	132	13.6	8	1aqb	175	15.8
4	1opd	85	11.6	6	1lit	131	13.4	8	1atl.A	200	15.9
4	1pih	73	10.9	6	1lou	97	13.2	8	1ble	161	15.1
4	1pou	71	11.2	6	1mai	119	13.7	8	1cex	197	15.2
4	1tpn	45	11.0	6	1pne	139	13.8	8	1cto	109	15.1
4	1ubi	71	10.9	6	1rie	123	13.6	8	1kid	189	16.2
4	1uxd	59	11.5	6	1sfp	111	13.4	8	1knb	186	16.1
4	1vif	60	10.9	6	1tit	89	12.9	8	1np4	184	15.5
4	1vig	67	11.2	6	1tul	102	13.5	8	1pkp	145	15.1
4	2ech	49	11.1	6	1whi	122	13.6	8	1ra9	159	15.5
4	2hqi	72	10.7	6	1wiu	93	13.0	8	1rlw	126	15.3
4	2igd	57	10.7	6	2bb8	71	12.9	8	1sfe	165	15.7
4	2sn3	65	10.8	6	2mcm	112	13.4	8	1std	162	16.0
5	1aba	87	12.5	6	2phy	125	13.3	8	1vhr.A	178	15.5
5	1ag4	103	12.5	6	2pld.A	101	13.7	8	1xnb	185	15.2
5	1aoy	74	12.0	6	2tbd	128	13.3	8	1yua	122	15.2
5	1awd	94	11.7	6	3chy	128	13.3	8	2gdm	149	15.1
5	1awj	77	11.7	6	3nll	138	13.6	8	2pth	193	15.4
5	1bdo	80	11.9	7	153l	185	14.9	8	2rn2	155	15.3
5	1bxa	105	12.6	7	1ahk	129	14.8	8	2sak	121	15.4
5	1cyo	88	12.6	7	1ax3	156	14.3	9	119l	162	16.5
5	1mb1	98	12.3	7	1ayo.A	125	14.9	9	1asx	152	16.6
5	1mzm	86	11.9	7	1b10	104	13.9	9	1gky	186	16.4
5	1put	106	12.2	7	1bc4	110	14.5	9	1pbw.B	195	17.3
5	1spy	85	12.2	7	1be1	137	13.9	9	2ucz	164	16.5

[a]The training set listed was used to derive the size-dependent potential. Size bins are defined in terms of radius of gyration (R_g) rather than number of residues (N_{res}).

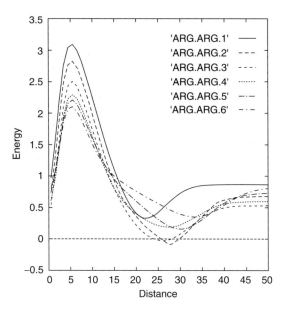

Figure 2. Size dependence of three representative terms in Eq. (4) for the amino acid pair arginine–arginine–arginine. Data for the first six size bins are shown.

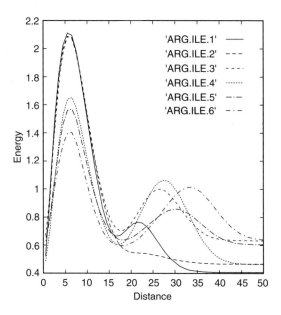

Figure 3. Size dependence of three representative terms in Eq. (4) for the amino acid pair arginine–isoleucine. Data for the first six size bins are shown.

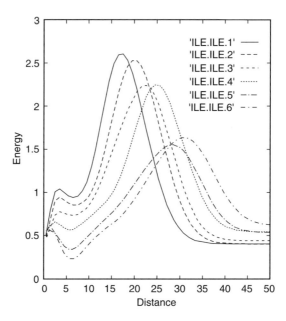

Figure 4. Size dependence of three representative terms in Eq. (4) for the amino acid pair isoleucine–isoleucine. Data for the first six size bins.

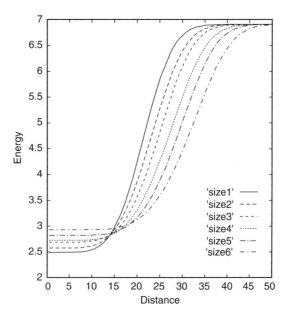

Figure 5. Density profiles for the first six size bins.

be easily rationalized: The arginine–arginine residues are pushed apart, while the isoleucine–isoleucine interaction is attractive. The arginine–isoleucine term is repulsive is well, but the minimum values occur at shorter distances than in the corresponding arginine–arginine plots, consistent with our intuitive picture of a spheroid with hydrophilic residues residing primarily on the surface. Not surprisingly, the basic effect of the density profile is to restrict the interresidue separation as a function of protein size. Note also that the density profile is the most sensitive to protein size (although the isoleucine–isoleucine pair potential clearly decreases with size).

Figure 6 illustrates the effect of adding the excluded volume and density profile to the arginine–arginine, arginine–isoleucine, and isoleucine-isoleucine potentials, respectively, for size bin 6. We see here that the linear portions of the potential are now restricted to a small range in distance (about 6–12 Å), outside of which the density profile and excluded volume become the dominant terms. The energies of each of the three residue pairs at large separation (e.g., 25 Å) relative to their minimum values increase in the expected order ($E_{\text{Ile-Ile}} > E_{\text{Arg-Ile}} > E_{\text{Arg-Arg}}$).

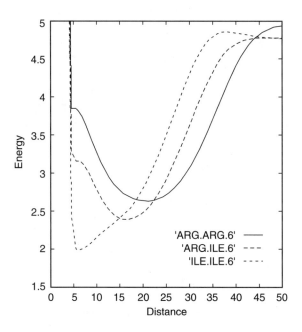

Figure 6. Total energy for three representative residue pairs: arginine–arginine, arginine–isoleucine, and isoleucine–isoleucine. The data corresponds to size bin 6.

III. TERTIARY FOLDING SIMULATIONS: PDB DERIVED
AND IDEAL SECONDARY STRUCTURES

A.   Physical Model

The physical model of the polypeptide chain we use has been described previously [2]; a few minor modifications are introduced as noted below. All bond angles and bond lengths are fixed at ideal values. The variables in the optimization are the torsional angles ϕ and ψ of the peptide backbone. Each residue is represented by a C_α atom and a C_β-like atom. The C_β atom position is given by the average projection of the side-chain center of mass onto the C_α–C_β bond vector.

We employ three different methods to describe the location and three-dimensional structure of secondary structure elements (i.e., α-helices and β-strands). The first is to take both the sequence location and backbone angles (which are frozen during the simulation) directly from the PDB entry. This is obviously not a realistic data set in a predictive situation, but is an essential computational experiment in that it indicates what level of accuracy is possible with "perfect" secondary structure information. The second is the replacement of PDB backbone angles with ideal backbone angles; this separates the effects of distortion of secondary structural elements from ideal geometries from errors in location in the sequence or in length. For these two types of calculations the correct size-dependent potential is selected by evaluating the radius of the gyration of the corresponding native structure. The third is to employ predicted, rather than PDB, secondary structure (along with the use of ideal geometries for the predicted elements) and to select the correct potential by predicting the radius of gyration from the number of residues of the target [22]. We have carried out an extensive investigation in this regard, using secondary structure prediction from various secondary structure prediction servers that are available over the Internet. These results are then combined to produce genuine *ab initio* structural prediction. The results, while far from a robust *ab initio* methodology over all protein types, yield important insights into the key obstacles to *ab initio* prediction and are in many cases surprisingly accurate. Predictions from the CASP3 contest are also included so that comparisons can be made with the work of others. While we are not generating these predictions as a blind test, it is the case that our CASP3 calculations were carried out using our software in a completely automated fashion, with no readjustment of parameters after obtaining results for the CASP3 targets.

B.   Simulation Methodology

Our simulation methodology is identical to that presented in previous publications [2], so we will describe it only briefly here. The algorithm is based on the Monte Carlo plus minimization (MCM) strategy proposed by Li and Scheraga [23]. This approach has proven to be extraordinarily efficacious in our previous work, and the present results reinforce our conclusions concerning its robustness and efficiency in enumerating the low-energy basins of attraction for low-resolution models such as those employed here. As in previous work [2], we have incorporated several key modifications of the algorithm, the most important of which is that the number of minimization steps is annealed as a function of the simulation temperature (i.e., more steps are taken later in the simulation), which yields a factor of 5–10 times reduction in computational effort. Finally, calculations are performed using a parallelized version of the code (an MPI implementation) on a network of PCs using Intel microprocessors and also on a large SGI Origin at the National Center for Supercomputing Applications.

The MCM procedure produces a large number of low-energy structures. The structurally unique predictions are extracted from the raw simulation data by a clustering algorithm. Figure 7 illustrates this process for the protein 1ACP. The raw simulation data (red dots) are combined into structurally similar clusters using a procedure discussed in Ref. 24. The criterion for separating structures into clusters is that the average RMSD between clusters (calculated over all structures in a particular cluster) be at least 5 Å. Clusters are represented by their lowest energy structure (black circles), which means that energies and RMSDs reported for clusters are based on their lowest-energy structure. The

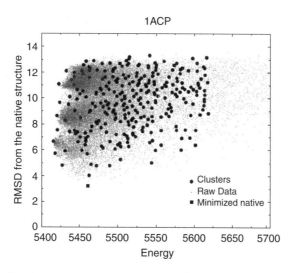

Figure 7. (See also color insert.) Comparison of raw data and clustered results (red dots: raw simulation data, black circles: cluster representatives, green square: locally minimized native structure).

TABLE III
RMSD with Respect to the Native of the ten Lowest Energy Clusters (Represented by Their Lowest Energy Member) for the Protein 1acp[a]

Cluster #	RMSD	Energy	N
1	6.65	5415.43	43
2	5.72	5417.83	58
3	8.64	5420.17	23
4	7.54	5420.80	25
5	11.08	5422.82	31
6	9.65	5424.04	34
7	4.79	5427.21	22
8	5.91	5431.10	16
9	8.27	5431.23	12
10	12.10	5432.68	19

[a]N gives the number of structures combined into a cluster.

RMSD between resulting representative structures is usually at least 5 Å, but this is not guaranteed by the clustering algorithm because we use the average RMSD as the clustering criterion. For the 10 representative structures lowest in energy we list energy, RMSD with respect to the native and number of structures combined into a cluster in Table III, and the RMSD between the representative structures themselves in Table IV. Derivation of the ranks of structures (discussed below) is straightforward given the data in Table III.

The lowest-energy structure obtained from the simulations is generally highly refined, meaning that its energy cannot be lowered significantly by performing more extensive searches. Refinement of higher-energy structures, structures that do not rank first, is possible though and in some of the cases,

TABLE IV
RMSD Between the Representative Structures from the Ten Lowest-Energy Clusters for the Protein 1acp

Cluster #	1	2	3	4	5	6	7	8	9	10
1	0.00	2.23	8.73	2.31	9.91	5.01	7.18	9.35	8.20	12.23
2		0.00	8.70	3.41	9.94	6.04	6.08	8.35	8.16	11.93
3			0.00	8.85	11.86	8.63	8.46	8.24	3.04	8.63
4				0.00	9.70	4.08	7.97	9.91	8.53	12.14
5					0.00	8.69	11.31	11.02	11.77	8.40
6						0.00	10.54	11.91	8.45	11.75
7							0.00	3.01	8.27	10.81
8								0.00	8.46	9.91
9									0.00	9.04
10										0.00

especially the larger proteins, actually results in improved ranks. We have not yet developed the optimal refinement strategy though and therefore do not report results for this approach.

C. Comparison of the Size-Dependent Potential with Previous Results Using PDB-Derived Secondary Structure

As a test set, we employed the subset of the 95 proteins used in Ref. 2 which are less than 100 residues and are not all β-strand. There is some overlap with the training set; but in tertiary folding, this is less of a concern than in secondary structure prediction because the three-dimensional phase space of the protein is so large that as long as an adequate number of proteins are used to generate the pair potential statistics, systematic bias of the results coming from the training set is unlikely to be large. In fact, we see little difference in performance for proteins depending upon whether they were included in the training set or not (or for the CASP3 targets we examined). By retaining the test set used in the previous chapter, we are able to directly compare our new potential with the older potential lacking size dependence, and thus assess the degree of progress that has been made by incorporating size dependence into the potential function.

As discussed above, after the tertiary folding simulations are completed, we group the resulting structures into clusters (without any reference to the native structure, which is presumed to be unknown during clustering) and report the highest-ranking clusters with RMSD from the native below 4 Å, 5 Å, 6 Å, and 7 Å, respectively.

In Table V, we compare these results for our test set with those obtained in Ref. 2. Note that Ref. 2 also included postsimulation screening algorithms; we have not developed such methods for the new potentials because some of the ideas have been incorporated directly into the energy function. Consequently we compare only with results taken directly from the simulations in Table V. However, we note that the overall quality of the results from the new potential is substantially better than those from the old, even when screening is employed in the latter. Table VI summarizes performance for various types of proteins and size classes.

The performance of the new potential function is particularly striking for proteins in the 50–100 residue size. For α-helical proteins in this category, the average rank of the best structure less than 7 Å is 3.6; furthermore, in the overwhelming majority of cases, the rank is 5 or better. This is a sufficient reduction in the number of possible structures that discrimination among the resulting structures via more expensive calculations at an atomic level of detail [25] becomes feasible. The reliability of the results demonstrates that the basic physics of the low-resolution model have been qualitatively improved as compared to previous efforts.

TABLE V
Comparison to Previous Results[a]

				"Old" Potential PDB—X-RAY			Size-dependent Potential PDB—X-RAY					PDB—IDEAL				
	N_{res}	N_α	N_β	<5Å	<6Å	<7Å	<4Å	<5Å	<6Å	<7Å	LER	<4Å	<5Å	<6Å	<7Å	LER
Alpha Proteins ($N_{res} < 50$)																
1ajj	17	6	0	—	1	1	—	1	1	1	4.0	—	1	1	1	4.9
1bgk	27	18	0	4	2	1	2	2	2	1	6.5	2	2	2	1	6.2
1erd	29	25	0	1	1	1	1	1	1	1	3.8	1	1	1	1	3.3
2erl	35	29	0	2	2	2	—	1	1	1	4.9	1	1	1	1	2.8
1res	35	27	0	3	1	1	1	1	1	1	3.5	1	1	1	1	3.8
1roo	17	14	0	1	1	1	1	1	1	1	3.7	1	1	1	1	3.7
1uxd	43	31	0	1	1	1	4	4	4	1	6.0	—	4	4	1	6.4
Mixed Alpha/Beta Proteins ($N_{res} < 50$)																
1aho	31	10	10	5	3	1	7	5	2	1	6.8	3	3	2	2	7.5
1ayj	46	11	15	33	1	1	—	—	2	2	7.7	—	3	3	2	8.6
1cmr	26	8	10	3	1	1	3	2	2	1	6.6	4	4	3	1	6.8
1gpt	47	13	19	23	2	2	13	13	12	3	8.1	—	—	2	2	8.9
1hev	25	7	11	1	1	1	3	1	1	1	5.0	—	3	2	2	7.1
2ktx	34	11	14	1	1	1	1	1	1	1	3.6	—	1	1	1	4.2
1pce	30	12	10	2	2	2	1	1	1	1	2.8	—	—	1	1	5.1
1ptq	43	6	8	732	21	18	—	—	20	11	8.6	—	—	16	1	6.8
2sn3	48	8	15	94	21	7	—	29	2	2	8.5	—	13	3	3	8.9
2vgh	34	6	12	126	61	21	—	—	—	4	7.1	—	—	—	3	8.2
1vtx	36	7	10	—	78	2	—	—	34	3	7.8	—	—	9	1	7.0
5znf	25	12	11	1	1	1	1	1	1	1	2.6	—	—	1	1	6.0
Alpha Proteins ($50 < N_{res} < 100$)																
1acp	73	45	0	256	115	30	—	7	2	1	6.7	—	—	11	11	11.3
1ail	67	60	0	5	5	2	1	1	1	1	3.0	1	1	1	1	3.9
1aj3	95	86	0	2	2	2	2	2	2	2	9.3	2	1	1	1	4.6
1am3	57	45	0	—	8	8	—	6	6	2	10.7	—	24	5	1	6.1
1c5a	62	49	0	1	1	1	—	3	3	2	8.2	10	3	3	3	8.0
1cc5	76	41	0	—	78	21	—	6	6	2	8.5	—	18	6	3	7.2
1ddf	87	66	0	—	7	7	—	63	3	2	12.7	—	58	8	8	7.1
2ezh	59	45	0	16	5	2	1	1	1	1	3.8	3	3	3	2	9.7
2ezk	76	64	0	28	8	1	—	—	1	1	5.7	—	—	1	1	5.9
2hp8	56	44	0	—	4	2	—	2	2	2	9.7	—	2	2	2	7.1
1hsn	62	46	0	88	88	67	—	—	19	19	11.4	—	—	98	17	8.3
1jvr	74	59	0	5	5	5	31	31	1	1	5.3	—	10	9	7	10.4
1lfb	69	48	0	—	94	94	—	—	5	5	10.4	—	15	11	11	10.6
1mzm	71	54	0	—	8	8	—	5	4	4	10.7	—	3	2	2	11.0
1nkl	70	56	0	—	—	2	1	1	1	1	3.9	2	2	2	2	9.6
1nre	66	55	0	22	22	22	22	1	1	1	4.9	19	1	1	1	4.6
2pac	77	26	0	—	—	136	—	—	53	1	6.4	—	—	76	5	11.2
1pou	70	57	0	—	6	6	1	1	1	1	2.3	4	4	4	4	11.2
1r69	61	41	0	46	9	8	—	6	6	3	11.3	—	23	12	5	10.7

TABLE V   (*Continued*)

			"Old" Potential			Size-dependent Potential									
			PDB—X-RAY			PDB—X-RAY					PDB—IDEAL				
	N_{res}	N_α	N_β	<5Å	<6Å	<7Å	<4Å	<5Å	<6Å	<7Å	LER	<4Å	<5Å	<6Å	<7Å	LER
1utg	62	53	0	4	2	1	—	21	1	1	5.6	—	14	1	1	5.3
5icb	72	52	0	—	—	—	8	8	2	1	6.1	—	—	8	1	6.2

Mixed Alpha/Beta Protein ($50 < N_{res} < 100$)

	N_{res}	N_α	N_β	<5Å	<6Å	<7Å	<4Å	<5Å	<6Å	<7Å	LER	<4Å	<5Å	<6Å	<7Å	LER
1aa3	56	31	8	—	—	—	19	19	6	3	8.4	7	7	7	5	9.4
2acy	92	24	41	—	—	16	—	—	5	5	12.0	—	—	—	—	13.0
1ag2	97	58	8	—	—	349	—	—	—	87	10.9	—	—	—	187	12.3
1bor	52	9	14	187	22	8	—	—	17	6	7.2	—	—	40	12	8.3
1btb	89	45	19	—	274	24	1	1	1	1	3.8	—	—	31	28	8.1
1ctf	67	38	19	15	12	4	1	1	1	1	3.0	—	—	4	4	11.1
2fdn	53	8	6	123	4	4	—	—	38	6	8.1	—	—	—	30	10.3
2fow	66	29	8	181	56	8	—	—	23	8	10.6	—	—	69	4	7.9
1fwp	66	22	17	484	2	2	—	3	3	3	10.3	—	42	10	10	10.3
1gbl	54	13	16	1	1	1	—	—	15	1	6.5	—	—	2	1	6.5
1pgx	57	15	33	4	4	4	2	2	2	2	9.5	—	35	28	11	8.1
1leb	63	36	6	142	27	4	—	3	3	3	10.9	—	6	6	6	8.7
1orc	56	25	17	2	2	1	8	6	6	6	7.1	46	2	2	1	6.2
5pti	55	16	14	109	16	16	—	—	14	4	10.1	—	—	47	14	7.1
2ptl	60	15	34	1	1	1	1	1	1	1	3.4	—	35	4	4	8.2
1ris	92	25	42	1	—	180	11	9	9	9	11.1	—	—	129	11	11.7
1svq	90	22	34	—	—	—	—	119	117	32	12.5	—	—	462	43	9.0

[a]Following global energy minimization, structures are clustered without reference to the native; the energetic ranks of clusters that have an RMSD close to the native (for old results, three RMSD cutoffs—5 Å, 6 Å, and 7 Å—were used; for new results, four RMSD cutoffs—4 Å, 5 Å, 6 Å, and 7 Å—were used). Energetic rank was defined so that the lowest-energy structure ranks 1, the second-lowest ranks 2, and so on. LER refers to the RMSD of the lowest-energy structure. The column "PDB—X-Ray" list's results of runs using location and configuration of secondary structure derived from the PDB entry. Column "PDB-Ideal" lists results for calculations where the location of secondary structure was derived from the PDB, but configuration of secondary structural elements was assumed to be ideal.

For mixed α/β-proteins, the absolute quality of the results is somewhat diminished, but the improvement as compared to previous work is even larger. There are two cases, 1ag2 and 1svq, where the rank obtained for the best low RMSD structure is above 10, with the 1ag2 result being particularly problematic. We have investigated this case further and show improved results for 1ag2 below. On the other hand, there is a significant number of cases for which no reasonable structures were recovered previously which now rank in the top 10.

The energies of structures located by the global optimization algorithm are lower than the native and locally minimized native structures in all cases, a

TABLE VI
Summary of Ranks Listed in Table V[a]

(a)

Class	N_{prot}	N_{conv}	Ave Rank	Max Rank	N_{conv}	Ave Rank	Max Rank	N_{conv}	Ave Rank	Max Rank	N_{conv}	Ave Rank	Max Rank
		RMSD < 4 Å			RMSD < 5 Å			RMSD < 6 Å			RMSD < 7 Å		
Small α	7	—	—	—	6	2	4	7	1	2	7	1	2
Small α/β	12	—	—	—	11	93	732	12	16	78	12	5	21
Medium α	21	—	—	—	11	43	256	18	26	115	20	21	136
Medium α/β	17	—	—	—	11	114	484	13	46	274	15	30	349

(b)

Class	N_{prot}	N_{conv}	Ave Rank	Max Rank	N_{prot}	Ave Rank	Max Rank	N_{conv}	Ave Rank	Max Rank	N_{prot}	Ave Rank	Max Rank
		RMSD < 4 Å			RMSD < 5 Å			RMSD < 6 Å			RMSD < 7 Å		
Small α	7	5	2	4	7	2	4	7	2	4	7	1	1
Small α/β	12	7	4	13	8	7	29	11	7	34	12	3	11
Medium α	21	8	8	31	17	10	63	21	6	53	21	3	19
Medium α/β	17	7	6	19	10	16	119	16	16	117	17	10	87

(c)

Class	N_{prot}	N_{conv}	Ave Rank	Max Rank	N_{conv}	Ave Rank	Max Rank	N_{conv}	Ave Rank	Max Rank	N_{conv}	Ave Rank	Max Rank
		RMSD < 4 Å			RMSD < 5 Å			RMSD < 6 Å			RMSD < 7 Å		
Small α	7	5	1	2	7	2	4	7	2	4	7	1	1
Small α/β	12	2	4	4	6	5	13	11	4	16	12	2	3
Medium α	21	7	6	19	16	11	58	21	13	98	21	4	17
Medium α/β	17	2	26	46	6	21	42	14	60	462	16	23	187

[a]Part a lists old results; part b lists results using the size-dependent potential and X-ray-derived secondary structure; part c lists results using the size-dependent potential and ideal secondary structure. The number of proteins N_{prot} is listed in column 2; the number of cases that converged within a specified RMSD from the native (< 4 Å, < 5 Å, < 6 Å, or < 7 Å) N_{conv} is listed in columns 3, 6, 9, and 12. (Note that the rank < 4 Å was not calculated for the old results, so a "—" is shown). Also listed are the average and maximum rank of converged clusters within each RMSD range.

feature that other groups using similar approaches have also observed [25]. A very important aspect of the results though, not apparent in the data presented here, is that for all simulations discussed above, the energy gap between the lowest-energy misfolded structures and low-energy native-like structures is quite small, on the order of 5–30 energy units where the energy scale is

TABLE VII
Comparison of Rankings for PDB Secondary Structure and DSSP Secondary Structure for Several
Cases from the Test Set

Protein	<4 Å	<5 Å	<6 Å	<7 Å	Comments
1ag2	—	—	—	87	PDB secondary structure, terminal loops deleted
1ag2	—	—	—	11	DSSP secondary structure, terminal loops included
1hsn	—	—	19	19	PDB secondary structure, terminal loops deleted
1hsn	—	—	23	10	DSSP secondary structure, terminal loops deleted
1orc	8	6	6	6	PDB secondary structure, terminal loops deleted
1orc	1	1	1	1	DSSP secondary structure, terminal loops included
1ris	9	9	9	9	PDB secondary structure, terminal loops deleted
1ris	—	—	3	3	DSSP secondary structure, terminal loops included

typically thousands of energy units. This is in sharp contrast to the results obtained with our previous tertiary folding potential, which routinely generated energy gaps between misfolded and native-like structures that were 5–10 times larger than those seen here.

D. Effects of Secondary Structure Definition and Truncation of Terminal Loops

The results presented above employ PDB-defined secondary structure and in some cases involve truncation of terminal loops, primarily carried out here to facilitate direct comparisons with the results of Ref. 2. However, the process of defining secondary structure even with X-ray crystallographic or NMR coordinates in hand is not entirely unambiguous, and the effects of terminal loops could be favorable or unfavorable. To examine these issues, we selected several proteins in Table V for which the results with the new potential appeared less accurate than would have been expected given the difficulty of the case being considered. Table VII presents results for these selected cases, listing the protein and identifying what experiments were carried out. Most of the cases examined are mixed α/β because these displayed the most significant problems. It can be seen that in some cases the use of a different secondary structure definition (e.g., DSSP rather than PDB) and the inclusion or deletion of a terminal loop has a substantial effect on the ranking of low RMSD structures. Clearly, more work needs to be done in understanding these effects.

E. Effects of Using Ideal Rather than PDB-Derived Three-Dimensional Topologies for Secondary Structure Elements

Having established that our new size-dependent potential is quite effective for generating low-resolution structures of proteins below 100 residues using secondary structure derived from PDB coordinates, we next ask what the effect is of using ideal torsional angles for helices and strands as opposed to PDB-derived torsion angles. Tables V and VIc summarize results for the entire test set of proteins utilizing ideal secondary structure elements. The results are surprisingly good; while there are certainly cases in which quantitative degradation of the rank of the best low-RMSD structure occurs (particularly with α/β-proteins—for example, the proteins 2fdn, 1fwp, and 5pti), in general the simulations are able to find such structures successfully and to rank them reasonably well in terms of total energy. Even in the case of 5pti, where there is severe distortion of the β-strands in the native structure, the use of ideal strands produces reasonable results. While incorporation of strand distortion is possible in our methodology [4], the reasonable predictive capability using ideal elements is likely to save considerable computational effort because one can carry out such simulations initially and then use the results as a starting point from which to incorporate distortions and other detailed effects.

IV. USE OF PREDICTED RATHER THAN PDB-DERIVED SECONDARY STRUCTURE ELEMENTS

A. Overview

Secondary structure prediction methods, while they have improved significantly over the past decade (principally via the use of multiple sequence analysis), still have nontrivial error rates. The best method at present appears to be the PSIPRED approach developed by Jones [26], which is claimed to achieve an accuracy between 76% and 78% on a reasonably large training set (it also outperformed other methods in the CASP3 contest). This level of reliability appears to be sufficient for low-resolution *ab initio* structure prediction and suggested to us that it was now worth experimenting with tertiary folding calculations based entirely on predicted, rather than PDB-derived, secondary structure [27–32]. Using servers set up on the World Wide Web, we are able to obtain predictions from PSIPRED and other secondary structure prediction algorithms for proteins in our test set. We have obtained results from a variety of servers to see what happens in cases where their predictions disagree; it is likely that *ab initio* prediction will involve trying a number of secondary structures, because in some cases the tertiary fold will be critical in selecting among plausible secondary structures predicted exclusively from sequence data.

Our calculations in this section endeavor to answer the following questions:

1. Can we for some percentage of cases make a successful *ab initio* prediction? We explore two different approaches below.

2. What are the effects of small errors in secondary structure—for example, elimination or addition of small elements, incorrect lengths of major elements and so on?

3. What is the impact of a major error—for example, replacing a long helix by a similar strand or missing an important loop?

In the present chapter, we have chosen to focus our *ab initio* prediction efforts primarily on α-helical proteins, although one mixed α/β-protein is also examined. The *ab initio* prediction calculations presented below are considerably more computationally intensive than those using PDB-derived secondary structure, because we have investigated a substantial number of secondary structure predictions for each protein. By studying helical systems intensively, we are able to draw conclusions concerning the necessary and sufficient conditions for success for such systems from a significant database of results. In addition to the α-helical proteins in the 50- to 100-residue range from the data set above, we also include two helical proteins from the CASP3 prediction contest. Our results for the CASP3 test cases are similar to those from the PDB-derived test suite.

B. Secondary Structure Prediction Methods

We use the following secondary structure prediction methods in our *ab initio* predictions:

- PSIPRED [26]: A two-stage neural network that predicts protein secondary structure based on the position specific scoring matrices generated by PSI-BLAST (available at *http://insulin.brunel.ac.uk/psipred/*). Average three-state prediction accuracy is between 76.5% and 78.3%. Currently the most accurate method.

- PhD [33,34]: Secondary structure is predicted by a system of neural networks (available at *http://cubic.bioc.columbia.edu/pp/*). Overall three-state prediction accuracy is 72.1%. The default secondary structure prediction settings were used in all predictions.

- JPRED [35,36]: A methodology that combines a total of six secondary structure predictions into one consensus prediction (available at http://jura.ebi.ac.uk:8888/ at the time of this writing). Average "real world" accuracy is 72.9%. Note that the PhD predictions generated by JPRED differ from the original PhD predictions (denoted: orig_phd) mentioned above. In addition to using the consensus prediction, we also report results for the six individual prediction methods included in the JPRED server.

By default, secondary structure prediction accuracies reported here are determined with DSSP as the reference (for details see Ref. 26). The secondary structure assignments used in the actual calculation differ from the original predictions in that helices and strands of less than three residues are eliminated. N- and C-terminal loops are deleted.

C. Simulation Protocols

The amino acid sequence of the target represents the only input data for our methodology. We do not carry out explicit database searches (i.e., threading) of any sort. Secondary structure predictions from the sources listed above are parsed and used directly in the structure predictions. In the case of JPRED we examine individually the results of all predictions that contribute to the consensus prediction (DSC [37], PhD [33,34], PREDATOR [38,39], NNSSP [40], Mulpred, and Zpred [41]). Because we do not assume any knowledge of approximate radius of gyration of the target, which is important for the selection of the correct potential energy parameters, we predict the radius of gyration via a simple formula [22] and use this prediction to assign the size bin for the tertiary folding simulation.

The first stage of our prediction algorithm applies the MCM-based approach described above to each of the nine secondary structure predictions for each target. Simulations are usually carried out on two to four nodes of a multiprocessor machine and take between 12 and 24 hours depending on protein size. To extract the structurally unique predictions, we apply the clustering algorithm discussed above. Table VIII shows the results of this procedure for the three targets discussed in more detail below. We list results for every secondary structure prediction (unless predictions consist only of loop or coil, in which case we did not believe it worthwhile to carry out the simulation).

Because it is quite possible that simulations utilizing different secondary structure predictions results in very similar representative low energy structures, we apply a second level of filtering which basically tries to eliminate structurally similar predictions and ranks the resulting "unique" predictions on an absolute energy scale. The first step in this process is the determination of the subset of residues common to all predictions (regardless of whether they belong to helices or strands). Secondary structure predictions for which the number of residues included in the simulation is substantially smaller than the average (due

TABLE VIII
Individual Clustering Results for the *ab initio* prediction Targets Discussed in More Detail in the Text (Stage 1 of the Composite Prediction Method)[a]

Protein	SSP	Q3	N_{res}	$<4\,\text{Å}$	$<5\,\text{Å}$	$<6\,\text{Å}$	$<7\,\text{Å}$
1aj3	cons	94.90	93	—	—	—	—
1aj3	dsc	87.76	93	—	—	—	—
1aj3	mul	86.73	92	—	—	—	—
1aj3	nnssp	95.92	98	—	—	2	2
1aj3	orig_phd	89.80	89	—	—	3	2
1aj3	phd	88.78	88	—	—	—	—
1aj3	pred	88.78	92	—	—	—	—

TABLE VIII (*Continued*)

Protein	SSP	Q3	N_{res}	<4 Å	<5 Å	<6 Å	<7 Å
1aj3	psipred	93.88	94	—	—	1	1
1aj3	zpred	93.88	98	—	—	2	2
1am3	cons	92.86	58	—	—	24	3
1am3	dsc	88.57	57	—	—	11	2
1am3	mul	77.14	59	—	—	11	11
1am3	nnssp	88.57	60	—	8	8	8
1am3	orig_phd	92.86	58	—	22	8	6
1am3	phd	94.29	58	—	4	4	2
1am3	pred	72.86	57	—	—	—	—
1am3	psipred	88.57	57	—	35	1	1
1am3	zpred	67.14	68	—	—	—	—
1mzm	cons	44.09	44	—	—	2	2
1mzm	dsc	66.67	67	—	—	26	3
1mzm	mul	37.63	68	—	—	243	4
1mzm	nnssp	38.71	93	—	—	—	—
1mzm	orig_phd	59.14	74	—	—	50	3
1mzm	phd	38.71	49	—	—	9	1
1mzm	pred	55.91	44	—	—	18	6
1mzm	psipred	78.49	78	—	1	1	1
1mzm	zpred	34.41	82	—	—	—	—
1eh2	cons	87.37	68	—	12	5	5
1eh2	dsc	80.01	73	—	—	—	—
1eh2	mul	74.74	43	—	3	3	3
1eh2	nnssp	86.32	67	—	6	3	1
1eh2	orig_phd	86.32	68	—	15	1	1
1eh2	phd	85.26	67	15	4	4	4
1eh2	pred	88.42	68	—	1	1	1
1eh2	psipred	95.79	72	—	3	2	1
1eh2	zpred	66.32	72	—	—	32	13
1bg8.A	cons	57.89	57	—	—	11	3
1bg8.A	dsc	42.11	55	—	—	64	6
1bg8.A	mul	51.32	67	—	—	—	24
1bg8.A	nnssp	63.16	67	—	—	31	9
1bg8.A	orig_phd	57.89	57	—	—	52	5
1bg8.A	phd	57.89	57	—	—	34	8
1bg8.A	pred	38.16	56	—	—	321	3
1bg8.A	psipred	50.01	52	—	92	17	5
1bg8.A	zpred	46.05	68	—	—	—	—

[a]Here N_{res} refers to the number of residues actually considered for every prediction. (cons: JPRED consensus prediction; dsc: DSC; mul: MULPRED; nnssp: NNSSP; orig_phd: PhD in its most current implementation; phd: PhD as run by JPRED; pred: PREDATOR; psipred: PSIPRED; zpred: ZPRED). Q3 refers to the three-state accuracy of a given prediction.

to deletion of terminal loops) are not considered at this stage. This set of residues is then extracted from the 50 clusters lowest in energy for every secondary structure prediction, and the energies of the resulting substructures are evaluated. After a second round of clustering, we obtain the final set of clusters (Table IX). At this point the RMSDs with respect to the native structures are reevaluated over the subset of common residues to allow a fair comparison of the tertiary folding results obtained from different secondary structure predictions. We refer to this method below as the composite energy prediction method.

D.   Final Rankings of Structures for Fully *Ab Initio* Predictions

We examine the use of two different approaches for producing fully *ab initio* predictions for the 22 proteins studied in this section. One approach is simply to use the secondary structure prediction with the highest calibrated prediction—accuracy—in this case, PSIPRED. Results for this approach are summarized in

TABLE IX

Final Clustering Results for the Subset of Common Residues for All *Ab Initio* Prediction Targets
(Stage 2 of the Composite Energy Prediction Method)[a]

Protein	N_{res}	$<4\,\text{Å}$	$<5\,\text{Å}$	$<6\,\text{Å}$	$<7\,\text{Å}$
1acp	70	—	—	10	5
1aj3	88	—	89	89	89
1am3	56	—	17	17	2
1bg8.A	52	—	—	92	1
1c5a	57	—	—	4	4
1cc5	68	—	22	12	2
1ddf	85	—	—	—	7
1eh2	65	—	4	4	3
1hsn	61	—	—	—	46
1jvr	66	39	12	12	2
1lfb	55	—	114	22	4
1mzm	66	—	—	65	4
1nkl	63	—	—	31	1
1nre	65	—	89	50	50
1pgx	53	—	—	—	35
1pou	64	30	3	3	1
1r69	57	—	5	5	1
1utg	56	—	—	4	3
2ezh	57	—	—	7	4
2ezk	67	—	—	—	—
2hp8	49	—	58	8	7
2pac	53	—	25	7	3

[a]Here N_{res} refers to the number of residues for which RMSD and energy are evaluated. We omitted predictions that were too short as compared to all others and the length of the sequence (1eh2: mul; 1jvr: psipred; 1mzm: cons, phd, pred; 1r69: mul, orig_phd, pred; 2ezh: orig_phd).

TABLE X

Individual Clustering Results for All *Ab Initio* Prediction Targets Using the PSIPRED Secondary Structure Predictions[a]

Protein	Q3	N_{res}	$<4\,\text{Å}$	$<5\,\text{Å}$	$<6\,\text{Å}$	$<7\,\text{Å}$
1acp	83.12	72	—	—	17	5
1aj3	93.88	94	—	—	1	1
1am3	88.57	57	—	35	1	1
1bg8.A	50.01	52	—	92	17	5
1c5a	93.94	63	4	1	1	1
1cc5	74.7	75	—	6	4	4
1ddf	81.1	86	—	—	43	9
1eh2	95.79	72	—	3	2	1
1hsn	87.34	62	—	—	—	20
1jvr	72.26	3	—	—	—	—
1lfb	58.97	59	—	—	—	34
1mzm	78.49	78	—	1	1	1
1nkl	94.87	71	—	—	4	1
1nre	60.49	65	—	—	—	380
1pgx	77.14	60	—	—	—	—
1pou	73.24	67	7	5	5	5
1r69	84.13	59	—	6	6	3
1utg	85.71	62	—	32	1	1
2ezh	81.54	57	—	15	1	1
2ezk	51.61	77	—	—	—	—
2hp8	64.71	53	—	6	2	1
2pac	70.73	77	—	19	2	2

[a]Here N_{res} refers to the number of residues actually considered for every prediction.

Table X. As above we list the rank of structures below a certain RMSD cutoff. The second is the composite energy prediction method discussed above. We summarize statistics for the success rate of each of these two approaches on the entire test set and CASP3 prediction targets below.

E. Results

1. Summary and Overall Success of Fully Ab Initio Prediction

We begin by summarizing the results for all of the secondary structure prediction methods (including the composite energy prediction method described above) and all of the target proteins. As in previous sections of this chapter, the ranks of the lowest-energy cluster with RMSDs from the native structure of 4 Å, 5 Å, 6 Å, and 7 Å are reported for both approaches. The first, and most striking, observation is that both approaches provide a surprisingly good success rate for *ab initio* prediction based on criteria used in CASP3. We have observed that for proteins in the 50–100 residue range, an RMSD below 7 Å typically provides a qualitatively reasonable folding topology at low resolution. Similar conclusions have been reached by Skolnick and co-workers [42] and by Cohen and Sternberg [43], whose estimates show that the probability of achieving a structure below 6 Å RMSD by chance is vanishingly small. Note also that for a significant fraction of proteins, structures below 6 Å are found; at this level, the correspondence with the native structure is quite satisfactory in agreement with the chapters cited above.

Both proposed fully *ab initio* prediction methods (composite energy method and exclusive use of PSIPRED predictions) yield a number of cases in which a low RMSD structure is ranked first; this would count as a successful prediction under any criterion. Using the assessment criteria of CASP3—that is, a maximum of five predictions—the composite energy method would achieve an RMSD of less than 7 Å in 68% of the cases; there are also four cases where the RMSD is less than 6 Å. Reliance entirely on PSIPRED would lead to an RMSD under 7 Å in 64% of the cases; however, 11 of those would have an RMSD under 6 Å. Thus, the use of the composite energy method appears to succeed slightly more often, however, the use of PSIPRED exclusively generates highly accurate predictions in significantly more cases.

We have employed the protocol described above in a completely automated fashion; but only in an actual blind test can one be sure that the results suffer from no unconscious bias. If these results hold up under truly blind test conditions, this would represent a significant advance in *ab initio* prediction methodology as judged by other *ab initio* efforts in CASP3.

While our new potential energy function certainly represents a step forward, there are also obviously areas where more work needs to be done. Primarily, the causes of failure to routinely achieve a low-RMSD structure in the top five predictions in some cases must be analyzed and understood. These failures are thus more interesting at this point than the successes because they point the way to development of an improved methodology. We therefore analyze a number of these cases in detail below so as to reveal the underlying difficulties and directions in which solutions must be developed.

2. Detailed Analysis of Specific Cases

Figure 8 presents the detailed secondary structure predictions for each of the cases that we analyze below. In conjunction with the tertiary folding results summarized in Table VIII, as well as the results using PDB-derived secondary structure presented above, we can extract insight into how various types of errors in secondary structure prediction affect tertiary folding accuracy. Due to the large amount of data, we have selected a subset of interesting examples to analyze in detail, however, the conclusions, summarized in the discussion following consideration of individual examples, reflect an examination of the results for all 22 of the proteins studied.

1aj3:

AA: |HQFFRDMDDEESWIKEKKLLVSSEDYGRDLTGVQNLRKKHKRLEAELAAHEPAIQGVLDTGCKKLSDDNTIGKEEIQQRLAQFVDHWKELKQLAAARGQ| DSSP AA

SS: (secondary structure prediction rows)

cons 94.90
dsc 87.76
mul 86.73
nnssp 95.92
orig_phd 89.80
phd 88.78
pred 88.78
psipred 93.88
zpred 93.88

(a)

1am3:

AA: |MDIRQGPKEPFRDYVDRFYKTLRAEQASQEVKNWMTETLLVQNANPDCKTILKALGPGATLEEMMTACQG| DSSP AA

cons 92.86
dsc 88.57
mul 77.14
nnssp 88.57
orig_phd 92.86
phd 94.29
pred 72.86
psipred 88.57
zpred 67.14

(b)

Figure 8(a–e). Secondary structure predictions for 1aj3, 1am3, 1mzm, 1eh2, 1bg8 Chain A (cons: JPRED consensus prediction; dsc: DSC; mul: MULPRED; mssp: NNSSP; orig_phd: PhD in its most current implementation; phd: PhD as run by JPRED; pred: PREDATOR; psipred: PSIPRED; zpred: ZPRED). References for the secondary prediction algorithms are given in the text.

253

1mzm:

AA: IAISCGQVASAIAPCISYARGQGSGPSAGCCSGVRSLNNAARTTADRRAACNCLKNAAAGVSGLNAGNAASIPSKCGVSIPYTISTSTDCSRVN | DSSP AA
    1234567890123456789012345678901234567890123456789012345678901234567890123456789012345678901234567890123

SS:	IIIIIIIIIIIIIIIIIIIII IIIIIIIIIIIIIIIIII IIIIIIIIIIIIIII	DSSP
SS:	HHHHHH HHHHHH HHHH	cons 44.09
SS:	EEEEE HHHHHHHHHHHHH EEEE	dsc 66.67
SS:	EEE E EEEHHHHH EEEE E	mul 37.63
SS:		nnssp 38.71
SS:	HHHHHHHHHH HHHHHHHHHHHHHHH EEEEEEEEE	orig_phd 59.14
SS:	EE HHHHH HHHHH EEEE EEE	phd 38.71
SS:	EEE HHHHHHHH HHHHHHHHHHH EEEEE	pred 55.91
SS:	HHHHHHHHHHH HHHHHHHHHH HHHHHH	psipred 78.49
SS:	EEEEEEEEEEEEEEE EEEEEEEE HHHHHEEHHHHEE EEEEEEEEEE E	zpred 34.41

(c)

1eh2:

AA: IPWAVKPEDKAKYDAIFDSLSPVNGFLSGDKVKPVLLNSKLPVDILGRVWELSDIDHDGMLDRDEFAVAMFLVYCALEKEPVMSLPPALVPPSKR | DSSP AA
    1234567890123456789012345678901234567890123456789012345678901234567890123456789012345

SS:	HHHHHHHHHH EHHHHHHHHH HHHHHHHHH E HHHHHHHHHHHHHHHHHH	DSSP
SS:	E HHHHHHHHH HHHHHHHHH HHHHHHHHHH	cons 87.37
SS:	HHHHHHHHHHHHHH HHHHHHHHHHHHHHHHHHHHHH	dsc 80.00
SS:	HHEE HHEEEH HHHHHHHH HHHHHHHHHHHHHH	mul 74.74
SS:	HHHHHHHHH HHHHHHHH HHHHHHHHHHHH	nnssp 86.32
SS:	E HHHHHHHHH EE HHHHHHHHH HHHHHHHHHHHH	orig_phd 86.32
SS:	EE HHHHHHHHH HHHHHHHHH HHHHHHHHHHHH	phd 85.26
SS:	HHHHHHH HHHHHHHHHH HHHHHHHHHHHH	pred 88.42
SS:	HHHHHHHHHHH HHHHHHHHHH HHHHHHHHHHHHHHH	psipred 95.79
SS:	HHHHHHH EEEE EE HHHEEEE EEEHHHHHEH HHHHHHHHHHHHHH	zpred 66.32

(d)

Figure 8(a–e) (Continued)

```
1bg8 - Chain A:

AA: |KKPVNSWTCEDFLAVDESFQPTAVGFAEALNNKDKPEDAVLDVQGIATVTPAIVQACTQDKQANFKDKVKGEWDKI| DSSP AA

     1234567890123456789012345678901234567890123456789012345678901234567890123456
SS: | HHH EHHHHH     HHHHHHHHHH       HHHHHHHHHHHHHHHH       EHHHHHHHHHH        | DSSP
SS: |     HHHHH      HHHHHHHH          EE     HHHHHHHHHHHHHH                    |          cons     57.89
SS: |     HHHHH      HHHHHHHH          HHEEE    HHHH                           |          dsc      42.11
SS: |     HHHHHHH      HHHH H        HH   H EEEE EEEEEEH   H HHHH HHHHHHHH      |          mul      51.32
SS: |     HHHHHHHH     HHHHHHHH         EE      HHHHHHHHH HHHHHHHHHHHHHHHHHH    |          nnssp    63.16
SS: |     HHHHHH       HHHHHHHH         EE  E EE   HHHHHHHHHHHHHH               |          orig_phd 57.89
SS: |     HHHHH        HHHHHHHH         EE  E EE   HHHHHHHHHHHHH                |          phd      57.89
SS: |     EEE          HHHHHHHH        EEEE      EEEE HHHHHHHHHHHHH             |          pred     38.16
SS: |     HHHHH        HHHHHHHH        EEEEE    EEEHHHHHHHHHHHHHH               |          psipred  50.00
SS: | EHHHHHHHHHHH     HHHHHHHH        HHHHHHHHEEEEEEEEEEEEEHHHHHHHHHHHHHHHHHHH |          zpred    46.05
```

(e)

Figure 8(a–e) (Continued)

255

1aj3: This is a case for which the average three-state prediction accuracy of all of the secondary structure prediction methods is quite good, typically in excess of 85%. However, only four of the secondary structure predictions yield reasonable tertiary folding results (NNSSP, original PhD, PSIPRED, and ZPRED). The reason in this case is quite obvious; the successful methods correctly predict that the region between residues 29 and 67 is a single long helix, whereas the remaining predictions insert a short loop in the middle of this part of the sequence. The short loop allows the two helical pieces surrounding it to fold, producing a very different shape than is enforced by the single long helix.

As we shall see below, in many cases the composite energy scoring method is capable of selecting the better tertiary architecture where there are qualitative differences between predictions. In the present example, however, the simple algorithm that we use to combine the predictions does not work well, for a completely understandable reason. By introducing a loop into the long helix, the protein is given greater flexibility. Because we have not explicitly included any sort of scoring function for secondary structure [44], the only discriminant is the energy of the tertiary fold, which in this case must favor the more flexible structure. In the present system, the non-native structures have energies far below the native-like and native structures.

The problem observed here will be potentially significant whenever the correct secondary structure is a long helix, and prediction methods have trouble distinguishing this from a pair of helices with a short loop in the middle—a very common motif in secondary structure prediction codes. In order to rectify this problem, it will probably be necessary to combine local energies, which determine secondary structures, with long-range energy terms. One approach is to replace fixed secondary structures by torsion angle energy wells, the depth and breadth of which are functions of the secondary structure prediction confidences. It may be possible to optimize the balance of torsion and long-range energy parameters such that correct helix assignments are favored. An alternative approach is to use an atomic level potential function and continuum solvation model to compare the energies of the predictions with different secondary structure assignments. We intend to explore both of these strategies is future work.

1am3: This example contains the other side of the long helix problem observed in 1aj3. Again, all of the secondary structure prediction three-state percentages are reasonable. However, three of the methods (PRED, and Zpred) predict a single long helix between residues 11 and 42, whereas the DSSP-derived secondary structure (and the remaining predictions) specify two short helices. In this case, the methods that incorrectly predict the long helix are unable to obtain reasonable RMSD structures from the native structure. However, here the composite prediction method easily eliminates the qualitatively incorrect predictions, in this case benefiting from the lower energies obtained due to greater flexibility of the two helical segments as opposed to a single long helix. Also of interest here is the result obtained from the Mulpred prediction, which inserts an incorrect short loop splitting the single helix between residues 12 and 26 into two shorter helices. This leads to a degradation in the rank of the best native-like structure, but does not eliminate the possibility of obtaining a reasonable prediction. Presumably, the magnitude of the effects of this sort of insertion are qualitatively larger when the size of the helix in question is large compared to the radius of gyration of the protein (as is the case in the two instances discussed here). It is also interesting that this error does not qualitatively degrade the results of the composite prediction method; it may be that structures with a significant bend at the short loop are energetically disfavored in this specific case.

1mzm: This protein is a startling example indicating that in some cases the tertiary folding potential can survive very large qualitative errors in secondary structure prediction. The only prediction that is satisfactory in terms of predicting major elements correctly is that of PSIPRED (and even here, a β-strand is incorrectly added on at the end), and indeed the PSIPRED results are certainly the best, particularly in terms of the RMSD of the low-energy structure which is below 5 Å. However, numerous other predictions are capable of achieving reasonable results, despite gross errors in the secondary structure of many different types. We have not analyzed in detail why this is the case; an initial speculation would be that this protein does not have a large number of alternative approaches to forming a good hydrophobic core. Also, because the potential energy function does not include explicit β-strand pairing terms, incorrect prediction of a strand is a local effect.

3.   Summary of the Results for All Proteins

The following is a brief analysis of how the various types of errors identified in the secondary structure predictions affected the proteins in the test set:

1. *Incorrect Prediction of Long Helices.* This problem, which amounts to missing a critical loop, affected at least some predictions in most of the proteins studied. Fortunately, in most cases at least one of the secondary structure prediction methods correctly identified the loop in question. Because the composite energy ranking protocol favors flexibility over long helices, the presence of several incorrectly predicted long helices was not, in general, a fatal error.

2. *Incorrect Replacement of a Helix by a Strand.* This problem most significantly affected the proteins 1jvr and 1lfb. In some cases, good low-energy tertiary folds are obtained despite the replacement of a helix with a

strand; in other cases, the replacement eliminates any good predictions. More work is needed to determine under what conditions this type of error can be overcome, and when it is fatal.

3. *Incorrect Replacement of an Important Helix by Loop.* Given our current composite energy ranking scheme, which favors flexibility, this error is in general fatal. Fortunately, in all but one case (1nre) at least one (and usually more than one) secondary structure prediction method correctly identified the important helix. As discussed above, a composite energy that combines local and long-range energy terms appears necessary in order to treat long helices. In the short term, simply preventing one secondary structure assignment from dominating the composite ranking may sufficiently diversify the resulting low-energy structures.

4. *Small Errors in the Prediction (Incorrect Lengths of Secondary Structure, Small Helix, or Strand Incorrectly Present or Missing).* Generally, these types of errors led to quantitative degradation in the ranking of low-RMSD structures as opposed to complete elimination of these structures.

4. Results from the CASP3 Prediction Contest

In addition to the test cases discussed above, we have also studied two small helical proteins that were targets in the CASP3 prediction contest. These studies allow us to compare our results with those of other groups [11]. The two targets we have investigated are target T0061 (PDB-code: 1bg8) and target T0074 (PDB code: 1eh2). Each is a helical protein between 50 and 100 residues and hence is part of the same general category as most of the proteins in the test set. The results for these two proteins are presented in detail in Tables VIII and IX and discussed below. We make explicit comparisons with the results of the Scheraga [25,45] and Samudrala [29] groups, both of whom carried out *ab initio* folding on these targets and used methods similar in spirit to what we present here. Those of the latter group are in fact quite analogous, because prediction methods are used to determine secondary structure, followed by tertiary folding simulations to generate a three-dimensional topology.

It should be noted that a nontrivial aspect of making these comparisons is that the proteins were truncated differently in the various calculations; we present all of the relevant information below so that the reader can draw his or her own conclusion. We do, however, wish to make one point with regard to the manner in which the comparison sequence is truncated. In our approach, truncation of terminal loop regions is done automatically using the secondary structure prediction, without reference to the native structure. In several of the comparisons we report below, truncation was carried out with the native structure in hand, presumably to minimize the RMSD obtained. While such results do indicate partial success of the folding algorithm, from a statistical point of view it is much easier to achieve an RMSD of 6–7 Å when an extensive choice of fragments are available to be optimized as opposed to when a single fragment is chosen *a priori*. This is particularly the case when the fragment is relatively small compared to the total length of the sequence.

1eh2: The secondary structure prediction methods generally performed well on this protein. The tertiary folding simulations were also quite effective, with the best results yielding an RMSD of less than 5 Å as the lowest energy prediction. The composite energy method provides a prediction ranking 3 with an RMSD of 6.02 Å, a respectable result for a protein in the 50 to 100-residue range. If the PSIPRED secondary structure method were to be used exclusively, the best prediction among five submitted predictions would be 4.84 Å; this is an excellent result, competitive with the best results obtained from threading methods [46]. We note that in both predictions, a long terminal loop of the protein was truncated, so that the total number of residues predicted was 72 in the PSIPRED simulation and 65 in the composite energy method.

In CASP3, results for 1eh2 varied greatly with prediction method. Several groups were able to identify a remote homolog and hence utilize threading approaches to structure prediction [46], whereas others use methods based more on *ab initio* approaches. When only ∼80% of the protein structure was predicted, the best results were in the 5 Å RMSD range; as the percentage of the protein predicted increased to 100%, the prediction accuracy degraded to 6.01 Å. Our results using PSIPRED secondary structure are comparable to the former results; in this case 74% of the residues were predicted to an accuracy of 4.84 Å.

The Scheraga group submitted a prediction for this target; however, they included the long terminal loop in their prediction which it is extremely difficult to predict correctly with *ab initio* methods. Consequently, their reported RMSD of 9.99 Å for the entire protein does not constitute a fair evaluation of the capabilities of their methodology. They also report a 5.8 Å RMSD for a 53-residue fragment of the protein. The calculations would most likely have been more successful had the terminal loop been deleted during the simulation, as was done in our approach. The Samudrala group, who achieved an RMSD of 11.3 Å, also included the terminal loop in their calculations. Their post-CASP3 analysis yielded an optimal fragment prediction of 7.0 Å for a 60-residue fragment. The results reported above (4.84 Å RMSD for 72 residues predicted) is qualitatively superior to either of these results, particularly as the truncation was carried out prior to the simulation.

1bg8—Chain A: 1bg8 is a target for which none of the predictors successfully located a remote homolog. The best results (and indeed the only ones that could be considered even partially successful for a protein this size) were those of Scheraga and co-workers, who achieved an RMSD of 7.27 Å (for all 76 residues reported experimentally) as the best result of four submitted predictions (their remaining predictions had RMSDs of 8.91 Å, 9.08 Å, and 9.23 Å). Their best results for a postprocessed fragment are 4.2 Å for a 61-residue fragment. Using the composite energy method, our lowest energy prediction achieves an RMSD of 6.69 Å, but for only 52 residues obtained after truncating to allow energetic comparisons among all of the secondary structure predictions. The PSIPRED calculations yield a 6.07 Å RMSD, again for 52 residues (PSIPRED incorrectly predicts a long terminal loop, which we truncate). These results are respectable in terms of RMSD but involve significant truncation in a region where there is actual secondary structure.

The Samudrala group achieved an RMSD of 10.1 Å for all 76 residues and 7.4 Å for 66 residues after postprocessing. The Scheraga group results in this case have to be considered best. Much of their success can be attributed to an impressive 79% accuracy in the secondary structure assembled in their most successful simulation; in this case, the standard neural-network-based secondary structure prediction methods that we (and Samudrala and co-workers) employed have a much poorer performance than they do for the test set, with accuracies below 65% in all cases.

V. CONCLUSION

We have demonstrated that the inclusion of size dependence in the derivation of a statistical potential for tertiary protein folding yields substantially improved results, as compared to previous efforts, for a substantial number of proteins of less than 100 residues in size. The new potential reliably yields highly ranked structures with low RMSDs as compared to the native structure (in contrast to earlier results that displayed occasional failures in this regard) and also provides a significant quantitative improvement in the energetic ranking of the best low RMSD cluster. There remain in most cases a small number (5–10) of competing misfolded structures with low energies; discrimination of these from the native-like topology, necessary for truly reliable tertiary structure prediction, will be a major objective of subsequent work. The reduction of the huge phase space of possible tertiary assemblies to a short list of discrete alternatives does, however, clearly represent progress in the nature and parameterization of the potential function.

We next examined the effect of replacing secondary structure elements derived from the PDB with idealized strands and helices, at the same locations. This substitution examines the effects of helix and strand distortion from ideal geometry on the predicted tertiary fold. Our conclusion is that, while there are occasional cases where substantial effects are observed, particularly for β-strands where a major distortion in length is manifested, the quality of the results is in general comparable to that obtained using PDB-derived secondary structure elements. This suggests that a folding protocol that initially uses idealized geometries and subsequently refines these geometries by allowing distortions is likely to be successful; furthermore, even if it is necessary in some cases to incorporate distortion directly into the initial simulations, the perturbations induced are relatively small and hence handling them should be computationally tractable.

Finally, we attempted genuine *ab initio* prediction by using predicted, rather than PDB-derived (in either geometry or location), secondary structure, focusing on small helical proteins. Recent improvements in secondary structure prediction, as exemplified by the PSIPRED code of Jones [26], allowed impressively accurate secondary structure predictions to be generated in many cases. When errors in secondary structure were made, the most difficult to deal with were cases in which a long helix was incorrectly predicted to be two short helices, or when two short helices were incorrectly predicted to be a single long helix. Reliable prediction of tertiary structure for α-helical proteins will clearly require secondary structure prediction methods than can robustly discriminate these two cases. Other types of large errors, such as replacement of a helix by a strand or a loop, produced variable results; in some cases, the predictions were surprisingly good despite such major errors. Smaller errors—for example, in length or position of a predicted helix—generally led to relatively minimal quantitative degradations in accuracy as compared to the use of PDB-derived secondary structure. Results for two small, helical CASP3 targets were presented which compared well with the work of other groups [11], including those employing fold recognition methods [46].

While there is still clearly a lot of work to be done, the above results are encouraging with regard to the possibility of developing reliable *ab initio* methods for protein structure prediction to low resolution, at least for small helical proteins. A different direction to pursue is the combination of these methods with fold recognition techniques (threading) and with experimental data, specifically NMR and X-ray crystallographic information. We have demonstrated in previous work [16] that the combination of a tertiary folding potential with sparse NMR constraints can successfully produce structures in the 2–4 Å resolution regime even for large systems; improvements in the folding potential will enhance the utility of such methods.

Acknowledgments

This work was supported in part by grants from the NIH: National Institute of General Medical Sciences (GM-52018) and National Center for Research Resources (P-41 RR06892). We also thank the Intel Corporation for donation of a large cluster of high-end Pentium-based workstations, without which many of these calculations could not have been carried out. Finally, many other computations were performed using SGI Origin machines at NCSA, via the NPACI program run by the NSF.

References

1. V. A. Eyrich, D. M. Standley, A. K. Felts, and R. A. Friesner, *Proteins Struct. Funct. Genet.* **35**, 41–57 (1999).

2. V. A. Eyrich, D. M. Standley, and R. A. Friesner, *J. Mol. Biol.* **288**, 725–742 (1999).

3. J. R. Gunn, A. Monge, R. A. Friesner, and C. H. Marshall, *J. Phys. Chem.* **98**, 702–711 (1994).

4. D. M. Standley, J. R. Gunn, R. A. Friesner, and A. E. McDermott, *Proteins Struct. Funct. Genet.* **33**, 240–252 (1998).

5. E. E. Abola, F. C. Bernstein, S. H. Bryant, T. F. Koetzle, and J. Weng, in *Crystallographic Databases—Information Content, Software Systems, Scientific Applications*, Data Commission of the International Union of Crystallography, Bonn/Cambridge/Chester, 1987, pp. 107–132.

6. F. C. Berstein, T. F. Koetzle, G. J. B. Williams, E. F. j. Meyer, M. D. Brice, J. R. Rodjers, O. Kennard, T. Shimanouchi, and M. Tasumi, *J. Mol. Biol.* **112**, 535–542 (1977).

7. G. Casari, and M. J. Sippl, *J. Mol. Biol.* **224**, 725–732 (1992).

8. P. D. Thomas and K. A. Dill, *J. Mol. Biol.* **257**, 457–469 (1996).

9. K.-C. Chou, M. Pottle, G. Nemethy, Y. Ueda, and H. Scheraga, *J. Mol. Biol.* **162**, 89–112 (1982).

10. A. Kolinski, A. Godzik, and J. Skolnick, *J. Chem. Phys.* **98**, 7420–7433 (1993).

11. C. A. Orengo, J. E. Bray, T. Hubbard, L. LoConte, and I. Sillitoe, *Proteins Struct. Funct. Genet.* **34**, 149–170 (1999).

12. L. Wang, T. Oconnell, A. Tropsha, and J. Hermans, *J. Mol. Biol.* **262**, 283–293 (1996).

13. J. Moult, T. Hubbard, K. Fidelis, and J. T. Pedersen, *Proteins Struct. Funct. Genet.* **34**, 2–6 (1999).

14. J. P. A. Kocher, M. J. Rooman, and S. J. Wodak, *J. Mol. Biol.* **235**, 1598–1613 (1994).

15. K. Yue and K. A. Dill, *Protein Sci.* **5**, 254–261 (1996).

16. D. M. Standley, V. A. Eyrich, A. K. Felts, R. A. Friesner, and A. E. McDermott, *J. Mol. Biol.* **285**, 1691–1710 (1999).

17. M. E. Mortenson, *Geometric Modeling*, 2nd ed., Wiley, New York, 1997.

18. P. D. Thomas and K. A. Dill, *Proc. Nat. Acad. Sci. USA* **93**, 11628–11633 (1996).

19. W. Kabsch and C. Sander, *Biopolymers* **22**, 2577–2637 (1983).

20. M. H. Hao and H. A. Scheraga, *Curr. Opin. Struct. Biol.* **9**, 184–188 (1999).

21. U. Hobohm and C. Sander, *Protein Sci.* **3**, 522–524 (1994).

22. J. Kuszewski, G. A. M., and G. M. Clore, *J. Am. Chem. Soc.* **121**, 2337–2338 (1999).

23. Z. Q. Li and H. A. Scheraga, *Proc. Natl. Acad. Sci. USA* **84**, 6611–6615 (1987).

24. H. C. Romesburg, *Cluster Analysis for Researchers*, Lifetime Learning Publications, Belmont, CA, 1984.

25. J. Lee, A. Liwo, D. R. Ripoll, J. Pillardy, and H. A. Scheraga, *Proteins Struct. Funct. Genet.* **34**, 204–208 (1999).

26. D. T. Jones, *J. Mol. Biol.* **292**, 195–202 (1999).

27. A. Kolinski and J. Skolnick, *Proteins Struct. Funct. Genet.* **18**, 353–366 (1994).

28. A. R. Ortiz, A. Kolinski, and J. Skolnick, *J. Mol. Biol.* **277**, 419–448 (1998).

29. R. Samudrala, Y. Xia, E. Huang, and M. Levitt, *Proteins Struct. Funct. Genet.* **34**, 194–198 (1999).

30. K. T. Simons, C. Kooperberg, E. Huang, and D. Baker, *J. Mol. Biol.* **268**, 209–225 (1997).

31. K. T. Simons, R. Bonneau, I. Ruczinski, and D. Baker, *Proteins Struct. Funct. Genet.* **34**, 171–176 (1999).

32. J. Skolnick, A. Kolinski, and A. R. Ortiz, *J. Biomol. Struct. Dyn.* **16**, 381–396 (1998).

33. B. Rost and C. Sander, *J. Mol. Biol.* **232**, 584–599 (1993).

34. B. Rost, C. Sander, and R. Schneider, *J. Mol. Biol.* **235**, 13–26 (1994).

35. J. A. Cuff, M. E. Clamp, A. S. Siddiqui, M. Finlay, and G. J. Barton, *Bioinformatics* **14**, 892–893 (1998).

36. J. A. Cuff and G. J. Barton, *Proteins Struct. Funct. Genet.* **34**, 508–519 (1999).

37. R. D. King and M. J. E. Sternberg, *Protein Sci.* **5**, 2298–2310 (1996).

38. D. Frishman and P. Argos, *Protein Eng.* **9**, 133–142 (1996).

39. D. Frishman and P. Argos, *Proteins Struct. Funct. Genet.* **27**, 329–335 (1997).

40. A. A. Salamov and V. V. Solovyev, *J. Mol. Biol.* **247**, 11–15 (1995).

41. M. J. Zvelebil, G. J. Barton, W. R. Taylor, and M. J. E. Sternberg, *J. Mol. Biol.* **195**, 957–961 (1987).

42. B. A. Reva, A. V. Finkelstein, and J. Skolnick, *Fold. Des.* **3**, 141–147 (1998).

43. F. E. Cohen and M. J. E. Sternberg, *J. Mol. Biol.* **138**, 321–333 (1980).

44. D. J. Osguthorpe, *Proteins Struct. Funct. Genet.* **34**, 186–193 (1999).

45. A. Liwo, J. Lee, D. R. Ripoll, J. Pillardy, and H. A. Scheraga, *Proc. Natl. Acad. Sci. USA* **96**, 5482–5485 (1999).

46. A. G. Murzin, *Proteins Struct. Funct. Genet.* **34**, 88–103 (1999).

DETERMINISTIC GLOBAL OPTIMIZATION AND *AB INITIO* APPROACHES FOR THE STRUCTURE PREDICTION OF POLYPEPTIDES, DYNAMICS OF PROTEIN FOLDING, AND PROTEIN–PROTEIN INTERACTIONS

JOHN L. KLEPEIS, HEATHER D. SCHAFROTH,
KARL M. WESTERBERG, AND CHRISTODOULOS A. FLOUDAS

*Department of Chemical Engineering, Princeton University,
Princeton, NJ, U.S.A.*

CONTENTS

*Computational Methods for Protein Folding: A Special Volume of Advances in Chemical Physics,
Volume 120,* Edited by Richard A. Friesner. Series Editors I. Prigogine and Stuart A. Rice.
ISBN 0-471-20955-4.   © 2002 John Wiley & Sons, Inc.

I. INTRODUCTION

Proteins are some of the most complex and vital molecules in nature. Their complexity arises from the intricate balance of intra- and intermolecular interactions that define their native three-dimensional structures and biological functionalities. Recent advances in genetic engineering and genome projects have heightened interest in predicting the folding dynamics and equilibrium structures of proteins and protein–protein complexes. This prediction ability is of great theoretical interest, especially in the fields of biophysics and biochemistry. The applications of these predictions promise to be especially valuable. The ability to predict the structure of individual and complexed protein molecules would increase our understanding of disease, aid in the interpretation of genome data, and revolutionize the process of *de novo* drug design.

Anfinsen's thermodynamic hypothesis [1] suggests that the native structure of a protein system is in a state of thermodynamic equilibrium corresponding to the system with the lowest free energy. Experimental studies have shown that, under native physiological conditions and after denaturation, globular proteins spontaneously refold to their unique, native structure [2]. Understanding the transition of a protein from a disordered state to its native state defines the protein folding problem. A natural extension of the protein folding problem is the related problem of predicting protein–protein interactions, also known as peptide docking. Prediction of protein–protein interactions requires the identification of equilibrium structures for protein–protein complexes. One part of this prediction challenge involves identifying the conformation of the binding sites through which complexed proteins interact, which can be accomplished experimentally or approached as an independent protein folding problem. Another part of the peptide docking prediction challenge involves identifying equilibrium structures for a number of candidate "docking" molecules complexed with a target macromolcule and then quantifying and comparing their relative binding affinities.

The use of computational techniques and simulations in addressing the protein folding and peptide docking problems became possible through the introduction of qualitative and quantitative methods for modeling these systems. The development of realistic energy models also established a link to the field of global optimization, where, based on Anfinsen's hypothesis, the quantity to be optimized is the free energy of the system. Because the number of local minima is vast, the corresponding problem formulation has earned the simple yet suggestive title of the "multiple-minima" problem. The basis for these difficulties is best summarized by Levinthal's paradox [3]. This paradox suggests a contradiction between the almost infinite number of possible stable states that the system may sample and the relatively short time scale required for actual protein folding. Levinthal's observations suggest that the native state is the lowest kinetically accessible free energy minimum, which may be different from the true global minimum. These principles have been used to develop computational techniques for predicting protein folding pathways [4–8]. Such techniques attempt to map the shape of the energy hypersurface and determine whether this surface "funnels" a protein toward a dominant conformational basin. By invoking the thermodynamic hypothesis, the overall shape of the energy hypersurface is neglected and the problem can be formulated in terms of global minimization, which requires the use of effective global optimization techniques. If this formulation is to reproduce the behavior of realistic systems, the folding of actual proteins should not be kinetically hindered. This has been verified for various systems by performing denaturation–refolding experiments. In addition, by introducing structural characteristics whose formation may act as kinetic barriers, such as the formation of disulfide bonds, the performance of the thermodynamic equilibrium model should be improved.

To better understand the dynamics of protein folding, it is also necessary to examine a protein's energy hypersurface. The characterization of the energy surface must include the identification of other stable and metastable configurations. Mathematically, these structures correspond to stationary points of the energy function. In particular, local minima represent stable conformations, while (first-order or higher-order) saddle points constitute transition states that connect two stable structures. A folding pathway defines the connection between two stable conformations (local minima) through a series of transition states (saddle points). Because the folding pathway may include a number of intermediates, a rigorous description of the energy surface would require the identification of all local minima and saddle points of the energy function.

Based on the complexity of the energy hypersurface, there is an obvious need for the development of efficient global optimization techniques. Although the energy can be expressed analytically, exhaustive searches are possible for only the smallest of systems. These observations, along with the importance of the protein folding and peptide docking problems, have propelled the introduction of new global search strategies specifically designed for these problems.

In the sequel, we first outline the basics of the deterministic global optimization approach, αBB, which has been used extensively to study the protein structure prediction, dynamics of protein–protein folding, and protein docking problems. This is followed by a comprehensive study of *ab initio* modeling for structure prediction of single-chain polypeptides in Section III. An extensive comparison of energy modeling, including solvation, entropic effects, and free energy calculations, is provided for the oligopeptides. The related problem of restrained structure refinement in the presence of sparse experimentally derived restraints is also discussed. Section IV moves beyond the static structure prediction problem toward an understanding of the dynamics of protein folding. An in-depth analysis of the coil-to-helix transition is provided for the alanine tetrapeptide. This analysis includes the elucidation of folding pathways and the identification of plausible reaction coordinates. Section V addresses the peptide docking problem. First, an approach for the determination of binding site structure is introduced. This is followed by a decomposition-based approach for the prediction of relative binding affinities. Both approaches are applied to peptide docking in HLA molecules.

II.   DETERMINISTIC GLOBAL OPTIMIZATION

A.   Twice Continuously Differentiable NLPs

The generic optimization problem to be addressed has the following form:

$$
\begin{aligned}
\min_{\mathbf{x}} \quad & f(\mathbf{x}) \\
\text{subject to} \quad & \mathbf{g}(\mathbf{x}) \leq 0 \\
& \mathbf{h}(\mathbf{x}) = 0 \\
& \mathbf{x} \in [\mathbf{x}^L, \mathbf{x}^U]
\end{aligned}
\tag{1}
$$

where $\mathbf{x}$ is a vector of n continuous variables, $f(\mathbf{x})$ is the objective function, $\mathbf{g}(\mathbf{x})$ is a vector of inequality constraints, and $\mathbf{h}(\mathbf{x})$ is a vector of equality constraints. Both the objective function and constraint equations are assumed to be twice continuously differentiable. $\mathbf{x}^L$ and $\mathbf{x}^U$ denote the lower and upper bounds on the $\mathbf{x}$ variables, respectively. The constraints define the feasible region for the problem.

Two main classes of global optimization techniques have been developed to address problem (1), namely, stochastic and deterministic approaches. Stochastic methods, such as those based on genetic algorithms [9] and simulated annealing [10], can be used to treat unconstrained nonconvex problems. However, the stochastic nature of the search strategy invalidates any claims regarding global optimality because it is impossible to obtain valid bounds on the solution of the problem. The addition of nonconvex constraints further complicates these solution schemes. In contrast, deterministic methods rely on a theoretically based search of the domain space to guarantee the identification of the global optimum solution.

A common characteristic of deterministic global optimization algorithms is the progressive reduction of the domain space until the global solution has been found with arbitrary accuracy. The solution is approached from above and below by generating converging sequences of upper and lower bounds, and the generation of these bounds on the global optimum solution is an essential part of all deterministic global optimization algorithms [11–13].

The αBB algorithm has been developed to address general twice continuously differentiable models of type (1) [14–18]. The algorithm is built on a branch-and-bound framework and can handle generic nonconvex optimization problems represented by formulation (1). ε-Convergence to the global optimum solution is guaranteed when the functions $f(\mathbf{x})$, $\mathbf{g}(\mathbf{x})$, and $\mathbf{h}(\mathbf{x})$ are twice continuously differentiable. The algorithm has been shown to terminate in a finite number of iterations for this broad class of problems [16,17,19,20].

The αBB global optimization approach is based on the convex relaxation of the original nonconvex formulation (1). This requires convex lower bounding of all expressions, and these expressions can be classified as (i) convex terms, (ii) nonconvex terms of special structure, and (iii) nonconvex terms of general structure. Obviously, convex lower bounding functions are not required for original convex expressions (e.g., linear terms). Certain nonconvex terms, including bilinear, trilinear and univariate concave functions, possess special structure that can be exploited in developing lower bounding functions. All other nonconvex terms can be underestimated using a general expression [18].

When applying the αBB approach to the protein folding problem, formulation (1) involves only nonconvex expressions of general structure. For this reason, the following exposition will briefly cover underestimation for terms of special structure and then focus on the development of a convex lower bounding formulation for global optimization involving generic nonconvex objective and constraint functions.

1. Underestimating Terms of Special Structure

In the case of a bilinear term xy, Ref. 21 showed that the tighte·t convex lower bound over the domain $[x^L, x^U] \times [y^L, y^U]$ is obtained by introducing a new variable w_B that replaces every occurrence of xy in the problem and satisfies the following relationship:

$$w_B = \max\{x^L y + y^L x - x^L y^L; \; x^U y + y^U x - x^U y^U\} \tag{2}$$

This lower bound can be relaxed and included in the minimization problem by adding two linear inequality constraints:

$$\begin{aligned} w_B &\geq x^L y + y^L x - x^L y^L \\ w_B &\geq x^U y + y^U x - x^U y^U \end{aligned} \tag{3}$$

Moreover, an upper bound can be imposed on w to construct a better approximation of the original problem [22]. This is achieved through the addition of two linear constraints:

$$w_B \leq x^U y + y^L x - x^U y^L$$
$$w_B \leq x^L y + y^U x - x^L y^U \tag{4}$$

A trilinear term of the form xyz can be underestimated in a similar fashion [23]. A new variable w_T is introduced and bounded by the following eight inequality constraints:

$$
\begin{aligned}
w_T &\geq xy^L z^L + x^L yz^L + x^L y^L z - 2x^L y^L z^L \\
w_T &\geq xy^U z^U + x^U yz^L + x^U y^L z - x^U y^L z^L - x^U y^U z^U \\
w_T &\geq xy^L z^L + x^L yz^U + x^L y^U z - x^L y^U z^U - x^L y^L z^L \\
w_T &\geq xy^U z^L + x^U yz^U + x^L y^U z - x^L y^U z^L - x^U y^U z^U \\
w_T &\geq xy^L z^U + x^L yz^L + x^U y^L z - x^U y^L z^U - x^L y^L z^L \\
w_T &\geq xy^L z^U + x^L yz^U + x^U y^U z - x^L y^L z^U - x^U y^U z^U \\
w_T &\geq xy^U z^L + x^U yz^L + x^L y^L z - x^U y^U z^L - x^L y^L z^L \\
w_T &\geq xy^U z^U + x^U yz^U + x^U y^U z - 2x^U y^U z^U
\end{aligned}
\tag{5}
$$

Fractional terms of the form x/y are underestimated by introducing a new variable w_F and two new constraints [23] which depend on the sign of the bounds on x:

$$
w_F \geq \begin{cases} x^L/y + x/y^U - x^L/y^U & \text{if } x^L \geq 0 \\ x/y^U - x^L y/y^L y^U + x^L/y^L & \text{if } x^L < 0 \end{cases}
$$
$$
w_F \geq \begin{cases} x^U/y + x/y^L - x^U/y^L & \text{if } x^U \geq 0 \\ x/y^L - x^U y/y^L y^U + x^U/y^U & \text{if } x^U < 0 \end{cases}
\tag{6}
$$

For fractional trilinear terms, eight new constraints are required [23]. The fractional trilinear term xy/z is replaced by the variable w_{FT} and the constraints for $x^L, y^L, z^L \geq 0$ are given by

$$
\begin{aligned}
w_{FT} &\geq xy^L/z^U + x^L y/z^U + x^L y^L/z - 2x^L y^L/z^U \\
w_{FT} &\geq xy^L/z^U + x^L y/z^L + x^L y^U/z - x^L y^U/z^L - x^L y^L/z^U \\
w_{FT} &\geq xy^U/z^L + x^U y/z^U + x^U y^L/z - x^U y^L/z^U - x^U y^U/z^L \\
w_{FT} &\geq xy^U/z^U + x^U y/z^L + x^L y^U/z - x^L y^U/z^U - x^U y^U/z^L \\
w_{FT} &\geq xy^L/z^U + x^L y/z^L + x^U y^L/z - x^U y^L/z^L - x^L y^L/z^U \\
w_{FT} &\geq xy^U/z^U + x^U y/z^L + x^L y/z - x^L y^U/z^U - x^U y^U/z^L \\
w_{FT} &\geq xy^L/z^U + x^L y/z^L + x^U y^L/z - x^U y^L/z^L - x^L y^L/z^U \\
w_{FT} &\geq xy^U/z^L + x^U y/z^L + x^U y^U/z - 2x^U y^U/z^L
\end{aligned}
\tag{7}
$$

Univariate concave functions are trivially underestimated by their linearization at the lower bound of the variable range. Thus the convex envelope of the concave function $ut(x)$ over $[x^L, x^U]$ is the linear function of x:

$$ut(x^L) + \frac{ut(x^U) - ut(x^L)}{x^U - x^L}(x - x^L) \tag{8}$$

The generation of the best convex underestimator for a univariate concave function does not require the introduction of additional variables or constraints.

2. Underestimating General Nonconvex Terms

A general nonconvex term $f(\mathbf{x})$ belonging to the class of twice continuously differentiable functions can be underestimated over the entire domain $\mathbf{x} \in [\mathbf{x}^L, \mathbf{x}^U]$ by the function $\hat{f}(\mathbf{x})$ defined as

$$\hat{f}(\mathbf{x}) = f(\mathbf{x}) + \sum_{i=1}^{n} \alpha_i(x_i^L - x_i)(x_i^U - x_i) \tag{9}$$

where the α_i's are nonnegative scalars.

$\hat{f}(\mathbf{x})$ is a guaranteed underestimator of $f(\mathbf{x})$ because the original nonconvex expression is augmented by the addition of separable quadratic functions that are negative over the entire domain $[\mathbf{x}^L, \mathbf{x}^U]$. Furthermore, because the quadratic term is convex, all nonconvexities in the original term $f(\mathbf{x})$ can be overpowered by using sufficiently large values of the α_i parameters.

The convex lower bounding function $\hat{f}(\mathbf{x})$, defined over the rectangular domain of $\mathbf{x}^L \le \mathbf{x} \le \mathbf{x}^U$, possesses a number of important properties that guarantee the convergence of the αBB algorithm to the global optimum solution:

(i) $\hat{f}(\mathbf{x})$ is a valid underestimator of $f(\mathbf{x})$. That is,

$$\forall \mathbf{x} \in [\mathbf{x}^L, \mathbf{x}^U] \text{ it can be shown that } \hat{f}(\mathbf{x}) \le f(\mathbf{x})$$

(ii) $\hat{f}(\mathbf{x})$ matches $f(\mathbf{x})$ at all corner points.

(iii) $\hat{f}(\mathbf{x})$ is convex in $\mathbf{x} \in [\mathbf{x}^L, \mathbf{x}^U]$.

(iv) The maximum separation between the nonconvex term of generic structure, $f(\mathbf{x})$, and its convex relaxation, $\hat{f}(\mathbf{x})$, is bounded and also proportional to the positive α parameters and to the square of the diagonal of the current box constraints:

$$\max_{\mathbf{x}^L \le \mathbf{x} \le \mathbf{x}^U} [f(\mathbf{x}) - \hat{f}(\mathbf{x})] = \frac{1}{4}\sum_{i}^{n} \alpha_i(x_i^U - x_i^L)^2 \tag{10}$$

(v) The underestimators constructed over supersets of the current set are always less tight than the underestimator constructed over the current box constraints for every point within the current box constraints.

The key development in the convex lower bounding formulation is the definition of the α parameters. Specifically, the magnitude of the α parameters may be related to the minimum eigenvalue of the Hessian matrix of the nonconvex term $f(x)$:

$$\alpha \geq \max\left\{0, -\frac{1}{2}\min_{i,x^L \leq x \leq x^U} \lambda_i(x)\right\} \tag{11}$$

where $\lambda(x)$ represent the eigenvalues of the Hessian matrix $(H_f(x))$ for the nonconvex term. An explicit minimization problem can be written to find the minimum eigenvalue (λ_{min}):

$$\min_{x,\lambda} \quad \lambda$$

$$\text{subject to} \quad \det(H_f(x) - \lambda I) = 0$$

$$x \in [x^L, x^U]$$

The solution of this problem is a nontrivial matter for arbitrary nonconvex functions.

One method for the rigorous determination of α parameters for general twice differentiable problems involves interval analysis of Hessian matrices to calculate bounds on the minimum eigenvalue [14,15]. The difficulties arising from the presence of the variables in the convexity condition can be alleviated through the transformation of the exact x-dependent Hessian matrix to an interval matrix $[H_f]$ such that $H_f(x) \subseteq [H_f]$, $\forall x \in [x^L, x^U]$. The elements of the original Hessian matrix are treated as independent when calculating their natural interval extensions [24,25]. The interval Hessian matrix family $[H_f]$ is then used to formulate a theorem in which the α calculation problem is relaxed [15]. In other words, a valid lower bound on the minimum eigenvalue can be used to calculate rigorous α values:

$$\alpha \geq \left\{0, -\frac{1}{2}\lambda_{min}([H_f])\right\} \tag{12}$$

where $\lambda_{min}([H_f])$ is the minimum eigenvalue of the interval matrix family $[H_f]$.

An $\mathcal{O}(n^2)$ method to calculate these α values is the straightforward extension of Gerschgorin's theorem [26] to interval matrices. For a real matrix $A = (a_{ij})$, the well-known theorem states that the eigenvalues are bounded below by λ_{min}

such that

$$\lambda_{\min} = \min_i \left(a_{ii} - \sum_{j \neq i} |a_{ij}| \right) \quad (13)$$

For an interval matrix $[A] = ([\underline{a}_{ij}, \bar{a}_{ij}])$, a lower bound on the minimum eigenvalue is given by

$$\lambda_{\min} \geq \min_i \left[\underline{a}_{ii} - \sum_{j \neq i} \max(|\underline{a}_{ij}|, |\bar{a}_{ij}|) \right]$$

This procedure provides a single α value that is valid for all variables.

Nonuniform diagonal shift matrices can be used to calculate a different α value for each variable in order to construct an underestimator of the form shown in Eq. (9). The nonzero elements of the diagonal shift matrix can no longer be related to the minimum eigenvalue of the interval Hessian matrix $[H_f]$. If all elements of the scaling vector are set to 1, the equation for the α_i values becomes

$$\alpha_i = \max\left\{ 0, -\frac{1}{2} \left(\underline{a}_{ii} - \sum_{j \neq i} |a|_{ij} \right) \right\}$$

However, the choice of scaling is arbitrary, and different α_i parameters can be estimated through various scaling techniques.

3. Convexification of Feasible Region

To obtain a valid lower bound on the global solution of the nonconvex problem, the lower bounding problem generated in each domain must have a unique solution. This implies that the formulation includes only convex inequality constraints, linear equality constraints, and an increased feasible region relative to that of the original nonconvex problem. The left-hand side of any nonconvex inequality constraint, $g(\mathbf{x}) \leq 0$, in the original problem can simply be replaced by its convex underestimator $\hat{g}(\mathbf{x})$, constructed according to Eq. (9), to yield the relaxed convex inequality $\hat{g}(\mathbf{x}) \leq 0$.

For an equality constraint containing general nonconvex terms, the equation obtained by simple substitution of the appropriate underestimators is also nonlinear. Therefore, the original equality $h(\mathbf{x}) = 0$ must be rewritten as two inequalities of opposite signs:

$$\begin{aligned} h^+(\mathbf{x}) &= h(\mathbf{x}) \leq 0 \\ h^-(\mathbf{x}) &= -h(\mathbf{x}) \leq 0 \end{aligned} \quad (14)$$

These two inequalities must then be underestimated independently to give $\hat{h}^+(\mathbf{x})$ and $\hat{h}^-(\mathbf{x})$.

4. Convex Lower Bounding Problem Formulation

Summarizing the concepts introduced so far, a convex relaxation for any nonconvex problem of type (1) belonging to the broad class of twice continuously differentiable continuous NLPs can be constructed as

$$\min_{\mathbf{x}} \quad \hat{f}(\mathbf{x})$$

$$\text{subject to} \quad \hat{\mathbf{g}}(\mathbf{x}) \leq 0$$

$$\hat{\mathbf{h}}^{+}(\mathbf{x}) \leq 0 \qquad \qquad (15)$$

$$\hat{\mathbf{h}}^{-}(\mathbf{x}) \leq 0$$

$$\mathbf{x} \in [\mathbf{x}^{L}, \mathbf{x}^{U}]$$

where ^ denotes the convex underestimator of the specified function over the domain $\mathbf{x} \in [\mathbf{x}^{L}, \mathbf{x}^{U}]$. Because the inclusion of convex terms and nonconvex terms of special structure has been neglected, these functions involve only α-type underestimating expressions. These underestimators are functions of the size of the domain under consideration, and because the αBB algorithm follows a branch-and-bound approach, this domain is systematically reduced at each new node of the tree. Tighter lower bounding functions can therefore be generated by updating the underestimating equations. The lower bounds on the problem form a nondecreasing sequence, and the underestimating strategy is therefore consistent, as required for convergence.

5. Variable Bound Updates

The quality of the convex lower bounding problem can also be improved by ensuring that the variable bounds are as tight as possible. These variable bound updates can be performed either at the onset of an αBB run or at each iteration.

In both cases, the same procedure is followed in order to construct the bound update problem. Given a solution domain, the convex underestimator for every constraint in the original problem is formulated. The bound problem for variable x_i is then expressed as

$$x_{i}^{L,\text{NEW}}/x_{i}^{U,\text{NEW}} = \begin{cases} \min/\max \limits_{\mathbf{x} \quad \cdot \mathbf{x}} & x_{i} \\ \text{subject to} & \hat{\mathbf{g}}(\mathbf{x}) \leq 0 \\ & \mathbf{x}^{L} \leq \mathbf{x} \leq \mathbf{x}^{U} \end{cases} \qquad (16)$$

where $\hat{\mathbf{g}}(\mathbf{x})$ are the convex underestimators of the constraints, and the bounds on the variables $\mathbf{x}^{L}$ and $\mathbf{x}^{U}$ are the best calculated bounds. Thus, once a new lower bound $x_{i}^{L,\text{NEW}}$ on x_i has been computed via a minimization, this value is used in the formulation of the maximization problem for the generation of an upper bound $x_{i}^{U,\text{NEW}}$.

Because of the computational expense incurred by an update of the bounds on all variables, it is often desirable to define a smaller subset of the variables on which this operation is to be performed. The criterion devised for the selection of the branching variables can be used in this instance, because it provides a measure of the sensitivity of the problem to each variable.

6. The αBB Algorithm

The global optimization method αBB deterministically locates the global minimum solution of (1) based on the refinement of converging lower and upper bounds. The lower bounds are obtained by the solution of (15), which is formulated as a convex programming problem. Upper bounds are based on the solution of (1) using local minimization techniques.

As previously mentioned, the maximum separation between the generic nonconvex terms and their respective convex lower bounding representations is proportional to the square of the diagonal of the current rectangular partition. As the size of the rectangular domains approach zero, this separation also become infinitesimally small. That is, as the current box constraints $[\mathbf{x}^L, \mathbf{x}^U]$ collapse to a point, the maximum separation between the original objective function of (1) and its convex relaxation in (15) becomes zero. This implies that for the positive numbers ϵ and $\mathbf{x}$ there always exists another positive number δ which, by reducing the rectangular region $[\mathbf{x}^L, \mathbf{x}^U]$ around $\mathbf{x}$ so that $\|\mathbf{x}^U - \mathbf{x}^L\| \leq \delta$, cause the difference between the feasible region of the original problem (1) and its convex relaxation (15) to become less than ϵ. Therefore, any feasible point $\mathbf{x}$ of problem (15), including the global minimum solution, becomes at least ϵ-feasible for problem (1) by sufficiently tightening the bounds on $\mathbf{x}$ around this point.

Once the solutions for the upper and lower bounding problems have been established, the next step is to modify these problems for the next iteration. This is accomplished by successively partitioning the initial rectangular region into smaller subregions. The number of variables along which subdivision is required is equal to the number of variables $\mathbf{x}$ participating in at least one nonconvex term of the (1) formulation. The default partitioning strategy used in the algorithm involves successive subdivision of the original rectangle into two subrectangles by halving on the midpoint of the longest side of the initial rectangle (bisection). Therefore, at each iteration a lower bound of the objective function (1) is simply the minimum over all the minima of problem (15) in each sub-rectangle of the initial rectangle. In order to ensure lower bound improvement, the subrectangle to be bisected is chosen by selecting the subrectangle that contains the infimum of the minima of (15) over all the subrectangles. This procedure guarantees a nondecreasing sequence for the lower bound. A nonincreasing sequence for the upper bound is found by solving the original nonconvex problem (1) locally and selecting it to be the minimum over all the previously recorded upper bounds. Obviously, if the single minimum of (15) for any subrectangle is greater than the current upper bound, this subrectangle can be discarded because the global minimum cannot lie within this subdomain (fathoming step).

Because the maximum separation between the nonconvex terms and their respective convex lower bounding functions is both a bounded and a continuous function of the size of rectangular domain, arbitrarily small feasibility and convergence tolerance limits are attained for a finite-sized partition element.

The basic steps of the αBB global optimization algorithm are as follows:

1. *Initialization.* A convergence tolerance, ϵ_c, and a feasibility tolerance, ϵ_f, are selected and the iteration counter, I, is set to one. The current variable bounds $[\mathbf{x}_I^L, \mathbf{x}_I^U]$ for the first iteration are set equal to the global ones $[\mathbf{x}_0^L, \mathbf{x}_0^U]$. Lower and upper bounds $[f^L, f^U]$ on the global minimum of (1) are initialized and an initial current point is selected from the domain.

2. *Local Solution of Nonconvex Problem.* The nonconvex optimization problem (1) is solved locally within the current variable bounds $[\mathbf{x}_I^L, \mathbf{x}_I^U]$. If the solution is ϵ_f-feasible, the upper bound f^U is updated as follows:

$$f^U = \min(f^U, f_I^U)$$

where f_I^U is the objective function value for the current ϵ_f-feasible solution.

3. *Partitioning of Current Rectangle.* The current rectangle, $[\mathbf{x}_I^L, \mathbf{x}_I^U]$, is bisected into two subrectangles $(r = a, b)$ for the variable (l) with the longest side of the initial rectangle:

$$l_I = \arg \max_i (x_{i,I}^U - x_{i,I}^L)$$

4. *Solution of Underestimating Problems.* The parameters $\alpha_{i,I,r}$ are updated for both rectangles $(r = a, b)$. The convex optimization problem (15) is solved inside both subrectangles $(r = a, b)$ using a nonlinear solver (e.g., MINOS5.4 [27], NPSOL [28]). If a solution $f_{I,r}^L$ is less than the current upper bound, f^U, then it is stored.

5. *Update of Lower Bound.* The iteration counter is increased by one, and the lower bound, f^L, is updated to be the minimum solution over the stored solutions from previous iterations. The selected region is erased from the stored set.

$$f^L = \min_{I',r} f_{I',r}^L, \qquad r = a, b, \quad I' = 1, \ldots, I - 1$$

6. *Update Bounds.* The bounds of the current rectangle are updated to those of the sub-rectangle containing the previously found solution (f^L).

7. *Check for Convergence.* If $\left(f^U - f^L\right) > \epsilon_c$, then return to Step 2. Otherwise, ϵ_c-convergence has been reached, and the global minimum solution corresponds to point providing f^U.

Figure 1 diagrams an unconstrained one-dimensional example of the approach. The mathematical proof that the αBB global optimization algorithm

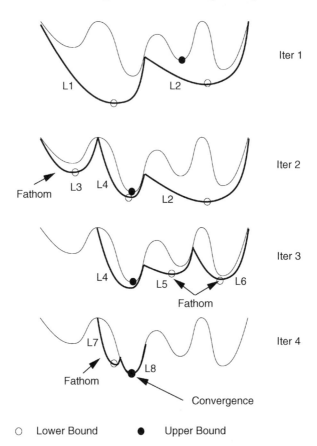

Figure 1. One-dimensional illustrative example of the αBB approach. In iteration 1 the overall domain is bisected, the two convex lower bounding functions are created, and their unique minima (*L*1 and *L*2) are identified. An upper bound is also identified. Because *L*1 is less than *L*2, the region containing *L*1 is further bisected in iteration 2, whereas the other region is stored. The minimum of one region (*L*3) is greater than the new upper bound, so this region can be fathomed. The other region is stored. In iteration 3 the region with the next lowest lower bound (*L*2) is bisected and because both new lower bound minima (*L*5 and *L*6) are greater than the current best upper bound, the entire region is fathomed. Finally, by iteration 4, the region containing *L*4 is bisected, which results in a region that can be fathomed (containing *L*7) and a convex region whose minimum (*L*8) equals the current upper bound and is the global minimum.

converges to the global optimum solution is presented in Ref. 19. In addition to computational chemistry related problems, the αBB approach has been applied to a variety of constrained optimization problems [15–18].

B. Enclosure of All Solutions

The αBB algorithm discussed in the previous section was originally designed to solve global optimization problems. However, this algorithm has also proven to be effective in the solution of non-linearly constrained systems of algebraic equations [23], provided only that the constraints are twice continuously differentiable. The key idea is to reformulate the algebraic system of equations as a global optimization problem that exhibits multiple global solutions and then use the αBB approach as a basis for the enclosure of all solutions. In the following sections, we discuss the enclosure of all solutions.

1. Problem Formulation

In general, a non-linearly constrained system of algebraic equations can be expressed in the form

$$
\begin{aligned}
f_i(\mathbf{x}) &= 0, & i &= 1, \ldots, N_f \\
g_j(\mathbf{x}) &\leq 0, & j &= 1, \ldots, N_g \\
\mathbf{x}^L &\leq \mathbf{x} \leq \mathbf{x}^U
\end{aligned}
\tag{17}
$$

where $f_i(\mathbf{x})$ represent the equality constraints (N_f is the number of such constraints) and $g_j(\mathbf{x})$ represent the inequality constraints (N_g is the number of such constraints).

In order to apply the αBB algorithm to (17), we must reformulate it as a global optimization problem. This is accomplished by introducing a slack variable s and minimizing its value over an augmented variable set $(\mathbf{x}, s)$ subject to a set of relaxed constraints:

$$
\begin{aligned}
&\min_{\mathbf{x}, s} s \\
\text{subject to} \quad & f_i(\mathbf{x}) - s \leq 0, & i &= 1, \ldots, N_f \\
& -f_i(\mathbf{x}) - s \leq 0, & i &= 1, \ldots, N_f \\
& g_j(\mathbf{x}) \leq 0, & j &= 1, \ldots, N_g \\
& \mathbf{x}^L \leq \mathbf{x} \leq \mathbf{x}^U
\end{aligned}
\tag{18}
$$

In comparing the two formulations, the following two facts are self-evident:

- If $s < 0$, the constraints in (18) are infeasible.
- If $s = 0$, the constraints in (18) reduce to the original problem (17).

It follows that $s = 0$ is the *global minimum* of (26)—provided that (17) has solutions—and that there is a one-to-one correspondence between global minima $(\mathbf{x}^*, s^*)$ of (18) and solutions $\mathbf{x}^*$ of the original problem (17). Therefore, the problem of finding all solutions to (17) can be reformulated as the problem of finding all global minima of (18).

In the next section, we will explain how the αBB global optimization algorithm can be used to enclose all global minima of (18), and hence, all solutions to (17).

2. *Framework for Enclosing All Solutions*

In this section, we describe the αBB global optimization algorithm as it is applied to the general problem of determining all solutions to a system of algebraic constraints (17). This adaptation is based on the correspondence between solutions of (17) and global minima of (18) with $s = 0$. Since the αBB algorithm can be applied to any problem involving constraints which are twice continuously differentiable (C^2), the only necessary assumptions we need to make are that $f_i(\mathbf{x})$ and $g_j(\mathbf{x})$ are C^2 functions for $i = 1, \ldots, N_f$ and $j = 1, \ldots, N_g$, respectively.

The algorithm proceeds by exploring the configuration space for solutions to (17). We begin with the full region $\mathbf{x} \in [\mathbf{x}^L, \mathbf{x}^U]$ and subdivide regions into smaller regions. Each region is tested before it is divided to see if a solution to (17) can possibly exist there. This is accomplished by finding a lower bound of the global minimum of (18) over the region in question. If the lower bound is positive, then $s = 0$ cannot lead to a feasible point of (18), and hence no solution to (17) can exist in the given region. The region will be fathomed (i.e., eliminated from further consideration). On the other hand, if the lower bound is negative or zero, there may or may not be a solution to (17) in that region. In this case, further subdivision and testing will be necessary. If the region size becomes small enough and the region is still active (i.e., its lower bound is negative or zero), then a solution to (17) is obtained within that region by a local search. The algorithm terminates when all regions have been fully processed.

Note that upper bounds of the global minimum need not be determined. Since we are assuming that the global minimum of (18) is zero, we can set the upper bound to this value from the start, and thus avoid the effort of solving an upper bounding problem.

Lower bounds of the global minimum of (18) are determined by solving the lower bounding problem over the given region:

$$\min_{\mathbf{x}, s} s$$

$$\text{subject to} \quad \hat{f}_i^{+}(\mathbf{x}) - s \leq 0, \quad i = 1, \ldots, N_f$$
$$\hat{f}_i^{-}(\mathbf{x}) - s \leq 0, \quad i = 1, \ldots, N_f \tag{19}$$
$$\hat{g}_j(\mathbf{x}) \leq 0, \quad j = 1, \ldots, N_g$$
$$\mathbf{x}^L \leq \mathbf{x} \leq \mathbf{x}^U$$

where $\hat{f}_i^+(\mathbf{x})$, $\hat{f}_i^-(\mathbf{x})$, and $\hat{g}_j(\mathbf{x})$ are convex functions which underestimate $f_i(\mathbf{x})$, $-f_i(\mathbf{x})$, and $g_j(\mathbf{x})$, respectively. Because the constraints are all convex functions, any local optimization package should be able to locate its global minimum. Furthermore, every feasible point of (18) is also a feasible point of (19) because these functions are underestimators of the original functions. It follows that the global minimum of (19) is a valid lower bound of the global minimum of (18).

The crux of the αBB algorithm is finding valid convex underestimators, $\hat{f}_i^\pm(\mathbf{x})$ and $\hat{g}_j(\mathbf{x})$, for the functions $\pm f_i(\mathbf{x})$ and $g_j(\mathbf{x})$, respectively, over a given region. An important consideration is that the convex underestimators be as tight (i.e., close in value to the original constraint functions) as is reasonably possible, because tighter underestimators lead to better lower bound estimates. It is important to be able to fathom regions as quickly as possible if they do not contain any solutions to (17). However, this cannot always be done: It frequently occurs that a region contains no solution to (17) [i.e., the global minimum of (18) over that region is positive], but the lower bound obtained from (19) for that region is negative. Such regions obviously must be explored further, until positive lower bounds are obtained. A better lower bound estimate can lead to significant improvement in the efficiency of the algorithm.

When applying this algorithm to the problem of finding all stationary points of a potential energy surface, the constraint functions, $\pm f_i(\mathbf{x})$ and $g_i(\mathbf{x})$, are general nonconvex functions. Whenever these constraint functions are C^2, they can be underestimated using the α underestimation described in Section II.A.2. In this context, the underestimators take the form

$$\hat{f}_i^+(\mathbf{x}) = f_i(\mathbf{x}) - \alpha_i^{f,+} \sum_k (x_k^U - x_k)(x_k - x_k^L)$$

$$\hat{f}_i^-(\mathbf{x}) = -f_i(\mathbf{x}) - \alpha_i^{f,-} \sum_k (x_k^U - x_k)(x_k - x_k^L) \tag{20}$$

$$\hat{g}_j(\mathbf{x}) = g_j(\mathbf{x}) - \alpha_j^g \sum_k (x_k^U - x_k)(x_k - x_k^L)$$

where the α parameters satisfy the convexity conditions

$$\alpha_i^{f,+} \geq -\frac{1}{2} \min_{\mathbf{x} \in [\mathbf{x}^L, \mathbf{x}^U]} \{\lambda_k(H_{f_i}(\mathbf{x})), 0\}$$

$$\alpha_i^{f,-} \geq +\frac{1}{2} \max_{\mathbf{x} \in [\mathbf{x}^L, \mathbf{x}^U]} \{\lambda_k(H_{f_i}(\mathbf{x})), 0\} \tag{21}$$

$$\alpha_j^g \geq -\frac{1}{2} \min_{\mathbf{x} \in [\mathbf{x}^L, \mathbf{x}^U]} \{\lambda_k(H_{g_j}(\mathbf{x})), 0\}$$

The discussion in Section II.A.2 applies equally well in this situation.

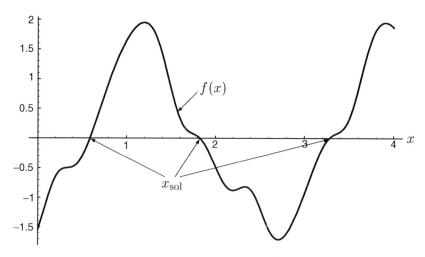

Figure 2. Plot of $f(x)$ for $x \in [0, 4]$.

3. Geometrical Interpretation

In this section, we give a geometric illustration of how the αBB algorithm works by showing how it would locate all of the solutions of a single equation $f(x) = 0$ over the interval $x \in [0, 4]$. The function we use for our illustration is

$$f(x) = -2 \cos \frac{\pi}{3}(x+0.05) + e^{-20(x-0.2)^2} - e^{-20(x-1.6)^2} + e^{-20(x-2.4)^2} - e^{-20(x-3.5)^2}$$

A graph of this function is given in Fig. 2. There are three solutions to $f(x) = 0$ in this interval. They are

$$x_{\text{sol}} \in \{0.59014, 1.82399, 3.27691\}$$

The corresponding global optimization problem is obtained by introducing a slack variable s and minimizing s subject to the constraints

$$f(x) - s \leq 0 \leq f(x) + s$$

The feasibility region for fixed s is determined by intersecting the region of space between $f(x) - s$ and $f(x) + s$ with the x-axis. This procedure is shown graphically in Fig. 3. For $s > 0$, the feasibility region forms intervals around the actual solutions to $f(x) = 0$. Minimizing s subject to the constraints above has the effect of pushing the two graphs together until they both meet at $f(x)$ (at $s = 0$). At $s = 0$, the feasibility region reduces to the solution set for $f(x) = 0$

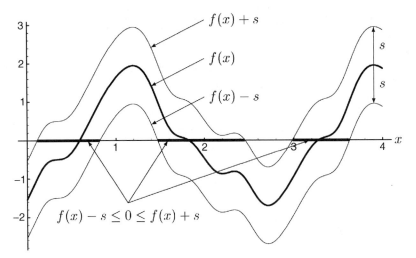

Figure 3. $f(x)$ is shifted by a positive slack variable $s = 1$. Note that the feasibility region of $f(x) - s \leq 0 \leq f(x) + s$ forms intervals around the solutions to $f(x) = 0$.

(each interval reduces to a point). For $s < 0$, the graphs cross and the feasibility region is empty. $s = 0$ is clearly the global minimum whenever $f(x) = 0$ has solutions.

In order to set up the lower bounding problem, we need to find convex underestimators for $\pm f(x)$ for each interval under consideration. We begin with the complete interval $[0, 4]$. The function $f(x)$ and a valid set of convex underestimators $\hat{f}^{\pm}_{[0,4]}(x)$ are plotted in Fig. 4. The convex underestimators

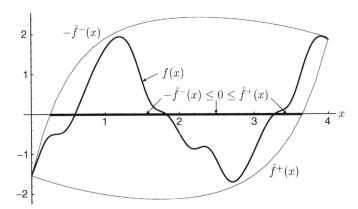

Figure 4. The functions $\hat{f}^{\pm}(x)$ are convex underestimators of $\pm f(x)$ over the interval $[0, 4]$. Note how $\hat{f}^{+}(x)$ and $-\hat{f}^{-}(x)$ form a convex envelope around $f(x)$.

$\hat{f}^{\pm}_{[0,4]}(x)$ essentially envelop the graph of $f(x)$ in a convex region. This convex region contains all the points $\hat{f}^{+}(x) \leq y \leq -\hat{f}^{-}(x)$, and its intersection with the x-axis is given by $\hat{f}^{+}(x) \leq 0 \leq -\hat{f}^{-}(x)$. All solutions to $f(x) = 0$ in the region $x \in [0,4]$ must lie in this intersection region because $\hat{f}^{+}(x)$ and $-\hat{f}^{-}(x)$ surround the function $f(x)$ (see Fig. 4). If this region had been empty, then no solution to $f(x) = 0$ could possibility exist in the interval $[0,4]$. This is not the case, but see later on when we discuss the interval $[2,3]$.

Determining whether or not the feasibility region of $\hat{f}^{+}(x) \leq 0 \leq -\hat{f}^{-}(x)$ is empty involves introducing a slack variable and minimizing it subject to

$$\hat{f}^{+}(x) - s \leq 0 \leq -\hat{f}^{-}(x) + s \tag{22}$$

This is the lower bounding problem. For $s = 0$, (22) reduces to $\hat{f}^{+}(x) \leq 0 \leq -\hat{f}^{-}(x)$. For $s \neq 0$, the feasibility region of (22) is determined by shifting the enveloping functions $\hat{f}^{+}(x)$ and $-\hat{f}^{-}(x)$ by an amount s—away from each other if $s > 0$, and toward each other if $s < 0$ (see Fig. 5). Graphically, minimizing s subject to (22) involves expanding or shrinking the region between the underestimators by adjusting s until the region between $\hat{f}^{+}(x) - s$ and $-\hat{f}^{-}(x) + s$ intersects the x-axis at a single point. For the interval $[0,4]$, this requires moving $\pm\hat{f}^{\pm}(x)$ toward each other, implying $s_{\min} < 0$. The fact that $s_{\min} < 0$ indicates that there might be solutions to $f(x) = 0$ in this interval: we will be forced to explore this region further. Note that the lower bounding problem is a *convex problem*, and so any local optimization package should reach this unique global minimum.

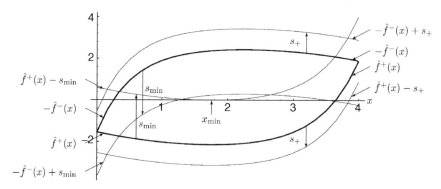

Figure 5. During the solution to the lower bounding problem, the convex underestimators $\hat{f}^{\pm}(x)$ are shifted by a slack variable. Two different shifts are shown above: One is positive, $s_{+} = 1$; and the other is negative, $s_{\min} = -2.135$. $s_{\min}$ represents the global minimum to the lower bounding problem: The feasibility region of the lower bounding problem is reduced to a single point $x_{\min} = 1.754$, shown above.

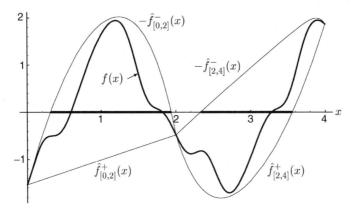

Figure 6. The interval [0, 4] has been subdivided into [0, 2] and [2, 4]. The convex underestimators $\hat{f}^{\pm}(x)$ for each subinterval, shown above, form a convex envelope around $f(x)$. As the intervals get smaller, the envelope gets tighter.

We therefore subdivide the interval $[0, 4]$ into two subintervals, $[0, 2]$ and $[2, 4]$, and explore each interval for solutions just as we did for $[0, 4]$. The convex underestimators for each interval, $\hat{f}^{\pm}_{[0,2]}(x)$ and $\hat{f}^{\pm}_{[2,4]}(x)$, are shown in Fig. 6. Note that each pair of underestimators envelops the corresponding portion of the function $f(x)$, and that the underestimators have improved: They are closer to the function $f(x)$. This will continue to happen as the intervals become narrower.

Again, the question we ask in each interval is: Can a solution to $f(x) = 0$ exist there? The question is answered by solving the lower bounding problem. In both cases, the region $\hat{f}^{+}(x) \leq 0 \leq -\hat{f}^{-}(x)$ does intersect the x-axis (see Fig. 6), indicating possible solutions in each interval. This fact is established by minimizing s subject to (22) within each interval. In both cases, $s_{min} < 0$, suggesting that $\hat{f}^{\pm}(x)$ must move toward each other to reduce the feasibility region to a point (see Fig. 7 and 8). Both intervals must be explored further.

So we subdivide again, and look at the intervals $[0, 1]$, $[1, 2]$, $[2, 3]$, and $[3, 4]$. The underestimators $\hat{f}^{\pm}_{[n,n+1]}(x)$ are plotted in Fig. 9. For the intervals $[0, 1]$, $[1, 2]$, and $[3, 4]$, the story is the same: $s = 0$ yields feasible points, s_{min} is negative, and so we must subdivide those intervals further. But something new happens for $[2, 3]$. The convex envelope $\hat{f}^{\pm}_{[2,3]}(x)$ completely isolates $f(x)$ from the x-axis. The lower bounding problem (22) is *infeasible* for $s = 0$. The region between $\hat{f}^{+}_{[2,3]}(x)$ and $-\hat{f}^{-}_{[2,3]}(x)$ must be *expanded* before it touches the x-axis (see Fig. 10), and thus s_{min} will be greater than zero. We have rigorously concluded that no solution to $f(x) = 0$ can exist in the interval $[2, 3]$, and so we

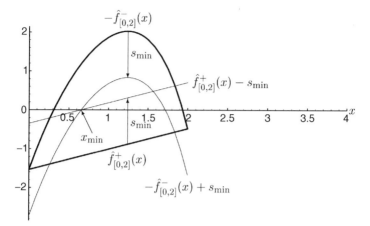

Figure 7. This figure represents the solution to the lower bounding problem in the interval $[0, 2]$. $(x_{min}, s_{min}) = (0.656, -1.189)$.

do not need to explore this interval any further. The ability to fathom regions like this is what distinguishes αBB from a straight gridsearch.

Exploration will continue with the intervals $[0, 1]$, $[1, 2]$, and $[3, 4]$. These intervals will be subdivided and further tested. As the algorithm progresses, most intervals will eventually be fathomed. A few intervals (three, in fact) will survive. Each of these intervals surrounds a solution point, which will be located by a local search once the interval size is small enough.

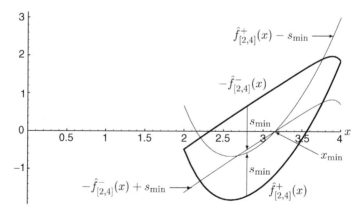

Figure 8. This figure represents the solution to the lower bounding problem in the interval $[2, 4]$. $(x_{min}, s_{min}) = (3.154, -1.150)$.

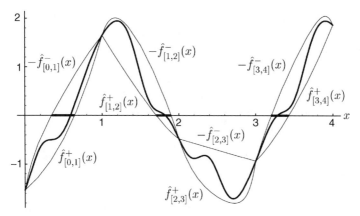

Figure 9. The intervals [0, 2] and [2, 4] have been further subdivided into [0, 1], [1, 2], [2, 3], and [3, 4]. Shown above are the convex envelopes around $f(x)$ formed by convex underestimators in each of these intervals. Note that the convex envelopes for [0, 1], [1, 2], and [3, 4] intersect the x-axis, but the convex envelope for [2, 3] does not. This will allow us to conclude rigorously that no solutions to $f(x) = 0$ exist in [2, 3].

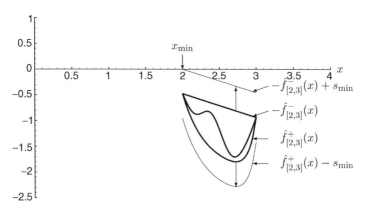

Figure 10. The lower bounding problem for the interval [2, 3] is solved. Note that the convex envelope must be *expanded* before it touches the x-axis, resulting in a positive value for s_{min}. This interval will be fathomed. $(x_{min}, s_{min}) = (2, +0.479)$.

III. STRUCTURE PREDICTION OF POLYPEPTIDES

A. Structure Prediction of Oligopeptides

The use of computational techniques and simulations in addressing the protein folding problem became possible through the introduction of qualitative and quantitative methods for modeling these systems. Given a sufficiently accurate description of the intramolecular forces, it is in principle possible to predict the folded conformation by optimization. In our work, we have focused not only on the development of global optimization methods, but also on the verification of energy modeling techniques.

In the area of energy modeling, our work has involved the investigation of numerous detailed representations of protein systems. In addition to the traditional all-atom potential energy models, our work has explored the effects of solvation contributions. In fact, although the problem of considering solvation effects in global conformational energy searches has been made tractable by the development of implicit solvation models, results for such formulations are essentially nonexistent, and those that have appeared are for limited searches only. In our work, both solvent accessible area and volume effects have been considered in the context of global searches for oligopeptides. In addition, we have examined the effects of several parameterizations for these models and have been able to identify those that provide the best correspondence between computational and experimental results.

1. Potential Energy Models

There are a number of approaches that may be used to model protein interaction energies. In reality, the dynamics of atoms are governed by the quantum theory of their participating electrons. Using the Born–Oppenheimer approximation, one can determine the energy for fixed atomic nuclei from the smallest eigenvalue of the Hamiltonian of the electron system. These approximations and their derivatives are calculated using *ab initio* methods. However, due to their computational complexity, such calculations are limited to extremely small molecules. Less detailed, semiempirical methods are based on all atom representations of the peptide. In general, these models, also known as force fields, are expressed as summations of empirically derived potential functions, with the mathematical form of individual energy terms based on the phenomenological nature of that term. Other simplified models have been used to reduce the degrees of freedom associated with the conformational energy expressions.

A number of empirically based molecular mechanics models have been developed for protein systems, including AMBER [29–31], CHARMM [32], DISCOVER [33], ECEPP [34–36], ECEPP/2 [37], ECEPP/3 [38], ENCAD [39,40], GROMOS [41], MM2 [42], and MM3 [43–45]. A general total energy equation, such as Eq. (23), includes terms for bond stretching (E_{bond}), angle bending (E_{angle}), torsion (E_{tor}), nonbonded (E_{nb}) and coupled (E_{cross}) interactions.

$$E_{tot} = E_{bond} + E_{angle} + E_{tor} + E_{nb} + E_{cross} \qquad (23)$$

Bond stretching and angle bending energies are included in those force fields that allow flexible geometries. A simple representation for both terms is based on the harmonic approximation, which corresponds to the classical description of the movement of a spring (by Hooke's law). The simplest approach, based on the fact that most bonds are near the minimum of their respective energy well, employs a quadratic term to model bond stretching and angle bending energies, as shown in Eqs. (24) and (25):

$$E_{bond} = \frac{k_{bond}}{2}(l - l_0)^2 \tag{24}$$

$$E_{angle} = \frac{k_{angle}}{2}(\theta - \theta_0)^2 \tag{25}$$

These equations act as penalty functions to force bond distances and bond angles, l and θ, to reference bond lengths and distances, l_0 and θ_0, whose values depend on the specific atoms involved. In actuality, these energy terms are more complicated. For bond energies cubic terms are often introduced, and angle energy terms usually include higher power expansions.

Torsional terms are used to describe the internal rotation energy of torsion angles, which exist between all atoms with a 1–4 relationship (separated by two other atoms). For rigid geometry force fields, these torsion angles can be used to define a set of independent variables that effectively describe any protein conformation. This approximately reduces the number of variables by a factor of 10 over those force fields that use a Cartesian coordinate system to describe flexible molecular geometries. In addition, bond and angle energies can be neglected for rigid geometry force fields. The torsion energy expression is typically represented by a Fourier series expansion that, as shown in Eq. (26), includes three terms:

$$E_{tor} = E_1(1 - \cos \phi) + E_2(1 - \cos 2\phi) + E_3(1 - \cos 3\phi) \tag{26}$$

The parameters involved in this expansion—namely E_1, E_2, and E_3—are torsional barriers that are usually specified for the pair of atoms around which the torsion occurs. Each term can be interpreted physically. The $1 - x$ $(\cos \phi)$ symmetry term accounts for those nonbonded interactions not included in general nonbonded terms. The $2 - x$ $(\cos 2\phi)$ symmetry term is related to the interactions of orbitals, while the $3 - x$ $(\cos 3\phi)$ symmetry term describes steric contributions.

Nonbonded energy terms attempt to model electrostatic and van der Waals interactions between those atoms that are not connected to each other or through a common atom. Typically, a Coulombic term is used to represent electrostatic energies based on atomic point charges, as shown in Eq. (27):

$$E_{\text{elec}} = \frac{Q_i Q_j}{\epsilon R_{ij}} \tag{27}$$

Here Q_i and Q_j represent the two point charges, while R_{ij} equals the distances between these two points. In some force fields, Coulombic interactions are modified by changing the dependence of the dielectric constant, ϵ. In general, van der Waals interactions are modeled using a 6–12 Lennard-Jones potential energy term. This expression, shown in Eq. (28), consists of a repulsion and attraction term.

$$E_{vdw} = \epsilon_{ij} \left[\left(\frac{R_{ij}^*}{R_{ij}} \right)^{12} - 2 \left(\frac{R_{ij}^*}{R_{ij}} \right)^{6} \right] \tag{28}$$

The energy minimum for a given atomic pair is described by the potential depth, ϵ_{ij}, and position, R_{ij}^*. Other force fields model van der Waals interactions using a modified Hill equation, which replaces the twelfth power term in Eq. (28) with an exponential term [42,43]. Different approaches are also used to describe nonbonded interactions between those atoms that may form hydrogen bonds. Some force fields model these interactions using only Coulombic terms, whereas other force fields employ special functions, such as a modified 10–12 Lennard-Jones-type potential term [46], as shown in Eq. (29).

$$E_{\text{hbond}} = \epsilon_{ij} \left[5 \left(\frac{R_{ij}^*}{R_{ij}} \right)^{12} - 6 \left(\frac{R_{ij}^*}{R_{ij}} \right)^{10} \right] \tag{29}$$

The cross term, shown in Eq. (23), accounts for interactions due to the inherent coupling between bonds, angles and torsions. Generally, these terms are small, and in many force fields they are neglected. Correction terms, which vary for each force field, are also typically added to the general energy equation. For example, the formation of disulfide bridges can be enforced by adding a penalty term to constrain the values of specified bond angles and bond lengths. Correction terms have also been used to adjust conformational energies according to the configurations of proline and hydroxyproline residues [38].

For a significant portion of this work, the ECEPP/3 (Empirical Conformational Energy Program for Peptides) [38] potential model is utilized. In this force field, it is assumed that the covalent bond lengths and bond angles are fixed at their equilibrium values. Then, the conformation is only a function of the independent torsional angles of the system, also known as dihedral angles. The total conformational energy is calculated as the sum of the electrostatic, nonbonded, hydrogen bonded, and torsional contributions. There is also a pseudo-potential for loop closing if the polypeptide contains two or more sulfur-containing residues. More recent work includes a revised treatment of prolyl and hydroxyprolyl residues [38]. For each prolyl or hydroxyprolyl residue contained in the polypeptide a fixed internal conformational energy for the pyrolidine ring is added. The main energy contributions (electrostatic, non-bonded, hydrogen bonded) are computed as the sum of terms for each atom pair (i,j) whose interatomic distance is a function of at least one dihedral angle. The general potential energy terms of ECEPP/3 are shown in Fig. 11, while the development of the appropriate parameters is discussed and reported elsewhere [38].

2. Solvation Energy Models

Solvation contributions are generally believed to be a significant force in stabilizing the native conformations of proteins. Explicit methods can be used to include solvation effects by actually surrounding the polypeptide with solvent

$$
\begin{aligned}
E = & \sum_{(i,j)\in ES} 332.0 \frac{q_i \, q_j}{D r_{ij}} && \text{(Electrostatic)} \\
+ & \sum_{(i,j)\in NB} F \frac{A}{r_{ij}^{12}} - \frac{C}{r_{ij}^6} && \text{(Nonbonded)} \\
+ & \sum_{(hx)\in HX} F \frac{A'}{r_{hx}^{12}} - \frac{B}{r_{hx}^{10}} && \text{(Hydrogen bonded)} \\
+ & \sum_{k\in TOR} (\frac{E_0}{2})(1 \pm \cos n_k \theta_k) && \text{(Torsional)} \\
+ & \sum_{l\in LOOP} B_L \sum_{il=1}^{il=3} (r_{il} - r_{io})^2 && \text{(Cystine Loop-Closing)} \\
+ & \sum_{l\in LOOP} A_L (r_{4l} - r_{4o})^2 && \text{(Cystine Torsional)}
\end{aligned}
$$

Figure 11. Potential energy terms in ECEPP/3 force field. r_{ij} refers to the interatomic distance of the atomic pair (ij). Q_i and Q_j are dipole parameters for the respective atoms, in which the dielectric constant of 2 has been incorporated. F_{ij} is set equal to 0.5 for 1–4 interactions and equal to 1.0 for 1–5 and higher interactions. A_{ij}, C_{ij}, A'_{ij}, and B_{ij} are nonbonded and hydrogen bonded parameters specific to the atomic pair. $E_{o,k}$ and $E_{o,l}$ are parameters corresponding to torsional barrier energies for a given dihedral angle. θ_k represents any dihedral angle. c_k takes the values -1, 1, and n_k refers to the symmetry type for the particular dihedral angle. The cystine loop-closing term is calculated as a penalty term of three distances involved in loop-closing, where r_{il} represents the actual distance and r_{io} represents the required distance. B_i, the penalty parameter, is set equal to 100. Finally, E_p is a fixed internal energy that is added for each proline residue in the protein. Energy units are kcal/mol and distance units are Å.

molecules and calculating solvent–peptide and solvent–solvent interactions. Although these methods are conceptually simple, explicit inclusion of solvent molecules greatly increases the computational time needed to simulate the polypeptide system. Therefore, most simulations of this type are limited to restricted conformational searches. In addition, it is difficult to quantify the effect of hydrophobic interactions that result from the ordering of water molecules.

Methods for estimating solvent free energies have also been developed using both integral equations and continuum models. Integral equation methods can be used to evaluate solvent structure and thermodynamic properties. Typically, molecular dynamics or Monte Carlo simulations are used to calculate ensemble averages from which free energy differences can be obtained. A number of methods have been proposed to estimate these solvation free energies from simulations based on molecular dynamics and Monte Carlo averages [47–49]. The integral equation method has also been used to analyze the solvent structure of a protein system [50]. In contrast, continuum models use a simplified representation of the solvent environment by neglecting the molecular nature of the water molecules. Calculations of solvation free energies using electrostatic continuum models rely on numerical solutions to the Poisson–Boltzmann equation from which dielectric and ionic strength effects are obtained [51]. Other continuum models estimate free energies of solvation as a function of surface areas and volumes.

In this work, solvation contributions are included implicitly using empirical correlations with both surface area [52] and volume [53]. The main assumption of these models is that, for each functional group of the peptide, a hydration free energy can be calculated from an averaged free energy of interaction of the group with a layer of solvent known as the hydration shell. In addition, the total free energy of hydration is expressed as a sum of the free energies of hydration for each of the functional groups of the peptide; that is, an additive relationship is assumed.

Accessible surface area methods assume that the free energy of hydration is proportional to the solvent-accessible surface area of the peptide, as described by the following equation:

$$E_{\text{HYD}} = \sum_{i=1}^{N} (A_i)(\sigma_i)$$ (30)

In Eq. (30), an additive relationship for N individual functional groups is assumed. (A_i) represents the solvent-accessible surface area for the functional group, and (σ_i) is an empirically derived free energy density parameter.

There are a number of ways to define the surface of a peptide. In developing these surfaces the peptide is represented by a union of spheres, with the radii of the spheres set by the van der Waals radii of the constituent atoms. A spherical test probe is then rolled over these spheres, thereby tracing out a surface. The molecular surface is set by direct contact between the probe sphere and the peptide spheres. In areas where the probe cannot make direct contact, the closest part of the probe is used. The solvent-accessible surface is defined by the surface traced by the center of the probe as the probe rolls over the peptide spheres. These areas depend on the radius of the probe sphere; when this radius is set to zero, both the molecular and solvent–accessible surface areas become the van der Waals surface of the peptide.

Solvent-accessible surface areas are calculated using the MSEED [52] program, which employs algorithms developed by Connolly [54]. MSEED eliminates many unnecessary computations by considering only those convex faces that are on the accessible surface. Rigorous implementation of Connolly's method requires the calculation of interior surface areas, which are ultimately found to be zero. A full description of the MSEED algorithm is given elsewhere [52]. A number of other methods for calculating surface areas are also available [55–57].

One potential problem that may arise when calculating accessible surface areas is the appearance of gradient discontinuities. This may occur when a new vertex or edge appears on the surface. If the discontinuity is large, minimization techniques requiring gradients may fail to converge to the local minimum conformation. A complete analysis of all situations for which the gradient of the molecular surface area becomes discontinuous has been reported [58].

Once the solvent–accessible surface areas have been calculated, these values must be multiplied by the appropriate (σ_i) parameters as shown in Eq. (30). A variety of parameter sets have been developed to model the transfer of atoms from a gaseous to a hydrated environment. The parameter values for the five ASP sets used in this study are given in Table I.

The ASP sets WE1 and WE2, are taken from Table 3 of Ref. 59. These parameters are both derived from Wolfenden's measured free energies of transfer of amino acid side-chain analogs from vapor to water [60]. Both sets have been adjusted to correct for entropy of mixing effects based on solute and solvent size differences [61,62], although the applicability of these corrections has been criticized [63,64]. The parameters for these two sets are negative for all atoms excluding carbon. Qualitatively, this means that the nitrogen, oxygen, and sulfur atoms are considered hydrophilic; that is, they favor solvent exposure. Comparatively, the WE1 and WE2 parameters are similar, with the largest relative difference being a 3 : 1 ratio (WE1 : WE2) for the σ_C parameter. Therefore, the hydrophobic character of these carbon atoms is stronger for the WE1 ASP set.

The OONS parameter set has been specifically developed to supplement the ECEPP/2 force field [65]. These parameters were derived by a least squares

TABLE I
Free Energy Density of Solvation Parameters for the ASP Set Employed with the Solvent-Accessible Surface Area Model[a]

Atom Type	WE1	WE2	OONS	SCKS	JRF
C aliphatic	12.0	4.0	8.0	32.5	216.0
C carboxyl, carbonyl	12.0	4.0	427.0	32.5	−732.0
C aromatic	12.0	4.0	−8.0	32.5	−678.0
N noncharged	−116.0	−113.0	−132.0	−17.5	−312.0
N charged	−186.0	−169.0	−132.0	−217.5	−312.0
O carboxyl, carbonyl	−116.0	−113.0	−38.0	−17.5	−262.0
O hydroxyl	−116.0	−113.0	−172.0	−17.5	−910.0
O charged	−175.0	−169.0	−38.0	−280.0	−910.0
S all	−18.0	−17.0	−21.0	−9.0	−281.0

[a]The first column describes the atom type, whereas the remaining columns provide the solvation parameters in cal/(mol $Å^2$) for the corresponding ASP set.

fitting to experimental free energies of gas to water transfer of small aliphatic and aromatic molecules. The most significant difference from the two previous ASP sets is a substantial increase in hydrophobic character for carboxyl (carbonyl) carbon atoms, which corresponds to a hundredfold increase when compared to the same WE2 parameter. In addition, the free energy parameter becomes negative for aromatic carbons, which indicates a hydrophilic tendency. The threefold decrease of the OONS values for carboxyl (carbonyl) and charged oxygen atom parameters, as compared to both the WE1 and WE2 ASP sets, is also significant.

Unlike the aforementioned models, the SCKS ASP set is not directly based on experimental free energies [66]. Instead, it is an optimized parameter set developed to complement the CHARMM [32] molecular mechanics force field. Specifically, through the use of experimental and molecular dynamics information, the relative weightings of solvation parameters were refined to provide the best correspondence between minimized and experimental structures. In comparing the individual free energy parameters, it is evident that the hydrophobic character of the carbon atoms is increased approximately three- and eightfold over the WE1 and WE2 values, respectively. In contrast, the uncharged oxygen and nitrogen atom parameters are 6.5 times smaller (less hydrophilic) than those for the WE1 and WE2 ASP sets. This decrease does not apply to charged oxygen and nitrogen atoms, which possess extremely hydrophilic values.

The JRF ASP set was derived from NMR studies of low energy solvated configurations of 13 tetrapeptides [67]. This represents an important difference from other derivations because actual peptides, rather than simple model compounds, were used to develop the JRF parameters. An ensemble of low-energy structures for these tetrapeptides was also produced using the ECEPP/2 potential function. Then, a nonlinear least-squares system was optimized for the best set of atomic solvation parameters. Although the parameters for oxygen, nitrogen, and sulfur atoms are negative, their large absolute values indicate much larger hydrophilicities than corresponding atoms of any other ASP set. In addition, both the carboxyl (carbonyl) and aromatic carbon atoms possess strong hydrophilic parameters, which contradicts other free energy parameter values for these atoms. The single positive value belongs to the aliphatic carbon atom type, which, although larger than any other parameter for this atom type, possesses the smallest magnitude for the JRF ASP set. Furthermore, because it was developed from minimum energy conformations of peptides, the JRF ASP set has been shown to produce undesirable perturbations during local minimizations if the solvation energy contributions are added at every iteration. Therefore, unlike the aforementioned ASP sets, the JRF solvation energy effects are only included at local minimum conformations.

For volume shell models, the free energy of hydration is assumed to be proportional to the water-accessible volume of a hydration layer surrounding the peptide. This can be represented in the form

$$E_{\text{HYD}} = \sum_{i=1}^{N}(VHS_i)(\delta_i) \tag{31}$$

An additive relationship for the N individual atoms of the peptide is assumed, and (VHS_i) represents the solvent-accessible volume of hydration shell for each atom i that is exposed to water. The (δ_i) parameters are empirically determined free energy of hydration densities for these atoms.

The hydration shell is defined by the volume inside a sphere of radius R_i^h but outside a sphere of radius R_i^v, with both radii centered on atom i. The larger radius, R_i^h, corresponds to the radius of the first hydration shell of atom i, while R_i^v is equal to the van der Waals radius. In order to calculate (VHS_i), the volume of a collection of overlapping hard spheres must be computed using:

$$V(\mathbf{R}) = \sum_{i} a_i S_i - \sum_{ij} b_{ij} D_{ij} + \sum_{ijk} c_{ijk} T_{ijk} - \sum_{ijkl} d_{ijkl} Q_{ijkl} \tag{32}$$

In Eq. (32), S_i signifies the volume of a single sphere, while D_{ij}, T_{ijk} and Q_{ijkl} represent the volume of intersection of two, three, and four spheres, respectively. This is sufficient because all higher-order overlaps can be decomposed into the three types of intersections included in Eq. (32). Therefore, the solvent-accessible volume of hydration can be written as

$$(VHS_i) = V(R_i^h) - V(R_i^v) \tag{33}$$

The first term in Eq. (33) is calculated using Eq. (32) with the radii of all atoms set equal to their van der Waals radii, whereas the second term is calculated with the radius of atom i equal to R_i^h and the van der Waals radii of all the other atoms. A number of methods to compute hydration shell volumes have been proposed [53,68,69].

The form of Eq. (32) is not suitable for force-field models using pairwise intramolecular potential, such as ECEPP/3. Furthermore, direct truncation at the double-overlap term would lead to large errors. In this work, the RRIGS (reduced-radius independent Gaussian sphere) approximation is used to efficiently calculate the exposed volume of the hydration shell [53]. This method uses a truncated form of Eq. (32) but also artificially reduces the van der Waals radii of all atoms other than atom i when calculating (VHS_i). These reductions effectively decrease the contribution of the double-overlap terms, leading to a cancellation of the error which results from neglecting the triple and higher overlap terms. In addition, the characteristic density of being inside the overlap volume of two intersecting spheres is not represented as a step function, but as a Gaussian function; this provides continuous derivatives of the hydration potential. Therefore, the solvation energy contributions can easily be added at every step of local minimizations because the RRIGS approximation has the same set of interactions as the ECEPP/3 potential.

Free energy density parameters for solvent accessible volumes have been developed for nonionic and charged organic solute molecules [70–72]. In this work, RRIGS specific (δ_i) parameters, which were developed by a least-square fitting of experimental free energy of solvation data for 140 small organic molecules [53], are used (Table II). The classification of the RRIGS atom types is more fragmented than for the solvent accessible surface area ASP sets. The most hydrophilic values belong to the nitrogen and selected oxygen and hydrogen atom types. In addition, aromatic carbons tend to possess slightly hydrophilic values, whereas the carbonyl and aliphatic carbon atoms exhibit the most hydrophobic parameter values.

3. Global Optimization Framework

The energy minimization problem is formulated as a unconstrained nonconvex global optimization problem, which is fashioned after the general formulation given in problem (1). Let $i = 1, \ldots, N_{RES}$ be an indexed set describing the sequence of amino acid residues in the peptide chain. There are ϕ_i, ψ_i, ω_i, $i = 1, \ldots, N_{RES}$, dihedral angles along the backbone of this peptide. In addition, let K^i denote the number of dihedral angles for the side chain of the ith residue and let J^N and J^C denote the number of dihedral angles for the amino and carboxyl end groups, respectively. Using these definitions the optimization

TABLE II
Free Energy Density of Solvation Parameters Employed in the RRIGS Model[a]

Atom Type	δ	R^v	R^h
H hydroxyl, amino	− 10.35	1.415	4.17
H acid	− 3.206	1.415	4.17
H amide	− 7.714	1.415	4.17
H thiol	2.709	1.415	4.17
C aliphatic CH_3	1.319	2.125	5.35
C aliphatic CH_2	0.2374	2.225	5.35
C aliphatic CH	− 1.271	2.375	5.35
C other aliphatic	− 2.297	2.060	5.35
C cyclic CH	0.2890	2.375	5.35
C aromatic CH	− 0.2137	2.100	5.35
C aromatic CR	− 1.713	1.850	5.35
C branched aromatic C	− 1.910	1.850	5.35
C aromatic COH	− 0.6063	1.850	5.35
C carbonyl	2.696	1.870	5.35
N primary amine	− 1.149	1.755	5.05
N secondary amine	− 10.28	1.755	5.05
N aromatic	− 10.48	1.755	5.05
N amide	− 7.332	1.755	5.05
O hydroxyl, ether	− 7.396	1.620	4.95
O acid, ester	0.07897	1.620	4.95
O ketone, carbonyl	− 15.70	1.560	4.95
O acid, amide carbonyl	− 15.56	1.560	4.95
S thiol, disulfide	− 4.706	2.075	5.37

[a]The second column provides the solvation parameters in cal/(mol Å²), and the last two columns correspond to the van der Waals and hydration radii (Å), respectively.

problem takes the following form:

$$\min \quad E(\phi_i, \psi_i, \omega_i, \chi_i^k, \theta_j^N, \theta_j^C)$$

$$\text{subject to} \quad \begin{aligned} -\pi \leq \phi_i \leq \pi, & \quad i = 1, \ldots, N_{\text{RES}} \\ -\pi \leq \psi_i \leq \pi, & \quad i = 1, \ldots, N_{\text{RES}} \\ -\pi \leq \omega_i \leq \pi, & \quad i = 1, \ldots, N_{\text{RES}} \\ -\pi \leq \chi_i^k \leq \pi, & \quad i = 1, \ldots, N_{\text{RES}}, \quad k = 1, \ldots, K^i \\ -\pi \leq \theta_j^N \leq \pi, & \quad j = 1, \ldots, J_N \\ -\pi \leq \theta_j^C \leq \pi, & \quad j = 1, \ldots, J_C \end{aligned} \qquad (34)$$

In general, E represents the total potential energy function and the free energy of solvation. However, in the case of the JRF ASP set, the potential energy function is minimized before adding the hydration energy contributions for this ASP set. In other words, gradient contributions from solvation are not considered. This approach is represented by the following equation:

$$E_{\text{JRF}}^{\text{Total}} = E_{\text{Min}}^{\text{Unsol}} + E_{\text{JRF}}^{\text{Sol}} \tag{35}$$

Even after reducing this optimization problem to a function of internal variables (dihedral angles), the multidimensional surface that describes the energy function has an astronomically large number of local minima. A large number of techniques have been developed to search this nonconvex conformational space. In general, the major limitation is that these methods depend heavily on the supplied initial conformation. As a result, there is no guarantee for global convergence because large sections of the domain space may be bypassed. To overcome these difficulties, the αBB global optimization approach [15,18,73] has been extended to identifying global minimum energy conformations of solvated peptides. The αBB global optimization algorithm effectively brackets the global minimum solution by developing converging sequences of lower and upper bounds. These bounds are refined by iteratively partitioning the initial domain. Upper bounds on the global minimum are obtained by local minimizations of the original energy function, E. Lower bounds belong to the set of solutions of the convex lower bounding functions, which are constructed by augmenting E with the addition of separable quadratic terms. The lower bounding functions, L, of the energy hypersurface can be expressed in the following manner:

$$
\begin{aligned}
L = E \ &+ \sum_{i=1}^{N_{\text{RES}}} \alpha_{\phi,i}(\phi_i^L - \phi_i)(\phi_i^U - \phi_i) + \sum_{i=1}^{N_{\text{RES}}} \alpha_{\psi,i}(\psi_i^L - \psi_i)(\psi_i^U - \psi_i) \\
&+ \sum_{i=1}^{N_{\text{RES}}} \alpha_{\omega,i}(\omega_i^L - \omega_i)(\omega_i^U - \omega_i) + \sum_{i=1}^{N_{\text{RES}}} \sum_{k=1}^{K^i} \alpha_{\chi,i,k}(\chi_i^{k,L} - \chi_i^k)(\chi_i^{k,U} - \chi_i^k) \\
&+ \sum_{j=1}^{J^N} \alpha_{j,\theta^N}(\theta_j^{N,L} - \theta_j^N)(\theta_j^{N,U} - \theta_j^N) + \sum_{j=1}^{J^C} \alpha_{j,\theta^C}(\theta_j^{C,L} - \theta_j^C)(\theta_j^{C,U} - \theta_j^C)
\end{aligned}
\tag{36}
$$

Here $\phi_i^L, \psi_i^L, \omega_i^L, \chi_i^{k,L}, \theta_j^{N,L}, \theta_j^{C,L}$ and $\phi_i^U, \psi_i^U, \omega_i^U, \chi_i^{k,U}, \theta_j^{N,U}, \theta_j^{C,U}$ represent lower and upper bounds on the dihedral angles $\phi_i, \psi_i, \omega_i, \chi_i^k, \theta_j^N, \theta_j^C$. The α parameters represent nonnegative parameters that must be greater or equal to the negative one-half of the minimum eigenvalue of the Hessian of E over the defined domain. The computational requirement of the αBB algorithm depends on the number of variables (global) on which branching occurs. Therefore, these global variables need to be chosen carefully.

The determination of the global minimum energy conformation using αBB requires the interfacing of a number of programs (αBB [15–18,73], PACK [74], NPSOL [28] and potential and solvation energy modules). PACK, a peptide generation program, is called once directly by αBB in order to initialize the current problem. In subsequent steps PACK is called through NPSOL [28], a local nonlinear optimization solver used to solve both the upper and lower bounding problems. PACK internally transforms to and from Cartesian and internal coordinate systems, and provides potential energy and gradient contributions for the ECEPP/3 potential model at every step of the local minimizations. When considering surface-accessible solvation, surface areas are calculated using MSEED [52]; whereas volumes of hydration shells are determined using the RRIGS module [53]. Finally, an additional module, UBC (upper bound check), is used to verify the quality of the upper bound solutions. The entire suite of programs has been combined to form the GLO-FOLD software package for the prediction of protein structure, as shown in Fig. 12.

The basic steps of the algorithm are as follows:

1. The initial best upper bound is set to an arbitrarily large value. The original domain is partitioned along one of the global variables.

2. A convex function (L) is constructed in each hyper-rectangle and minimized using NPSOL, with calls (through PACK) to both ECEPP/3 and one of the two solvation modules. If a solution is greater than the current best upper bound, the entire subregion can be fathomed; otherwise the solution is stored.

3. The local minima solutions for L are used as initial starting points for local minimizations of the upper bounding function (E) in each hyper-rectangle. Again, the appropriate calls are made to PACK and the potential and solvation energy modules. In solving the upper bounding problems, all variable bounds are expanded to $[-180, 180]$. These solutions are upper bounds on the global minimum solution in each hyper–rectangle.

4. The current best upper bound is updated to be the minimum of those thus far stored. If a new upper bound (from Step 3) is selected, the upper bound check, UBC, module is called. UBC checks that the absolute value of each gradient in the objective function gradient vector is below a specified tolerance (kcal/mol/deg). If a gradient does not satisfy this check the corresponding variable bounds are incrementally increased and the problem is solved with the previous point used as the initial starting point. This process is repeated until the gradient constraints are satisfied or an iteration limit is exceeded. UBC also employs algorithms to

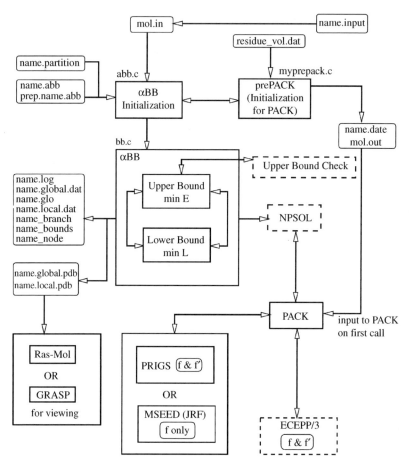

Figure 12. Interface for αBB within GLO-FOLD. The arrows indicate the direction of information flow. The names of the input, output, and intermediate files are indicated, in addition to selected source code files. References to "*f & f'*" and "*f* only" describe whether gradient evaluations or only function evaluations are used in the respective modules.

calculate the second derivative matrix [75], which is used to verify that the upper bound solution is a local minimum; that is, the Hessian matrix is positive semidefinite. If the matrix is not positive semidefinite or the gradient checks are not satisfied, the upper bound solution is rejected.

5. The hyper-rectangle with the current minimum value for L is selected and partitioned along one of the global variables.

6. If the best upper and lower bounds are within the ϵ tolerance, the program will terminate; otherwise it will return to Step 2.

4. Computational Studies

Single-residue examples were defined as terminally blocked by using acetyl (amino) and methyl (carboxyl) end groups. All dihedral angles were treated as global variables, excluding the three θ angles of the end groups. The relative convergence was set to 10^{-2}. For these examples, all dihedral angles, excluding those of the end groups, were treated as global variables. The remaining variables were treated locally; that is, they were allowed to vary during minimization, but their domain space was not partitioned. When using the RRIGS and JRF models, the global variables were assigned initial α values of 3.0. For the other solvation models, the α values were increased to 5.0.

For a number of residues, the JRF global minimum solutions possess ω angles in the range of $[-30, 30]$ with the corresponding ϕ and ψ angles near the $[-150, 80]$ region. Additional runs were conducted in which the ω angles were constrained to the range of $[160, 200]$. In all cases, with the exception of serine, this constraint led to increases in solvation energies and decreases in potential energy terms while the structures became either β-sheet-like or α-helical. Without exception, the ω angles for the all other global minimum energy solutions were within the $[160, 200]$ range. The remaining analysis in this section refers to these constrained (ω within $[160, 200]$) minima for the JRF ASP set. This is appropriate not only in comparing the JRF results with other solvation results, but it also makes the analysis relevant for the oligopeptide studies because similar ω bounds are typically used.

The results of the solvation models are more clearly evaluated when examining energy differences. For example, ΔE^{POT} ($\Delta E^{POT} = E^{POT}_{ASP} - E^{POT}_{RRIGS}$) refers to the change in potential energy of an area based global minimum (E^{POT}_{ASP}) and the RRIGS global minimum (E^{POT}_{RRIGS}) solution for a given terminally blocked residue. This difference is positive in almost all cases, which indicates that the potential energy of the RRIGS structure is always lower and provides more stabilization at the corresponding global minimum solution. In most cases, this difference is very small, especially for the OONS and SCKS ASP sets. In fact, for both, of these ASP sets, several residues, most noticeably phenylalanine and tyrosine, have more potential energy stabilization at their corresponding global minima. However, for five peptides, namely phenylalanine, serine, threonine, tyrosine and leucine, the JRF potential energy is more than 10 kcal/mol less stabilizing. This set of residues includes the three residues (serine, threonine, and tyrosine) that contain hydroxyl groups among the side-chain atoms, as well as the two regular aromatic residues (phenylalanine and tyrosine). It is also interesting to note that these atom types (i.e., hydroxyl oxygen and aromatic carbon) correspond to two of the most hydrophilic type atoms in the JRF ASP set. Finally, the leucine ΔE^{POT} seems to be abnormally high because of the large torsional contribution at the JRF global minimum conformation.

The results for ΔE^{HYD} ($\Delta E^{HYD} = E_{ASP}^{HYD} - E_{RRIGS}^{HYD}$), which refers to the change in hydration energy between an area based global minimum (E_{ASP}^{HYD}) and the RRIGS global minimum (E_{RRIGS}^{HYD}) solutions, are especially interesting. These differences are positive in most cases, which indicates that the hydration energy of the RRIGS structure is generally lower. However, when examining the JRF results, ΔE^{HYD} is negative for four examples, namely histidine, phenylalanine, tryptophan, and tyrosine. Excluding the special case of proline, these four residues correspond to the naturally occurring residues which possess ringed side-chain structures. Other trends are also apparent. The most positive ΔE^{HYD} values for the JRF ASP set are provided by the aliphatic residues. In addition, the acidic residues, glutamic and aspartic acid, and the amide forms of these residues, glutamine and asparagine, have comparable values for ΔE^{HYD}.

For the other (gradient inclusive) ASP sets, the ΔE^{HYD} of different residues are less varied. However, it is important to consider that for all residues, excluding tyrosine, the ASP sets follow a WE2, WE1, OONS, and SCKS order when ranked beginning with the most stabilizing hydration energy. Low hydration energies are expected for WE2 because of the consistently small hydrophobic and relatively large hydrophilic parameters. In most cases, the WE1 ΔE^{HYD} are only slightly larger than for WE2. This can be directly attributed to the increased hydrophobicity of the free energy parameter for the carbon atoms of the WE1 ASP set. When comparing the OONS and WE1 ASP sets, the increased ΔE^{HYD} is more noticeable, which is most likely a result of the combined effects of the strong hydrophobic value of the carboxyl (carbonyl) carbon parameter and the decreased hydrophilic value of the carboxyl (carbonyl) oxygen parameter for the OONS ASP set. However, for aromatic residues (i.e., phenylalanine, tryptophan and tyrosine), these effects are partially offset by introducing a hydrophilic character for aromatic carbons in the OONS ASP set. In fact, for tyrosine this change is strong enough to cause the OONS ASP set to produce a more stabilizing hydration energy than the WE1 ASP set. A comparison between the OONS and SCKS reveals the largest increase in ΔE^{HYD} values. This can partly be attributed to the relatively large value of the free energy parameters for carbon atoms of the SCKS ASP set. The increase is also enhanced for aromatic residues because of the hydrophilic nature of the aromatic carbon atoms for the OONS ASP set. In addition, for residues with nitrogen-containing side chains, the ΔE^{HYD} increase is heightened because of a subsequent decrease in the value of the free energy parameter for nitrogen atoms in the SCKS ASP set. Finally, a comparison of other surface accessible solvation results to the JRF results is qualitatively similar to those made between the RRIGS and JRF models. Specifically, the strong hydration energy stabilization of ring-containing residues, as well as the decreased stabilization provided by aliphatic residues, is evident.

A more detailed analysis was performed by generating adiabatically relaxed ϕ–ψ maps for N–acetyl–N'–methyl–alanineamide. The adiabatic curves define regions within a given energy of the global minimum value. The first map corresponds to an adiabatically relaxed map for the unsolvated form of the peptide. This was calculated by fixing the ϕ and ψ angles at $3°$ increments and using a local minimization solver to minimize the ECEPP/3 potential energy by varying the remaining dihedral angles. The other maps were constructed by a similar procedure, although the minimized energy now included both ECEPP/3 and the appropriate hydration free energy. In generating the data for the JRF, the ECEPP/3 energy was first minimized in the absence of solvent at each point and the map was generated by adding the solvation free energy for the JRF model at the minimized conformation.

These maps reveal several important effects of including solvation. Experimental data for the alanine peptide suggests that more than one conformation is present in solution, and NMR coupling constants indicate a large population of conformations with $-70 > \phi > -80$ [76]. It is also expected that hydration may weaken intrapeptide hydrogen bonding. The unsolvated map indicates well-defined regions for intramolecular hydrogen bonding (C_7) and for right-handed α-helices (α_R). The global minimum occurs within the C_7 region. The RRIGS map retains some features of the unsolvated map, with the global minimum in the C_7 region and a very strong α_R region. However, there is a broadening of the β-sheet (C_5) region as well as a less distinct C_7 minimum. This can be contrasted with both the WE1 and WE2 adiabatic maps, which exhibit large C_5 regions and significant decreases in the size of both the C_7 and α_R regions. The OONS map contains an even larger low-energy region that connects the C_5 and C_7 domains. The α_R low-energy region is also broader than either of those indicated by the WE1 or WE2 map. In all three cases (WE1, WE2, and OONS) the global minimum is shifted to the β-sheet domain. In contrast, the SCKS adiabatic map is more similar to the RRIGS map because of its smaller and disjoint C_5 and C_7 regions, as well as the location of the global minimum in the C_7 well. The largest disparity between these maps exists with the JRF adiabatic map, which indicates a complete shift away from the C_7 minimum toward the C_5 region.

Qualitatively, similar trends are observed for the ϕ–ψ distribution of other terminally blocked amino acids. The RRIGS model predicts a majority of global minima in the C_7 region, which indicates a tendency to preserve certain potential energy effects. As expected, the majority of WE1 and WE2 global minima lie within the C_5 domain, with the same distribution for each parameter set. The most uniform distribution of global minima belongs to the OONS ASP set, for which there are an almost equal number of C_5 and C_7 global minimum structures. This agrees with the large low-energy regions displayed on the N-acetyl-N'-methyl-alanineamide adiabatic map. The large population of C_7 global minima for the SCKS ASP set is also suggested by the strong C_7 region on the N-acetyl-N'-methyl-alanineamide map. In accordance with the distinct implementation of the JRF model, these results are less predictable. Specifically, although almost half of the JRF global minima lie in the C_5 domain, a significant number also exhibit α-helical type structures, which contrasts with the φ–ψ map of N-acetyl-N'-methyl-alanineamide.

Met-enkephalin (H–Tyr–Gly–Gly–Phe–Met–OH) is an endogenous opioid pentapeptide found in the human brain, pituitary, and peripheral tissues and is involved in a variety of physiological processes. The peptide consists of 24 independent torsional angles and a total of 75 atoms and has played the role of a benchmark molecular conformation problem. The energy hypersurface is extremely complex with the number of local minima estimated on the order of 10^{11} [77]. Based on a previous study, the unsolvated global minimum potential energy conformation, with an ECEPP/3 energy of -11.707 kcal/mol, was shown to exhibit a type II' β-bend along the N-C' peptidic bond of Gly^3 and Phe^4 [78].

In studying the effects of solvation on the structure of met-enkephalin, the results for the unsolvated structure were verified by employing the algorithm outlined in Section III.A.2. A major difference from the previous implementation [78] is the addition of the UBC module, as well as the expansion of all variable bounds (to $[-180, 180]$) when solving the upper bounding problems. Because the backbone dihedral angles (i.e., φ and ψ) are the most influential variables in defining the backbone structure, the corresponding 10 backbone dihedral angles were treated as global variables for the enkephalin problems. Although they were not partitioned during the global search, all other variables (i.e., ω and all χ) were allowed to vary during local minimizations. The global variables were assigned initial α values of 5.0 when using the unsolvated, RRIGS, and JRF models and were assigned values of 10.0 for all other models. In the case of unsolvated met-enkephalin, the structural and energetic results of the previously identified global minimum energy structure [78] were confirmed.

Experimental results have indicated that met-enkephalin in aqueous solution does not possess an unique structure [79]. In general, experimentally determined aqueous conformations are found to exhibit characteristics of extended random-coil polypeptide with no discernible secondary structure. When considering the effects of hydration, the competition for backbone hydrogen bonding (with water), which contributes to the bending of the unsolvated conformation, should result in a more extended structure.

The RRIGS model predicts a more extended structure than the global minimum structure reported for the unsolvated case [78]. In fact, although a slight turn occurs near the N-terminus, the structure possesses no hydrogen bonds (<2.2 Å) and an overall end-to-end C^α distance of 10.16 Å. In addition, there exists close proximity of the Tyr and Phe aromatic rings, as shown in

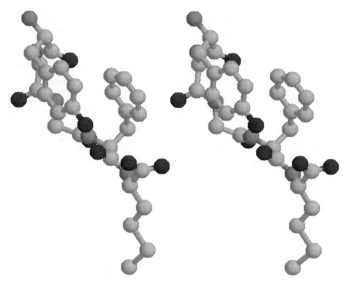

Figure 13. Plot of met-enkephalin conformation (in stereo). Global minimum energy of -50.01 kcal/mol using the RRIGS model for hydration.

Fig. 13. The centroids of these rings are separated by 4.16 Å, which is slightly closer than the preferential aromatic–aromatic interaction distance of 4.5 to 7 Å [80]. Furthermore, the aromatic rings are essentially in a parallel, as opposed to the more common orthogonal, orientation. This suggests an attempt to balance the slightly hydrophilic nature of the aromatic carbon atoms, as given by the RRIGS δ_i, and the favorable hydrophobic interactions between the two rings. The values of the dihedral angles for the global minimum energy conformation are given in Table III.

TABLE III
Dihedral Angles at the Global Minimum Energy Conformation of Met-enkephalin, Using the RRIGS Model for Hydration

	ϕ	ψ	ω	χ_1	χ_2	χ_3	χ_4
Tyr	-168.32	-30.81	178.52	-173.58	-101.26	18.83	
Gly	78.83	-86.96	182.73				
Gly	162.94	91.72	172.83				
Phe	-150.72	162.32	181.50	66.66	92.68		
Met	-77.80	106.79	181.63	-67.82	178.91	180.01	-60.01

The global minimum structures for the area-based hydration models (gradient inclusive) are less extended, as exhibited by Figs. 14 and 15. The lowest energy structures for the WE1 and WE2 models are very similar, with an the end-to-end C^α distance of 5.85 Å for both solvation models. In addition, the bend near the N-termini is stabilized by a hydrogen bond between the CO of the tyrosine residue and the NH proton of the phenylalanine residue (approximately 1.98 Å). This bend is similar to the type II' β-bend of the unsolvated global minimum energy structure, although it is shifted to the Gly^2–Gly^3 backbone region. The aromatic ring separation is wider (approximately 6.48 Å for both models) than for the RRIGS global minimum structure, although the side-chain orientations are similar. The values of the dihedral angles for the WE1 and WE2 global minimum structures are given in Tables IV and V, respectively.

The lowest energy conformation for the OONS ASP is also similar to the WE1 and WE2 structures. In this case, the end-to-end C^α distance is again 5.85 Å. The bending near the N-termini is again similar to a type II' β-bend along the Gly^2–Gly^3 backbone, although in this case it is stabilized by a slightly weaker hydrogen bond between the CO of the tyrosine residue and the NH proton of the phenylalanine residue (approximately 2.01 Å). The 6.60 Å aromatic ring separation is also slightly larger, which may be attributed to the slightly hydrophilic character of the aromatic carbon parameters as compared to the WE1 and WE2 ASP sets. The values of the dihedral angles for the global minimum structure are given in Table VI.

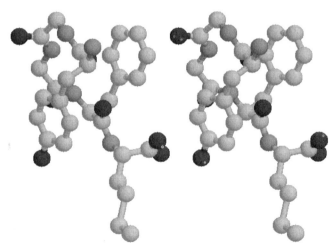

Figure 14. Plot of met-enkephalin conformation (in stereo). Global minimum energy of −30.31 kcal/mol using the WE1 model for hydration.

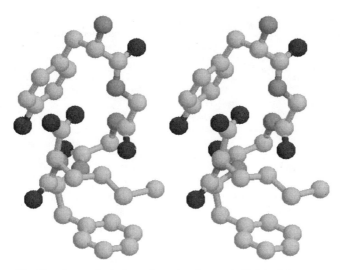

Figure 15. Plot of met-enkephalin conformation (in stereo). Global minimum energy of −0.62 kca/mol using the SCKS model for hydration.

TABLE IV

Dihedral Angles at the Global Minimum Energy Conformation of Met-enkephalin, Using the WE1 Model for Hydration

	ϕ	ψ	ω	χ_1	χ_2	χ_3	χ_4
Tyr	−162.65	−43.34	−177.43	−173.76	−90.62	2.61	
Gly	66.15	−86.62	172.92				
Gly	−152.31	32.40	−178.49				
Phe	−157.59	154.87	179.36	52.02	−96.19		
Met	−90.62	128.89	−179.18	−169.29	176.88	180.14	−59.99

TABLE V

Dihedral Angles at the Global Minimum Energy Conformation of Met-enkephalin, Using the WE2 Model for Hydration

	ϕ	ψ	ω	χ_1	χ_2	χ_3	χ_4
Tyr	−162.70	−43.23	−177.47	−173.94	−90.83	2.63	
Gly	66.15	−86.59	173.03				
Gly	−152.49	32.41	−178.55				
Phe	−157.84	154.97	179.26	52.12	−96.11		
Met	−89.96	129.19	−179.17	−169.47	176.75	180.13	−59.99

TABLE VI

Dihedral Angles at the Global Minimum Energy Conformation of Met-enkephalin,
Using the OONS Model for Hydration

	ϕ	ψ	ω	χ_1	χ_2	χ_3	χ_4
Tyr	−166.11	−50.84	−176.25	−188.97	−102.81	2.45	
Gly	63.86	−86.04	175.39				
Gly	−151.94	33.86	−178.80				
Phe	−159.47	153.41	179.46	50.93	−96.43		
Met	−79.75	148.31	−178.93	−68.16	181.45	178.08	59.70

The SCKS global minimum structure is even less extended, as shown in Fig. 15. Although the aromatic ring separation becomes wider (8.13 Å), the overall end-to-end C^α distance decreases to 5.80 Å. In this structure, there are two stabilizing hydrogen bonds—a 1.86 Å hydrogen bond between the NH proton of the first glycine residue and the CO of the methionine residue, and a 2.02 Å hydrogen bond between the CO of the first glycine residue and the NH proton of the phenylalanine residue. This backbone structure exhibits a type II′ β-bend around the Gly³ and Phe⁴ residues, which is similar to the global minimum energy conformation for unsolvated met-enkephalin. This compact structure is consistent with the relatively strong hydrophobic values of all carbon atom free energy parameters, as well as the relatively weak hydrophobic values of the oxygen and nitrogen atoms for the SCKS ASP set. The values of dihedral angles corresponding to the global minimum energy structure are given in Table VII.

In contrast, the JRF global minimum energy structure resembles a more extended conformation, with an overall end-to-end C^α distance of 9.56 Å. The plot of this structure, given in Fig. 16, shows that the residues near the N-terminus are almost fully extended, although there is slight turn near the C-terminus. This bending is stabilized by the formation of 2.10 Å hydrogen bond between the CO of the second glycine residue and the NH proton of the

TABLE VII

Dihedral Angles at the Global Minimum Energy Conformation of Met-enkephalin,
Using the SCKS Model for Hydration

	ϕ	ψ	ω	χ_1	χ_2	χ_3	χ_4
Tyr	−82.91	154.09	−176.27	−172.88	79.47	−166.08	
Gly	−151.61	81.91	168.71				
Gly	84.09	−72.41	−169.54				
Phe	−137.07	18.52	−173.06	57.94	−86.04		
Met	−162.71	158.63	−179.76	51.94	173.67	179.21	−58.18

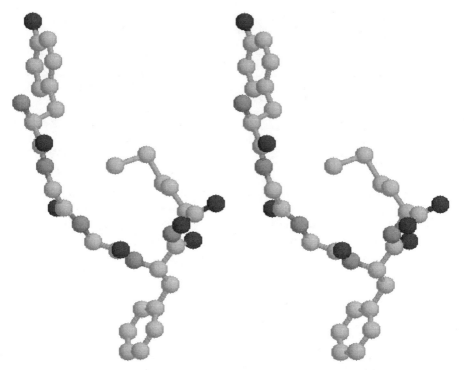

Figure 16. Plot of met-enkephalin conformation (in stereo). Global minimum energy of −283.76 kcal/mol using the JRF model for hydration.

methionine residue. In addition, the structure displays a large 14.87 Å separation between the centroids of the Phe and Tyr aromatic rings. This can be partly attributed to the strongly hydrophilic character of the aromatic and carboxyl (carbonyl) carbons parameters for the JRF ASP set. The values of dihedral angles corresponding to the global minimum energy are given in Table VIII.

The structures were further analyzed by comparing energy evaluations at corresponding global minimum solutions. This information is given in Tables IX and X. In all cases, excluding the SCKS model, the JRF global minimum energy structure provides that most stabilizing values for the hydration energy. However, these stabilizing hydration energies are generally offset by the relatively high value for potential energy at the JRF global minimum conformation (5.06 kcal/mol, obtained by calculating $E_{TOT} - E_{HYD}$ from Tables IX and X). In fact, the high potential energy causes the JRF structure to exhibit the highest values for overall energy, excluding the case of the JRF model. Only when considering the JRF model do these stabilizing hydration free energies tend to dominate the prediction of the global minimum structure. This is

TABLE VIII
Dihedral Angles at the Global Minimum Energy Conformation of Met-enkephalin,
Using the JRF Model for Hydration

	ϕ	ψ	ω	χ_1	χ_2	χ_3	χ_4
Tyr	− 84.96	160.74	179.09	− 59.83	100.80	− 179.29	
Gly	− 160.26	151.83	− 177.53				
Gly	159.50	− 157.94	178.71				
Phe	− 76.55	76.23	− 178.05	− 61.87	108.68		
Met	− 132.90	147.47	− 179.83	− 65.17	− 175.99	− 84.91	59.38

evidenced by the fact that the JRF structure provides an overall energy, more
than 100 kcal/mol lower than any other total energy, which can be directly
attributed to the differences in hydration energy. When using the SCKS model,
the only case for which the JRF conformation does not produce the most
stabilizing hydration energy, the JRF structure provides the least stabilizing

TABLE IX
Comparison of Hydration Energies for Met-enkephalin[a]

	Global of	E_{TOT}	E_{HYD}	E_{NB}	E_{ES}	E_{TOR}	(RMSD)
RRIGS	RRIGS	− 50.01	− 41.42	21.84	− 31.46	1.02	0.00
	WE1	− 47.87	− 38.12	22.09	− 32.61	0.78	2.83
	WE2	− 47.91	− 38.14	22.09	− 32.63	0.76	2.83
	OONS	− 47.17	− 37.95	22.25	− 32.13	0.66	2.66
	SCKS	− 47.24	− 35.61	21.47	− 35.40	2.30	4.04
	JRF	− 41.63	− 46.69	23.29	− 19.13	0.90	4.83
WE1	RRIGS	− 26.60	− 18.00	21.84	− 31.46	1.02	2.83
	WE1	− 30.31	− 20.56	22.09	− 32.61	0.78	0.00
	WE2	− 30.31	− 20.53	22.09	− 32.63	0.76	0.01
	OONS	− 29.01	− 19.79	22.25	− 32.13	0.66	0.80
	SCKS	− 27.80	− 16.17	21.47	− 35.40	2.30	3.33
	JRF	− 19.49	− 24.55	23.29	− 19.13	0.90	4.33
WE2	RRIGS	− 29.87	− 21.27	21.84	− 31.46	1.02	2.83
	WE1	− 33.26	− 23.52	22.09	− 32.61	0.78	0.01
	WE2	− 33.27	− 23.49	22.09	− 32.63	0.76	0.00
	OONS	− 32.01	− 22.79	22.25	− 32.13	0.66	0.80
	SCKS	− 30.77	− 19.15	21.47	− 35.40	2.30	3.33
	JRF	− 22.93	− 27.99	23.29	− 19.13	0.90	4.32

[a]The first column refers to the hydration model used in the function evaluations, which are
performed at the global solutions for the hydration model listed in the second column. The total
energy, E_{TOT}, is provided along with the contributions from hydration, E_{HYD}, nonbonded interactions
(including hydrogen bonding), E_{NB}, electrostatic interactions, E_{ES}, and torsion, E_{TOR}. The last
column provides the heavy-atom root-mean-squared deviation between the global minimum energy
structures of the hydration models listed in the first two columns.

TABLE X
Comparison of Hydration Energies for Met-enkephalin[a]

	Global of	E_{TOT}	E_{HYD}	E_{NB}	E_{ES}	E_{TOR}	(RMSD)
OONS	RRIGS	− 24.18	− 15.59	21.84	− 31.46	1.02	2.66
	WE1	− 31.08	− 21.33	22.09	− 32.61	0.78	0.80
	WE2	− 31.09	− 21.31	22.09	− 32.63	0.76	0.80
	OONS	− 31.45	− 22.23	22.25	− 32.13	0.66	0.00
	SCKS	− 29.57	− 17.95	21.47	− 35.40	2.30	3.38
	JRF	− 19.60	− 24.66	23.29	− 19.13	0.90	4.12
SCKS	RRIGS	3.43	12.02	21.84	− 31.46	1.02	4.04
	WE1	0.90	10.65	22.09	− 32.61	0.78	3.33
	WE2	0.89	10.67	22.09	− 32.63	0.76	3.33
	OONS	1.66	10.88	22.25	− 32.13	0.66	3.38
	SCKS	− 0.62	11.00	21.47	− 35.40	2.30	0.00
	JRF	17.44	12.38	23.29	− 19.13	0.90	3.78
JRF	RRIGS	− 139.36	− 130.77	21.84	− 31.46	1.02	4.83
	WE1	− 180.59	− 170.84	22.09	− 32.61	0.78	4.33
	WE2	− 180.57	− 170.79	22.09	− 32.63	0.76	4.32
	OONS	− 181.70	− 172.48	22.25	− 32.13	0.66	4.12
	SCKS	− 171.67	− 160.04	21.47	− 35.40	2.30	3.78
	JRF	− 283.76	− 288.82	23.29	− 19.13	0.90	0.00

[a]The first column refers to the hydration model used in the function evaluations, which are performed at the global solutions for the hydration model listed in the second column. The total energy, E_{TOT}, is provided along with the contributions from hydration, E_{HYD}, nonbonded interactions (including hydrogen bonding), E_{NB}, electrostatic interactions, E_{ES}, and torsion, E_{TOR}. The last column provides the heavy-atom root-mean-squared deviation between the global minimum energy structures of the hydration models listed in the first two columns.

hydration energy. This indicates that unlike the other hydration models, the SCKS model does not provide more hydration energy stabilization for extended conformations. This agrees with the prediction of the SCKS global minimum energy structure, which exhibits the most folded conformation. The SCKS structure also closely resembles the unsolvated global minimum energy structure and it exhibits the lowest potential energy contribution, − 11.63 kcal/mol, which is only 0.08 kcal/mol higher than the global minimum potential energy. This suggests that low potential energy conformations are not only favored but also enhanced by hydration effects for the SCKS model. Excluding the SCKS model, the other models predict relatively large hydration energies at the SCKS structure. In fact, for the RRIGS, WE1 and WE2 models, the SCKS structure produces the highest values for the hydration energies. For the OONS and JRF model, the hydration energies are only smaller than those for the RRIGS structure. This is consistent with the hydrophilic nature of the aromatic carbons for the OONS and JRF models. Specifically, because the aromatic ring separation is smallest for the RRIGS structure, the OONS and JRF hydration models tend to provide higher hydration energies for this structure. Although hydration energies for the RRIGS structure are typically high, the RRIGS model predicts a stabilizing hydration energy for this structure, second only to the JRF structure. It is this hydration energy contribution, when coupled with a relatively low potential energy (-8.59 kcal/mol), that sets the RRIGS global minimum energy structure. For the other hydration models, low potential energy contributions (-9.77, -9.75, and -9.22 kcal/mol for WE2, WE1, and OONS, respectively) seem to be more important in the prediction of relatively compact structures. In these cases the relative weighting of the hydration energy contributions does not favor extended conformations. However, these models also do not provide low hydration energies at the most compact structures, such as the SCKS global minimum energy structure. This indicates an interplay of hydration and potential energy contributions, although the prediction of relatively compact structures suggest the importance of low potential energy contributions.

Like met-enkephalin, leu-enkephalin (H–Tyr–Gly–Gly–Phe–Leu–OH) is an endogenous pentapeptide in which the methionine residue has been replaced by a leucine residue. Qualitatively, the results for the hydrated forms of leu-enkephalin are similar to those for met-enkephalin [81].

5. Free Energy Modeling

Locating the global minimum *potential* energy or the global minimum *potential plus solvation* energy conformation is not sufficient because Anfinsen's thermodynamic hypothesis requires the minimization of the conformational free energy. Specifically, potential energy minimization neglects the entropic contributions to the stability of the molecule. An approximation to these entropic contributions can be developed by using information about low-energy conformations. That is, once a sufficient ensemble of low-energy minima has been identified, a statistical analysis can be used to estimate the relative entropic contributions, and thus the relative free energy, for conformations in the ensemble.

Therefore, the analysis of the free energy of peptides requires efficient methods for locating not only the global minimum energy structure but also large numbers of low-energy conformers. A variety of methods have been used to find such stationary points on potential energy surfaces. For example, periodic quenching during a Monte Carlo or molecular dynamics trajectory can be used to identify local minima [82]. However, a drawback of these approaches is their inherent stochastic nature. In its original form, the αBB *deterministic* global optimization algorithm [15–18,73] has been shown to be an efficient method for finding the global minimum energy conformation for both unsolvated and solvated peptide systems [78,81,83]. Here, novel methods are proposed within the framework of the αBB algorithm to optimize the free energy of peptide systems. These modifications facilitate the generation of ensembles of low-energy conformers, which can be used to identify the global minimum free energy conformation, as well as perform detailed free energy rankings.

In peptide systems, this entropic contribution arises from fluctuations around a local conformational state. There exist a number of procedures, including both exact and approximate calculations, that can be used to determine the entropic contributions, and thus the free energy, of peptide systems.

First, assume that the full conformational space R can be considered as the union of disjoint basins of attraction, and the conformational space associated with a given basin (denoted by γ) is defined by R_γ. The energy, E, is a function of the variable set θ, which corresponds to the set of dihedral angles used to describe the conformational state of the system. Each basin of attraction is characterized by a unique local minimum at position θ_γ^*, with a corresponding energy E_γ^*. That is, local minimization starting at any point in R_γ will lead to the local minimum at θ_γ^*. It should be noted that this approximation of the conformational space excludes all maxima and saddle point conformations.

For a given temperature, T, the probability that a peptide occupies the conformational space of a given basin (R_γ) can be described by a Gibbs–Boltzmann distribution:

$$p_\gamma = \frac{\int_{R_\gamma} \exp(-\beta E(\theta))\, d\theta}{\int_R \exp(-\beta E(\theta))\, d\theta} \tag{37}$$

Here β is equivalent to $1/k_B T$. If the numerator is redefined as the partition function (Z_γ) for the basin, Eq. (37) can be rewritten as

$$p_\gamma = \frac{Z_\gamma}{Z} \tag{38}$$

The total partition function for the entire conformational space is represented by Z. Because this function is described by a disjoint set of basins (R_γ), it is equivalent to the following form:

$$Z = \sum_\gamma Z_\gamma \tag{39}$$

Once the probability is known, the corresponding free energy, G_γ, associated with each basin can also be calculated:

$$G_\gamma = -\frac{\ln p_\gamma}{\beta} \tag{40}$$

Using these definitions, a rigorous procedure can be envisioned for calculating the exact probability associated with a given basin. First, a sample of conformations must be generated with initial starting energies E_i, as defined by the total set I. Each structure is minimized to identify its corresponding basin minimum (θ_γ^*). These structures define the set $I(\gamma)$ (i.e., those structures associated with basin γ). As the sampling goes to infinity, the probability associated with basin γ can be calculated by the following expression:

$$p_\gamma^{\text{exact}} = \frac{\sum_{i(\gamma) \in I(\gamma)} \exp(-\beta E_{i(\gamma)})}{\sum_{i \in I} \exp(-\beta E_i)} \tag{41}$$

Obviously, such a method is intractable for large systems, and this is the impetus for developing approximate methods.

6. Harmonic Approximation

A tractable method for including entropic effects for proteins relies on the concept of the harmonic approximation. Initially, the theoretical development of this approximation for polymer systems generated debate in the literature [84–86]. In the work of Goldberg [84] a classical rigid model was used to characterize a partition function based on the fixed bond length and bond angle assumptions. In contrast, Flory [86] derived a different partition function using a classical flexible model. Later analysis by Gō and Scheraga [85] actually showed that the flexible model was also applicable to the fixed bond length and bond angle system (i.e., a peptide described by the internal coordinate system).

In either case (i.e., rigid or flexible), entropic contributions can be calculated by employing an harmonic approximation [85]. The fundamental concept is to characterize the basin of attraction (γ) by the properties of its corresponding local minimum (θ_γ^*), and not by a random sampling of conformations. These properties include the local minimum energy value, E_γ^*, and the convexity around the local minimum. Essentially, the convexity measure is used to approximate the basin of attraction region as a hyperparabola centered at the local minimum. Therefore, the anharmonic nature of the true basin, which defines the deviation from approximated harmonic behavior, controls the error associated with this assumption.

At each minimum (θ_γ^*) the harmonic approximation to the entropy can be evaluated using the following expression:

$$S_\gamma^{\text{approx}} = -\frac{k_B}{2} \ln\left[\text{Det}(H_\gamma)\right] + \hat{f}(T) \tag{42}$$

Here $\text{Det}(H_\gamma)$ refers to the determinant of the Hessian (second derivative matrix) evaluated at the local minimum θ_γ^*. The function $\hat{f}(T)$ is an additive term that is only dependent on temperature. The approximated free energy can then be calculated by combining the energetic and entropic contributions through the follow expression:

$$G_\gamma^{\text{approx}} = E_\gamma^* - TS_\gamma^{\text{approx}} + \bar{f}(T) \tag{43}$$

By substituting the harmonic entropic approximation from Eq. (42), Eq. (43) becomes

$$G_\gamma^{\text{approx}} = E_\gamma^* + \frac{1}{2\beta} \ln\left[\text{Det}(H_\gamma)\right] + \tilde{f}(T) \tag{44}$$

In this equation, it becomes evident that the free energy for a given basin is estimated using only the properties of the corresponding local minimum—that is, the local minimum energy (E_γ^*) and a measure of local convexity ($\text{Det}(H_\gamma)$). A temperature-dependent term, $\tilde{f}(T)$, is included, although it does not affect relative free energy comparisons.

Expressions for the probabilities and partition functions can also be developed. By combining Eqs. (38), (40), and (44), an approximation for the partition function of a given basin can be written as:

$$\ln Z_\gamma^{\text{approx}} = -\beta E_\gamma^* - \frac{\ln\left[\text{Det}(H_\gamma)\right]}{2} - \beta\tilde{f}(T) + \ln Z \tag{45}$$

A further simplification can be made by realizing that $-\beta\tilde{f}(T)$ and $\ln Z$ are constant for a given temperature (i.e., $f(T) = -\beta\tilde{f}(T) + \ln Z$). Equation (45) can be rewritten as

$$Z_\gamma^{\text{approx}} = \left[\frac{1}{[\text{Det}(H_\gamma)]}\right]^{1/2} \exp(-\beta E_\gamma^*) f(T) \tag{46}$$

Finally, by using Eq. (39), an approximate probability associated with a given basin (γ) can be calculated using the following equation:

$$p_\gamma^{\text{approx}} = \frac{[\text{Det}(H_\gamma)]^{-1/2} \exp(-\beta E_\gamma^*)}{\sum_{i=1}^{N} [\text{Det}(H_i)]^{-1/2} \exp(-\beta E_i^*)} \tag{47}$$

As expected, the $f(T)$ term disappears, and the statistical weight becomes a function of only the temperature (through β), the local minimum energy value, and the measure of convexity. In order to develop a meaningful comparison of relative free energies, the total partition function [i.e., the denominator of Eq. (47)] must include an adequate ensemble of low-energy local minima, as well as the global minimum energy conformation.

These probabilities can be used to estimate the occupancy of each individual basin, or summed in order to calculate cumulative probabilities for an ensemble of structures exhibiting similar physical or energetic properties. It should be noted that the determination of free energy using the harmonic approximation does not require the explicit inclusion of a contribution based on the density of states. That is, the harmonic approximation decomposes the energetic states within a basin of attraction into one energetic value represented by the local minimizer of the basin. In contrast to counting methods, which estimate probabilities based on the density of states, the contribution of each structure should be accounted for only once. Therefore, using the harmonic approximation requires a structural comparison of all local minimizers.

The probabilities obtained through the harmonic approximation can also be used to calculate thermodynamic quantities. Once the set of unique minimizers has been identified, these structures can be ranked according to their free energy values and then divided into bins of a specified energy width. Probabilities for each bin can be calculated by summing the individual probabilities [as defined in Eq. (47)]:

$$P_j^{\text{approx}} = \sum_{\gamma=1}^{n_j} p_\gamma^{\text{approx}} \tag{48}$$

Here P_j^{approx} signifies the probability for energy bin j. The summation includes the n_j individual probabilities (p_γ^{approx}) belonging to bin j. Average thermodynamic quantities can now be estimated using equations with the following form:

$$\langle E \rangle_T = \sum_j P_j^{\text{approx}} \langle E \rangle_j \tag{49}$$

Here the total average energy, $\langle E \rangle_T$, is calculated by summing the bin probabilities multiplied by the mean energy of bin j, $\langle E \rangle_j$.

7. Free Energy Problem Formulation

As before, the energy minimization problem for proteins is formulated as a nonconvex nonlinear optimization problem. The inclusion of free energy modeling into the protein folding problem does not change the general formulation. However, an additional condition must be satisfied; that is, an ensemble of local minimum low-energy conformations must be generated along with the global minimum energy conformation. Once this ensemble has been compiled, a free energy ranking can be performed using the harmonic approximation presented in the previous section.

Several rigorous methods can be envisioned for locating local minimum energy conformations using the αBB deterministic global optimization approach. As an introduction to the ideas used here, two rigorous approaches for finding all local minimum energy conformations are discussed.

The first method relies on the introduction of a single inequality constraint to the problem formulation given by (34). The new formulation is:

$$
\begin{aligned}
\min \quad & E(\phi_i, \psi_i, \omega_i, \chi_i^k, \phi_j^N, \phi_j^C) \\
\text{subject to} \quad & (E^* - E) + \epsilon^* < 0 \\
& -\pi \leq \phi_i \leq \pi, \quad i = 1, \ldots, N_{\text{RES}} \\
& -\pi \leq \psi_i \leq \pi, \quad i = 1, \ldots, N_{\text{RES}} \\
& -\pi \leq \omega_i \leq \pi, \quad i = 1, \ldots, N_{\text{RES}} \\
& -\pi \leq \chi_i^k \leq \pi, \quad i = 1, \ldots, N_{\text{RES}}, \quad k = 1, \ldots, K^i \\
& -\pi \leq \phi_j^N \leq \pi, \quad j = 1, \ldots, J^N \\
& -\pi \leq \phi_j^C \leq \pi, \quad j = 1, \ldots, J^C
\end{aligned}
\tag{50}
$$

The additional constraint requires that the objective function values be larger than the energy value at some local (or global) minimum, as denoted by E^*, plus a positive parameter, ϵ^*. When $\epsilon^* = 0$, the solution of the corresponding global optimization problem will give the best local minimum energy conformation with an energy larger than E^*. The original formulation given by (34) is actually a special case of this problem in which $E^* = -\infty$ and $\epsilon^* = 0$. That is, in (34) no bounds are placed on the value of the objective function, E. The global minimum energy conformation is only required to take some finite value. In order to locate all local minima, a set of global optimization problems must be solved iteratively with updating of the parameter E^*.

The problem of finding all local minimum energy conformations can also be formulated as a single global optimization problem, which can be deterministically solved using the αBB algorithm [23]. This method stems from the idea that all stationary points (i.e., minima, maxima, and transition states) of the energy hypersurface satisfy the constraint $\nabla E(\theta) = \mathbf{0}$. This can be written as:

$$
\frac{\partial E(\theta)}{\partial \theta_i} = 0, \quad i = 1, \ldots, N_\theta
\tag{51}
$$

Here N_θ represents the total number of dihedral angles defined by the variable set θ. The problem of finding local minima is equivalent to finding all solutions of Eq. (51) for which the Hessian of E is positive definite.

The problem posed in Eq. (51) involves the solution of a system of nonlinear equations. The identification of all multiple global solutions requires the use of a deterministic global optimization method, as outlined in Section II.B. The application of this method to protein systems will be described fully in Section IV.B.

Both methods for rigorously locating all local minimum energy conformations have some disadvantages. On one hand, the first approach should effectively locate low energy conformers in order of increasing energy. However, locating each minimum requires the solution of a full global optimization problem. The second approach avoids this drawback because it can be solved as a single global optimization problem. However, when dealing with a high-dimensional search space, the number of necessary subdivisions may be computationally inhibitive. In addition, this method will potentially locate stationary points other than local minima. Therefore, the development of other methods for locating low-energy local minimum energy conformations were pursued.

8.   Ensemble of Local Minimum Energy Conformations

Because the number of local minima on a given energy hypersurface may become astronomically large (e.g., the number of local minima for met-enkephalin is estimated to be on the order of 10^{11} [77]), methods that do not necessarily provide all local minima were developed. Specifically, it was determined that the generation of ensembles of low-energy conformers is possible through algorithmic modifications of the general αBB procedure. Rigorous implementation of the global optimization algorithm requires the minimization of a *convex* lower bounding function in each domain. The unique solution θ for each lower bounding minimum can then used as a starting point for the minimization (or function evaluation) of the original energy function in the current domain. In the case of local minimization, each partitioned region provides a single minimum energy conformation as the algorithm proceeds. Using this information, along with the global minimum energy conformation, a list of low-energy conformers can be constructed.

A method for increasing the number of local minima produced within each subdomain would involve the selection of multiple random starting points for minimizing the upper bounding function. At first, this approach appears to be equivalent to choosing random points for local minimization. Initially, when the subdomains constitute significant portions of the original domain space, this is the case. However, as the separation between lower and upper bounds decreases, the subdomains are localized in regions of low energy. Therefore, the random point selection is localized in regions that contain low-energy local minima.

However, this approach does not take advantage of the information provided by the lower bounding functions. Rigorously, these functions possess a single minimum in each subdomain. Because the choice of α affects the convexity of the lower bounding functions, the α values can be modified to ensure a certain nonconvexity in these functions. In this case, the lower bounding functions possess multiple minima, and these functions can be minimized several times in each domain. In addition, because the lower bounding functions smooth the original energy hypersurface, the location of these multiple minima provide information on the location of low-energy minima for the upper bounding function. Therefore, by using the location of the minima of the lower bounding function as starting points for local minimization of the upper bounding function, an improved set of low-energy conformations can be identified. As before, these conformations are also localized in those domains with low-energy as the subdomains decrease in size. This Energy-Directed Approach (EDA) is represented schematically in Fig. 17.

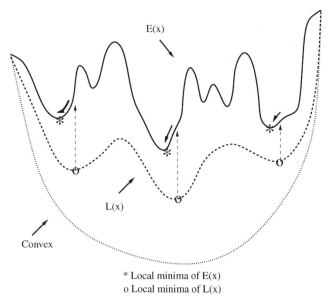

E(x)

L(x)

Convex

* Local minima of E(x)
o Local minima of L(x)

Figure 17. Using multiple lower bound minima to find low-energy conformers of the upper bounding function.

The basic steps of the algorithm, which are qualitatively similar to those outlined in Fig. 12, are as follows:

1. The initial best upper bound is set to an arbitrarily large value. The original domain is partitioned along one of the global variables. α values are initially chosen to be constant ($\alpha = \alpha_0$) for all global variables.

2. The lower bounding function (L) is constructed in each hyper-rectangle. Three local minimization are performed using the following procedure:

   a. Fifty random points are generated and used for function evaluations.

   b. The point with the minimum value is used as a starting point for local minimization of L using NPSOL, with calls (through PACK) to ECEPP/3 and possibly the RRIGS solvation module.

   c. The unique solutions are stored.

   If the minimum valued solution (of all local minima of L in this subdomain) is greater than the current best upper bound the subdomain is fathomed.

3. The unique local minima (points) for L are used as initial starting points for local minimizations of the upper bounding function (E) in each hyper-rectangle. Again, the appropriate calls are made to PACK and the potential and solvation energy modules. Two additional minimizations are performed using the following procedure:

   a. Fifty random points are generated and used for function evaluations.

   b. The point with the minimum value is used as a starting point for local minimization of E using NPSOL, with calls (through PACK) to ECEPP/3 and possibly the RRIGS solvation module.

   In all cases, the UBC (upper bound check) module is also called. UBC checks that the absolute value of each gradient in the objective function gradient vector is below a specified tolerance (10^{-6} kcal/mol/deg). If a gradient does not satisfy this check, the corresponding variable bounds are incrementally increased and the problem is solved with the previous point used as the initial starting point. This process is repeated until the gradient constraints are satisfied or an iteration limit is exceeded. UBC also employs algorithms to calculate the second derivative matrix [75], which is used to verify that the upper bound solution is a local minimum; that is, the Hessian matrix must be positive semidefinite. If the matrix is not positive semidefinite or the gradient checks are not satisfied, the upper bound solution is rejected. All local minima are stored.

4. The current best upper bound is updated to be the minimum of those thus far stored.

5. The hyper-rectangle with the current minimum value for L (this is the minimum value of all local minima of L in each subdomain) is selected and partitioned along one of the global variables. All α values are updated according to the following rule:

$$\alpha = \alpha_0 R^L \qquad (52)$$

In this equation α_0 refer to the initial values from Step 1. R is a reduction parameter $(0 < R \leq 1)$, and L refers to the current level in the branch and bound tree. For $R = 1$ the α values are kept constant at the initial value, α_0.

6. If the best upper and lower bounds are within the ϵ tolerance, or a maximum iteration limit has been exceeded, the program will terminate, otherwise it will return to Step 2.

A second approach incorporates free energy information into the branch and bound algorithm. Specifically, harmonic entropic contributions are calculated and included at each minima of the upper and lower bounding functions. In this way, the progression of lower and upper bounds includes a temperature-dependent entropic term. A similar modification to the Monte Carlo minimization method has also been proposed [87] and has been shown to be effective in locating low-energy conformers of peptides [88,89].

The problem formulation is identical to the one given in (34). That is, the minimization of E and L are still performed using only potential and solvation energy contributions. However, once local minima have been located, the free energy is calculated by the following expression:

$$G = U_{\text{Min}} + \frac{1}{2\beta} \ln \left[\text{Det}(H_{\text{Min}}) \right] \qquad (53)$$

This equation is similar to Eq. (44), although the additive term $f(T)$ has been omitted because it is a function of temperature only. U_{Min} represents the local minimum energy of E or L, and $\text{Det}(H_{\text{Min}})$ is the determinant of the Hessian evaluated at this local minimum. The specification of a thermodynamic temperature $(\beta = 1/k_B T)$ is required as an additional input parameter.

A single rigorous application of the αBB algorithm to this problem will result in the identification of the global minimum free energy at a given temperature. However, the goal is to identify an ensemble of low energy and, in this case, low free energy conformers so that a free energy ranking and comparison can be made. Therefore, the algorithmic steps for the Free Energy-Directed Approach (FEDA) are similar to those for EDA, with the additional evaluation of the free energy (G) at each local minima of E and L. The thermodynamic temperature used in Eq. (53) must be specified as an additional input parameter.

9. Free Energy Computational Studies

The EDA was first applied to the isolated form of met-enkephalin. All 24 dihedral angles were considered variable, with the 10 dihedral angles of the backbone residues acting as global variables (variables on which branching occurs). For both peptides, the EDA algorithm detailed above was applied 10 times. The input conditions correspond to initial α values of 5 and 10, with a subsequent reduction of these values based on the current level in the branch and bound tree.

Once the ensemble of local minima had been compiled, a set of distinct conformations was identified by checking for repeated and symmetric conformations. In addition, a conformation was only considered unique if at least one dihedral angle differed by at least 50° when comparing each pair of conformations. These conformations were then used to generate results and distributions according to energy and free energy values. Energy bins were used to characterize a group of distinct structures between a range of energy values (every 0.5 kcal/mol) relative to the global minimum energy structure. For example, Bin 1 contains structures that are 0.0–0.5 kcal/mol above the global minimum energy structure, Bin 2 contains structures that are 0.5–1.0 kcal/mol above the global minimum energy structure, and so on.

In the case of isolated met-enkephalin, the 10 (EDA) runs generated a total of 83,908 distinct local minima. The potential energy global minimum (PEGM) conformation for met-enkephalin possesses an energy of -11.707 kcal/mol. This conformation exhibits a type II' β-bend along the N–C' peptidic bond of Gly^3 and Phe^4. Essentially, this structure corresponds to the free energy global minimum (FEGM) conformation for a temperature of 0 K—that is, when entropic contributions are not included. When considering the harmonic free energy, the prediction of the FEGM can be calculated over a range of temperatures. Table XI provides information on the FEGM for temperatures ranging from 100 K to 500 K.

As Table XI shows, the PEGM persists as the FEGM at a temperature of 100 K. However, at the next three temperature points (i.e., 200 K, 300 K, 400 K) the FEGM exhibits a potential energy contribution 1.808 kcal/mol higher than the PEGM. The ϕ and ψ values for this structure are also significantly different than those for the PEGM. In fact, the conformational code (B*AAAE) indicates that the central residues display an α helical configuration. At a temperature of 500 K, the FEGM structure changes again, while the potential energy difference between the FEGM and PEGM increases to 5.369 kcal/mol. These differences suggest that the inclusion of entropic contributions greatly affects the relative stability of individual low energy structures. In addition, as the temperature increases, the stability offered by entropic contributions offsets substantial differences in potential energy.

TABLE XI

Dihedral Angle Values for PEGM and FEGM Structures of Isolated Met-enkephalin Using EDA[a]

Residue	DA	PEGM	100 K	200 K	300 K	400 K	500 K
Tyr$_1$	ϕ	-83.4	-83.4	179.8	179.8	179.8	90.2
	ψ	155.8	155.8	-18.2	-18.2	-18.2	149.1
	ω	-177.1	-177.1	-178.1	-178.1	-178.1	177.5
	χ_1	-173.2	-173.2	178.2	178.2	178.2	169.8
	χ_2	79.3	79.3	81.3	81.3	81.3	-108.2
	χ_3	-166.3	-166.3	177.3	177.3	177.3	177.6
Gly$_2$	ϕ	-154.3	-154.3	-59.8	-59.8	-59.8	-66.1
	ψ	85.8	85.8	-37.6	-37.6	-37.6	87.5
	ω	168.5	168.5	-178.8	-178.8	-178.8	-173.4
Gly$_3$	ϕ	83.0	83.0	-67.0	-67.0	-67.0	147.2
	ψ	-75.0	-75.0	-40.1	-40.1	-40.1	-36.7
	ω	-170.0	-170.0	179.7	179.7	179.7	175.1
Phe$_4$	ϕ	-136.9	-136.9	-70.9	-70.9	-70.9	-92.5
	ψ	19.1	19.1	-39.5	-39.5	-39.5	-34.7
	ω	-174.1	-174.1	-179.8	-179.8	-179.8	-179.1
	χ_1	58.9	58.9	173.9	173.9	173.9	179.1
	χ_2	94.5	94.5	-102.6	-102.6	-102.6	74.7
Met$_5$	ϕ	-163.5	-163.5	-161.0	-161.0	-161.0	-154.7
	ψ	160.9	160.9	122.1	122.1	122.1	135.3
	ω	-179.8	-179.8	-178.0	-178.0	-178.0	179.9
	χ_1	52.9	52.9	-174.7	-174.7	-174.7	-172.6
	χ_2	175.3	175.3	174.0	174.0	174.0	175.1
	χ_3	-179.9	-179.9	179.0	179.0	179.0	179.9
	χ_4	-178.6	-178.6	-60.1	-60.1	-60.1	-60.0
G		-11.707	-2.499	6.151	14.175	22.200	29.592
E		-11.707	-11.707	-9.899	-9.899	-9.899	-6.338

[a]The temperatures are provided in the first row. The last two rows indicate the harmonic free energy (kcal/mol) and the potential energy value (kcal/mol), respectively.

Table XII provides information on the distribution of distinct low free energy minima within 8.0 kcal/mol of the FEGM for a range of temperatures. For a given temperature the general trend indicates a large increase in the number of minima as the free energy increases above the FEGM. Several exceptions to this trend occur at high temperature and large bin number. In these cases, the number of minima remains constant or even decreases slightly. This is most likely due to an inadequate sampling of higher potential energy minima. For a given bin, it is also apparent that the clustering of low free energy structures increases with temperature. This increased density of the free energy bins indicates that increases in energy are offset by entropic contributions.

TABLE XII
Number of Distinct Minima in Bins for Isolated Met-enkephalin Using EDA[a]

Bin	0 K	50 K	100 K	150 K	200 K	250 K	300 K	350 K	400 K	450 K	500 K
1	2	1	2	10	6	3	3	4	16	16	8
2	3	5	13	22	12	9	15	24	18	21	31
3	12	25	36	58	52	42	40	40	59	69	77
4	45	48	55	105	105	100	101	115	164	184	184
5	49	69	120	233	199	206	213	249	309	397	475
6	90	125	263	451	435	403	410	491	726	893	918
7	166	292	467	806	763	765	848	1,043	1,438	1,655	1,687
8	303	497	766	1,250	1,297	1,362	1,524	1,906	2,464	2,821	2,695
9	552	776	1,233	1,929	2,079	2,247	2,601	3,069	3,932	4,284	4,111
10	840	1,177	1,710	2,915	3,168	3,475	3,927	4,707	5,774	6,030	5,562
11	1,121	1,675	2,681	3,879	4,355	4,899	5,708	6,655	7,573	7,775	7,116
12	1,618	2,467	3,526	5,303	5,935	6,572	7,364	8,333	9,437	9,448	8,721
13	2,331	3,223	4,491	6,821	7,619	8,360	9,203	10,228	10,730	10,473	9,719
14	2,973	4,050	6,037	8,058	8,834	9,712	10,598	11,244	11,651	11,285	10,630
15	3,747	5,250	7,258	9,031	9,821	10,585	11,504	11,939	11,915	11,396	10,745
16	4,588	6,422	8,053	8,587	9,687	10,958	11,563	11,432	9406	8,482	8,338

[a]Each bin represents a 0.5 kcal/mol range above the previous bin. The temperatures are given in the first row.

These observations are also supported by the information shown in Fig. 18. This plot displays the range of potential energy in free energy bins at temperatures of 250 and 500 K, with the potential energy bins included for comparison. As expected, the potential energy values for the free energy bins increase with increasing temperature. In addition, the range of potential energy values increases in higher free energy bins. It is interesting to note that this occurs because the minimum potential energy is relatively (i.e., within a few kcal/mol of the PEGM) low for each bin, whereas the maximum potential energy value increases in higher bins. The corresponding differences are also greater at higher temperature. For example, at 500 K some bins exhibit a 20-kcal/mol range in potential energy. These trends explain the increased number of low free energy conformers. That is, bins of low free energy contain conformers of relatively high potential energy because of their more stabilizing entropic contributions. The plot also implies that the PEGM appears in bins 3 and 10 for temperatures of 250 and 500 K, respectively.

Relative free energies were also calculated for clusters of low-energy conformers. This analysis is useful because it is difficult to capture the true accessibility of individual structures based on a pointwise approximation of entropic effects. That is, the harmonic free energy approximation does not provide a continuous free energy landscape. By clustering structures into larger

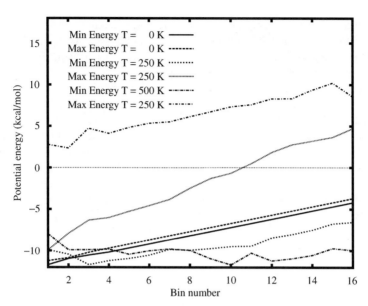

Figure 18. Potential energy comparison for isolated met-enkephalin using EDA. Minimum and maximum potential energies versus bin number are plotted for three temperatures: T = 0 K, 250 K, 500 K.

groups, it is hoped that the error associated with these estimates will average out. Typically, structures are clustered by calculating and comparing root-mean-squared deviations. Because the enkephalin peptide is relatively small, structures were grouped based on the Zimmerman codes for the central residues of the peptide [90]. Specifically, for met-enkephalin, structures were said to belong to the same cluster if the central three residues possessed the same three code letters based on the Zimmerman classification [90]. The relative free energy of a cluster was calculated by the following equation:

$$G_{\text{cluster}} = -\frac{\ln \sum_{i \in C} p_i^{\text{approx}}}{\beta} \tag{54}$$

In Eq. (54) the individual p_i^{approx}, which refers to the statistical weight based on the harmonic approximation, are summed for the set of conformations belonging to a particular cluster (C). These individual probabilities were calculated by normalizing each probability with respect to the overall probability at a given temperature:

$$p_i^{\text{approx}} = \frac{\exp[-\beta(G_0^{\text{approx}} - G_i^{\text{approx}})]}{\sum_j \exp[-\beta(G_0^{\text{approx}} - G_j^{\text{approx}})]} \tag{55}$$

A reference free energy, G_0^{approx}, was used to normalize the probabilities at each temperature point. All free energies, G_0^{approx}, G_i^{approx} and G_j^{approx}, refer to the harmonic approximation of the free energy as calculated using Eq. (44). The denominator, which represents the total probability at a given temperature, is calculated by summing over the set of all conformers.

The relative free energies for clusters of met-enkephalin structures are given in Table XIII. At each temperature point the Zimmerman code and correspond-ing data for the top three clusters are listed. The results indicate that the structure exhibiting the individual lowest free energy does not always belong to the cluster with lowest free energy. At 100 and 200 K the DC*B and AAA clusters are consistent with the structures of the FEGM. However, although the FEGM retains the AAA structure at 300 and 400 K, the group of structures possessing the lowest $G_{cluster}$ at these temperatures exhibits a CD*A Zimmerman code. This is, at least in part, attributable to the large number of structures grouped in this cluster. In contrast to the α-helical-type structure for the FEGM, the CD*A structures possess elements of a β-turn conformation. Specifically the lowest free energy conformer exhibiting a CD*A structure at 300 and 400 K, possesses a type II β-bend along the Gly^2–Gly^3 backbone.

TABLE XIII
Clustered Relative Free Energies for Isolated Met-enkephalin Using the EDA[a]

Temperature (K)	Code	Number	$\sum_i p_i^{approx}$	$G_{cluster}$
	DC*B	113	0.636	0.0899
100	CC*B	136	0.0794	0.503
	C*DE	557	0.0765	0.511
	AAA	323	0.230	0.585
200	DC*A	1828	0.213	0.615
	C*DE	676	0.192	0.656
	CD*A	2685	0.297	0.723
300	DC*A	1843	0.100	1.372
	AAA	328	0.0990	1.379
	CD*A	2654	0.219	1.209
400	DC*A	1799	0.0452	2.461
	AAA	329	0.0380	2.600
	CD*A	2449	0.112	2.174
500	C*C*A	1361	0.0256	3.640
	C*AE	1463	0.0229	3.752

[a]From left to right, the information provided in this table includes temperature, Zimmerman code[i], number of individual structures in cluster, total probality ($\sum_i p_i^{approx}$), and free energy of cluster ($G_{cluster}$).

TABLE XIV
Input Parameters Used for FEDA Runs[a]

Run No.	α_0	R	T (K)	Run No.	α_0	R	T (K)
1	5	0.90	50	6	5	0.90	300
2	5	0.90	100	7	5	0.90	350
3	5	0.90	150	8	5	0.90	400
4	5	0.90	200	9	5	0.90	450
5	5	0.90	250	10	5	0.90	500

[a]Here α_0 refers to the initial α values used for all global variables. R refers to the reduction rate applied at each level of the branch and bound tree. T refers to the thermodynamic temperature at which the free energy was calculated.

FEDA was also applied to the isolated form of met-enkephalin. For this approach, the thermodynamic temperature appears as an input parameter, and these values had to be specified along with initial α values. Several methods can be envisioned for initializing the FEDA. For example, if the goal is to characterize the low free energy conformers at a single temperature, a full set of FEDA runs could be performed for that temperature. This type of search should efficiently locate the global and many low free energy conformers for that temperature. However, the goal was to effectively characterize the FEGM and low free energy conformers over a range of temperatures. Therefore each of the 10 (FEDA) runs were conducted at a unique temperature point in the range of 50 to 500 K. The details of the conditions for these runs are given in Table XIV.

In total, 87,974 distinct local minima were found after compiling the results from the 10 (FEDA) runs for isolated met-enkephalin. The PEGM and FEGM found using the FEDA are displayed in Table XV. It should be noted that when comparing PEGM for the EDA and FEDA, both structures possess the same potential energies, but a different set of dihedral angles. However, these structures are actually the same. That is, the different values of χ_2 and χ_3 for Tyr$_1$ represent a degenerate state for tyrosine, which is generated by rotating both of these dihedral angles by 180°. An important observation is that at 200 K the FEDA method predicts a slightly lower FEGM. The structure possesses a lower potential energy (-10.547 vs. -9.899 kcal/mol) and exhibits a free energy value that is 0.044 kcal/mol lower than the EDA predicted FEGM. The remaining FEGM predictions are consistent for the two approaches.

An analysis of the distribution of distinct minima, as given by Table XVI, reveals that the results are qualitatively consistent with those produced by the EDA. It should be noted that in all cases the lowest free energy bin is as densely populated as the corresponding EDA bins, which indicates that each run using the FEDA was able to find a better distribution of low free energy conformers near the FEGM. This is not unexpected, considering that the FEDA runs were

TABLE XV
Dihedral Angle Values for PEGM and FEGM Structures of Isolated Met-enkephalin Using FEDA[a]

Residue	DA	PEGM	100 K	200 K	300 K	400 K	500 K
Tyr$_1$	ϕ	-83.4	-83.4	-163.1	179.8	179.8	-90.2
	ψ	155.8	155.8	-40.5	-18.2	-18.2	149.1
	ω	-177.1	-177.1	-177.7	-178.1	-178.1	177.5
	χ_1	-173.2	-173.2	-172.2	178.2	178.2	169.8
	χ_2	-100.7	-100.7	93.2	81.3	81.3	71.8
	χ_3	13.7	13.7	-177.2	177.3	177.3	-2.4
Gly$_2$	ϕ	-154.3	-154.3	65.1	-59.8	-59.8	-66.1
	ψ	85.8	85.8	-89.7	-37.6	-37.6	87.5
	ω	168.5	168.5	174.1	-178.8	-178.8	-173.4
Gly$_3$	ϕ	83.0	83.0	-152.6	-67.0	-67.0	147.2
	ψ	-75.0	-75.0	34.4	-40.1	-40.1	-36.7
	ω	-170.0	-170.0	-178.9	179.7	179.7	175.1
Phe$_4$	ϕ	-136.8	-136.8	-155.4	-70.9	-70.9	-92.5
	ψ	19.1	19.1	159.8	-39.5	-39.5	-34.7
	ω	-174.1	-174.1	179.2	-179.8	-179.8	-179.1
	χ_1	58.9	58.9	52.1	173.9	173.9	179.1
	χ_2	-85.5	-85.5	82.9	-102.6	-102.6	74.7
Met$_5$	ϕ	-163.5	-163.5	-79.3	-161.0	-161.0	-154.7
	ψ	160.9	160.9	130.4	122.1	122.1	135.3
	ω	-179.8	-179.8	-178.7	-178.0	-178.0	179.9
	χ_1	52.9	52.9	-66.8	-174.7	-174.7	-172.6
	χ_2	175.3	175.3	179.8	174.0	174.0	175.1
	χ_3	-179.9	-179.9	-179.9	179.0	179.0	179.9
	χ_4	-178.6	-178.6	-60.0	-60.1	-60.1	180.0
G		-11.707	-2.499	6.107	14.175	22.200	29.592
E		-11.707	-11.707	-10.547	-9.899	-9.899	-6.338

[a]The temperatures are provided in the first row. The last two rows indicate the harmonic free energy (kcal/mol) and the potential energy value (kcal/mol), respectively.

conducted at the same discrete temperature points used in the analysis. However, when comparing the populations of higher energy bins at low temperatures, the number of minima is larger for the EDA. Some of this variation, especially near the 150 to 200 K range, is probably due to the lower FEGM found by the FEDA. In general, the FEDA seems to provide a denser distribution of distinct minima at higher temperatures and large bin number.

A comparison of the relative efficiencies for the EDA and FEDA to generate low-energy local minima can also be made by examining Fig. 19. In this plot the cumulative fraction of conformers, which is equal to the total number of unique conformers within the first 8, 12, and 16 energy bins over the total number of unique conformers, is given as a function of temperature. It is apparent that both approaches are highly efficient. For example, at 400 K approximately 90% of

TABLE XVI
Number of Distinct Minima in Bins for Isolated Met-enkephalin Using FEDA[a]

Bin	0 K	50 K	100 K	150 K	200 K	250 K	300 K	350 K	400 K	450 K	500 K
1	2	1	3	10	8	5	5	6	17	15	8
2	3	6	14	9	10	11	16	23	19	23	30
3	12	26	38	52	53	43	42	41	56	63	86
4	46	48	55	87	91	100	97	107	156	188	193
5	47	69	116	180	189	205	208	249	324	407	478
6	87	122	259	373	400	391	403	481	721	898	988
7	161	290	470	654	730	758	846	1,051	1,476	1,801	1,756
8	297	488	760	1,063	1,246	1,368	1,524	1,936	2,576	2,966	3,052
9	543	762	1,182	1,637	1,918	2,188	2,597	3,181	4,136	4,618	4,538
10	828	1,140	1,624	2,413	2,996	3,511	4,032	4,863	6,033	6,481	6,070
11	1,066	1,560	2,569	3,542	4,193	4,852	5,726	6,791	8,047	8,466	7,832
12	1,527	2,404	3,433	4,735	5,785	6,616	7,499	8,630	9,989	10,069	9,426
13	2,244	3,070	4,470	6,288	7,382	8,341	9,315	10,632	11,286	11,130	10,484
14	2,818	4,004	5,833	7,451	8,649	9,727	10,862	11,833	12,430	11,937	11,102
15	3,657	5,064	7,075	8,723	9,617	10,818	12,004	12,606	12,358	11,968	11,238
16	4,472	6,257	7,848	8,718	10,108	11,295	12,167	12,003	9,952	8,640	8,576

[a]Each bin represents a 0.5 kcal/mol range above the previous bin. The temperatures are given in the first row.

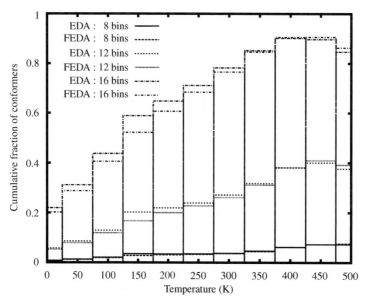

Figure 19. Plot of cumulative fraction of low energy conformers for isolated met-enkephalin, which is equal to the number of unique conformers within the first 8, 12, and 16 energy bins over the total number unique conformers, versus temperature. Both EDA and FEDA data are plotted.

TABLE XVII
Clustered Relative Free Energies for Isolated Met-enkephalin Using the FEDA[a]

Temperature (K)	Code	Number	$\sum_i p_i^{approx}$	$G_{cluster}$
	DC*B	107	0.532	0.125
100	C*DE	990	0.232	0.291
	CC*A	1604	0.0636	0.547
	C*DE	1275	0.331	0.439
200	AAA	322	0.209	0.623
	DC*A	1729	0.174	0.694
	CD*A	2128	0.263	0.796
300	C*DE	1360	0.125	1.239
	AAA	327	0.111	1.309
	CD*A	2116	0.192	1.313
400	C*DE	1362	0.0464	2.440
	DC*A	1714	0.0429	2.502
	CD*A	1966	0.0922	2.368
500	C*AE	2088	0.0308	3.459
	C*C*A	1900	0.0279	3.555

[a]From left to right, the information provided in this table includes temperature, Zimmerman code[i], number of individual structures in cluster, total probability $(\sum_i p_i^{approx})$, and free energy of cluster $(G_{cluster})$.

the total unique conformations identified are in the top 16 free energy bins, which ranges up to 8 kcal/mol above the FEGM. The lower fractions at lower temperatures indicate that a relatively large number of conformations have high potential energies and that these energetic differences are not offset by entropic effects at low temperatures. A more subtle comparison can be made by observing that the EDA cumulative fractions are generally higher for temperatures lower than 400 K. Although the total number of unique conformations is slightly lower for the EDA, this trend indicates that the EDA is more efficient at filling low-energy bins, especially at lower temperatures.

The results for the cluster analysis of the FEDA met-enkephalin structures are given in Table XVII. There are some differences between the EDA and FEDA cluster free energies, although the overall trend is the same. At all temperatures, excluding 200 K, the cluster exhibiting the lowest cluster free energy is the same as in the EDA analysis. At 200 K, the FEDA predicts the AAA cluster as having a slightly higher free energy than the C*DE cluster, which only appears as the third cluster in Table XIII. In both analyses, the transition from the ground-state DC*B cluster to the CD*A cluster as temperature increases is evident.

Because both the EDA and FEDA provide large amounts of statistical information for the peptide system, these data were used to perform a simple thermodynamic analysis of the folding process. It is widely accepted that the folding of peptides progresses successively. The first step of this process is typically associated with a structural collapse—that is, a transition from random extended structures to an ensemble of compact structures. This transition should also be associated by significant changes in the description of the ensemble as temperature changes. For example, a peak in the specific heat at the transition temperature indicates a steep decrease in average potential energy of the ensemble. In order to verify that such a transition occurs for met-enkephalin, the specific heat was calculated using the following expression:

$$C = \frac{\beta^2(\langle E^2 \rangle_T - \langle E \rangle_T^2)}{N} \tag{56}$$

Here N refers to the number of amino acid residues in the peptide. The average energy and squared energy ($\langle E \rangle_T$ and $\langle E^2 \rangle_T$, respectively) were calculated at 10 temperature points using expressions of the form given in Eq. (49). The bin probabilities were based on an energy width of .015625 kcal/mol. In addition, a reference free energy, G_0^{approx} (the lowest free energy), was used to normalize the probabilities at each temperature point.

The results for isolated met-enkephalin are shown in Fig. 20. Both the EDA and FEDA predict a transition temperature in the 250–275 K temperature range. This is consistent with the increase in bin density and structural diversity at higher temperatures, and it suggests a sharp increase in the average potential energy of the system at this temperature. It also supports the transition from the DC*B ground-state (PEGM) cluster to the higher potential energy CD*A cluster in this temperature range.

Similar results for characterizing the folding transitions of enkephalins have also been obtained by multicanonical simulations [91]. This is encouraging because the two methods possess fundamental differences. In contrast to this work, the multicanonical approach does not rely on the identification of low-energy local minima or the concepts of the harmonic approximation. Instead, thermodynamic quantities are developed by first generating large ensembles of structures with wide ranging energies and then employing reweighting techniques. In addition, although the multicanonical simulations included detailed atomistic level modeling, only unsolvated systems were considered.

The EDA was then applied to the RRIGS solvated form of met-enkephalin using the same protocol and conditions as detailed above. Qualitatively, the PEGM (in this case, PEGM refers to potential + solvation) for solvated met-enkephalin exhibits a more extended conformation than that which is observed for the isolated form. As detailed in Table XVIII, the PEGM structure persists as

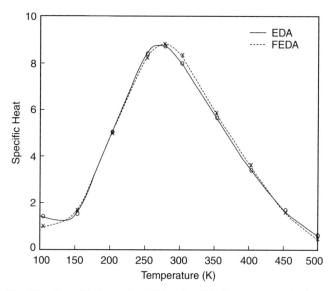

Figure 20. Plot of specific heat using EDA and FEDA free energy results for isolated met-enkephalin.

the FEGM at 100 K. However, at each subsequent temperature, the FEGM structure changes, and this change is accompanied by an increase in total energy (potential and solvation). As with isolated met-enkephalin, the difference in total energy between the PEGM and FEGM at 500 K is greater than 5 kcal/mol. This suggests that entropic effects are important in defining the predicted native structure. When considering individual structures, entropic effects tend to produce more extended FEGM conformations at higher temperatures, especially with regard to the placement of the aromatic rings. It is interesting to note that in a previous study the positioning of aromatic rings was found to be a major difference when considering the ability of solvation models to predict extended PEGM conformations for the solvated enkephalin peptides [83]. The sequence of FEGM structures is illustrated in Fig. 21.

The distribution of the 72784 distinct minima for solvated met-enkephalin exhibits some important differences from those results obtained for the isolated form of the peptide. This is evidenced by the information presented in Table XIX and the plot in Fig. 22. In particular, the low- and intermediate-energy bins are much denser than the corresponding bins for isolated met-enkephalin, especially within 4 kcal/mol (8 bins) of the FEGM. In addition, some higher-energy bins are actually more populated at lower temperatures. One obvious reason for these differences is the high density of conformers for the original system (at 0 K). This high density of states causes the original energy differences to be relatively

TABLE XVIII

Dihedral Angle Values for PEGM and FEGM Structures of Solvated Met-enkephalin[a]

Residue	DA	PEGM	100 K	200 K	300 K	400 K	500 K
Tyr$_1$	ϕ	− 168.2	− 168.2	− 170.9	− 168.4	− 168.4	− 152.5
	ψ	− 30.9	− 30.9	− 28.5	− 34.3	− 34.3	153.2
	ω	178.6	178.6	177.5	− 178.9	− 178.9	178.5
	χ_1	− 173.5	− 173.5	178.8	178.7	178.7	− 179.0
	χ_2	− 100.9	− 100.9	61.3	− 100.8	− 100.8	− 101.2
	χ_3	19.3	19.3	− 4.1	179.0	179.0	− 179.9
Gly$_2$	ϕ	78.5	78.5	73.8	177.8	177.8	− 173.9
	ψ	− 86.5	− 86.5	47.6	− 179.9	− 180.0	177.1
	ω	− 177.3	− 177.3	− 179.2	180.0	180.0	− 179.8
Gly$_3$	ϕ	162.4	162.4	167.6	− 180.0	− 180.0	179.6
	ψ	92.2	92.2	− 145.2	179.9	179.9	− 179.3
	ω	172.6	172.6	175.2	179.7	179.7	179.6
Phe$_4$	ϕ	− 150.3	− 150.3	− 149.3	− 155.3	− 155.4	− 155.4
	ψ	159.8	159.8	135.8	147.2	149.5	149.3
	ω	− 178.1	− 178.1	− 176.6	− 176.8	− 178.3	− 178.3
	χ_1	65.8	65.8	177.3	− 179.5	− 179.5	− 179.7
	χ_2	− 87.4	− 87.4	− 108.1	− 111.7	− 105.6	74.4
Met$_5$	ϕ	− 75.0	− 75.0	− 85.5	− 78.7	− 78.7	− 78.9
	ψ	113.9	113.9	− 41.1	− 51.1	113.4	113.5
	ω	− 178.4	− 178.4	179.9	179.7	− 179.1	− 179.1
	χ_1	− 172.3	− 172.3	− 65.6	− 67.2	− 67.4	− 67.4
	χ_2	176.1	176.1	− 179.6	− 178.8	− 178.8	− 178.8
	χ_3	− 180.0	− 180.0	− 179.4	− 179.9	− 179.9	− 179.9
	χ_4	60.0	60.0	179.5	− 180.0	60.0	− 60.0
G		− 50.060	− 41.896	− 34.566	− 28.604	− 22.828	− 17.166
E		− 50.060	− 50.060	− 48.676	− 46.030	− 45.780	− 44.797

[a]The temperatures are provided in the first row. The last two rows indicate the harmonic free energy (kcal/mol) and the potential energy value (kcal/mol), respectively.

small, and the entropic correction tends to induce an even stronger equalization of the free energy values. This equalization is best illustrated by the data plotted in Fig. 22, which indicate that the efficiency of locating low-free-energy conformers is relatively high at all temperatures. In fact, the highest density of states occurs near the middle of the temperature range, rather than at high temperatures as predicted for the isolated peptide. This behavior may be due to a lack of much-higher-energy local minima that would probably populate these high-temperature, high-energy bins.

Similar conclusions can be drawn by examining the data presented in Fig. 23, which provides information on the energy extrema for free energy bins at temperatures of 0, 250, and 500 K. As expected, for both 250, and 500 K, the

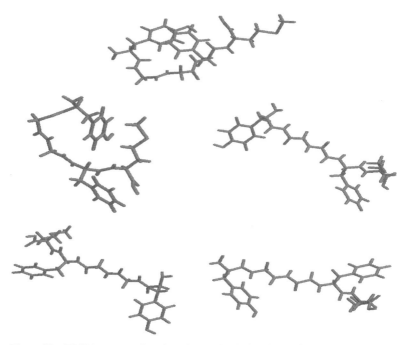

Figure 21. FEGM structures for solvated met-enkephalin. The top figure is the PEGM and the FEGM for 100 K. The structures at other temperatures (200 K, 300 K, 400 K, 500 K) are shown left to right, top to bottom.

range of energy values increases for higher-free-energy bins. In addition, for all bins, the minimum energy is relatively low and generally within a few kcal/mol of the PEGM. However, unlike the isolated met-enkephalin results, the maximum energy values do not become larger at higher temperatures. In fact, the curves for maximum energy at 250 and 500 K are almost identical. This indicates that relatively high energy minima may be needed in order to fill out these high-temperature bins.

A clustering analysis of the low-free-energy conformers was also performed for solvated met-enkephalin, and the results are shown in Table XX. At 100 K, the lowest free energy cluster included the FEGM structure, which is also the PEGM structure. At higher temperatures, the correlation between the extended FEGM structures and the lowest-free-energy cluster was also evident. In fact, all low energy clusters at 300, 400, and 500 K possess highly extended backbone conformations, with nearly all geometries within the E and E* regions on the Zimmerman conformational map. In fact, although the number of individual structures in each cluster is not excessively large, many of these extended conformers reside in the lowest free energy bins.

TABLE XIX
Number of Distinct Minima in Bins for Solvated Met-enkephalin[a]

Bin	0 K	50 K	100 K	150 K	200 K	250 K	300 K	350 K	400 K	450 K	500 K
1	10	11	16	17	21	18	19	22	21	21	13
2	14	17	35	122	236	149	98	95	97	94	79
3	34	66	299	542	896	607	378	283	223	195	166
4	117	296	668	1589	2075	1496	885	635	520	412	343
5	326	626	1907	3163	3636	2644	1730	1175	814	678	548
6	717	1582	3324	4902	5438	4256	2812	1957	1418	1047	762
7	1440	2865	5393	6733	6816	5790	4451	3061	2172	1623	1202
8	2611	4521	6906	7692	7569	6730	5390	4376	3123	2299	1705
9	3891	6337	7857	7952	7650	7221	6301	4972	4073	3132	2263
10	5567	7342	8094	7304	6858	7158	6736	5925	4699	3788	2903
11	6677	8090	7193	6612	6320	6374	6675	6232	5426	4453	3501
12	7624	7483	6618	5915	5645	6028	6295	6270	5754	5015	4161
13	7650	6920	5726	4864	4582	5279	5756	5972	5822	5328	4577
14	7047	6106	4680	3875	3645	4280	5113	5546	5689	5387	4879
15	6375	5066	3710	3086	2978	3449	4361	4973	5376	5271	5012
16	5534	4090	2848	2237	2140	2796	3437	4233	4809	5141	4964

[a]Each bin represents a 0.5 kcal/mol range above the previous bin. The temperatures are given in the first row.

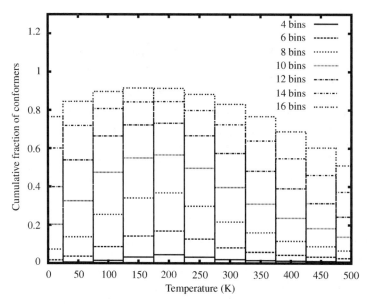

Figure 22. Plot of cumulative fraction of low-energy conformers for solvated met-enkephalin, which is equal to the number of unique conformers within the first 4, 6, 8, 10, 12, 14, and 16 energy bins over the total number unique conformers, versus temperature.

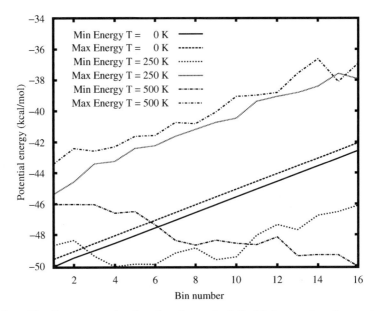

Figure 23. Energy comparison for solvated met-enkephalin. Minimum and maximum potential energies versus bin number are plotted for three temperatures: T = 0 K, 250 K, 500 K.

A specific heat profile was also derived for solvated met-enkephalin in order to understand how the dominance of these extended cluster geometries affect the folding transition. These results are shown in Fig. 24. As with isolated met-enkephalin, a folding transition is indicated by the peak in the specific heat, which, in this example, occurs between 275 and 300 K. This represents a significant change in average energy, which accompanies the collapse from an ensemble of extended conformations (EE*E and E*EE clusters) to the more compact ground-state cluster. For the solvated met-enkephalin example, this transition is clearly illustrated by the cluster analysis and the structure plots given in Fig. 21.

B. Structure Refinement with Sparse Restraints

To effectively determine protein function, it is important to predict the three-dimensional structure of the macromolecule. Over the last several decades a number of experimental and theoretical approaches have been developed and refined in order to achieve this goal. Experimentally, there now exist two basic techniques used to perform protein structure refinement. The first, X-ray crystallography, relies on the ability to crystallize the protein so that diffraction patterns can be used for sufficient resolution. These requirements have limited

TABLE XX
Clustered Relative Free Energies for Solvated Met-enkephalin[a]

Temperature (K)	Code	Number	$\sum_i p_i^{\text{approx}}$	G_{cluster}
	C*H*E	139	0.466	0.152
100	C*DF	286	0.224	0.297
	C*G*A	205	0.0991	0.459
	C*A*E	1112	0.0521	1.174
200	A*E*E	393	0.0468	1.217
	E*EE	149	0.0421	1.259
	E*EE	148	0.0474	1.818
300	EE*E	152	0.0445	1.856
	D*E*E	149	0.0273	2.147
	EE*E	151	0.0476	2.419
400	E*EE	145	0.0391	2.575
	EEE	159	0.0266	2.883
	EE*E	149	0.0460	3.059
500	E*EE	142	0.0327	3.397
	EEE	156	0.0299	3.488

[a]From left to right, the information provided in this table includes temperature, Zimmerman code[i], number of individual structures in cluster, total probability ($\sum_i p_i^{\text{approx}}$) and free energy of cluster (G_{cluster}).

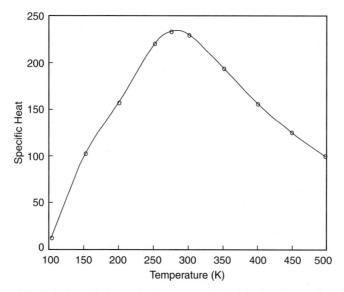

Figure 24. Plot of specific heat using free energy results for solvated met-enkephalin.

the applicability of this technique. A more powerful method, NMR (nuclear magnetic resonance) spectroscopy, is based on solution measurements of the system. Several key developments, including multidimensional NMR experiments, have resulted in the ability to determine solution structures for proteins consisting of over 200 residues.

This section focuses on the development of a novel approach for protein structure prediction via experimental NMR restraints. Traditionally, the protein folding global optimization problem involves a progression of unconstrained minimizations. However, the introduction of experimentally derived or artificial restraints can be used to recast the fundamental protein folding problem as a constrained global optimization problem. The constraints, through reduction of the feasible search space, serve two important purposes: (1) to attempt to correct any deficiencies of the energy model and (2) to focus the efforts of the global optimization algorithm.

This constrained approach is applied to the NMR structure prediction problem, although a variety of restraint information could be used. The proposed constrained formulation differs from traditional NMR approaches in several fundamental ways. First, the energy model is represented by a detailed full atom force field, rather than simplified nonbonded potential terms. This should make the approach especially effective when the number of NMR restraints per residue decreases; that is, the accuracy of the energy model becomes more significant. In addition, traditional solution approaches apply target function distance geometry or simulated annealing to unconstrained problem formulations in which restraints are represented by penalty function approximations. The solution of the constrained formulation requires the use of constrained local optimization solvers and an overall global optimization framework for nonlinearly constrained problems. This is accomplished through the application of the ideas of the αBB deterministic global optimization approach [15–18,73]. αBB-based global optimization techniques have also been applied to NMR-type structure prediction problems [92,93].

Because the location of the global minimum relies on effectively solving constrained local optimization problems, convergence to the global minimum can be enhanced by consistently identifying low-energy solutions. These observations illustrate the need for reliably locating low-energy feasible points for initializing the constrained local optimization routine. Along these lines, a combined torsion angle dynamics (TAD) and simulated annealing scheme has been implemented within the context of the global optimization framework. TAD has recently been shown to be more effective than Cartesian coordinate dynamics [94,95]. In this case, the degrees of freedom are rotations around single bonds, which reduces the number of variables by approximately tenfold because bond lengths, bond angles, chirality, and planarities are kept fixed at optimal values during the calculation.

1. Energy Modeling

Basic data obtained from NMR studies consist of distance and torsion angle restraints. Once resonances have been assigned, nuclear Overhauser effect (NOE) contacts are selected and their intensities are used to calculate interproton distances. Information on torsion angles are based on the measurement of coupling constants and analysis of proton chemical shifts. Together, this information is used to formulate a nonlinear optimization problem, the solution of which should provide the correct protein structure. Typically, a hybrid energy function of the following form is employed:

$$E = E_{\text{forcefield}} + W_{\text{NMR}} E_{\text{NMR}} \tag{57}$$

The energy, E, specified by this target function includes a chemical description of the protein conformation through the use of a force field, $E_{\text{forcefield}}$. The force field potentials are generally much simpler representations of all atom force fields. The distance and dihedral angle restraints appear as weighted penalty, E_{NMR}, terms that should be driven to zero.

The second term of Eq. (57) can be rewritten as

$$E_{\text{NMR}} = E_{\text{distance}} + E_{\text{dihedral}} \tag{58}$$

Here E_{distance} and E_{dihedral} represent the violation energies based on the distance and dihedral angle restraints, respectively. These functions can take several forms, although a simple square well potential is commonly used. The expressions also include a summation over both upper and lower distance violations; for example, $E_{\text{distance}} = E_{\text{distance}}^{\text{upper}} + E_{\text{distance}}^{\text{lower}}$. When considering upper distance restraints, this becomes

$$E_{\text{distance}}^{\text{upper}} = \sum_j \begin{cases} A_j(d_j - d_j^{\text{upper}})^2 & \text{if} \quad d_j > d_j^{\text{upper}} \\ 0 & \text{otherwise} \end{cases} \tag{59}$$

The squared violation energy is considered only when the calculated distance d_j exceeds the upper reference distance d_j^{upper}. This squared violation can then be multiplied by a weighting factor A_j. A similar contribution is calculated for those distances that violate a lower reference distance, d_j^{lower}:

$$E_{\text{distance}}^{\text{lower}} = \sum_j \begin{cases} A_j(d_j - d_j^{\text{lower}})^2 & \text{if} \quad d_j < d_j^{\text{lower}} \\ 0 & \text{otherwise} \end{cases} \tag{60}$$

For dihedral angle restraints the functional form is similar to that of Eqs. (59) and (60). As before, the total violation, E_{dihedral}, is a sum over upper and lower violations (i.e., $E_{\text{dihedral}} = E_{\text{dihedral}}^{\text{upper}} + E_{\text{dihedral}}^{\text{lower}}$). A dihedral angle ω_j can be restrained by employing a quadratic square well potential using upper (ω_j^{upper}) and lower (ω_j^{lower}) bounds on the variable values. However, due to the periodic nature of these variables, a scaling parameter must be incorporated to capture the symmetry of the system. Furthermore, by centering the full periodic region on the region defined by the allowable bounds, all transformed values will lie in the domain defined by $[\omega_j^{\text{lower}} - \Delta HW_{\omega_j}, \omega_j^{\text{upper}} + \Delta HW_{\omega_j}]$, where ΔHW_{ω_j} is equal to half the excluded range of dihedral angle values (i.e., $\Delta HW_{\omega_j} = \pi - (\omega_j^{\text{upper}} - \omega_j^{\text{lower}})/2$). This results in the following set of equations:

$$
E_{\text{dihedral}}^{\text{upper}} = \sum_j \begin{cases} A_j \left(1 - 2\left[\dfrac{\omega_j - \omega_j^{\text{upper}}}{2\pi - (\omega_j^{\text{upper}} - \omega_j^{\text{lower}})}\right]^2 \right)(\omega_j - \omega_j^{\text{upper}})^2 & \text{if} \quad \omega_j > \omega_j^{\text{upper}} \\ 0 & \text{otherwise} \end{cases}
$$

$$(61)$$

$$
E_{\text{dihedral}}^{\text{lower}} = \sum_j \begin{cases} A_j \left(1 - 2\left[\dfrac{\omega_j - \omega_j^{\text{lower}}}{2\pi - (\omega_j^{\text{upper}} - \omega_j^{\text{lower}})}\right]^2 \right)(\omega_j - \omega_j^{\text{lower}})^2 & \text{if} \quad \omega_j < \omega_j^{\text{lower}} \\ 0 & \text{otherwise} \end{cases}
$$

$$(62)$$

The force field energy term, $E_{\text{forcefield}}$ of Eq. (57), models the nonbonded interactions of the protein, which can consist of simple or more detailed energy functions. In practice, when considering NMR restraints, force field terms are often simplified to include only geometric energy terms such as quartic van der Waals repulsions. Such functions neglect rigorous modeling of energetic terms in order to ensure that experimental distance violations are minimized. In fact, a basic representation for this target function would be

$$
E_S = E_{\text{distance}} + E_{\text{dihedral}} \tag{63}
$$

Here the E_{distance} function includes additional distance restraints to avoid van der Waals contacts. Notice that when all restraints are satisfied, the objective function is driven to zero.

A detailed modeling approach is proposed by using the ECEPP/3 force field [38]. When considering an unconstrained minimization, this approach provides the following objective function:

$$
E_D = E_{\text{distance}} + E_{\text{dihedral}} + E_{\text{ECEPP/3}} \tag{64}
$$

In contrast to Eq. (63), the detailed energy modeling greatly increases the complexity of the objective function. It should also be noted that the transformation from Cartesian to internal coordinate space results in highly nonlinear functions. That is, there is not a one-to-one correspondence between distances and internal coordinates. The advantage for working in dihedral angle space is that the variable set decreases, with the disadvantage being the increased nonconvexity of the energy hypersurface.

2. Global Optimization

The determination of a three-dimensional protein structure defines an optimization problem in which the objective function may correspond to one of the target functions outlined in the previous section. For the simple case, the formulation becomes

$$\min_{\phi} \quad E_S(\phi) = E_{\text{distance}} + E_{\text{dihedral}} \qquad (65)$$

A standard procedure for addressing this global optimization problem consists of a combination of simulated annealing and molecular or torsional angle dynamics [96]. Generally, multiple initial conformers are generated and optimized to provide a set of acceptable structures. Typically, a set containing on the order of 100 acceptable conformers may be identified, from which a subset of similar structures (approximately 20) are used to characterize the system. The simulated annealing protocol is incorporated in an attempt to reduce trapping in local minimum energy wells.

However, the minimization of complex target functions necessitates the use of rigorous global optimization techniques. For the detailed target function, given by Eq. (64), the unconstrained formulation is similar to formulation (65). Through the use of the constrained optimization approach, the dihedral angle bounds are implicitly included as box constraints. Furthermore, distance restraints are treated explicitly. This reformulation removes both E_{dihedral} and E_{distance} from the target function, leaving only $E_{\text{forcefield}}$:

$$\min_{\phi} \quad E_{\text{ECEPP/3}}$$
$$\text{subject to} \quad E_l^{\text{distance}}(\phi) \leq E_l^{\text{ref}}, \quad l = 1, \ldots, N_{\text{CON}} \qquad (66)$$
$$\phi_i^L \leq \phi_i \leq \phi_i^U, \quad i = 1, \ldots, N_\phi$$

Here $i = 1, \ldots, N_\phi$ corresponds to the set of dihedral angles, ϕ_i, with ϕ_i^L and ϕ_i^U representing lower and upper bounds on these dihedral angles. In general, the lower and upper bounds for these variables are set to $-\pi$ and π, although appropriate bounds derived from NMR data are also suitable.

3. Torsion Angle Dynamics

Standard unconstrained molecular dynamics simulations have been used extensively to model the folding and unfolding of protein systems [97–99]. In addition, several methods for NMR structure calculation have been based on molecular dynamics in Cartesian space [96]. Torsion angle dynamics differs from traditional molecular dynamics in that bond lengths and bond angles are fixed at equilibrium values, thereby allowing for a transformation from the Cartesian to the internal coordinate system. The constraints on the systems also dampen high-frequency motions, which permits the use of longer time steps during the numerical integration of the equations of motion. The use of TAD in place of conventional MD has been found to improve the efficiency of NMR structure prediction [94,95].

A major disadvantage for employing TAD in place of Cartesian MD is that the equations of motion become much more complex for the constrained system. For unconstrained Cartesian MD the accelerations of the atoms can be calculated independently due to the decoupled nature of the equations of motion. The addition of constraints to the Cartesian system transforms the equations from a system of ODEs to a system of differential algebraic equations (DAEs). The alternative to solving this system of DAEs is to transform the equations of motion to the internal coordinate reference frame. In this case, the solution of a linear matrix equation in each time step is required, which, due to the highly coupled structure of the equations, scales as a cubic function of the number of degrees of freedom (torsion angles). To avoid the potentially prohibitive computational cost required for the solution of the equations of motion, a fast recursive algorithm, which scales linearly with the number of torsion angles, was implemented. The algorithm is based on spatial operator algebra that has been used to simulate the dynamics of astronautical and robotic equipment [100].

The algorithm solves for the torsional accelerations, $\ddot{\theta}$:

$$M(\theta)\ddot{\theta} + C(\theta, \dot{\theta}) = 0 \qquad (67)$$

In this equation, M is an $N \times N$ nonlinear mass matrix and C is the N-dimensional vector of velocity-dependent (Coriolis and other) forces. θ, $\dot{\theta}$, and $\ddot{\theta}$ represent the torsional position, velocities and accelerations, respectively. The ability to calculate the accelerations recursively relies on the chainlike structure of the protein, in which each node of the chain represents a rigid body. These rigid bodies consist of one atom or a cluster of atoms whose relative positions are fixed. To simplify the explanation of the algorithm, an unbranched chain will be considered, although the approach can be easily extended to branched systems. For this simple case, the first rigid body, at one end of the chain, defines the base ($k = 0$), while the last rigid body, at the other end of the chain, defines the tip ($k = N$). The rotatable torsion angle between bodies k and $k - 1$ is defined as θ_k.

The framework of the algorithm to calculate $\ddot{\theta}$ can be broken down into three steps:

Step 1. A recursion from the base to the tip is required to calculate the positions, spatial velocities, Coriolis and gyroscopic terms for each of the rigid bodies. To proceed, the 6×6 spatial transformation matrix, ϕ_k, between rigid bodies k and $k - 1$ must first be defined:

$$\phi_k = \begin{bmatrix} I_3 & \tilde{l}(\mathbf{r_k} - \mathbf{r_{k-1}}) \\ O_3 & I_3 \end{bmatrix} \tag{68}$$

Here I_3 and O_3 denote the 3×3-dimensional identity and zero matrices, while the $\tilde{l}$ operator refers to the cross-product tensor associated with $\mathbf{r_k} - \mathbf{r_{k-1}}$, where $\mathbf{r_k}$ is the position vector that defines the reference frame for rigid body k. The spatial velocity, $\mathbf{V_k}$, can be computed from the following relation:

$$\mathbf{V_k} = \phi_k^T \mathbf{V_{k-1}} + H_k^T \dot{\theta}_k \tag{69}$$

The spatial velocity is a six-dimensional vector that combines both the three-dimensional angular, ω, and linear, $\mathbf{v}$, velocities:

$$\mathbf{V}_k \equiv \begin{pmatrix} \omega_k \\ \mathbf{v_k} \end{pmatrix} \tag{70}$$

$\mathbf{H_k}$ is also a six-dimensional vector with the first three elements corresponding to the unit vector, $\mathbf{e_k}$, in the direction of the bond forming the connection between rigid bodies k and $k - 1$:

$$\mathbf{H_k} \equiv \begin{pmatrix} e_k \\ 0 \end{pmatrix} \tag{71}$$

The Coriolis and gyroscopic terms, $\mathbf{a_k}$ and $\mathbf{b_k}$, respectively, can then be calculated using the following relationships:

$$\mathbf{a_k} = \begin{pmatrix} 0 \\ \tilde{\omega}_{k-1}[\mathbf{v_k} - \mathbf{v_{k-1}}] \end{pmatrix} + \begin{pmatrix} \tilde{\omega}_k & 0 \\ 0 & \tilde{\omega}_k \end{pmatrix} H_k^T \dot{\theta}_k \tag{72}$$

$$\mathbf{b_k} = \begin{pmatrix} \tilde{\omega}_k J_k \tilde{\omega}_k \\ m_k \tilde{\omega}_k \tilde{\omega}_k \mathbf{Y_k} \end{pmatrix} \tag{73}$$

Both $\mathbf{a_k}$ and $\mathbf{b_k}$ are six-dimensional vectors. m_k, $\mathbf{Y_k}$, and $\mathbf{J_k}$ represent the mass, the center-of-mass vector, and the 3×3 inertia matrix for the rigid body, respectively. Finally, the spatial inertia, $\mathbf{L_k}$, of the rigid body about the reference frame is given by the following 6×6 matrix:

$$\mathbf{L_k} = \begin{pmatrix} \mathbf{J_k} & m_k \tilde{\mathbf{Y}}_k \\ -m_k \tilde{\mathbf{Y}}_k & m_k I_3 \end{pmatrix} \tag{74}$$

Step 2. The next step requires a backward recursion from the tip, $k = N$, to the base, $k = 1$. The recursion is used to store a number of auxiliary quantities needed for the final forward recursion to calculate the accelerations. In addition, the gyroscopic terms, $\mathbf{b_k}$, and the spatial inertia terms, $\mathbf{L_k}$, calculated in step 1 can be used to initialize two auxiliary quantities, $\mathbf{z_k}$ and $\mathbf{P_k}$, respectively. Both $\mathbf{P_k}$ and $\mathbf{z_k}$ are updated recursively using the following intermediate terms:

$$D_k = \mathbf{H_k P_k H_k^T} \tag{75}$$

$$\mathbf{G_k} = \mathbf{P_k H_k^T D_k^{-1}} \tag{76}$$

$$\epsilon_k = -\mathbf{H_k}(\mathbf{z_k} + \mathbf{P_k a_k}) - \nabla \mathbf{E_k} \tag{77}$$

Here D_k and ϵ_k denote scalar quantities, whereas $\mathbf{G_k}$ is a six-dimensional vector. The final equation requires the gradient of the potential function, ∇E_k. The recurrence relationships for $\mathbf{P_{k-1}}$ and $\mathbf{z_{k-1}}$ are given by:

$$\mathbf{P_{k-1}} \leftarrow \mathbf{P_{k-1}} + \phi_k(\mathbf{P_k} - \mathbf{G_k H_k^T P_k})\phi_k^T \tag{78}$$

$$\mathbf{z_{k-1}} \leftarrow \mathbf{z_{k-1}} + \phi_k(\mathbf{z_k} + \mathbf{P_k a_k} + \mathbf{G_k}\epsilon_k) \tag{79}$$

Step 3. A final forward recursion from the base to the tip is used to obtain the $\ddot{\theta}$ values. The six-dimensional vector α_k is used to store intermediate quantities, with α_k equal to a vector of zeroes for $k = 0$.

$$\alpha_k = \phi_k^T \alpha_{k-1} \tag{80}$$

$$\ddot{\theta}_k = \epsilon_k D_k^{-1} - G_k \alpha_k \tag{81}$$

The following recursion relation is used to update the values of α_k:

$$\alpha_k \leftarrow \alpha_k + \mathbf{H_k}\ddot{\theta}_k + \mathbf{a_k} \tag{82}$$

For branched molecular structures, each node can potentially spawn more than one child so that both the inward and outward recursions must be modified. In the case of an inward recursion, the results from each of the child nodes must be summed up before moving up one level. In the case of the outward recursion, each of the node branches requires a separate recursion.

The TAD is carried out using simulated annealing, with temperature control provided by coupling to an external bath [101]. This coupling provides a means for forcing or damping the torsional velocities using the following scaling factor at time t:

$$f_T = \sqrt{1 - \frac{1}{\beta} + \frac{T_0}{\beta T(t)}} \tag{83}$$

In this equation, β is a force constant, while T_0 is the bath temperature and $T(t)$ is the actual temperature. The actual temperature is calculated from the kinetic energy, $E_{kinetic}$, with the following relationship:

$$T(t) = \frac{2E_{kinetic}(t)}{N_\phi k_B} \tag{84}$$

where k_B is the Boltzmann constant. The value for f_T is used to scale the torsional velocities:

$$\dot{\theta}(t) \leftarrow f_T \dot{\theta}(t) \tag{85}$$

Once torsional velocities have been determined, the accelerations, $\ddot{\theta}$, can be calculated using the recursive algorithm outlined above. A basic leap-frog technique is then employed to calculate velocities at the half-time step, which can be used to calculate torsional positions, θ, and new estimated velocities at the full-time step.

4. Algorithmic Steps

The algorithmic steps for the constrained αBB approach can be generalized to any force field model or routine for solving constrained optimization problems. Here, the αBB approach is interfaced with PACK [74] and NPSOL [28]. PACK is used to transform to and from Cartesian and internal coordinate systems, as well as to obtain function and gradient contributions for the ECEPP/3 force field and the distance constraint equations. NPSOL is a local nonlinear optimization solver that is used to locally solve the constrained upper and lower bounding problems in each subdomain.

The implementation can be broken down into two main phases: initialization and computation. The basic steps of the initialization phase are as follows:

1. Choose the set of global variables. Because the bounds on these variables will be refined during the course of global optimization, they should be selected based on their overall effect on the structure of the molecule. In this work (and in general) the ϕ and ψ dihedral angles provide the largest structural variability and are chosen to constitute the global variable set.
2. Set upper and lower bounds on all dihedral angles (variables). If information is not available for a given dihedral angle, the variable bounds are set to $[-\pi, \pi]$. Because a constrained local optimization solver is used, these box constraints are strictly enforced.
3. Identify the set of NOE-derived distance restraints to be used in the constraints. In general, this set can include all intra- and inter-residue restraints. In this work, only backbone sequential and medium to

long-range information was used in developing the constraints, because intra-residue restraints are less likely to affect the overall fold. Although the formulation can handle multiple constraints, distance restraints were included as one constraint ($N_{CON} = 1$) for the computational studies.

4. Choose the value of E_l^{ref} to be used in the constraint equations. This can be determined by simply performing several local constrained optimizations or possibly a short global optimization run with simplified energy models. In this work, information based on X-PLOR [96] results was used to define the E^{ref} parameter (see below).

5. Identify initial α values for both the objective and constraint functions.

6. Set initial best upper bound to an arbitrarily large value.

The computation phase of the algorithm involves an iterative approach, which depends on the refinement of the original domain by partitioning along the global variables. In each subdomain, upper and lower bounding problems are solved locally and used to develop the sequence of converging upper and lower bounds. The basic steps are as follows:

1. The original domain is partitioned along one of the global variables.

2. Lower bounding functions for both the objective and constraints are constructed in both subdomains. A constrained local minimization is performed using the following procedure:

   a. 100 random points are generated and used for evaluation of the lower bounding objective function and constraints. The point with the minimum objective function value is used as a starting point for local minimization using NPSOL.

   b. If the minimum value found is greater than the current best upper bound, the subdomain can be fathomed (global minimum is outside region); otherwise the solution is stored.

3. The upper bounding problems (original constrained formulation) are then solved in both subdomains according to the following procedure:

   a. 100 random points are generated and used for evaluation of the objective function and constraints. The point with the minimum objective function value and feasible constraints is used as a starting point for local minimization using NPSOL. If a feasible starting point is not found, local minimization is not performed.

   b. All feasible solutions are stored.

4. The current best upper bound is updated to be the minimum of those thus far stored.

5. The subdomain with the current minimum value of $L_{forcefield}$ is selected and partitioned along one of the global variables.

6. If the best upper and lower bounds are within a defined tolerance, the program will terminate; otherwise it will return to Step 2.

To enhance the search for low-energy feasible points, the basic procedure described in Step 3a is modified to include TAD. The following protocol is used:

1. Set counter, $c = 1$. Perform TAD (1000 high-temperature steps followed by 3000 annealing steps) using E_S as the target function. The torsion angle bounds of the current subdomain determine the dihedral angle restraint functions. In addition to the NOE-derived distance restraints, sterically based distance restraints are added to prevent van der Waals overlaps.

   a. If the $E_l^{\text{distance}} < E_l^{\text{ref}}$ $\forall l = 1, \ldots, N_{\text{CON}}$, go to Step 2. Else go to Step 1b.

   b. Increment counter, $c = c + 1$. If $c < 5$, reduce weight of sterically based distance restraints, perform new TAD and go to Step 1a. Else go to Step 2.

2. Set counter, $c = 1$. Perform local minimization using NPSOL with dihedral angle box constraints to implicitly enforce bounds. The objective function is a weighted combination of forcefield energy and distance restraint terms:

$$E = E_{\text{ECEPP/3}} + \sum_l W_l E_l^{\text{distance}} \tag{86}$$

where the weights, W_l, are based on the violation of the distance constraints:

$$W_l = \sqrt{1 + \frac{E_l^{\text{distance}}}{E_l^{\text{ref}}}} \tag{87}$$

   a. If $E_l^{\text{distance}} < E_l^{\text{ref}}$ $\forall l = 1, \ldots, N_{\text{CON}}$, go to Step 3. Else go to Step 2b.

   b. Increment counter, $c = c + 1$. If $c < 5$, increase weight of distance restraint terms, perform TAD (100 high-temperature steps followed by 300 annealing Steps) and go to Step 2a. Else go to Step 3.

3. Solve the constrained minimization problem using NPSOL.

5. Computational Study

The global optimization algorithm was tested on Compstatin, a synthetic 13-residue (ICVVQD WGHHRCT) cyclic peptide (disulfide bridge between the Cys^2 and Cys^{12} residues) that binds to C3 (third component of complement) and inhibits complement activation [102]. Two-dimensional NMR techniques [103] yield a total of 30 backbone sequential (including H^β - backbone), 23 medium- and long-range (including disulfide), and 82 intra-residue NOE restraints. In addition, 7 ϕ angle and 2 χ_1 angle dihedral restraints are available. In previous work [103], traditional distance geometry-simulated annealing protocol was utilized to minimize the associated target function in the Cartesian coordinate space using the program X-PLOR [96]. NOE distance and dihedral angle restraints were modeled using a quadratic square well potential, while van der Waals overlaps were prevented through the use of a simple quartic potential function.

The NMR refinement protocols resulted in a family of 21 structures with similar geometries for the Gln^5–Gly^8 region. A representative structure was obtained by averaging the coordinates of the individually refined structures and then subjecting this structure to further refinement to release geometric strain produced by the averaging process. The formation of a type I β-turn was identified as a common characteristic for these structures.

The consistency of the ensemble of Compstatin solution structures was determined by evaluating distance restraints for each of the original 21 structures (accession number 1a1p at the RCSB, http://www.rcsb.org), as well as for the average Compstatin conformation. In considering distance restraints, only backbone sequential and medium/long-range NOEs were considered. That is, the 82 intra-residue restraints were neglected because they are less likely to effect the overall fold of the Compstatin peptide. This results in a total of 52 restraints, with an additional restraint on the distance between the sulfur atoms forming the disulfide bridge. In order to quantify these results, the violation energy, E_{VIO}, which can be calculated by summing Eqs. (59) and (60), was calculated for each of the original PDB structures. In these calculations, the value of the weighting factor (A_j) was assumed to be constant and set equal to 50 kcal/mol/$Å^2$.

The results of the analysis, shown graphically in Fig. 25 indicate that the average structure ($\overline{Compstatin}$) possesses the third largest violation energy, whereas the smallest value is provided by structure 8 ($\langle Compstatin \rangle_8$). These quantities provide a range of comparison for violation energies and were used to set the constraint parameter, E^{ref}, to 200 kcal/mol. This value is chosen so that the sum of the violation energies will necessarily result in an improvement over the violation energy for the average Compstatin structure.

To measure the performance of the proposed global optimization approach, the ensemble and average Compstatin structures ($\langle Compstatin \rangle$ and $\overline{Compstatin}$) were then used as starting points for local minimization. Because these calculations are performed in the torsion angle space, which requires fixing bond lengths and bond angles to equilibrium values, the corresponding Compstatin PDB structures could only be used to derive torsion angle values. These dihedral angles were then used as input to directly evaluate the corresponding force field energy. Differences in bond lengths and bond angles propagate through the generation of the corresponding ECEPP/3 structure,

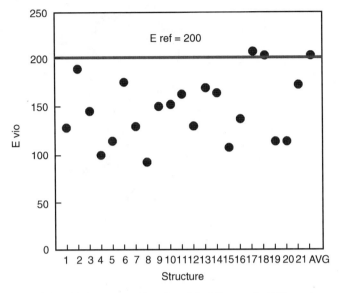

Figure 25.   Violation energy, E_{VIO}, for original Compstatin PDB structures.

which produces an inherent RMSD between the PDB structure and the ECEPP/3-generated structure. For example, when using the set of dihedral angles calculated from the $\overline{Compstatin}$ PDB, the ECEPP/3 structure possesses a 0.581 Å all atom RMSD (all heavy atoms in backbone and side chains) with respect to the original $\overline{Compstatin}$ structure, with a corresponding ECEPP/3 energy of 519.2 kcal/mol. In addition, due to the differences in bond lengths and angles, the total distance violation for the ECEPP/3 structure ($\overline{Compstatin}_{ECEPP}$) increases from 6.9 to 8.7 Å, which results in a subsequent increase in violation energy to 315 kcal/mol. The superposition of the original and ECEPP/3 $\overline{Compstatin}$ conformations is shown in Fig. 26.

Due to the relatively large distance violations and energies obtained after transformation of PDB to PACK (ECEPP/3) structures, the 22 structures were then subjected to local minimization. The problem formulation for local minimization uses the set of 53 restraints for the constraint function, a constant 50 kcal/mol/Å weighting factor (A_j), and a constraint parameter, E^{ref}, equal to 200 kcal/mol. In all cases, the corresponding violation energy reached the upper bound value of 200 kcal/mol. The corresponding total distance violations increased, with an average value of 6.766 Å. The smallest distance violation (5.873 Å) was reported for structure number 10 ($\langle Compstatin \rangle_{10}^{Local}$), whereas the corresponding energy for this structure (-41.685 kcal/mol) was only slightly above the average energy of -47.75 kcal/mol. The lowest energy structures

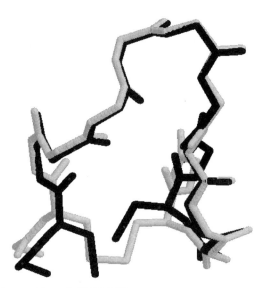

Figure 26. Superposition of ($\overline{Compstatin}_{\text{Orig}}$) structure (in light gray) and corresponding ECEPP/3 structure (in black) using calculated dihedral angles ($\overline{Compstatin}_{\text{ECEPP}}$).

(-71.613 for $\langle Compstatin \rangle_2^{Local}$, -68.704 kcal/mol for $\langle Compstatin \rangle_{21}^{Local}$, -67.653 kcal/mol for $\langle Compstatin \rangle_9^{Local}$) provided above average values for total distance violation (6.963 Å, 6.832 Å, 7.120 Å, respectively). In addition, the conformation obtained from the average Compstatin structure ($\overline{Compstatin}$) exhibited near average values for energy (-52.283 kcal/mol) and total distance violations (6.392 Å). The range of ECEPP/3 energies after local minimization are shown in Fig. 27.

The structural characteristics of these locally minimized structures were quantified using RMSD (root-mean-squared deviation) calculations. For the original PDB structures, comparison with the average Compstatin structure provided RMSD values between 1 and 2 Å for only backbone atoms. As expected, these structures possess common structural features. However, when comparing original PDB structures and their locally minimized counterparts, most RMSD values are larger than 2 Å, indicating that significant conformational changes occur during local minimization. This is due to both the reduced set of NOE restraints in the constraint function and the role of the detailed energy force field. In contrast, the RMSD values for the β-turn region remain consistently low when comparing the original PDB structures to their locally minimized counterparts. These results indicate that the β-turn is a conserved structural feature, even with the addition of the detailed energy model.

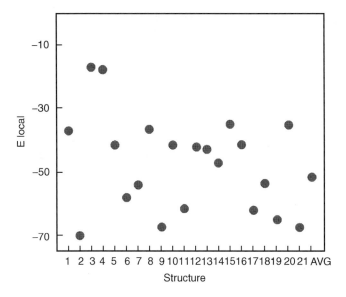

Figure 27. Locally minimized energy, $E_{ECEPP/3}$, for Compstatin structures.

The constrained global optimization approach was first applied to Compstatin structure prediction without the use of TAD. A subset of 26 (all ϕ and ψ) torsion angles, from a total of 73, were treated globally, whereas the remaining ones were allowed to vary locally. As was the case for local minimization, the same set of restraints were used to formulate the nonlinear constraint, with a constant 50 kcal/mol/Å weighting factor and a constraint parameter equal to 200 kcal/mol. The lowest-energy structure satisfying the constraint functions provided an ECEPP/3 energy of -85.71 kcal/mol, an energy value more than 15 kcal/mol lower than those values provided by local minimization. The global minimization required approximately 40 CPU hours on a HP C160. The total distance violation equaled 6.690 Å, which is near the average distance violation for the local minimum structures.

RMSD calculations were performed to again quantify the structural differences between the global minimum energy structure and the other Compstatin structures. RMSD values between the full backbone and the β-turn segments of the 22 locally minimized PDB structures and the global minimum energy structure are plotted in Figs. 28 and 29, respectively. When comparing full backbone RMSD vcalues, the $\langle Compstatin \rangle_9^{Local}$, $\langle Compstatin \rangle_{21}^{Local}$, $\langle Compstatin \rangle_{19}^{Local}$ and $\langle Compstatin \rangle_{17}^{Local}$ provide the best agreement with the global minimum energy structure. These structures also correspond to four of the lowest energy local minima, indicating that some of the lowest energy

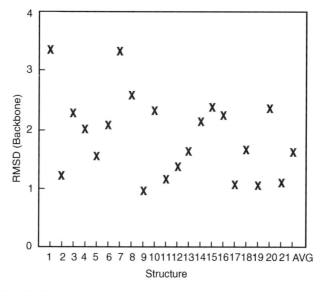

Figure 28. RMSD values for backbone when comparing global minimum energy structure to locally minimized PDB structures.

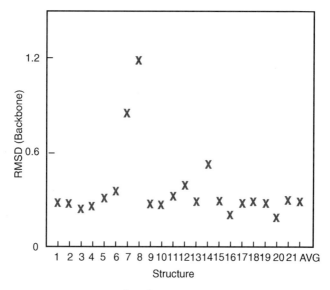

Figure 29. RMSD values for Gln^5–Gly^8 backbone when comparing global minimum energy structure to locally minimized PDB structures.

conformers exhibit similar backbone structural characteristics. In contrast, the lowest energy local minimum, $\langle Compstatin \rangle_2^{Local}$, is less similar to the global minimum energy structure. For the β-turn segment, the correlation between low RMSD values and low energy local minima does not exist. This observation, coupled with the relatively low RMSD values between all structures, indicates that the β-turn structure is a characteristic for all conformers, including the global minimum energy structure. Plots for superpositioning (backbone atoms) of the average local minimum energy structure $\overline{Compstatin}^{Local}$ and the global minimum energy structure are given in Fig. 30.

6. *Comparison with TAD: DYANA*

A comparison to an independent method for solving distance restraint problems was also made in order to gauge the performance of the proposed αBB constrained formulation. Specifically, a torsional angle dynamics (rather than a Cartesian coordinate dynamics such as X-PLOR) package was used [94]. The coupled simulated annealing/TAD protocol from DYANA was applied to a starting sample of 1000 randomly generated structures. The same dihedral angle constraints and 53 medium- and long-range distance constraints were considered; that is, no heuristic methods for reducing the variable space were employed. In the case of unspecified symmetric hydrogens, a pseudoatom approach, in which the restraint is based on a pseudoatom central to the symmetric hydrogen atoms, was used. A subset consisting of the 20 conformers

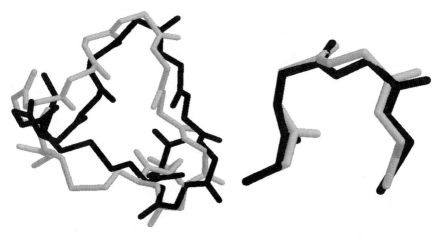

Figure 30.   Superposition of global minimum (in black) and $\overline{Compstatin}^{Local}$ (in light gray) structures. The left panel shows the full (backbone atom) structure, whereas the right panel compares only the β-turn region.

TABLE XXI
Local Minimization Results for the Best DYANA (TAD)-Generated Conformations[a]

Local Minimum	D_{VIO} (Å)	E_{VIO} (kcal/mol)	$E_{ECEPP/3}$ (kcal/mol)
$Compstatin_1^{DYANA}$	6.234	200.0	− 11.945
$Compstatin_2^{DYANA}$	6.538	200.0	6.782
$Compstatin_3^{DYANA}$	6.163	200.0	− 10.208
$Compstatin_4^{DYANA}$	5.476	200.0	− 14.516
$Compstatin_5^{DYANA}$	6.927	200.0	5.006

[a]Here D_{VIO} refers to the total distance violation, E_{VIO} is the corresponding violation, and energy and $E_{ECEPP/3}$ is the force field energy at the local minima.

exhibiting the best target values were then used as starting points for a second set of runs. Finally, a set of five conformations (with the smallest violations) were used for further analysis. Because each method (DYANA vs. ECEPP/3) employed different structural definitions, based on fixed bond lengths and bond angles, a direct comparison was not sufficient. Instead, the DYANA-generated structures were used as starting points for local minimizations using the local constrained formulation. In all cases, the violations reached the upper bound of 200 kcal/mol for E^{ref}. The corresponding violation values, including final local minimum energy values ($E_{ECEPP/3}$), are given in Table XXI.

The results given in Table XXI indicate that although the DYANA conformers satisfy the corresponding constraint, their energy values are significantly higher than that of the global minimum energy structure (more than 70 kcal/mol). This can be anticipated because the goal of the DYANA algorithm is to minimize distance restraint violations via penalty term optimization, while neglecting any detailed force field terms. In fact, an analysis of the structural characteristics indicate that the type I β-turn does not appear along the Gln^5–Gly^8 backbone in these structures. This is verified by the data in Table XXII, which gives the φ and ψ dihedral angle values for the central β-turn residues.

TABLE XXII
φ and ψ Values for Central Residues (Asp^6 and Trp^7) for Anticipated β-Turn Region[a]

Local minimum	ϕ_2 (°)	ψ_2 (°)	ϕ_3 (°)	ψ_3 (°)
$Compstatin_1^{DYANA}$	166.9	− 66.07	− 80.00	− 40.40
$Compstatin_2^{DYANA}$	165.9	− 65.55	− 81.02	− 33.99
$Compstatin_3^{DYANA}$	180.0	− 60.94	− 81.76	− 42.43
$Compstatin_4^{DYANA}$	168.8	− 50.32	− 80.00	− 42.22
$Compstatin_5^{DYANA}$	165.4	− 72.75	− 97.79	− 39.86

[a]The subscripts refer to the second and third residues in the Gln^5–Gly^8 sequence.

TABLE XXIII

Local Minimization Results for the Best DYANA (TAD)-Generated
Conformations Using All Restraints.[a]

Local minimum	D_{VIO} (Å)	E_{VIO} (kcal/mol)	$E_{ECEPP/3}$ (kcal/mol)
$Compstatin_{1c}^{DYANA}$	6.222	200.0	24.714
$Compstatin_{2c}^{DYANA}$	5.643	200.0	− 31.216
$Compstatin_{3c}^{DYANA}$	6.527	200.0	− 17.569
$Compstatin_{4c}^{DYANA}$	7.135	200.0	− 27.110
$Compstatin_{5c}^{DYANA}$	5.926	200.0	− 14.656

[a]Here D_{VIO} refers to the total distance violation, E_{VIO} is the corresponding violation, and energy and $E_{ECEPP/3}$ is the forece field energy at the local minima.

The problem is evidenced by the Asp[6] residue, which has ϕ–ψ values in a forbidden region of the Ramachandran plot. It appears that this may be related to clustering of the side chains in the DYANA-predicted structures.

In order to further examine this deviation from the previous results (which define a type I β-turn), the DYANA protocol was also tested on the full set of restraints, including intra-residue distances. The five DYANA-predicted structures exhibiting the lowest target function values were then subjected to local minimization using the constrained formulation. As before, only the 53 medium- and long-range distance restraints were included during the local minimizations. As the results in Table XXIII show, the average energy has decreased for this set of conformers. However, the structural analysis of the Gln[5]–Gly[8] region, given in Table XXIV still indicates that a type I β-turn is not preferred.

An additional comparison between the structural characteristics of these (DYANA) local minima and the global minimum was also performed using RMSD calculations, as given in Tables XXV and XXVI. These values are consistently larger than those between the average ($\overline{Compstatin}^{Local}$) and local

TABLE XXIV

ϕ and ψ Values for Central Residues (Asp[6] and Trp[7]) for Anticipated β-Turn Region[a]

Local Minimum	ϕ_2 (°)	ψ_2 (°)	ϕ_3 (°)	ψ_3 (°)
$Compstatin_{1c}^{DYANA}$	− 180.0	− 58.61	− 80.00	− 47.72
$Compstatin_{2c}^{DYANA}$	177.5	− 63.77	− 82.74	− 33.53
$Compstatin_{3c}^{DYANA}$	180.0	− 63.98	− 82.18	− 23.32
$Compstatin_{4c}^{DYANA}$	163.0	− 58.56	− 109.2	− 4.53
$Compstatin_{5c}^{DYANA}$	− 180.0	− 70.46	− 92.40	− 41.22

[a]The subscripts refer to the second and third residues in the Gln[5]– Gly[8] sequence.

TABLE XXV
RMSD Values for Full Compstatin Structures[a]

Local Minimum	Heavy Atoms	Backbone Atoms
$Compstatin_{1c}^{DYANA}$	4.117	2.812
$Compstatin_{2c}^{DYANA}$	4.866	3.893
$Compstatin_{3c}^{DYANA}$	5.243	3.943
$Compstatin_{4c}^{DYANA}$	4.892	2.654
$Compstatin_{5c}^{DYANA}$	4.506	3.180

[a]Column 2 reports RMSD using all heavy atoms, while column 3 accounts for only backbone atoms (N, C^α, C'). Both columns compare the DYANA local minimum structures ($Compstatin_i^{DYANA}$) to the global minimum Compstatin PDB structure ($Compstatin^{Global}$).

minimum solutions structures ($\langle Compstatin \rangle_i^{Local}$) and global minimum energy structure. The RMSD values indicate not only that there is significant structural difference over the entire structure (Table XXV), but also that the β-turn region (Table XXVI) is not a structural characteristic of the DYANA local minima. This is evidenced by the superpositioning of the lowest-energy DYANA structure and the global minimum energy structure, given in Fig. 31.

7. Global Optimization and Torsion Angle Dynamics

The modified constrained global optimization was also applied to the Compstatin structure prediction problem using the same constraint function and parameters [104]. The goal of introducing TAD as a component of the upper bound solution approach is to increase the number of feasible points available for initialization of the constrained local minimization. Initially, TAD is used in combination with simple van der Waals overlap restraints to drive the distance violations to zero.

TABLE XXVI
RMSD Values for the β-Turn Regions (Residues 5 through 8)[a]

Local Minimum	Heavy Atoms	Backbone Atoms
$Compstatin_{1c}^{DYANA}$	1.163	0.625
$Compstatin_{2c}^{DYANA}$	1.473	0.732
$Compstatin_{3c}^{DYANA}$	1.607	0.721
$Compstatin_{4c}^{DYANA}$	1.327	0.721
$Compstatin_{5c}^{DYANA}$	1.277	0.781

[a]Column 2 reports RMSD using all heavy atoms, while column 3 accounts for only backbone atoms (N, C^α, C'). Both columns compare the DYANA local minimum structures ($Compstatin_i^{DYANA}$) to the global minimum Compstatin PDB structure ($Compstatin^{Global}$).

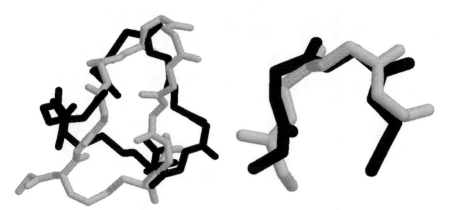

Figure 31. Superposition of global minimum (in black) and $Compstatin_{1c}^{DYANA}$ (in gray) structures. The left panel shows the full (backbone atom) structure, whereas the right panel compares only the β-turn region.

Taken independently, this methodology is comparable to the typical implementation of TAD for NMR structure prediction [94]. Although there are potential deficiences in the independent TAD algorithm; that is, the simplified force field term is insufficient for sparse sets of distance restraints.

The use of TAD in the context of the global optimization approach surmounts this difficulty by using an iterative TAD scheme with two forms of the target function. The first set of TAD runs focuses on the reduction of the distance violations, while employing a simplified forcefield in the form of additional distance restraints to avoid atomic overlaps. This approach mimics the effects of a typical TAD approach for structure prediction. To ensure that these conformers provide low energy, this step is then followed by unconstrained minimization with a hybrid distance and ECEPP/3 energy objective function. If the ECEPP/3 energy is acceptably low, the algorithm proceeds to the constrained local minimization step, otherwise an iterative set of TAD runs are performed with readjustment of the relative weight of the distance and ECEPP/3 terms. Fig. 32 shows a typical sequence for both the ECEPP/3 and distance violations energy during one solution of the upper bounding problem for Compstatin.

The results of the combined constrained global optimization and TAD algorithm can be assessed by examining the sequence of ECEPP/3 energies obtained from the solution of the upper bounding problems, as depicted in Fig. 33. When compared to the original algorithm, the TAD implementation augments the number of feasible starting points by more than a factor of two. This enhancement leads to earlier identification of low-energy conformers.

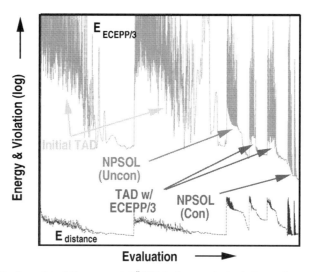

Figure 32. Log plot of $E_{ECEPP/3}$ and $E^{distance}$ during a typical solution to the upper bounding problem for C3.

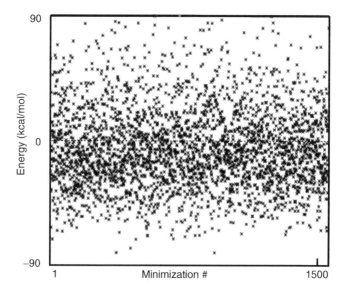

Figure 33. Energy values for Compstatin conformers obtained from combined constrained global optimization and TAD algorithm.

In particular, conformers with energies less than -70 kcal/mol, and thus lower in energy than the locally minimized PDB structures, are identified within the first 10 iterations of the global optimization approach. This property has important algorithmic implications, including the ability to fathom regions based on the current estimate of the global minimum. In general, the TAD-enhanced search provides more consistent and denser population of low-energy conformers.

Both experimental and theoretical methods exist for the prediction of protein structures. In both cases, additional restraints on the molecular system can be derived and used to formulate a nonconvex optimization problem. Here, the traditional unconstrained problem was recast as a constrained global optimization problem and was applied to protein structure prediction using NMR data. Both the formulation and solution approach of this method differ from traditional techniques, which generally rely on the optimization of penalty-type target function using SA/MD protocols.

As a first step, the penalty-type restraint functions were replaced by nonlinear constraints, which can be individually enumerated for all or subsets of the distance restraints. In addition, the objective function was transformed to a full atom force field potential, a modification that should be particularly useful for systems possessing sparse set of restraints. To solve this reformulated molecular structure prediction problem, the concepts of a deterministic global optimization approach, αBB, were applied. This methodology can be used to develop theoretical guarantees for convergence to the global minimum of nonconvex constrained problems. The algorithm was further enhanced by modifying the upper bounding solution approach to include an iterative scheme involving TAD.

The approach was applied to the Compstatin structure prediction problem using both the original TAD approach and the coupled αBB-TAD approach. When considering basic structural features, such as the formation of a type I β-turn, the predicted structure was found to agree with results based on X-PLOR [96]. However, constrained global optimization was able to identify conformers with significantly lower energies than those obtained from either local minimization or independent TAD algorithms. In particular, the coupled αBB-TAD implementation consistently produced dense populations of low-energy conformers.

C. Perspectives and Future Work

1. Structure Prediction of Polypeptides

In spite of pioneering contributions and decades of effort, the *ab initio* prediction of the folded structure of a protein remains a very challenging problem. The approaches for the structure prediction of polypeptides can be classified as (i) homology or comparative modeling methods, (ii) fold recognition or threading methods, (iii) *ab initio* methods that utilize knowledge-based information from structural databases (e.g., secondary and/or tertiary structure restraints), and (iv) *ab initio* methods without the aid of knowledge-based information.

Knowledge-based *ab initio* methods exploit information available from protein databases regarding secondary structure, introduce distance constraints, and extract similar fragments from multiple sequence alignments in an attempt to simplify the prediction of the folded three-dimensional protein structure. Significant contributions include the work of Levitt and co-workers [40,105], Skolnick and co-workers [106,107], Baker and co-workers, [108,109], Dill and co-workers, [110], and Friesner and co-workers, [93,111,112]. *Ab initio* methods that are not guided by knowledge-based information represent the most challenging category. Important advances include the pioneering work of Scheraga and co-workers [113–115], Rose and co-workers [116], and Dill and co-workers [117,118]. Orengo et al. (1999) [119] provide a recent assessment of the current status of both types of *ab initio* protein structure prediction approaches.

We have recently developed the novel ASTRO-FOLD approach for the *ab initio* prediction of the three-dimensional structures of proteins [120]. The four stages of the approach are outlined in Fig. 34. The first stage involves the identification of helical segments and is accomplished by: partitioning the amino acid sequence into pentapeptides such that consecutive pentapeptides possess an overlap of four amino acids; atomistic level modeling using the selected force field; generating an ensemble of low-energy conformations; calculating free energies that include entropic, cavity formation, polarization and ionization contributions for each pentapeptide; and calculating helix propensities for each residue using equilibrium occupational probabilities of helical clusters.

In the second stage, β-strands, β-sheets, and disulfide bridges are identified through a novel superstructure-based mathematical framework originally established for chemical process synthesis problems [121]. Two types of superstructure are introduced, both of which emanate from the principle that hydrophobic interactions drive the formation of β-structure. The first one, denoted as *hydrophobic residue-based superstructure*, encompasses all potential contacts between pairs of hydrophobic residues (i.e., a contact between two hydrophobic residues may or may not exist) that are not contained in helices (except cystines that are allowed to have cystine–cystine contacts even though they may be in helices). The second one, denoted as *β-strand-based superstructure*, includes all possible β-strand arrangements of interest (i.e., a β-strand may or may not exist) in addition to the potential contacts between hydrophobic residues. The hydrophobic residue-based and β-strand-based superstructures are

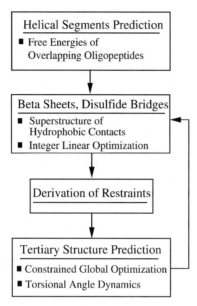

Figure 34. Overall flowchart for the *ab initio* structure prediction using ASTRO-FOLD. The first stage addresses the prediction of helical segments based on free energy calculations of overlapping oligopeptides. The second stage introduces a superstructure-based framework coupled with integer-linear optimization for the prediction of a rank-ordered list of β-sheets and disulfide bridges. The third stage derives lower and upper bounds on the (ϕ, ψ) dihedral angles of the secondary structure residues, the distances between pairs of contacts of hydrophobic residues, and the (ϕ, ψ) angles of the loop/turn residues. The fourth stage introduces a constrained formulation for the tertiary structure prediction and its solution via the αBB global optimization approach enhanced by torsion angle dynamics. An iterative loop over the final three stages allows for analysis of multiple β-sheet and disulfide bridge configurations.

formulated as mathematical models that feature three types of binary variables: (i) representing the existence or nonexistence of contacts between pairs of hydrophobic residues; (ii) denoting the existence or nonexistence of the postulated β-strands; and (iii) representing the potential connectivity of the postulated β-strands. Several sets of constraints in the model enforce physically legitimate configurations for antiparallel or parallel β-strands and disulfide bridges, while the objective function maximizes the total hydrophobic contact energy. The resulting mathematical models are Integer Linear Programming (ILP) problems that not only can be solved to global optimality, but also can provide a rank ordered list of alternate β-sheet configurations.

The third stage serves as a preparative phase for atomistic-level tertiary structure prediction, and therefore it focuses on the determination of pertinent information from the results of the previous two stages. This involves the introduction of lower and upper bounds on dihedral angles of residues belonging to predicted helices or β-strands, as well as restraints between the C^α atoms for residues of the selected β-sheet and disulfide bridge configuration. Furthermore, for segments that are not classified as helices or β-strands, free energy runs of overlapping heptapeptides are conducted to identify tighter bounds on their dihedral angles.

The fourth and final stage of the approach involves the prediction of the tertiary structure of the full protein sequence. The problem formulation, which relies on dihedral angle and atomic distance restraints acquired from the previous stage, is equivalent to the problem outlined in Section III.B. The generation of low-energy starting points for constrained minimization is enhanced by introducing torsion angle dynamics [94] within the context of the αBB global optimization framework, as described in Section III.B.7.

An important question regarding the prediction of the native folded state of a protein is how the formation of secondary and tertiary structure proceeds. Two common viewpoints provide competing explanations to this question. The classical opinion regards folding as hierarchic, implying that the process is initiated by rapid formation of secondary structural elements, followed by the slower arrangement of the tertiary fold. The opposing perspective is based on the idea of a hydrophobic collapse, and it suggests that tertiary and secondary features form concurrently. This work bridges the gap between the two viewpoints by introducing a novel *ab initio* approach for tertiary structure prediction in which helix nucleation is controlled by local interactions, while nonlocal hydrophobic forces drive the formation of β-structure. The agreement between the experimental and predicted structures validates the use of the ASTRO-FOLD method for generic tertiary structure prediction of polypeptides.

2. Parallelization Issues

The extension of our global optimization approaches to larger protein systems requires the use of distributed computing environments. Such implementations have been developed independently of system architecture, and the code has been compiled and optimized using the MPI (message passing interface) standard.

On a fundamental level, these parallel implementations exploit the inherent branch-and-bound structure of the αBB algorithm. A major characteristic of a branch and bound framework is that as the size of the domain decreases, the quality of the representation improves, which implies that finer initial domains result in better approximations. This is equivalent to simultaneously exploring multiple domains in order to perform a more efficient search, which is the rationale behind advocating the development of a parallel algorithm.

Distributed frameworks for branch-and-bound algorithms can rely on two basic protocols. The most simplistic structure consists of a tree hierarchy in which a master processor directs the overall flow of the algorithm. In this case, global communication constructions can be maintained in order to control termination and domain processing. The second alternative relies on a ring structure in which all processors act locally and utilize predetermined communication patterns to relay information and detect termination.

Initial implementations of the αBB algorithm have employed the tree hierarchy through a master–slave decomposition approach. This requires the creation of only one communication group in which a single master processor maintains the list of lower bounds. The initial domains for the slave nodes are determined by the master through partitioning of the global domain to the appropriate level in the branch-and-bound tree, and these regions are sent to the nodes for further processing. Once the upper and lower bounding problems have been solved, the relevant information is returned to the master, which extracts and sends to the idle node the next region from the lower bound list. The local processing of each domain can also encompass several levels in the branch-and-bound tree depending on the computational requirements for solving one node in the tree. This procedure can be efficient for treating large protein systems because of low communication time overhead. That is, the time spent in solving the lower and upper bounding problems for each region is long relative to the time required for communication.

The overall protein folding solution approach also affords other levels of parallelism. For example, during the helix prediction phase, the full protein is decomposed into smaller segments. This decomposition allows us to identify the major secondary structural components (α-helical, β-sheet) of the protein by solving smaller global optimization subproblems (using αBB) in parallel. The extent of parallelism depends on the length of these subsegments and the parallelism of the underlying αBB algorithm.

IV. DYNAMICS OF PROTEIN FOLDING

A. Background

The protein folding problem is a very important problem in computational chemistry and molecular biology. The ability of a protein to function properly within the cell depends on its tertiary structure. Considering how precisely and reliably a protein shapes itself to perform its specific task, very little is understood about the mechanism of protein folding. Better understanding and insight on the mechanism of protein folding are of major importance.

In Section III, we discussed the structure prediction problem, in which the native conformation is sought. In this section, we pursue the protein-folding problem further by studying the *folding mechanism*—that is, the pathways followed by a protein as it proceeds from its initial (extended) conformation to its native state, as well as the rates associated with these folding processes.

1. Studying the Dynamics of Secondary Structure Formation

According to the hierarchic model of protein folding, the time scale of formation of secondary structures, such as α-helices and β-sheets, within a given protein occurs on a much shorter time scale than the formation of tertiary structure. Whether this is true or not, there is much evidence—both theoretical and experimental—that the folding of large proteins begins with the formation of these secondary structure elements [122,123]. Therefore, an initial step in understanding protein folding is understanding the folding process of the secondary structures such as α-helices and β-sheets. Insights can be gained into the folding mechanism of these structures by studying short peptides that exhibit the structure we wish to study.

Alpha helices have been studied for a relatively long time [124,125]. Numerous short peptides have been observed in the lab to form α-helices in solution, and have been the subject of many experimental and theoretical studies [4,5,126–131]. Our recent analysis of tetra-alanine, the shortest peptide to form an α-helix, has provided us with enormous insights into the folding mechanism of these structures and will be presented in Section IV.C.

The situation is very different for β-sheet structures. Until recently, experimental studies of these structures have been mostly unsuccessful, largely due to the fact that short peptides which fold into a β-sheet conformation tend to aggregate in solution [125]. These difficulties have finally been overcome with the recent discoveries of designed sequences, such as Beta-nova [132] and others [133,134], as well as the second β-hairpin fragment of Protein G (residues 41–56) [135], thus opening the door to a proper study of β-hairpin and β-sheet formation [136–144]. Our ongoing efforts to analyze the Protein G fragment (41–56) will be discussed in Section IV.E.

2. Searching for Stationary Points

A promising approach to understanding protein folding is the study of its potential energy surface. The first step in the study of any potential energy surface is the identification of stationary points (local minima and saddle points), because these points play a crucial role in defining the topography of the surface. The local minima represent stable configurations of the protein molecule, and the first-order saddle points generally correspond to transition states that connect two such configurations. A protein-folding process can be thought of as a transition between two local minima through a transition state, or a series of such transitions.

The problem of finding stationary points of a potential energy surface is an old one, and numerous methods have been developed to solve it. The most obvious method is applying the Newton–Raphson method to the equation $\nabla V = 0$. The Newton–Raphson method tends to yield a solution whenever the initial guess is close to a stationary point and the Hessian matrix has the appropriate signature for the type of stationary point desired (minima, first-order saddle, etc). It cannot be used, for example, to walk away from a local minimum towards a first-order saddle point.

The various "eigenmode-following" methods are sophisticated variants of the Newton–Raphson method [145–150]. The Hessian is diagonalized, and a modified Newton–Raphson step is generated by "shifting" some of the eigenvalues of the Hessian, from positive to negative or vice versa, before applying its inverse. These methods allow one to step away from local minima in search of transition states, and vice versa.

There are a number of stochastic methods used to find stationary points [151]. Local minima can be obtained by frequent quenching of a constant energy (or temperature) trajectory [82]. Simulated annealing by running a constant temperature trajectory simulation, slowly reducing the temperature to zero in the process, can sometimes lead to good candidates for the global minimum. The method of "slowest slides" [152] has been used to search for transition states connecting two given local minima: A constant energy trajectory is followed during a transition from one local minimum to the other, and the maximum along that trajectory is taken as an initial guess for the transition state.

The global minimum can also be found by use of genetic algorithms, in which new conformations are generated from old conformations by random mutations in the hope of eventually lowering the potential energy. Of particular interest to us is the Conformation Space Annealing (CSA) algorithm [115,153, 154], which is a combination of genetic, annealing, and buildup methods. This algorithm can also be used to generate a variety of low-energy conformations.

Other methods exist for searching for the global minimum of a potential energy surface. Diffusion equation and distance scaling methods have been applied to the problem of finding the global minimum of a potential energy surface [155]. Smoothing transformations are applied to the potential energy surface to eliminate the irrelevant local minima. The remaining minima are tracked back to the original potential energy surface as the transformations are gradually removed. Another method involves obtaining a large sample of local minima and forming a "convex global underestimator" of the potential energy surface based on those sample points [156]. The global minimum of the original potential energy surface is sought in the vicinity of the global minimum of the convex global underestimator.

Many dynamical studies of protein folding are carried out these days by performing molecular dynamics simulations, in which the time evolution of the protein's configuration is determined directly by solving Newton's equations of motion. Not only is it possible to obtain numerous low-energy minima in the vicinity of the starting point, but rate and pathway information can also be inferred directly from the trajectories generated by these simulations. A major drawback of these simulations, however, is their computational expense. Current limitations on simulations are on the order of a few hundred nanoseconds of real time (a 1-µs simulation has been reported recently [97]), which is far too short to enable a full simulation of the folding process of even a modest sized protein.

All of these methods, good in their own right, share one very important drawback: There is no guarantee that all (or even the most important) local minima and first- or higher-order transition states will be found. In this chapter, we propose a method of finding all stationary points of a given potential energy surface in which we apply the αBB deterministic branch-and-bound global optimization algorithm to the system of equations $\partial V/\partial x_i = 0$. The general algorithm is discussed in Section II.B, and its specific application to the stationary point search is discussed in Section IV.B. We have successfully applied this method to small systems, such as triatomic molecules, alanine, alanine dipeptide, and tetra-alanine [130,131]. We will discuss tetra-alanine in Section IV.C.

3. Analyzing the Potential Energy Surface

Once the minima and first-order saddles are determined, the potential energy surface can be analyzed. The folding mechanism of the protein can be understood by enumerating the reaction pathways from the extended conformations to the native state. The first step in constructing the pathways is to determine for each transition state which two minima it connects. This is accomplished by performing a downhill search from the transition state along each of the two reaction coordinate directions. The result is a list of minimum–saddle–minimum "triples." The reaction pathways can then be enumerated by joining these triples together in chains using graph theory techniques.

Transition rates can be calculated using Rice–Ramsperger–Kassel–Marcus (RRKM) theory [157]. The basic assumptions of RRKM theory is that the protein can be treated thermodynamically in the vicinity of the minima as well as the transition state, and that the transition is completed once the transition state is crossed (i.e., there are no re-crossings). Once the transition rates have been determined, the Master equation can be solved for the occupation probabilities of each state as functions of time. This gives us a direct indication of how long it takes for a protein prepared in a given unfolded state to reach its native state. It is also possible to use this information to calculate the time evolution of other quantities, such as (ensemble-averaged) energies, atomic distances, and dihedral angles.

Becker and Karplus [4] proposed a graphical representation of the topography of a potential energy surface based on the connectivity tree originally introduced by Czerminiski and Elber [5]. They define a finite energy (temperature) generalization of the "catchment region." As the energy (temperature) is increased, regions that were once disconnected by high barriers begin to merge. This coalescence process is described by means of a "energy (temperature) disconnectivity graph." The shape of the disconnectivity graph reveals an enormous wealth of dynamical information. We extended this idea by constructing a "rate disconnectivity graph" that is based on transition rates, rather than energy levels or barrier heights.

We have applied these methods to tetra-alanine (an α-helix), which we discuss in Section IV.C, and to the 41–56 fragment of Protein G (a β-hairpin), which we discuss in Section IV.E.

B. The αBB Global Optimization Approach

Stationary points of all orders (i.e., minima, maxima, first-order and higher-order transition states) of a given potential energy surface $V(\mathbf{x})$ are determined by the constraints

$$\frac{\partial V}{\partial x_i} = 0 , \qquad i = 1, \ldots, N_x \tag{88}$$

where N_x is the number of variables: $\mathbf{x} = (x_1, \ldots, x_{N_x})$. Equation (88) is an example of a nonlinearly constrained system of algebraic equations. Indeed, (88) can be obtained from (17) in Section II.B.1 by assigning $f_i(\mathbf{x}) = \partial V/\partial x_i$ for $i = 1, \ldots, N_f = N_x$, and $N_g = 0$.

In Section II.B., we explained how such systems of equations can be solved using the αBB global optimization algorithm. This algorithm applies whenever the constraint functions $\partial V/\partial x_i$ are twice continuously differentiable (C^2)—in other words, whenever the potential energy function itself is C^3. Unlike other methods of locating stationary points, the αBB provides a rigorous theoretical guarantee of finding *all* of the stationary points on a potential energy surface.

According to the αBB algorithm, the original problem (88) is first reexpressed as a global optimization problem by introducing a slack variable:

$$\min_{\mathbf{x},s} s$$

subject to
$$\partial V/\partial x_i - s \leq 0 , \qquad i = 1, \ldots, N_x$$
$$-\partial V/\partial x_i - s \leq 0 , \qquad i = 1, \ldots, N_x \tag{89}$$
$$\mathbf{x}^L \leq \mathbf{x} \leq \mathbf{x}^U$$

The global minima of (89) with $s = 0$ correspond to solutions to the original problem (88).

Configuration space is searched for stationary points by subdividing the full conformational space into smaller and smaller regions. At each stage, the

current region is tested for possible stationary points by solving the lower
bounding problem:

$$\min_{\mathbf{x},s} s$$

subject to
$$\partial V/\partial x_i - \alpha_i^+ \sum_k (x_k^U - x_k)(x_k - x_k^L) - s \leq 0$$

$$-\partial V/\partial x_i - \alpha_i^- \sum_k (x_k^U - x_k)(x_k - x_k^L) - s \leq 0 \tag{90}$$

$$\mathbf{x}^L \leq \mathbf{x} \leq \mathbf{x}^U$$

The left-hand side of each constraint in (90) is a convex underestimator of the
corresponding term in (89), and it is obtained by subtracting off a sufficiently
large quadratic term. The lower bounding problem (90) is indeed convex,
provided that the coefficients $\alpha_i^\pm$ satisfy

$$\alpha_i^+ \geq -\frac{1}{2} \min_{\mathbf{x} \in [\mathbf{x}^L, \mathbf{x}^U]} \{\lambda_k(H_{\partial V/\partial x_i}(\mathbf{x})), 0\}$$

$$\alpha_i^- \geq +\frac{1}{2} \max_{\mathbf{x} \in [\mathbf{x}^L, \mathbf{x}^U]} \{\lambda_k(H_{\partial V/\partial x_i}(\mathbf{x})), 0\} \tag{91}$$

Assuming that (91) is satisfied, the lower bounding problem is convex and can be
solved to global optimality by any commercial local optimization package. The
global minimum s_{LB} of (90) provides a valid lower bound of the global minimum
of (89), and thus it can be used to check if a stationary point can exist in the
current region $[\mathbf{x}^L, \mathbf{x}^U]$. If $s_{\mathrm{LB}} > 0$, no such solution exists, and the region can be
fathomed. If $s_{\mathrm{LB}} \leq 0$, then a solution may or may not exist in $[\mathbf{x}^L, \mathbf{x}^U]$, and so that
region will be subdivided and both subregions checked by the same procedure.
The αBB algorithm terminates when all regions have either been fathomed, or
reduced sufficiently in size at which point a solution to (88) is obtained by a local
search.

Calculating values of $\alpha_i^\pm$ according to (91) is difficult in general because the
Hessian matrices $H_{\partial V/\partial x_i}$ depend on $\mathbf{x}$. A simplified method of calculating $\alpha_i^\pm$ is
to start with small values of $\alpha_i^\pm$ (e.g., $\alpha_i^\pm = 5$) and increase the values of $\alpha_i^\pm$ until
no new solutions are found. This can be a practical solution to many problems
where the correct values of $\alpha_i^\pm$ are difficult to determine. However, this method
has the one serious drawback in that it sacrifices the theoretical guarantee of
finding *all* solutions. In spite of this fact, we were able to identify all minima
and first-order transition states using modest values of $\alpha_i^\pm$ for alanine, alanine
dipeptide, and tetra-alanine. Tetra-alanine will be discussed in Section IV.C.

A more robust method involves calculating the Hessian matrices $H_{\partial V/\partial x_i}$ at
various grid points to get a sample of required $\alpha_i^\pm$ values. First we select a grid,

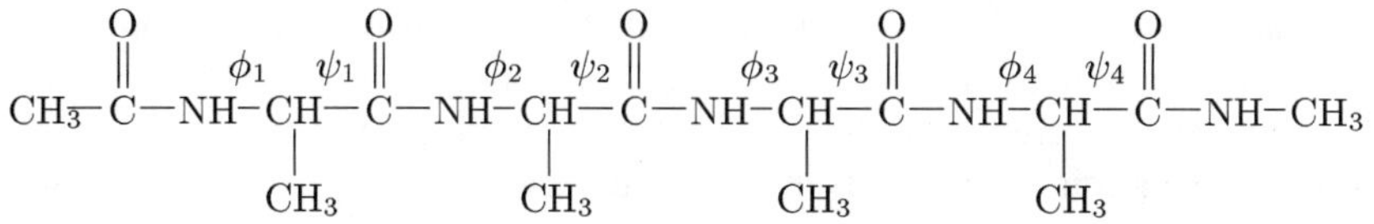

$$
\begin{array}{cccccccccc}
\mathrm{O} & & & \mathrm{O} & & & \mathrm{O} & & & \mathrm{O} & & & \mathrm{O} \\
\parallel & \phi_1 & \psi_1 & \parallel & \phi_2 & \psi_2 & \parallel & \phi_3 & \psi_3 & \parallel & \phi_4 & \psi_4 & \parallel \\
\end{array}
$$

$$\mathrm{CH_3\!-\!C\!-\!NH\!-\!CH\!-\!C\!-\!NH\!-\!CH\!-\!C\!-\!NH\!-\!CH\!-\!C\!-\!NH\!-\!CH\!-\!C\!-\!NH\!-\!CH_3}$$

$$
\begin{array}{cccc}
\mathrm{CH_3} & \mathrm{CH_3} & \mathrm{CH_3} & \mathrm{CH_3}
\end{array}
$$

Figure 35. Tetra-alanine.

$\{\mathbf{x}^k\}$. Then we evaluate the Hessian for each constraint at each grid point, $H_{\partial V/\partial x_i}(\mathbf{x}^k)$, and use (91) to determine precomputed values of $\alpha_i^{\pm}(\mathbf{x}^k)$ at each grid point. During the αBB run, appropriate values of $\alpha_i^{\pm}$ for a given region are determined by selecting the maximum $\alpha_i^{\pm}$ over all grid points contained in the region. This method of generating $\alpha_i^{\pm}$ was used when we studied triatomic molecules, which is discussed in Ref. 130.

C. Dynamics of Coil-to-Helix Transitions

In this section we attempt to elucidate the formation of α-helices by studying tetra-alanine, which is one of the smallest peptides that can exhibit a full α-helical turn. Tetra-alanine is depicted in Fig. 35.

In Sections IV.C.1–IV.C.6, we study tetra-alanine in vacuum. We use the ECEPP/3 potential energy surface [38] (see Section III.A.1 and Fig. 11), which is an all-atom potential energy function. In Section IV.C.7, we consider tetra-alanine in solution by adding a solvation free-energy term to the ECEPP/3 potential energy surface. The solvation free energy is modeled by the volume method using the Reduced Radius Independent Gaussian Sphere (RRIGS) approximation (see Section III.A.2). To simplify the calculations, we fix bond lengths and bond angles, allowing only the eight backbone (ϕ, ψ) dihedral angles to vary.

1. Stationary Points for Unsolvated Tetra-Alanine

The first step in elucidating the folding process of tetra-alanine is to determine the local minima and first-order saddles of its potential energy surface. We first obtained a testbed of minima and first-order saddles by applying a brute-force eigenmode-following search (Eigenmode III [145]) using a grid of starting points. Our search results are summarized in Table XXVII. For our initial attempt

TABLE XXVII
Eigenmode III Results for Unsolvated Tetra-alanine

	4^8 Grid	6^8 Grid
Local minima	16,125	62,373
First-order saddles	18,902	212,938

to analyze tetra-alanine [130], we generated a 4^8 grid of starting points and performed minimum and first-order saddle searches from each point. The transition states were then followed down to the minima they connect, resulting in additional minima found. Given the relative high percentage of starting points that resulted in unique stationary points, we decided to increase the grid to 6^8 and perform first-order saddle searches from each point. Additional minima were obtained by following each such transition state down to the minima they connect. After merging these new results with the results from the 4^8 grid, we had generated a total of 62,373 minima and 212,938 first-order saddle points [131].

Tetra-alanine is one of the smallest peptides that can exhibit an α-helical conformation as well as an extended conformation. These two conformation types can characterized by their (ϕ, ψ) angle values. Alpha-helical conformations tend to have (ϕ, ψ) angle values in the vicinity of $(300°, 300°)$. On the other hand, extended conformations tend to have (ϕ, ψ) values in the vicinity of $(300°, 120°)$.

Therefore, to facilitate the classification of tetra-alanine conformations, we subdivide the (ϕ, ψ) plane into regions and classify those regions according to Table XXVIII. Values of (ϕ, ψ) corresponding to α-helix formation are classified as "a," and values of (ϕ, ψ) corresponding to β-sheet formation are classified as "b." Each conformation of tetra-alanine is characterized by four (ϕ, ψ) pairs, and hence can be classified by a concatenation of four symbols.

Of the 62,373 minima, we found one α-helical conformation, min.1 (aaaa), and one extended conformation, min.1587 (bbbb). Their potential energy and free energy[1] values can be found in Table XXIX. The α-helix conformation is the lowest energy conformation of tetra-alanine. We will be concentrating on the folding process from the extended conformation to the ground state.

We checked the αBB algorithm described in Section IV.B against the Eigenmode III search for stationary points by conducting αBB runs on selected regions of the potential energy surface. Selected results are given in Table XXX.

TABLE XXVIII
Classification Scheme for (ϕ, ψ) Pair

Symbol	ψ	Decoration	ϕ
a	$270° \leq \psi \leq 335°$	No prime	$270° \leq \phi \leq 330°$
i	$335° \leq \psi$ or $\psi \leq 90°$	Prime	$180° \leq \phi \leq 270°$
b	$90° \leq \psi \leq 150°$	Double prime	Otherwise
j	$150° \leq \psi \leq 270°$		

[1]By "free energy," we mean potential energy plus the contributions from vibrational entropy. Free energy can be calculated using (93) in Section IV.C.2.

TABLE XXIX
Ground State and Extended Conformation of Unsolvated Tetra-alanine

Minimum	Classification	E (kcal/mol)	F (kcal/mol)
min.1	aaaa	− 6.643	− 11.798
min.1587	bbbb	4.916	− 5.549

We started with a constant value of $\alpha = 20$, and then increased α in subsequent runs until we found all stationary points located by the Eigenmode III search. In all cases, modest values of α (less than 100) were sufficient to locate all minima and first-order saddles found by Eigenmode III. In many cases, additional saddle points were located.

2. Transition Rates and the Master Equation

Having now identified the local minima and first-order transition states, we are now in a position to enumerate the reaction pathways between states and calculate transition rates. The connectivity between the various minima is determined by following each transition state back to the minima they connect.

TABLE XXX
Selected Results from αBB Tetra-alanine Runs

Region	Saddle Type	Eigenmode III	αBB	α
aaaa	min	1	1	25
bbbb	min	1	1	20
	1st	4	4	
	2nd	6	6	
	3rd	4	4	
	4th	1	1	
bibi	min	1	1	20
	1st	1	2	
	2nd	0	1	
bbbj′	min	2	2	20
	1st	8	9	
	2nd	4	17	
	3rd	3	16	
	4th	2	7	
	5th	0	1	
aai′i	min	2	2	80
	1st	1	1	

This is accomplished by perturbing the transition state in each of the two directions along the reaction coordinate and then using Eigenmode III to locate a local minimum from that starting point. This gives us a list of (minimum, transition state, minimum) triples.

We can then calculate the transition rate matrix using Rice–Ramsperger–Kassel–Marcus (RRKM) theory. According to RRKM theory [130,157,158], the transition rate for a single transition is given by

$$W_{j' \to \text{ts} \to j} = \frac{kT}{h} \frac{Q_{\text{ts}}}{Q_{j'}} \tag{92}$$

The partition functions at the minima and first-order saddles are related to the free energies of those stationary points, and they can be evaluated using the harmonic approximation

$$Q = e^{-F/kT} = e^{-E/kT} \prod_i \frac{kT}{h\nu_i} \tag{93}$$

where E and F are the potential energy and free energy, respectively, of the stationary point, and ν_i are the vibrational frequencies of the molecule around the stationary point. The product over frequencies takes into account the vibrational entropy of the system. Substituting (93) into (92) yields

$$W_{j' \to \text{ts} \to j} = \frac{\prod_i \nu_i^{j'}}{\prod_{i \neq \text{r.c.}} \nu_i^{\text{ts}}} e^{-(E_{\text{ts}} - E_{j'})/kT}$$

Summing over all transition states connecting two particular minima yields the transition rate matrix

$$W_{jj'} = \sum_{\text{ts}} W_{j' \to \text{ts} \to j}$$

The time evolution of occupation probabilities can be calculated by solving the Master equation

$$\frac{dP_j}{dt} = w_{jj'} P_{j'}(t) \tag{94}$$

where

$$w_{jj'} = \begin{cases} W_{jj'} & \text{if } j \neq j' \\ -\sum j'' W_{j''j} & \text{if } j = j' \end{cases}$$

Coupled differential equations like (94) are solved by diagonalizing the matrix $w_{jj'}$, so that

$$\sum_{j'} w_{jj'} u_{j'}^{(i)} = \lambda^{(i)} u_j^{(i)}$$

The general solution to (94) can be written in the form

$$P_j(t) = \sum_i a_i e^{\lambda^{(i)} t} u_j^{(i)} \tag{95}$$

where the coefficients a_i are determined by the initial probability distribution at $t = 0$.

One of the eigenvalues $\lambda^{(0)}$ is zero. The associated eigenvector corresponds to the equilibrium ($t = \infty$) probability distribution,

$$u_j^{(0)} = P_j(+\infty) = Q_j \Big/ \sum_{j'} Q_{j'}$$

All other eigenvalues are negative, and they correspond to transient probabilities with a decay time of $\tau^{(i)} = -1/\lambda^{(i)}$.

The time evolution of occupation probabilities for the extended conformation and the three lowest free energy states of unsolvated tetra-alanine at room temperature $T = 300\,\text{K}$, starting with the extended conformation at $t = 0$ (i.e., $P_{bbbb}(0) = 1$, all other $P_j(0) = 0$), is given in Fig. 36. It takes tetra-alanine about 10^{-10} sec to reach the ground state from the extended conformation.

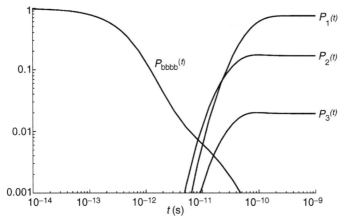

Figure 36.   Time evolution of the extended conformation and the three lowest free energy states of unsolvated tetra-alanine at $T = 300$ K.

3. Pathways

Details of the folding process can be determined by enumerating the pathways from the extended conformation to the ground state. A pathway is defined as a sequence of minima joined together by transition states:

initial state → ts → min → ts → min → ⋯ → min → ts → final state

Pathways between these two states can be enumerated using graph-theory techniques. We construct a graph where each node in the graph represents a minimum and each edge in the graph represents a transition state that connects two minima. The set of all pathways from one minimum to another can be generated by an exhaustive search.

If we conduct this exhaustive search without restriction, we would generate an enormous number of pathways. It is important to restrict the pathways we generate in a sensible manner. We selected pathways based on two criteria: (1) We restrict the length of the pathway (i.e., number of minima) to be less than or equal to some prescribed maximum length, and (2) we also apply a transition rate cutoff, effectively ignoring transitions whose rates fall below the cutoff value. The number of pathways from the extended conformation to the ground state of unsolvated tetra-alanine at $T = 300$ K for various length and rate cutoffs is given in Table XXXI. The total number of minima and transition states involved in such pathways are given in Table XXXII.

These two criteria were applied in an attempt to find the most relevant pathways. Because the faster pathways are likely to be the most important ones, it makes sense to eliminate pathways that involve one or more slow transitions (i.e., transitions which fail to meet the rate cutoff). The length cutoff is chosen for more practical reasons. Even with a transition rate cutoff, the number of pathways increases exponentially with the length cutoff (about a factor of 10 for

TABLE XXXI
Number of Pathways from Extended Conformation to Ground State with Given Length Restriction and Rate Cutoff

Maximum Length	No Rate Cutoff	10^6 Hz	10^7 Hz	10^8 Hz	10^9 Hz	10^{10} Hz	10^{11} Hz
6							
7	4						
8	38						
9	999	421	421	421	421	285	130
10	19963	10836	10828	10828	10733	7443	2099
11	297974	150831	150396	149391	146493	92216	21004
12	4132256	1868821	1859469	1832692	1768736	1002874	221592

TABLE XXXII

Number of Minima/Transition States Involved in Pathways from the Extended Conformation to Ground State with Given Length Restriction and Rate Cutoff

Maximum Length	No Rate Cutoff	10^6 Hz	10^7 Hz	10^8 Hz	10^9 Hz	10^{10} Hz	10^{11} Hz
6							
7	12/14						
8	26/42						
9	236/488	96/183	96/183	96/183	96/183	86/160	65/114
10	886/2339	339/952	339/951	339/951	332/932	287/790	188/466
11	2817/8341	664/2177	663/2173	657/2152	651/2120	526/1696	357/1044
12	6403/21316	943/3405	938/3388	922/3341	913/3291	754/2622	509/1699

each additional minimum). An exhaustive pathway search would be intractable if we did not impose a length cutoff. It is assumed that the fastest pathways are also among the shortest in length. Although we have no proof of this, we will see evidence later on that suggests that we have found the most relevant pathways.

We examined in detail the pathways of length 9 and 10 with a transition rate cutoff of 10^6 Hz. An example pathway of length 9 is given in Fig. 37. For each such pathway, we estimated the amount of time it would take for tetra-alanine to proceed from the extended conformation to the ground state along that particular pathway by solving the Master equation for a reduced system consisting only of the minima and transition states involved in the pathway. The decay time of the longest-lived transient probabilities was used as an estimate of the overall transition time. The fastest transition times were on the order of 5×10^{-11} sec, and most of the 10, 836 pathways we looked at had transition times less than 1×10^{-9} sec. Clearly, there is no single most important pathway: there are many pathways which are all equally important. We also found that the pathways of length 9 tended to be among the fastest of the pathways of length 10 or less, suggesting that shorter pathways tend to be faster.

We also studied the pathways of length 10 or less in terms of changes in the ϕ and ψ angles. Each (ϕ, ψ) pair is classified according to Table XXVIII. In proceeding from the extended conformation to the ground state, each of the four (ϕ, ψ) pairs must proceed from "b" to "a." We observed that this process tends to follow regular patterns.

We make the following general observations regarding the rotation of the ψ angles:

  1. Each ψ angle normally progresses in the sequence $b \rightarrow i \rightarrow a$ or $b \rightarrow j \rightarrow i \rightarrow a$.

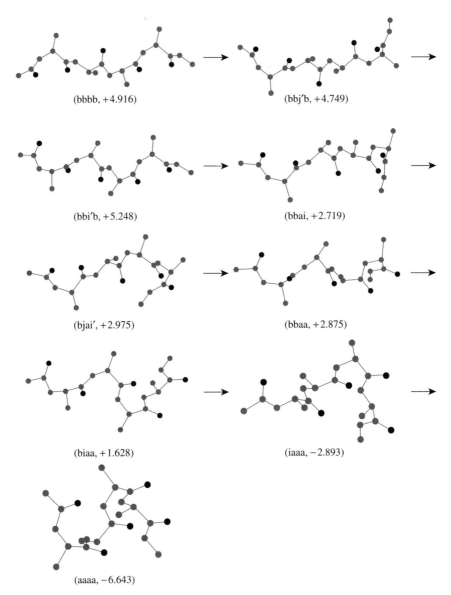

Figure 37. One possible pathway from the extended conformation to the ground state of unsolvated tetra-alanine.

2. No direct b → a transitions are observed,[2] indicating that a rotation of ψ from β-sheet to α-helical values is too large for a single transition.

3. Most pathways of length 10 or less involve at least one transition where more than one ψ angle changes (cooperative motion).

4. A wide variety of cooperative motion is possible, but the two most common types are as follows:

$$
\begin{aligned}
&\text{bi} \rightarrow \text{ia} \quad 36\% \\
&\text{bj} \rightarrow \text{ii} \quad 14\%
\end{aligned}
$$

5. There is a tendency for one-half of the molecule to fold (nearly) completely followed by the other half (e.g., bbbb → bbaa → aaaa).

We can analyze the pathway given in Fig. 37 in terms of these observations. The individual ψ angles proceed as follows:

$$
\begin{aligned}
\psi_1: &\quad b \rightarrow i \rightarrow a \\
\psi_2: &\quad b \rightarrow j \rightarrow b \rightarrow i \rightarrow a \\
\psi_3: &\quad b \rightarrow j \rightarrow i \rightarrow a \\
\psi_4: &\quad b \rightarrow i \rightarrow a
\end{aligned}
$$

Except for a slight backtrack in ψ_2, this pathway is consistent with (1) and (2). This pathway also exhibits three transitions that involve cooperative motion. Two of them are in the form bi → ia, which is the most common form observed. The other cooperative motion, ji → ba (nonadjacent alanines), has also been observed but is not nearly as common as the two forms listed above. Finally, it should be remarked that this pathway does pass through a bbaa minimum. In other words, the right side (the carboxyl terminus) folds completely before the left side (the amino terminus) folds at all. Not all pathways follow this rule strictly, although we have found that tetra-alanine tends to fold its right side most of the way before its left side makes significant progress.

The rotation of the φ angles plays less of a role in the folding process than rotation of ψ angles. φ takes on similar values for α-helical and β-sheet conformations. We found that the very slowest transitions (on the order of 100 Hz or less) tend to involve rotations of the φ angles from inside to outside of the range $180° \leq \phi \leq 330°$ and vice versa. In fact, none of the minima involved

---

[2]This has been checked rigorously for all pathways length 11 or less with a rate cutoff of 10^6 Hz. What we have in fact found is that there are transition states that connect two minima b → a, but either the transition itself is very slow or else the minima are so high in energy that it seems unlikely that a fast pathway (of *any* length) could pass through it. Our conclusion is that b → a is not observed for all but the very slow pathways.

in pathways of any length with a rate cutoff of 10^6 Hz involves ϕ angles outside this range (they would be indicated in our classification scheme by a double-prime). This can be proved rigorously by examination of the rate disconnectivity graph, which we will discuss next.

4. Rate Disconnectivity Graph

We constructed the rate disconnectivity graph for tetra-alanine at $T = 300$ K. It is shown in Fig. 38. The rate disconnectivity graph provides us with the rate-dependent connectivity of the potential energy surface [4,5,130,131]. If we begin at the top of the graph, with a very small rate cutoff, all of the minima fall into one group that is represented by a single node. As we increase the rate cutoff, transitions get eliminated. At some point, a critical transition gets eliminated which disconnects the minima into two groups. This is represented by the node splitting into two at the rate cutoff value. As the rate cutoff is increased further, more and more transitions are eliminated and the graph continues to bifurcate as

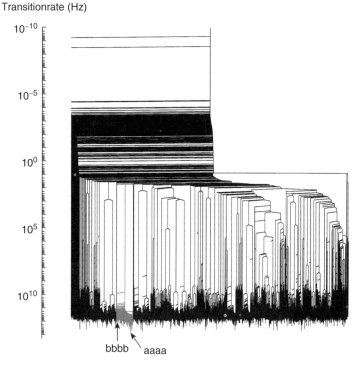

Figure 38. Complete rate disconnectivity graph for unsolvated tetra-alanine at $T = 300$ K. The α-helical ground state and the extended conformation both lie in the highlighted subtree.

the groups of minima further subdivide. At the base of the graph, no transitions remain, and each minimum falls into its own group. The minima can be identified at the base of the graph.

The rate disconnectivity graph for tetra-alanine shown in Fig. 38 covers 23 orders of magnitude in transition rates and contains 62,357 minima.[3] Starting at the top, we see that a relatively small number of minima break away as the rate cutoff is increased to around 10 Hz. Between 10 Hz and 100 Hz, a number of large groups of minima (several thousand minima each) break away from the main branch, indicating a great deal of interesting dynamics occurring on a time scale of about 0.1 sec. Between 10^2 Hz and 10^{10} Hz, relatively little happens. There seems to be two well-separated time scales with characteristic times roughly 0.1 sec and 10^{-10} sec.

The highlighted section of the rate disconnectivity graph contains a total of 3713 minima, including the extended conformation and the α-helical ground state. If we apply a transition rate cutoff anywhere between 10^2 Hz and 10^{10} Hz, we would find that all of the minima in the highlighted region would be connected to one another and disconnected from all of the rest. In other words, it would take about 10^{-10} sec to make transitions between two minima within this group and would take about 10^{-2} sec to make transitions out of this group. This is consistent with our solution of the Master equation (see Fig. 36).

We looked for a distinguishing characteristic of the minima within this group. We found that all 3713 minima in this group satisfy the constraints

$$180° \leq \phi_i \leq 330°$$

for all four ϕ angles. Conversely, we found that all except for one minimum which satisfies these constraints on all four ϕ angles lies within this group. This leads us to the following conclusions:

1. Transitions involving large changes in ϕ (from within $[180°, 330°]$ to outside this range, or vice versa) tend to be very slow, requiring longer than 0.01 sec (sometimes much longer). This is no doubt a result of very high barriers separating these two regions of configuration space.

2. Transitions involving small changes in ϕ (i.e., those that stay within the range $[180°, 330°]$) and arbitrary changes in ψ tend to be much faster, typically on the order of 10^{-10} sec. The folding of tetra-alanine from its extended conformation (bbbb) to the ground state (aaaa) falls into this catagory.

---

[3]The remaining 16 minima are not connected to the main group by any transition states.

5. Time Evolution of Quantities

Another way of obtaining an overall picture of the folding process of tetra-alanine is to study the time-evolution of averages of certain quantities, such as energy, dihedral angles, or distances between specific atoms. If q_j is the value of some quantity at minimum j, then $\langle q \rangle$, the average value of q, and σ_q, the standard deviation, can be calculated as a function of time with the help of the Master equation:

$$\langle q \rangle(t) = \sum_j P_j(t)q_j = \sum_{i,j} a_i e^{\lambda^{(i)}t} u_j^{(i)} q_j$$

$$= \sum_i a_i \left(\sum_j u_j^{(i)} q_j \right) e^{\lambda^{(i)}t} \tag{96}$$

$$\langle q^2 \rangle(t) = \sum_i a_i \left(\sum_j u_j^{(i)} q_j^2 \right) e^{\lambda^{(i)}t} \tag{97}$$

$$\sigma_q(t) = \sqrt{\langle q^2 \rangle(t) - \langle q \rangle^2(t)} \tag{98}$$

Plots of $\langle q \rangle$ and $\langle q \rangle \pm \sigma_q$ as functions of time for $q = E, \phi_1$, and ψ_1 are given in Figs. 39–41.

To obtain the correct time evolution of $\langle q \rangle$ and σ_q, it is necessary to solve the Master equation over all of the minima.[4] We can also calculate the approximate time evolution of $\langle q \rangle$ and σ_q by restricting our attention to only a certain subset of pathways. This is accomplished by restricting the minima and transition states we use to solve the Master equation to those which are visited by the selected pathways.

In Figs. 39–41, we compare the overall time evolution of E, ϕ_1, and ψ_1 with the time evolution obtained by restricting our attention to pathways with various length restrictions. The deviations are rather large for a length cutoff of 10, but are much smaller for a length cutoff of 11 or 12 (the same holds true for the other ψ_i and ϕ_i angles, not shown). It appears that applying a length cutoff of 11 will yield most of the relevant pathways.

We can also determine the effect of various transition rate cutoffs on the time evolution of E, ϕ_i, and ψ_i. In Fig. 42, we compare the overall time evolution of E with that obtained by restricting our attention to pathways with a length cutoff

---

[4]Actually, we only solve the Master equation over the 3713 minima in the highlighted region of the rate disconnectivity graph shown in Fig. 38. This is necessary because solving the Master equation for all 62,373 minima would require diagonalizing a $62,373 \times 62,373$ matrix which does not fit in computer memory. Fortunately, it is also sufficient because the other minima are unreachable during times on the order of 10^{-9} sec.

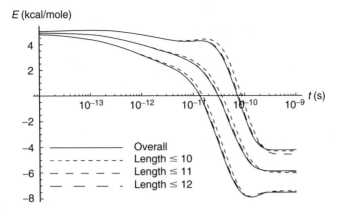

Figure 39. Time evolution of E as a function of time (average $\pm$ one standard deviation), given that the system occupies the extended conformation at $t = 0$ sec. Various pathway length cutoffs are employed.

of 11 and various transition rate cutoffs. We find significant deviation from the overall time evolution only when the transition rate cutoff is increased to 10^{11} Hz (the same holds for ϕ_i and ψ_i, not shown). It appears that the most significant pathways are those of length 11 or less which satisfy a transition rate cutoff of 10^{10} Hz. There are 92,216 such pathways, and they involve only 526 minima and 1696 transition states. This is significantly less than the 62,373 minima and 212,938 transition states that we started with.

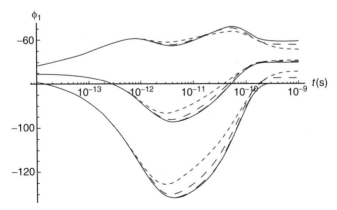

Figure 40. Time evolution of ϕ_1 as a function of time (average $\pm$ one standard deviation), given that the system occupies the extended conformation at $t = 0$ sec. Various pathway length cutoffs are employed.

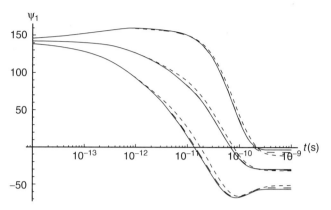

Figure 41. Time evolution of ψ_1 as a function of time (average ± one standard deviation), given that the system occupies the extended conformation at $t = 0$ sec. Various pathway length cutoffs are employed.

6. Reaction Coordinates

It would be useful to characterize the folding process by means of determining a viable reaction coordinate. A reaction coordinate is a quantity that accurately measures the progress from the initial state to the final state. Ideally, it should be monotonic and proceed at a uniform rate along each individual pathway. If we examine the time evolution of E, ϕ_i, and ψ_i (Fig. 39–41), we see that the energy and the ψ angles seem to make reasonable reaction coordinates, but the ϕ angles

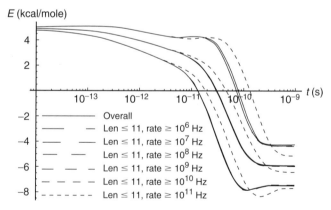

Figure 42. Time evolution of E as a function of time (average ± one standard deviation), given that the system occupies the extended conformation at $t = 0$ sec. A pathway length limit of 11, along with various transition rate cutoffs, are employed.

definitely do not. However, these plots only reveal the average progress of these quantities. What we would really like to know is which, if any, of these quantities proceeds monotonically and uniformly for *each* pathway.

To help answer this question, we developed two "reaction coordinate indicators"—one that measures the monotonicity of the reaction coordinate, and the other that measures the uniformity of the reaction coordinate. For a given pathway of length N

$$\text{min}_1 \rightarrow \text{min}_2 \rightarrow \cdots \rightarrow \text{min}_N$$

a certain quantity q takes on values

$$q_1 \rightarrow q_2 \rightarrow \cdots \rightarrow q_N$$

The two reaction coordinate indicators are d/D and D^2/S, where

$$d = \left| \sum_{i=1}^{N-1} (q_{i+1} - q_i) \right| \qquad \text{(displacement)}$$

$$D = \sum_{i=1}^{N-1} |q_{i+1} - q_i| \qquad \text{(distance)}$$

$$S = (N - 1) \sum_{i=1}^{N-1} |q_{i+1} - q_i|^2 \qquad \text{(squared distance)}$$

d/D measures the monotonicity of q along the given pathway, and D^2/S measures the uniformity of q along the given pathway. Both indicators take the value 1 in the ideal case.

For each of the quantities E, ϕ_i and ψ_i, we tabulated the average value and standard deviation of these two reaction coordinate indicators over the 92,216 relevant pathways in Table XXXIII. As expected, the ϕ angles perform poorly on the monotonicity test (d/D is very small), whereas the energy and the ψ angles perform reasonably well on the monotonicity test. However, none of the quantities do very well on the uniformity test: the average value of D^2/S is around 0.30 for each of the dihedral angles and around 0.48 for the energy. This suggests that changes in a given dihedral angle tend to occur in a small number of big steps, rather than in a large number of small steps. This is consistent with our earlier pathway analysis, where we found that the ψ angles tend to change one or two at a time.

It is clear that progress toward the α-helical ground state should not be measured in terms of a single ψ angle, but should reflect the progress of *all* ψ

TABLE XXXIII
Average and Standard Deviation Values of the Reaction Coordinate Indicators d/D and D^2/S for
Various Quantities Over All Pathways of Length 11 or Less with Transition Rates Exceeding 10^{10} Hz
from the Extended Conformation to the Ground State of Unsolvated Tetra-alanine

Quantity	d/D		D^2/S	
	Average	Standard	Average	Standard
E	0.796	0.099	0.482	0.144
ϕ_1	0.224	0.138	0.291	0.080
ψ_1	0.899	0.120	0.256	0.060
ϕ_2	0.032	0.034	0.304	0.077
ψ_2	0.850	0.100	0.283	0.051
ϕ_3	0.081	0.081	0.332	0.084
ψ_3	0.867	0.129	0.298	0.071
ϕ_4	0.046	0.038	0.302	0.075
ψ_4	0.849	0.132	0.293	0.059
$\sum_i \psi_i$	0.927	0.066	0.749	0.066
$d_{\alpha 1, \alpha 4}$	0.674	0.138	0.355	0.115
d_1	0.762	0.129	0.467	0.098
d_2	0.712	0.111	0.523	0.142
$d_1 + d_2$	0.818	0.103	0.587	0.133

angles. This suggests that we might look at $\sum_i \psi_i$ as a reaction coordinate. The time evolution of $\sum_i \psi_i$ is plotted in Fig. 43, and the average value and standard deviation of the reaction coordinate indicators are given in Table XXXIII. The average value of the reaction coordinate indicators, $d/D = 0.927$ and $D^2/S = 0.749$, both indicate very strongly that $\sum_i \psi_i$ makes a good reaction coordinate. To confirm this, we constructed a scatter plot of D^2/S vs. d/D for each of the 92,216 pathways, shown in Fig. 44. For most of the pathways, the reaction coordinate indicators are both near 1, further suggesting that $\sum_i \psi_i$ makes a good reaction coordinate.

Further insight into the folding process may be gained by looking for a more physically significant reaction coordinate. An α-helix is stabilized by the formation of hydrogen bonds between the i and $i + 3$ residues. Because these residues tend to be farthest apart in the extended conformation, and must be brought close together to form the hydrogen bond, it makes sense to use the hydrogen bonding distance as a reaction coordinate.

We first tried d_{α_1,α_4}, the distance between the first and fourth α-carbons. This distance is indicated in Fig. 45. This distance varies from 9.079 Å in the extended conformation to 4.998 Å in the ground state. The α-helical ground state is not the only conformation with $d_{\alpha_1,\alpha_4} < 5.0$ Å. Of the 526 minima involved in the 92,216 relevant pathways, 26 of them satisfy this inequality.

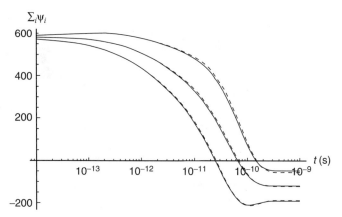

Figure 43. Time evolution of $\sum_i \psi_i$ as a function of time (average $\pm$ one standard deviation), given that the system occupies the extended conformation at $t = 0$ sec. Solid curve shows the overall time evolution, and dotted line shows time evolution with a pathway length limit of 11 and a transition rate cutoff of 10^{10} Hz.

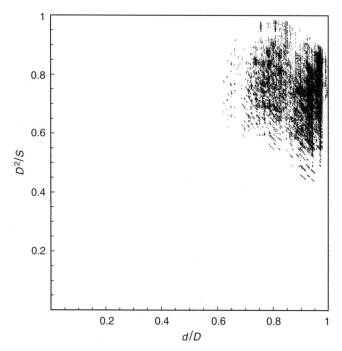

Figure 44. Scatter plot of reaction coordinate indicators for $\sum_i \psi_i$ for each pathway. Only pathways of length 11 or less with all transition rates exceeding 10^{10} Hz are used (92,216 pathways).

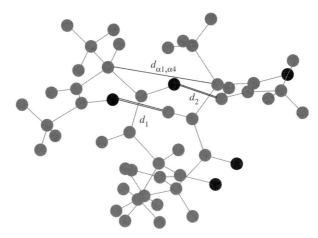

Figure 45. Alpha-helical ground state of unsolvated tetra-alanine, with the hydrogen bonds indicated.

The distance between α-carbons is only a crude measure of hydrogen bonding. A more direct measure is the distance between the nitrogen-bonded hydrogen atom and the oxygen atom that shares it. It turns out there are two candidate hydrogen bonding distances, as indicated in Fig. 45. These distances in the ground state are $d_1 = 1.934$ Å and $d_2 = 1.921$ Å. It turns out that neither distance alone uniquely determines the ground state. Of the 526 relevant minima, 9 of them satisfy $d_1 < 2$ Å and 7 of them satisfy $d_2 < 2$ Å. However, only the ground state satisfies both inequalities. Apparently there are two hydrogen bonds which stabilize the α-helix in tetra-alanine.

We tabulated the average value and standard deviation of the reaction coordinate indicators for d_{α_1,α_4}, d_1, d_2, and $d_1 + d_2$ in Table XXXIII. The motivation of including $d_1 + d_2$ among the distance parameters is similar to that of including $\sum_i \psi_i$. Because there are two hydrogen bonds to form, it makes sense that reaction progress should be measured by *both* hydrogen bond distances. Any of the four distance parameters would make a reasonable reaction coordinate, but $d_1 + d_2$ is clearly the best with $d/D = 0.818$ and $D^2/S = 0.587$. A scatter plot of D^2/S vs. d/D for $d_1 + d_2$ is given in Fig. 46.

7. Solvated Tetra-Alanine

We next studied tetra-alanine in solvation. We used the ECEPP/3 potential energy surface coupled with the volume method for calculating solvation energies using the Reduced Radius Independent Gaussian Sphere (RRIGS) approximation.

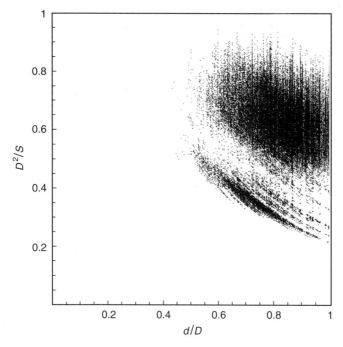

Figure 46. Scatter plot of reaction coordinate indicators for $d_1 + d_2$ for each pathway. Only pathways of length 11 or less with all transition rates exceeding 10^{10} Hz are used (92,216 pathways).

We determined the minima and first-order saddles by applying a brute force eigenmode-following search (Eigenmode III) with a 6^8 grid of start points, just as we did for unsolvated tetra-alanine. The results of this search can be found in Table XXXIV.

Of the 66,228 minima, we found one α-helical conformation, min.1 (aaaa), and one extended conformation, min.874 (bbbb). The potential energy (which includes the solvation energy) and free energy (which includes contributions from the vibrational entropy) of these two states can be found in Table XXXV.

TABLE XXXIV
Eigenmode III Results for Solvated Tetra-alanine

	6^8 Grid
Local minima	66,228
First-order saddles	195,639

TABLE XXXV
Ground State and Extended Conformation of Solvated Tetra-alanine

Minimum	Classification	E (kcal/mol)	F (kcal/mol)
min.1	aaaa	−35.249	−40.741
min.874	bbbb	−30.823	−41.194

The first thing to notice is that, although the α-helical conformation has the lowest potential energy (and hence the lowest free energy at $T = 0$ K), the extended conformation has a lower free energy at room temperature $(T = 300\,\text{K})$ than the ground state. The result of adding solvation energy reduces the energy gap from 11.6 kcal/mol to 4.4 kcal/mol. The entropic term in the free energy is more than enough to overpower this energy gap and reduce the free energy of the extended conformation below that of the α-helical ground state. This has significant implications.

We calculated the free energies of all the minima in order to determine the equilibrium probability distribution (see Section IV.C.2). We found that the several hundred lowest free energy minima have about the same free energy, and that no single minimum has an equilibrium occupation probability which exceeds 0.004. This is in stark contrast with unsolvated tetra-alanine, where the ground state had an equilibrium occupation probability of 0.748, and the lowest three potential energy states accounted for 0.936 of the total equilibrium probability.

As a check, we calculated the transition rate matrix for solvated tetra-alanine at $T = 300$ K, and we also solved the Master equation starting with the extended conformation at $t = 0$ sec. We plotted the time evolution of the occupation probabilities of the 300 lowest free energy states. That plot is given in Fig. 47. The equilibrium probability distribution is achieved in about 10^{-10} sec.

It seems likely that solvated tetra-alanine exhibits liquid-like behavior at $T = 300\,\text{K}$. To be sure, we need to verify that the several hundred minima that share the equilibrium probability distribution do not occupy the same region of configuration space. If that were the case, the potential energy surface would have one deep basin with a rough bottom. The true characteristics of a liquid-like molecule is that it randomly (and quickly) samples widely distinct configurations. By plotting the distribution of minima on four (ϕ, ψ) plots (not shown), we reached the conclusion that the minima that share the equilibrium probability distribution do occupy distinct regions of configuration space.

If solvated tetra-alanine is to be liquid-like at $T = 300$ K, then there must be a phase transition. This should show up as a peak in the heat capacity versus temperature plot. The heat capacity can be calculated by calculating energy

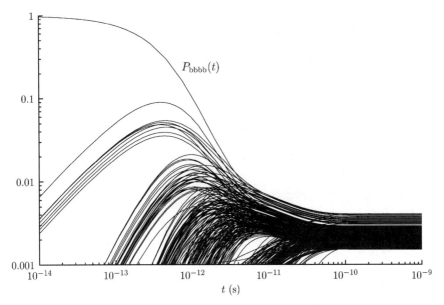

Figure 47. Time evolution of the extended conformation and the 300 lowest free energy states of solvated tetra-alanine at $T = 300$ K. No single state has an equilibrium probability that exceeds 0.004.

fluctuations at equilibrium

$$C_v = \frac{d}{dT}\langle E \rangle_{eq} = \frac{\langle E^2 \rangle_{eq} - \langle E \rangle_{eq}^2}{kT^2}$$

where equilibrium averages may be calculated from free energies

$$\langle q \rangle_{eq} = \frac{\sum_i q_i e^{-F_i/kT}}{\sum_i e^{-F_i/kT}}$$

We calculated C_v as a function of T for temperatures ranging from (just above) 0 K to 1000 K for both solvated and unsolvated tetra-alanine. The plots are given in Figs. 48 and 49. The transition temperatures are given by

$$T^{\text{solv}}_{\text{sol-liq}} = 130 \text{ K} \qquad T^{\text{unsolv}}_{\text{sol-liq}} = 395 \text{ K}$$

The lower transition temperature for solvated tetra-alanine can be traced back to the reduction in the energy gap between the α-helical ground-state conformation and the other higher-energy states, including the extended conformation, and

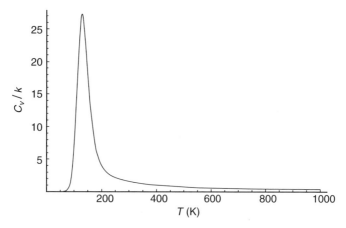

Figure 48. Heat capacity as a function of temperature for solvated tetra-alanine.

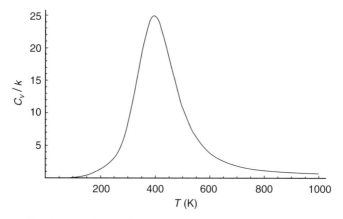

Figure 49. Heat capacity as a function of temperature for unsolvated tetra-alanine.

indeed does explain the appearance of liquid-like behavior for solvated tetra-alanine (but not for unsolvated tetra-alanine) at $T = 300$ K.

D. Overall Framework and Implementation

In this section we present the methods involved in the dynamical study of a particular peptide sequence, and we discuss the implementation details of those methods. The overall framework is summarized in Fig. 50. The dynamical study of a particular potential energy surface divides into two major parts: (1) the search for stationary points (minima and first- and higher-order transition states) and (2) the dynamics analysis.

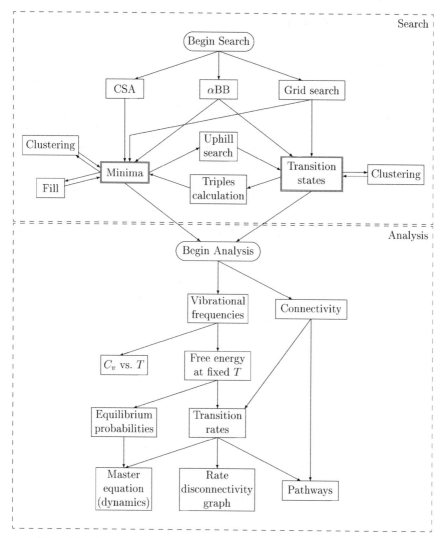

Figure 50. Overall framework for the dynamical study of a given peptide sequence.

The stationary point search generally proceeds as follows. First, an initial sample of minima and or transition states is generated using one of the global optimization methods (αBB, CSA, or grid search). Additional stationary points can be generated, if needed, by performing uphill searches from minima to saddle points, or downhill searches from saddle points to minima, or by interpolating widely separated minima to locate new minima in between ("fill"). Similar minima and transition states may be combined by clustering, if desired.

Once an adequate sample of minima and transition states has been found, we begin the dynamical analysis. Connectivity between minima and transition states has already been determined by the triples calculation (i.e., downhill searches). The free energy of each stationary point is calculated (using the vibrational frequencies), and from that the transition rates may be calculated. Then we can construct a C_v vs. T plot, determine equilibrium probability distributions, solve the Master equation, construct the rate disconnectivity graph, and perform a full pathway analysis.

1. Local Stationary Point Search Methods

Eigenmode-Following Search. Eigenmode-following search algorithms are essentially sophisticated variations of the Newton–Raphson method applied to the equations $\partial V/\partial x_i = 0$. We employ the version introduced by Tsai and Jordan [145]. At each iteration, the Hessian may be updated by direct calculation, or by BFGS (minimum search only) or Powell updating (with occasional direct calculation) [159–161]. We generally used the Powell updating for the uphill searches and a full Hessian calculation for the downhill searches, so as to ensure that the correct connectivity is determined.

The Newton–Raphson step is given by

$$\Delta \mathbf{x} = -H^{-1}\mathbf{g} = -\sum_i \frac{g_i}{b_i}\mathbf{e}_i$$

where $\mathbf{g}$ and H are the gradient and Hessian, respectively, b_i and $\mathbf{e}_i$ are eigenvalues and eigenvectors of H, respectively, and g_i is the component of the gradient in the direction of $\mathbf{e}_i$. The Newton–Raphson algorithm tends to locate stationary points that have the same signature (i.e., number of negative eigenvalues) as the Hessian matrix at the starting point. More specifically, potential energy tends to be minimized along modes for which $b_i > 0$, and it tends to be maximized along modes for which $b_i < 0$. Eigenmode-following algorithms circumvent this limitation by "shifting" some of the eigenvalues to change their sign, so that the "eigenvalues" used to construct the step have the desired signature. Thus, if a minimum is sought, then all eigenvalues are rendered positive by shifting the negative b_i's to positive values. If a first-order saddle is desired, then eigenvalues are shifted as needed so that one specific eigenvalue is negative. If the Hessian already has the required signature, no shifting takes place—the search essentially becomes a Newton–Raphson search in the vicinity of the saddle point. Eigenmode-following searches, when they converge, virtually always converge to a saddle point of the correct saddle order.

When searching for a first-order saddle from a starting point in the vicinity of a minimum, there is some question as to which eigenvalue should be shifted to a negative value (i.e., which eigenmode to follow "uphill"). There are two possible answers:

1. At each iteration, follow the mode with the smallest eigenvalue.
2. Choose a specific mode at the starting point, and continue to follow that "same" mode as each step is taken. Eigenmodes at each subsequent step are identified with eigenmodes at previous steps by maximum overlap.

In the first case, it is automatically the case that at any point where the Hessian has the correct signature, no eigenvalue shifting takes place at all, and the Newton–Raphson step is taken. In the second case, a specific mode is selected, which often does not start out as the lowest eigenvalue mode. However, as the selected mode is driven uphill, its eigenvalue decreases, eventually causing that mode to overtake the other modes in becoming the lowest eigenvalue mode. Eventually the eigenvalue is driven to a negative value, after which the first-order saddle will be found.

SUMSL.   The SUMSL algorithm [162], which is made available as part of ECEPP/3 [38], is designed to find local minima in the vicinity of a starting point. It employs the BFGS updating method for the Hessian. It is specifically designed for minimum searches and, as such, is generally much more efficient than the eigenmode-following algorithm.

2.   Methods for Finding Minima and First-Order and Higher-Order Transition States

αBB Stationary points of all orders are generated by solving the stationary conditions

$$\frac{\partial V}{\partial x_i} = 0 \qquad i = 1, \dots, N_x$$

using the αBB method described in Sections II.B and IV.B. This algorithm offers a theoretical guarantee of enclosing all solutions within the starting region in a finite amount of time.

CSA.   The Conformational Space Annealing (CSA) algorithm attempts to reach the global minimum (free) energy conformation by a combination of genetic, annealing, and buildup algorithms [115,153,154]. The user provides an initial bank of minima (usually by locally minimizing randomly selected points). Seed points are selected from the bank and modified according to prespecified rules. The modified points are then minimized by local search, and then considered for introduction into the bank, possibly replacing a point which is already there. If the candidate point falls within a certain "cutoff distance" from any other point in the bank, the candidate point and the bank point closest to it are compared. Otherwise, the candidate point is considered to be in its own "class," and it is compared with all other points in the bank. In either case, the highest (free) energy point is discarded. The "cutoff distance" is initialized to one-half the average distance between points in the initial bank, and it is annealed down by a fixed factor every iteration.

Termination conditions include one or more of the following:

1. Iteration count limit.
2. Round limit. Each round ends after every point in the current bank has been used as a seed point.
3. (Free) energy lower limit.
4. Update counter limit. The "update counter" is incremented whenever the fraction of candidate points which actually make it into the bank is sufficiently small for a given number of minimizations.
5. Stop file. The user can stop the algorithm by creating a special file whose existence is checked each iteration.

Virtually all of the effort is spent performing the local minimizations of the modified seed points. The parallel version of this algorithm divides the modified seed points among all of the processes (including the master process) to be minimized in parallel. The master process handles the rest of the algorithm.

Grid Search. A sampling of stationary points of a specified order (first-order saddle or minima, generally) is found by initiating an eigenmode-following search from each point on a specified grid to a saddle point of the specified order. After the searches are completed, duplicate points are thrown out. This algorithm requires one search for each grid point, and thus the time requirements depend exponentially on the number of variables for which alternative values are provided for the grid. It is therefore unsuited for large problems, but yields good results for small problems (e.g., tetra-alanine, discussed in Section IV.C).

The parallel version of this algorithm divides the gridpoints among all of the processes (including the master process), which then perform the searches. The results are sent back to the master process.

Uphill Search. First-order saddle points are found by performing eigenmode-following searches "uphill" from each minimum. For every minimum and every choice of eigenmode, an initial step is taken along that mode (in each of the two possible directions), followed by an eigenmode-following search along that mode to a first-order saddle. A total of $2N$ searches are required for each minimum, where N is the number of eigenmodes. One may alternatively restrict the number of modes followed for each minimum. The resulting saddles are collected and duplicates are removed.

The parallel version of this algorithm divides the minima among all of the processes (including the master process), which then perform all of the required searches. The results are sent back to the master process.

Triples Calculation. The connection between first-order saddles and the minima they connect are found by performing a minimization on each side of the saddle point. An initial step from the saddle point is taken in each of the two directions along the eigenmode corresponding to the negative eigenvalue, each followed by a minimization. Minima found this way are compared with minima that have been found previously, and duplicates are discarded in favor of the previously found minima. This algorithm may also be used to locate previously unknown minima by downhill search from a saddle point, in which case the new minima are retained.

The parallel version of this algorithm divides the saddle points among all of the processes, which then perform the necessary minimizations. As the saddles and minima are sent back to the master process, duplicate minima are discarded in favor of previously determined minima.

Fill. "Fill" refers to the act of filling in a "scaffolding" of minima (such as might be obtained by a CSA run) by searching for additional minima between pairs of minima found in the initial set. The reason why this may be necessary is because the minima generated by the CSA algorithm are often too far apart for connections between them to develop after a single uphill/downhill search (this is practically by design of the CSA algorithm, which spreads itself thin so as to sample a large portion of the conformational space).

A "distance cutoff" and a "coordination number" are provided, along with an initial set of minima. Ideally, this algorithm will first cluster the points according to the distance cutoff (i.e., split the points into equivalence classes, where two points are equivalent if they can be connected to one another by a "path" involving points which are within the cutoff distance from each other). Then each cluster will be paired with the N clusters closest to it in distance, where N is the coordination number. For every pair of clusters generated this way, additional points will be added along the line joining the two clusters (more specifically, along the line joining the two representative points in the two clusters which are the least distance apart). The points will be uniformly spaced, and the number of points chosen is the least number which results in each point being within the cutoff distance from its nearest neighbor. The new points can then be used as starting points for minimization.

For practical reasons, the algorithm actually proceeds as follows. Every point is considered in turn. Distances from that point to every other point are first determined and then sorted. Connections between this point and all points which are within the cutoff distance are noted, so that equivalence classes may be determined. Then the pairs consisting of this point and each of the next N points (N is the coordination number), along with their mutual distances, are added to a "pairs" list. After all points have been considered and the equivalence classes are determined (class-wise), duplicate pairs are discarded from the "pairs" list, and then the points are generated as described above.

The most CPU-intensive part of the algorithm is the generation of a list of distances between a given point and every other point, along with the sorting of that list. In the parallel version of this algorithm, the master process sends the set of points to each slave process so that they will know what to do. While the master process is carrying out the remainder of the algorithm, each slave process calculates the distances between a given point and every other point and then sorts the list. As each point is considered in turn, the master process cycles through the slave processes, receiving the needed distances from each one.

Clustering. The number of minima and transition states can be reduced by "clustering" them—that is, by identifying points that lie within a specified distance of one another with a single point. The first point in the set of points to be clustered is selected as a cluster center and compared with every other point. Points within a certain cutoff distance from the selected cluster center are identified as belonging to that cluster and taken out of circulation. The next point in the set that is not yet part of any cluster is selected as the next cluster center, and it is compared with all other points not yet part of any cluster. The algorithm continues this way until all points have been assigned to a cluster.

Note that the clusters generated by this algorithm have the property that the cluster centers used to generate them appear earlier than all of the other points in the cluster. Thus, by first sorting the set in increasing order of potential energy, we can guarantee that each cluster will be represented by its lowest-energy member and, in particular, that the global minimum energy point will be among the cluster centers.

Minima should be clustered first using the algorithm as described above. The connectivities between the transition states and the minima they connect should then be redefined so that transition states connect the cluster centers associated with the minima they actually connect. Then the transition states can be clustered using the algorithm as described above with one additional caveat: One transition state cannot be identified as belonging to a cluster centered by another transition state unless they connect the same two minima (clusters).

The most CPU intensive part of the algorithm is the calculation of distances between selected cluster centers and all other nonclustered points. The parallel version of this algorithm runs as follows. The points are first sorted, and then they are shipped from the master process to each slave process. The master process sends the first cluster center to one of the slave processes which begins comparing that point to all of the remaining points in the set. As cluster matches occur, results are reported back to the master process that records the clustering information. The master process continues sending cluster centers to available slave processes and awaits reports of clustering until all points have been clustered.

The situation is complicated by the fact that the master process cannot send a new cluster center to another slave process until it is established that the potential cluster center does not belong to a cluster defined by a previous cluster center. As long as each slave process performs the comparisons in order, the master process will be able to deduce that the next unclustered point should be regarded as a new cluster center as soon as all active slave processes have reported progress beyond that point. To facilitate this process, each slave process reports its progress back to the master process at well-defined intervals,[5] in addition to those instances where a cluster match is found.

3. Methods for Analyzing the Potential Energy Surface

Vibrational Frequencies Calculation. The vibrational frequencies are determined by solving the generalized eigenvalue problem

$$(H - (2\pi v)^2 I)x = 0$$

where H is the Hessian and I is the generalized inertia tensor, defined so that the kinetic energy of the system is given by

$$K = \frac{1}{2} \sum_{i,j} \frac{dx_i}{dt} I_{ij} \frac{dx_j}{dt}$$

The inertia tensor is calculated by first calculating dr_j/dx_i for $j = 1, \ldots, 3N_a$ and $i = 1, \ldots, N_x$ by finite differencing and then using the following formula:

$$I_{ii'} = \sum_{j=1}^{3N_a} m_j \frac{dr_j}{dx_i} \frac{dr_j}{dx_{i'}}$$

where m_j is the mass of the $j/3$-th atom.

This makes use of the Cartesian coordinate functions $\mathbf{r}(\mathbf{x})$. The formulae above depend on the Cartesian coordinates being physically correct.

---

[5]A geometric sequence is used so as to generate a number of early reports without generated an enormous number of total reports. Thus, reports are sent back after comparing the cluster center to the next $1, 5, 5^2, 5^3, \ldots$ points.

Unfortunately, most methods of generating Cartesian coordinates from generalized coordinates (in our case, dihedral angles) involve fixing the positions and orientations of specific atoms, which leads to the introduction of unphysical forces and torques being applied to the molecule. We eliminate these unphysical forces by augmenting the set of generalized coordinates to include overall translation and rotation coordinates, calculating the vibrational frequencies using the above methods, and then discarding the six zero-mode frequencies (which must exist). The resulting vibrational frequencies are physically correct.

Vibrational frequencies can be computed at the end of an eigenmode-following search at little cost, because the Hessian has already been generated. Alternatively, the vibrational frequencies can be calculated all at once after the minima and saddles have all been found. In the latter case, the calculation can be run in parallel by distributing the work to each process, having them calculate the frequencies, and then having them pass the results back to the master process.

Free Energy Calculation. The free energy for a given stationary point is defined as follows:

$$F = E - TS_{\text{vib}}$$

The vibrational entropy is calculated from the vibrational frequencies by employing the Classical Harmonic Oscillator approximation

$$S_{\text{vib}} = -k \ln \prod_i \frac{h\nu_i}{kT}$$

where the product is taken over all vibrational frequencies. For saddle points of order 1 or higher, the negative eigenvalue modes are not counted as "vibrational modes."

Other methods of calculating the vibrational entropy exist, but are not currently implemented. Perhaps the simplest is the Quantum Harmonic Oscillator approximation:

$$S_{\text{vib}} = -k \ln \prod_i 2 \sinh \frac{h\nu_i}{2kT}$$

Anharmonic methods exist in the literature [163,164].

Equilibrium Probabilities. Equilibrium probabilities are calculated from the contribution to the partition function from each minimum, which can be expressed in terms of its free energy:

$$P_i = \frac{e^{-F_i/kT}}{\sum_j e^{-F_j/kT}}$$

The minimum free energy over the entire system is first subtracted off in order to prevent overflow/underflow problems that could arise from modest nonzero free energies (positive or negative).

Average values (as well as standard deviations) of any quantity can now be computed at equilibrium:

$$\langle q \rangle = \sum_i q_i P_i$$

$$\sigma_q = (\langle q^2 \rangle - \langle q \rangle^2)^{1/2}$$

Temperature derivatives are also possible:

$$\frac{d\langle q \rangle}{dT} = \frac{\langle qE \rangle - \langle q \rangle \langle E \rangle}{kT^2}$$

assuming that q_i does not depend explicitly on temperature. In particular, the specific heat $C_v = d\langle E \rangle / dT$ can be calculated.

Transition Rates.   Transition rates are computed by Rice–Ramsperger–Kassel–Marcus (RRKM) theory. Each transition state is associated with two rates:

$$W_{i \to ts \to j} = \frac{kT}{h} e^{-(F_{ts} - F_i)/kT}$$

$$W_{j \to ts \to i} = \frac{kT}{h} e^{-(F_{ts} - F_j)/kT}$$

These rates are collected together in a (sparse) matrix:

$$W_{ij} = \sum_{ts} W_{j \to ts \to i}$$

Time-Dependent Probabilities (Master Equation).   The time development of occupation probabilities can be determined by solving the Master equation:

$$\frac{dP_i}{dt} = \sum_j W_{ij} P_j - \left(\sum_j W_{ji} \right) P_i = \sum_j w_{ij} P_j$$

where

$$w_{ij} = \begin{cases} W_{ij} & (\text{if } i \neq j) \\ -\sum_{i'} W_{i'i} & (\text{if } i = j) \end{cases}$$

Solving the Master equation involves determining the eigenvalues and eigenvectors of the (nonsymmetric, but easily symmetrizable) matrix w:

$$\sum_j w_{ij} u_j^{(k)} = \lambda^{(k)} u_i^{(k)}$$

The occupation probabilities as a function of time can be computed (and, e.g., plotted):

$$P_j(t) = \sum_k a_k e^{\lambda^{(k)} t} u_j^{(k)}$$

where the coefficients a_k are determined from the initial conditions $P_j(0)$. The time constants are determined from the eigenvalues

$$\tau_k = -1/\lambda^{(k)}$$

One of the eigenvalues is zero, which corresponds to the equilibrium probability distribution ($\tau = \infty$). The remaining eigenvalues will be negative.

Average values (as well as standard deviations) of any quantity can now be computed as a function of time (and, e.g., plotted):

$$\langle q \rangle(t) = \sum_j q_j P_j(t) = \sum_k a_k \left(\sum_j q_j u_j^{(k)} \right) e^{\lambda^{(k)} t}$$

$$\sigma_q(t) = (\langle q^2 \rangle(t) - \langle q \rangle(t)^2)^{1/2}$$

Solving the Master equation requires the diagonalization of a matrix whose size is the number of minima in the system. This is an extraordinarily expensive operation and may be prohibitive in both space and time resources required. A 4000×4000 matrix requires 128 megabytes of storage and generally requires about a day of CPU time to diagonalize. There is no parallel algorithm available for this operation.

Pathways. Each transition state connects two minima on the potential energy surface. A pathway between two minima is defined as a series of such connections:

initial state $\rightarrow$ ts $\rightarrow$ min $\rightarrow$ ts $\rightarrow \cdots \rightarrow$ ts $\rightarrow$ min $\rightarrow$ ts $\rightarrow$ final state

The set of all (nonlooping) pathways from one minimum to another can be found by an exhaustive search. We begin at the initial state and move to each minimum that is connected to the initial state. For each such minimum, we recursively explore all minima connected to that minimum, taking care not to visit a given minimum more than once along the same pathway. When the final state is reached, the pathway is reported. When all possible routes have been explored, the algorithm terminates.

For any reasonably sized system of minima and transition states, the number of possible nonlooping pathways between any two minima is likely to be prohibitively large. There are several criteria that can be applied to reduce the number of pathways:

1. Monotonicity in any specified quantity (such as energy or free energy). Transitions are ignored if they violate the proposed monotonicity.
2. Maximum length (i.e., maximum number of minima, including the initial and final state, visited along the pathway).
3. Minimum transition rate. Transitions are ignored if they are slower than this cutoff rate.

The following information is available during a pathway calculation:

1. The set of minima and/or transition states visited along the way by at least one of the pathways.
2. Transition rates for each transition taken along a given pathway.
3. An overall "transition time" for a given pathway. This is determined by (a) solving the Master equation over the minima and transition states involved in that one pathway alone and (b) using the lifetime of the longest-lived transient probability eigenvector.
4. Values of any number of quantities for each minimum visited along a given pathway.
5. The two reaction coordinate indicator values associated with any number of quantities along each given pathway (explained below).
6. The average value and standard deviation of any number of quantities over all pathways, at a fixed position along the pathways.
7. The average value and standard deviation of the two reaction coordinate indicators over all the pathways (explained below).

For a given pathway

$$\text{min}_1 \rightarrow \text{min}_2 \rightarrow \cdots \rightarrow \text{min}_N$$

a certain quantity q takes on values

$$q_1 \rightarrow q_2 \rightarrow \cdots \rightarrow q_N$$

To help determine if q would make a good reaction coordinate, we developed two "reaction coordinate indicators". They are d/D (monotonicity) and D^2/S

(uniformity), where

$$d = \left| \sum_{i=1}^{N-1} (q_{i+1} - q_i) \right|$$

$$D = \sum_{i=1}^{N-1} |q_{i+1} - q_i|$$

$$S = (N - 1) \sum_{i=1}^{N-1} (q_{i+1} - q_i)^2$$

An ideal reaction coordinate varies both monotonically (same direction) and uniformly (in equal steps) from its initial value to its final value. Both reaction coordinate indicators take on the value of 1 in this ideal case. Values less than 1 indicate nonideality.

Less detailed information about the connectivity of the minima is also available. The level of connection between two minima is defined as the minimum-length pathway that connects them. The level of connection between a given minimum and all other minima can be generated iteratively as follows. First, start off by marking the given minimum as level 1 with all other minima marked (temporarily) as unreachable. For each level n, starting with $n = 1$, we follow each minimum marked as level n to all the minima they are connected to. For each such connected minimum, if it is yet to be marked as reachable, it is marked as level $n + 1$ (if it is marked already, then a shorter pathway has already reached it). We continue on with level $n + 1$, stopping whenever no additional minima are marked for a given level.

This procedure may be used to determine the connection component which contains a given minimum (i.e., the set of minima connected to the given minimum by *any* length pathway). By iteratively applying this procedure, the minima can be divided into connection components.

It should be noted that pathway traversal can be substantially optimized when a length restriction is given. First, the level of connection between the final state and all other minima is determined. Then, for every transition considered during the pathway search, it is determined whether or not the final state could possibly be reached in the proper number of steps. If it is not possible according to the precalculated level of connection, the transition is avoided.

Rate Disconnectivity Graph. Minima can be classified into connection components. If a transition rate cutoff is applied, transition states may be eliminated if the transitions they represent occur too slowly. In this case, the number of connection components may increase. The rate-dependent connectivity information can be summarized by drawing a rate disconnectivity graph. One starts off at the top of the graph with a low rate cutoff, in which case the minima are separated into their connection components. As the rate cutoff is increased, transition states get eliminated from consideration. At some critical value of the transition rate cutoff, a critical transition state gets eliminated, causing one of the connection components to divide in two. As the rate cutoff is increased further, more and more transition states are eliminated from consideration, causing further bifurcation of connection components. At the highest rate cutoffs, no transition states remain, and all minima occupy their own connection component. Minima can be identified at the base of the graph.

The rate disconnectivity graph is built from the bottom up. Each minimum starts off by occupying the leaf node of its own tree. Connectivities between pairs of minima are sorted in decreasing order of transition rate, so that the highest transition rates will be considered first.[6] For each such pair of minima, we locate the subtrees generated so far which contain each of the two minima. If the two minima already belong to the same subtree, nothing happens. If the two minima belong to different subtrees, those two subtrees are joined by a bifurcation node, which is labeled with the transition rate. The rate disconnectivity graph will be completed after each transition has been considered, at which point there will be one tree for each connection component.

Once the rate disconnectivity graph is constructed, one can walk along the nodes in the tree, print a subtree in text format, or write Mathematica code which plots the rate disconnectivity graph in graphical form.

E. Perspectives and Future Work

In this section, we discuss our ongoing efforts to elucidate the folding mechanism of β-hairpin and β-sheet structures by studying one of the short peptides that has been recently discovered to form such structures in the native state.

Our first task centered on the selection of an appropriate peptide sequence and a potential energy surface. Our initial efforts were focused on a 12-residue designed sequence using the ECEPP/3 potential energy surface with an additional solvation term using the volume method. Unfortunately, we were unable to locate a low-energy hairpin structure and, upon further investigation, discovered that the lowest-energy state of this system was an α-helix. It seems that ECEPP/3 is unable to predict the β-hairpin structure of this peptide sequence. So we checked other peptide sequences as well as other potential energy surfaces to see if we could predict a β-hairpin fold. We eventually found success with the second β-hairpin segment of Protein G (residues 41–56) using

---

[6]The transition rate associated with a given pair of connected minima is by default the maximum of the two transition rates associated with that connection. The minimum transition rate can be selected instead.

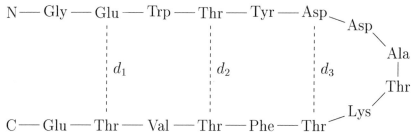

Figure 51. A schematic of Protein G (41–56) in its hairpin conformation. The dotted lines indicate hydrogen bonds, and the distances d_1, d_2, and d_3 refer to the distances between the C_α atoms.

the Effective Energy Function (EEF1) [165], which is the CHARMM potential plus a solvation term based on a Gaussian solvent exclusion model.

Segment 41–56 of Protein G is a 16-residue peptide that has been determined experimentally to fold into a β-hairpin in aqueous solution [135]. A schematic of this hairpin structure is depicted in Fig. 51. The hairpin structure is stabilized by the formation of three pairs of hydrogen bonds as indicated in Fig. 51. The corresponding distances between the C_α atoms are designated d_1, d_2, and d_3 and will play an important role in our analysis of this molecule.

The potential energy surface we employ for this peptide in aqueous solution is split into two terms:

$$E = E_{\text{pep}} + E_{\text{solv}}$$

where E_{pep} includes the peptide intramolecular interactions, and E_{solv} includes the peptide–solvent and solvent–solvent interactions.

The intramolecular interaction term, E_{pep}, is modeled with the CHARMM22 potential energy function, an all-atom potential that takes the general form [32,166]

$$E_{\text{pep}} = \sum_{\text{bonds}} K_b (b - b_0)^2 + \sum_{\text{Urey–Bradley}} K_{UB}(S - S_0)^2 + \sum_{\text{angles}} K_\theta(\theta - \theta_0)^2$$
$$+ \sum_{\text{dihedrals}} K_\phi(1 + \cos(n\phi - \delta)) + \sum_{\substack{\text{improper} \\ \text{dihedrals}}} K_\omega(\omega - \omega_0)^2$$
$$+ \sum_{\text{nonbond}} \left\{ \epsilon_{ij}[(R_{ij}^{\text{min}}/r_{ij})^{12} - 2(R_{ij}^{\text{min}}/r_{ij})^6] + (q_i q_j)/r_{ij} \right\} \tag{99}$$

The quantities b, S, θ, ϕ, ω are the bond length, Urey–Bradley distance, bond angle, dihedral angle, and improper dihedral angle, respectively, with the zero subscript representing equilibrium values. The parameters have been determined empirically and are given in Ref. [166].

The solvation term, E_{solv}, is based on the Gaussian solvent exclusion model, which takes the general form [165]

$$E_{solv} = \sum_i \Delta G_i^{solv}$$

$$= \sum_i \left\{ \Delta G_i^{ref} - \sum_{j \neq i} \frac{\alpha_i e^{-((r_{ij} - R_i)/\lambda_i)^2}}{4\pi r_{ij}^2} V_i \right\} \quad (100)$$

where r_{ij} is the distance between atoms i and j. The parameters ΔG^{ref}, α_i, R_i, λ_i, and V_i can be found in Ref. [165]. In addition, partial charges for several atoms in charged residues have been modified, effectively neutralizing the side chains in the CHARMM22 potential.

To simplify calculations, we fix the bond lengths and bond angles to their equilibrium values according to the CHARMM22 parameters, allowing only the ϕ, ψ, ω, and χ dihedral angles to vary. This reduces the number of degrees of freedom from $3N_a - 6 = 735$ to $N_h = 88$. Energy values, as well as the Cartesian gradient and Hessian matrix, were computed by the TINKER software package [167]. The Cartesian gradients and Hessians were converted to torsional gradients and Hessians by methods developed in our computer lab. All in all, one Hessian evaluation requires approximately 0.50 sec of CPU time on a 600-MHz pentium machine running linux, where the bulk of the calculations were performed.

In order to study the folding pathways of Protein G (41–56), we need to generate an adequate sample of stationary points of the potential energy surface. Not only do we need to generate conformations that resemble the hairpin native state, as well as extended conformations, but we also need to find conformations that lie along the low-lying pathways connecting these two regions of conformation space. Thus, we need to find low-lying conformations, as well as transition states, over a large region of conformation space.

The approach we have chosen to follow is to first generate an initial sampling of minima, forming a "scaffolding," and then building upon that scaffolding by performing uphill and downhill searches using an eigenmode-following algorithm. Before carrying out this search, however, we first want to identify the global minimum energy conformation on the potential energy surface, which will serve as the native structure. The results of the global minimum search are given in Fig. 52. The study of the pathways for the transitions from extended to β-sheet conformations is currently in progress.

V. PROTEIN–PROTEIN INTERACTIONS

Understanding protein–protein interactions, also known as peptide docking, is critically important for rational protein engineering and pharmaceutical design.

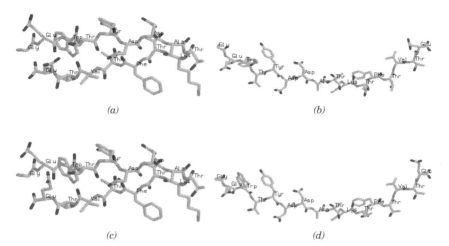

Figure 52. (a) Overall minimum energy conformation ($E = 653.020, F = 602.768$). (b) Overall minimum free energy conformation ($E = 654.139, F = 602.647$). (c) Minimum energy extended conformation ($E = 673.439, F = 605.342$). (d) Minimum free energy extended conformation ($E = 675.854, F = 604.347$). Energy and free energy values are expressed in kcal/mol. Free energy values are at $T = 300$ K.

Peptide docking is the binding of one protein to another protein, and such binding is essential to processes ranging from chemotherapy to the communications between cells. Advances in understanding and predicting how solvation, electrostatics, and other forces affect the strength, specificity, and kinetics of peptide docking interactions is vital for discovering new drugs, for developing tools for characterizing and treating disease, and for designing sensors and other molecular recognition devices. No comprehensive peptide docking prediction method yet exists. In this section we review the state of the art in peptide docking prediction methods.

A. Background

Predicting peptide docking and protein–protein interactions computationally involves predicting the shapes, characteristics, and interactions of "target" molecules and the "docking" ligand molecules that bind to them. One part of the prediction challenge involves determining the conformation or structure of the binding sites in the target molecule. The other part of the prediction challenge involves determining the binding affinity of different docking molecules for the target molecule. This includes identifying a set of equilibrium structures for complexes between different docking molecules and the target molecule and then quantifying and comparing, or "scoring," the binding affinity of docking molecule structures. The following two sections discuss methods used for binding site structure prediction and binding affinity prediction.

1. Prediction of Binding Site Structure

The identification of binding site conformations in target molecules usually requires experimental structure determination of the binding site. One class of protiens that has received particular attention is the class of proteins derived from the major histocompatibility complex (MHC), a set of genes critical in the immune response [168]. Crystallographic studies have been performed for the two major classes of MHC molecules, class I [169,170] and Class II [171]. Such crystallographic information is invaluable. For instance, it can define rigid binding sites for docking molecules and thus greatly reduce the conformational space being searched in computational searches for structures of target/docking molecule complexes.

The determination of high-quality models of protein structure for which no experimentally determined coordinates exist has received considerable attention in the literature. A commonly used approach is based on homology modeling, in which a model for a target protein is generated using the known structure of a homologous protein. Typically, a backbone model first is constructed for the structurally conserved regions, and then loops and side chains are added [172,173]. For the prediction of side-chain conformations, many approaches based on homology modeling are available. These approaches differ from each other in (a) the rotamer libraries used, (b) the energy function chosen, and (c) the search strategy employed. In sampling conformational space through rotamer libraries, many different approaches have been used, including backbone-independent rotamer libraries [174] or rotamer sets that incorporate backbone–side-chain interactions [175]. Also employed are extended rotamer libraries derived from cluster analyses of experimentally determined databases [176], as well as augmented libraries that use discrete values around observed χ angle values $\pm 10°$ [177]. Regarding the energy function used, simplistic local interactions typically are limited to van der Waals or hard-sphere energies [178,179]. Finally, the employed search strategies are mainly heuristic methods involving Monte Carlo techniques [180], genetic algorithms [181], neural networks [182], mean-field optimization [179], and combinatorial searches [175].

Recently, a novel decomposition-based approach has been proposed for predicting binding site structures in the MHC II HLA-DR1 protein [183]. In this approach, existing MHC II crystal structures are used to predict the binding site conformations of other MHC II molecules. The approach uses the detailed potential energy force field ECEPP/3 and an area-based solvation method. A global optimization search, based on the αBB algorithm, is used to identify the global minimum energy conformation of the binding sites. As discussed further in later sections, the predicted binding sites agree with available crystallographic data with only small rms deviations.

2. Prediction of Binding Affinity

The development of accurate "scoring" functions to identify and compare equilibrium structures of target/docking molecule complexes is a challenging and unsolved problem. A general scoring function is represented in Eq. (101):

$$\Delta G = \Delta G_{complex} - \Delta G_{ligand} - \Delta G_{pocket} \qquad (101)$$

Here $\Delta G_{complex}$, ΔG_{binder} and ΔG_{pocket} are the free energies of the target/docking molecule complex, the free docking ligand molecule, and the free target pocket or binding site, respectively. ΔG is then the free energy of binding or binding affinity.

Due to the computational complexity of rigorous energy calculations, many methods have relied on qualitative modeling of peptide docking interactions. As a first approximation, models have been developed which assume that the docking and target molecules are rigid. In this rigid binding approximation case, the use of shape complementarity has had some limited success [184]. Such algorithms model the ligand and target macromolecule according to their surface topology and attempt to identify which complexes exhibit the best "fit." Here, scoring functions are based on the complementarity of the molecules, which, in most cases, is related to their solvent accessible surface areas [54,185,186]. The strength of these methods is that they can be made computationally efficient and used to screen large databases of potential ligands. However, studies comparing the computational results of these methods to experimentally determined native complexes indicate that rigid models identify many non-native low-energy structures. The rigid-docking scoring function can be refined by adding additional components, such as conformational energy and solvation energy.

On the other end of the flexibility spectrum are fully flexibile, exact models. It has been demonstrated that exact modeling of binding free energies provides results in nearly exact quantitative agreement with experimental results [29,187,188]. In contrast to the rigid description of docking, these methods allow for flexibility of both the ligand and receptor molecules. However, for general peptide docking problems, these thermodynamic integration and free energy perturbation methods are computationally infeasible with current computing power. These problems are only tractable when approximate structures are known and relatively small. More detail on these methods can be found elsewhere [189,190]. A comprehensive theoretical treatment of the thermo-dynamics of binding processes in macromolecules is also available [191].

More computationally feasible methods are based on calculating binding free energies using empirically derived free energy functions. Some methods of approximating free energy functions involve structure-based potentials [192]. Other approximations utilize parameterization of experimental data to construct scoring functions based on conformational energy, hydrophobic and hydrophilic surface areas, and hydrogen bonding geometries [193,194]. However, these methods are generally not transferable from one docking system to another.

A more universal approach, applicable to flexible ligands, is to base free energy calculations on general force field models, which involve potential energy functions similar to those described in the preceding sections. This free energy function must also account for solvation energy, which can be calculated from structure-based solvation terms or continuum-based models of solvation. Rigorously, entropic effects of side-chain rotations should also be considered. Reviews of methods used to evaluate binding free energies can be found elsewhere [195,196].

Once a method for "scoring" the binding affinity has been selected, the exact form of the approach for determining and optimizing the target/docking molecule complex must be developed. Several general approaches have been employed. The most obvious and most difficult approach would be to optimize the entire system of the two interacting peptides. To accomplish this, the relative position of the two peptides, which is defined by six degrees of freedom (three translation and three rotation), along with the total number of internal degrees of freedom for the two molecules, must be considered. This problem becomes intractable for all but the smallest systems. Alternative approaches have decomposed the problem by considering the binding affinities of shorter subsequences at different binding sites of the target macromolecule. The full binding ligand can then be constructed based on the optimally docked subsequences. This approach relies on the ability to build a suitable ligand. Another alternative method is based on independently generating conformations of the isolated ligand. Binding affinities for a number of these rigid conforma-tions then can be calculated and compared, with the drawback that conforma-tions with higher binding affinities may be overlooked.

The following discussion classifies peptide docking approaches according to their treatment of the internal flexibility of the docking ligand molecule. Some approaches combine aspects of both rigid and flexible methods, and the choice of scoring function is often closely related to these classifications. For example, it is implicitly difficult for shape-based approaches to capture internal flexibility due to their simplified description of the molecular surface. Detailed energy-based approaches better represent the free energy of the system and can represent internal conformational changes, but their increased dimensionality makes these methods more computationally expensive. The complexity of these approaches indicates that rigorous global optimization methods are needed to address the peptide docking prediction challenge.

Rigid Models. The first, and most common, methods used to address the peptide docking problems were based on the concept of shape complementarity. These methods employ, at least initially, rigid approximations for both the docking ligand and target receptor molecules. In the most general case, six degrees of freedom—three translational and three rotational—must be optimized to determine the best "fit" for the receptor–ligand complex. Approximations often are used in practice to reduce the number of degrees of freedom. In addition, the alignment of each ligand must be optimized within the binding site. Typically, several screening stages are used to reduce these optimizations to a manageable number.

One shape-based method utilizes a simplified protein model, which is generated by representing each amino acid by a single sphere. The scoring function is based on interfacial areas and a simplified nonbonded potential energy term. Potential ligand structures are screened by systematically rotating the ligand and then translating the structure, along only one dimension, into the pocket [197–199]. These approximations and simplifications are necessary in order to make the problem tractable, especially in the context of a systematic search. A recent modification attempts to overcome these computational limitations by using a simulated annealing, rather than a systematic search, to screen the ligand structures [200].

Distinctive characteristics of molecular surfaces also have been used to reduce the number of degrees of freedom for shape-based docking problems. One study considers local shape functions, which are generated by placing spheres at surface points along the docking ligand and target receptor surfaces. The volume within the surface and the unit vector that extends from the center of the sphere to the surface characterize these functions. A combinatorial algorithm can then be used to compare these local shape functions at "knobs and holes" [201] on the ligand and receptor surfaces so that the best alignments of the two molecules can be identified [202].

More detailed descriptions of molecular surfaces also have been used in determining shape complementarity. One procedure creates a webbed surface for the ligand and receptor by using a local coordinate system to define the surface points for each molecule. After setting the ligand position, a least-squares method is used to align the surface points of the two molecules. The method also screens ligands according to a Coulombic scoring function [203].

An alternative approach transforms the problem from identifying complementary shapes for the receptor and ligand proteins into one of matching similar shapes for these two molecules. This is accomplished by (a) describing the binding site as a collection of spheres that lie on the outside of the receptor surface and (b) characterizing the ligand as a collection of spheres that lie on the inside of the ligand surface [84,204,205]. Potential matches are identified by grouping and comparing distances between the center of spheres for each molecule. Local refinement of translation and rotation vectors is used for the highest-ranking matches. The complexity of the problem is to some degree obscured, because it also depends on the choice of location, size, and number of spheres used to model the receptor molecule. Other modifications of this procedure include the addition of hydrogen bonding criteria and the use of local minimization of the potential energy in order to relax the rigidity of the ligand molecule [206,207].

The "soft docking" model represents the target and docking molecules as a collection of cubes rather than spheres. This method combines aspects of surface complementarity, grid search, and soft potential modeling. The "cubic" representation along with a grid search makes the translational and rotational searches much more efficient. In addition, the cubes implicitly allow for some volume overlap, which can be used in combination with surface complementarity to screen docked complexes [208].

In general, when considering a rigid receptor, the concept of a grid search can be used to reduce the computational requirements of evaluating scoring functions. This is accomplished by precomputing values for the receptor based on points of a three-dimensional grid [209]. The concept is similar to cubic lattice model approaches in molecular conformation problems, for which a recently proposed algorithm using a tabu search has been highly effective [210]. This approach has been the basis of a number of recent studies [211,212], including one that employs a Monte Carlo search in the context of "knobs and holes" docking [212].

Flexible Models.   In the most general case, flexible docking approaches attempt to optimize the free energy of the entire target/docking molecule complex, which is described by translational, rotational, and internal variables of the system. In contrast to most rigid modeling approaches, these methods typically do not require prior knowledge of ligand conformations. As a result, their success in predicting ligand binding is highly dependent on the use of detailed scoring functions to evaluate free energy changes. In addition, although some studies have considered full macromolecular–ligand systems, most approaches also depend on effective decomposition strategies of the overall docking problem.

Several simple approaches have been implemented in an attempt to model flexible docking. For example, a number of methods have incorporated ligand flexibility by considering databases of multiple ligand conformations [213,214]. However, these methods require reliable databases and methods for developing appropriate ligand conformations, and these typically are not available. On the other hand, thermodynamic integration and free energy perturbation methods allow for full flexibility and detailed modeling of binding free energies. However, these simulations, usually accomplished by molecular dynamics, effectively explore only a single low-energy minimum. This has led to the need for global optimization methods that efficiently search the conformational energy hypersurface associated with peptide docking problems.

One of the most common approaches is based on Monte Carlo (MC) simulated annealing algorithms. This method was first applied to flexible ligand docking using molecular affinity potentials [215]. Molecular affinity potentials increase the computational efficiency of the search by employing precomputed energy grids [209]. In this case, flexibility is introduced by allowing internal rotations of torsion angles, along with translational and rotational movement. However, for each docking example, a set of simulated annealing runs is necessary in order to increase the confidence of the reported structures.

A second method, also based on simulated annealing, involves a two-step procedure to dock flexible oligopeptide ligands [216]. In the first step, a modified potential energy force field is used to reduce unfavorable intermolecular contacts. This energy model is employed in local energy minimizations of arbitrarily docked ligands, which are needed in order to generate an initial set of ligand conformations. The scoring function for the second step describes energy interactions between the flexible ligand and rigid receptor molecules. The set of minimized conformations is then used to generate starting points for a Monte Carlo minimization procedure. Although experimental results were not initially available, later comparison has shown that this method does not correctly predict MHC binding. These discrepancies are most likely attributable to incorrect energy modeling (e.g., no inclusion of solvation), along with the inherent inefficiencies associated with simulated annealing searches.

Another MC-based method employs a multiple-start technique in an attempt to reproduce the results of a systematic search. The first step involves a Monte Carlo search with a grid-based scoring function in order to limit steric overlaps of the ligand and receptor molecules. A second, energy-directed, simulated annealing search uses a pairwise potential energy function. Rather than rely on a single search, this method employs a large number of short simulated annealing runs. Although initial results were based on both rigid receptor and ligand conformations [217], more recent work has addressed the issue of flexible ligand docking [218].

Another type of MC method is the scaled collective variable Monte Carlo method used in the software package PRODOCK [219]. This method performs energy minimizations after each MC step, which helps to distinguish native conformations from low-energy non-native conformations. Bezier splines and other techniques have been incorporated into the method to improve its efficiency. In addition, PRODOCK allows different amino acids in the docking complex to be defined as rigid or flexible.

In a similar way, genetic algorithms recently have been used to dock flexible ligands. In some cases, scoring functions have been based on potential energy force fields [220], although some modified potentials also have been used [221]. The results of one method [222], which includes solvation effects, emphasize the need for developing reliable scoring functions. In general, as with simulated annealing, the ability to model flexibility is limited as ligand size increases. The coupling of these effects with the implicit unreliability of both the genetic algorithm and simulated annealing search techniques must be closely considered when approaching large-scale docking problems such as de novo drug design.

Combinatorial methods also have been used to address the difficulties of modeling full ligand flexibility. In theory, these methods are similar to buildup methods used for the protein folding problem, although peptide docking also includes intermolecular interactions. An initial application to the peptide docking problem was based on rigid ligand models generated from a database of chemical structures [205]. A more detailed implementation uses libraries of low-energy conformations for single amino acid residues. These conformations subsequently are joined and grouped according to scoring functions based on the intra- and intermolecular energies of the target/docking ligand complex [223]. More recent methods have employed databases developed for smaller ligand fragments such as functional groups [224] or even atoms [225]. In general, these ligand buildups are initialized by selecting a starting point within the target binding site pocket. As with the protein folding approaches, such combinatorial techniques must employ effective reduction schemes in order to limit the number of generated conformations.

Similar approaches combine the ideas of fragment assembly and site mapping. In contrast to the single anchor requirement of simple buildup methods, these techniques attempt to identify a number of anchor fragments or residues that can be joined through a process of fragment assembly. The first step, site mapping, is equivalent to docking probe fragments at specific sites of the target macromolecule. Some methods have screened the binding affinities of these probes using shape-based modeling [226], whereas others have relied on other energy-based descriptions, such as hydrogen bonding interactions [227,228]. In general, these site maps are constructed by local minimization, grid, or library searches of the probe conformations. Other techniques employ a multiple copy simultaneous search [229,230]. Once anchor positions have been determined using one of these methods, the resulting segments must be joined by fragment assembly. Bridges can be formed by searching through molecular libraries, or in some cases using an exhaustive search over all connections [231]. A recently proposed technique applies a dynamic programming approach, as discussed above, to the fragment assembly phase of a nonameric ligand in an MHC HLA-A2 complex [232]. A molecular dynamics simulation also has been utilized for studying the binding afinity of the HLA-B*2705 protein [233].

Recently, a novel decomposition-based approach has been proposed for predicting the binding site structure of and peptide docking to the MHC II HLA-DR1 protein [234]. The approach performs site mappings of the five polymorphic pockets of MHC II molecules that accommodate peptide docking [171]. In one part of the approach, existing MHC II crystal structures are used to predict the binding site conformations of other MHC II molecules. In another part of the approach, each naturally occurring amino acid is treated as a probe molecule for each of the five pockets. The approach uses a deterministic global optimization search technique to identify the best conformation for each pocket or residue. The scoring function accounts for both intra- and intermolecular interactions using the detailed potential energy force field ECEPP/3 along with several solvation model approaches. The global optimization search, based on the αBB algorithm, is used to identify the global minimum energy conformation for the pockets and for both the bound and free residues. The corresponding energy differences are then used to provide rank-ordered lists of the best binders for each pocket. As discussed in later sections, results for pocket 1 of the HLA-DRB1 macromolecule have exhibited good agreement with experimental binding assays [234].

A recent review of approaches for peptide docking can be found in Floudas et al. [235]. The main disadvantages of most of these approaches are as follows:

(a) Only a very limited conformational space is considered because usually fewer than 10 rotamers are used for each residue.

(b) The simplicity of the energy functions is not able to give a realistic description of the molecular system.

(c) No systematic search methodology exists to guarantee the determination of the global optimal solution, even in methods using simplified energy functions.

Thus many current models of binding site structure prediction and binding affinity prediction in peptide docking are not able to guarantee that they have found the optimum docking solution because they consider only a few of the many conformations two docking partners may adopt, because they are not quantitative, or because they do not fully consider entropic, electrostatic, or other energetic effects.

B. Prediction of Binding Site Structure

We have developed a theoretical approach that, based on crystallographic data from MHC II molecules, determines the three-dimensional structure of MHC II molecule binding sites for which crystallographic data are not available. Class II histocompatibility molecules are cell surface molecules that form complexes with self and nonself peptides and then present them to T cells as part of the immune response. A number of class II histocompatibility molecules have been analyzed by crystallography, including HLA-DR1 [171], HLA-DR3 [236], and I-E^k [237]. Crystal structures are not available, however, for the vast majority of class II MHC molecules. MHC II molecules for which crystal structures are not available are important in autoimmune diseases such as diabetes and rheumatoid arthritis, and being able to predict such structures would advance the understanding and treatment of these diseases.

Our approach to binding site structure prediction uses the ECEPP/3 potential energy model for describing the energetics of atomic interactions (as described in Section III.A.1 above) and employs the rigorous deterministic global optimization algorithm αBB (as described in Section II.A.6 above) to obtain the global minimum energy conformation of the binding site. With this approach, we predicted the binding sites of HLA-DR3 and I-E^k based on the crystallographic structure of HLA-DR1 [171]. The root mean square differences (based on all atoms) between the structures we predicted and the actual crystal structures of the two molecules [236,237] are between 1.09 and 2.03Å. We also calculated the binding affinity of our predicted structures using the decomposition approach discussed in Section V.C.2. These binding affinities are in good agreement with the results obtained by applying the decomposition approach to the actual crystal structures.

1. Definition of Problem

The recent crystallographic studies of class II HLA molecules [171,236,237] suggest an overall similarity in their structures. The conformation of HLA-DR3 in the HLA-DR3-CLIP complex is only slightly different from that of HLA-DR1 in HLA-DR1-HA [236], and a comparison of two I-E^k structures with HLA-DR1 identifies that only a few differences in β-chain amino acids exist between I-E^k and both the HLA-DR1 and HLA-DR3 sequences. However, these few variable residues are sufficient to explain antigenic differences without recourse to allosteric transitions or novel conformations.

Consequently, specific information about the structure of the histocompatibility molecules is needed in order to be able to analyze their specificity. Because crystal structures of class II molecules are not available except for the human crystals of HLA-DR1-HA and HLA-DR3-CLIP and the murine crystals I-E^k-HB and I-E^k-Hsp, we propose a novel approach based on decomposition and deterministic global optimization that enables the prediction of the three-dimensional structure of the binding sites of class II molecules and can be used efficiently for the qualitative assessment of their binding affinities.

The question that is addressed is stated as follows: *Given the (x, y, z) coordinates of the atoms in pockets 1, 4, 6, 7, and 9 of HLA-DR1 [171], can we predict the three-dimensional structures of the corresponding pockets of HLA-DR3 and I-E^k?*

2. Approach

A systematic approach is presented below for the structure prediction of an antigen binding site based on the crystallographic data of the HLA-DR1 molecule [171]. The approach examines each of the binding sites separately and involves the following steps:

1. The binding sites of HLA-DR1 molecule are evaluated. All amino acids within a radius of $\mathscr{R} = 5.0$ Å of the atoms of the binding amino acid in the crystallographic studies [171] are identified as shown in Table XXXVI. The Program for Pocket Definition, as described in Ref. 234 and Section V.C.3, constructs these pockets through the selection of all residues that are within a radius $\mathscr{R}$ of the atoms of the crystallographic binder.

2. The amino acid substitutions between HLA-DR1 and the HLA-II molecule (e.g., HLA-DR3, I-E^k) are identified and are shown in Table XXXVII. Note that pocket 1 of HLA-DR1 requires only one substitution (Gly → Val in position β86) to give pocket 1 of HLA-DR3. Pockets 4, 6, and 7 involve three substitutions, whereas pocket 9 features only one substitution, in the representation of the corresponding pockets of HLA-DR3. Note also that all pockets of HLA-DR1 require three or four substitutions in order to give the corresponding pockets of I-E^k.

3. For each one of the substituted residues, the intra- and intermolecular energy interactions are modeled. Specifically, the electrostatic, nonbonded, torsional, and hydrogen bonding contributions [38] are considered for each

TABLE XXXVI
HLA-DR1 Pocket Compositions for $\mathscr{R} = 5.0$ Å

		Pocket		
1	4	6	7	9
pheα24	glnα09	gluα11	valα65	asnα69
ileα31	gluα11	asnα62	asnα69	leuα70
pheα32	asnα62	valα65	gluβ28	ileα72
trpα43	pheβ13	aspα66	tyrβ47	metα73
alaα52	leuβ26	leuβ11	trpβ61	argα76
serα53	glnβ70	pheβ13	leuβ67	trpβ09
pheα54	argβ71	argβ71	argβ71	aspβ57
gluα55	alaβ74			tyrβ60
asnβ82	tyrβ78			trpβ61
valβ85				
glyβ86				
pheβ89				
thrβ90				

TABLE XXXVII
Substitutions for HLA-DR3 and I-E^k Binding Sites

Pocket	Substitutions for HLA-DR3	Substitutions for I-E^k
1	β86: Gly →Val	β85: Val → Ile β86: Gly → Phe β90: Thr → Leu
4	β13: Phe → Ser β26: Leu → Tyr β74: Ala → Arg	β13: Phe → Ser β74: Ala → Glu β78: Tyr → Val β71: Arg → Lys
6	β11: Leu → Ser β13: Phe → Ser β71: Arg → Lys	β11: Leu → Ser β13: Phe → Cys β71: Arg → Lys
7	β28: Glu → Asp β47: Tyr → Phe β71: Arg → Lys	β28: Glu → Val β47: Tyr → Phe β67: Leu → Phe β71: Arg → Lys α72: Ile → Val
9	β9: Trp → Glu	β9: Trp → Glu β60: Tyr → Asn

substituted residue, as well as the interactions of the substituted residues with the rest of the amino acids that constitute the examined binding site. The solvation energy also is considered through solvent-accessible areas [52,238] as explained in Section III.A.2. The dihedral angles that define the three-dimensional structure of the substituted residues are considered explicitly as variables. The relative position of each amino acid also must be determined, and this is done through the determination of each amino acid's translation vector and Euler angles. Lower and upper bounds are considered for the N and C′ coordinates of the substituted amino acids, based on the available crystallographic data [171,236,237].

4. Having the mathematical model that includes the intra- and intermolecular energetic interactions and the solvation energy, and which has as variables the dihedral angles of the substituted amino acids as well as their translation vectors and Euler angles, we minimize the total potential energy by employing the αBB deterministic global optimization approach [14–18] as described in later sections below.

5. The resulting global minimum energy conformer provides information on the predicted (x, y, z) coordinates of the atoms of the substituted residues. Structure verification is made by superposition of all atoms of the predicted structure and the ones derived from crystallographic data. The superposition is based on the global minimization of the root mean square

differences of the distances between all the atoms involved in the pocket as described in the computational studies section below (Section III.C.5).

3. Modeling

When bond angles and bond lengths are assumed to be rigid, the geometric shape of a protein is uniquely determined by its dihedral angles. If more than one polypeptide is involved, the relative orientations and locations of these different chains also must be defined. This can most easily be accomplished by defining a translation vector and a rotation matrix. The translation vector is based on the Cartesian coordinates of the initial nitrogen atom of each independent chain. Euler angles specify the rotations necessary to orient a particular polypeptide and are defined as the angles between the coordinate axes defined by the initial hydrogen, nitrogen, and alpha carbon of each residue.

The system under study involves all the residues of the binding site. The substituted amino acids constitute the problem variables, whereas the residues that remain the same are treated as fixed based on the crystallographic data. Because there may be multiple amino acid substitutions, the problem variables are the amino coordinates (N_x, N_y, N_z), the Euler angles $(\varepsilon_1, \varepsilon_2, \varepsilon_3)$, and the dihedral angles $(\phi, \psi, \omega, \chi_k)$ of all substituted residues. In contrast to other existing approaches, the Euler angles and dihedral angles are considered to span the entire feasible range $[-180°, +180°]$ and are not restricted to specified discrete values.

Consequently, the total energy function is defined as

$$E_{\text{Total}} = E_{\text{Unsolvated}}^{\text{MIN}} + E_{\text{Solvated}} \tag{102}$$

where $E_{\text{Unsolvated}}^{\text{MIN}}$ is the potential energy of the system without considering solvation, E_{Solvated} is the solvation energy of the system, and E_{Total} is the potential and solvation energy of the system. Based on the above description the mathematical formulation can be posed in the following way:

$$\min \ E_{\text{Total}}(\phi^m, \psi^m, \omega^m, \chi_k^m, N_x^m, N_y^m, N_z^m, \varepsilon_1^m, \varepsilon_2^m, \varepsilon_3^m) \tag{103}$$

$$\text{subject to} \quad -\pi \leq \phi^m, \psi^m, \omega^m, \chi_k^m, \varepsilon_1^m, \varepsilon_2^m, \varepsilon_3^m \leq \pi \tag{104}$$

$$(N_x^m)^L \leq N_x^m \leq (N_x^m)^U \tag{105}$$

$$(N_y^m)^L \leq N_y^m \leq (N_y^m)^U \tag{106}$$

$$(N_z^m)^L \leq N_z^m \leq (N_z^m)^U \tag{107}$$

$$(C_x'^m)^L \leq C_x'^m(\phi^m, \psi^m, \omega^m, \chi_k^m, N_x^m, N_y^m, N_z^m, \varepsilon_1^m, \varepsilon_2^m, \varepsilon_3^m) \leq (C_x'^m)^U \tag{108}$$

$$(C_y'^m)^L \leq C_y'^m(\phi^m, \psi^m, \omega^m, \chi_k^m, N_x^m, N_y^m, N_z^m, \varepsilon_1^m, \varepsilon_2^m, \varepsilon_3^m) \leq (C_y'^m)^U \tag{109}$$

$$(C_z'^m)^L \leq C_z'^m(\phi^m, \psi^m, \omega^m, \chi_k^m, N_x^m, N_y^m, N_z^m, \varepsilon_1^m, \varepsilon_2^m, \varepsilon_3^m) \leq (C_z'^m)^U \tag{110}$$

where $m = 1, \ldots, M$ corresponds to total number of substitutions.

The additional constraints (105–110) represent the bounds on the N and C' coordinates and express the binding of the specific residue with the rest of the pocket [234], because the substituted residue is part of a longer polypeptide and consequently is not allowed to rotate freely. Because the C' coordinates can be evaluated as functions of the independent variables, the restrictions on the position of C' are implemented by the incorporation of a penalty function, P, in the objective function:

$$P = \beta\{\langle C_x'^l - C_x' \rangle + \langle C_x' - C_x'^u \rangle$$
$$+ \langle C_y'^l - C_y' \rangle + \langle C_y' - C_y'^u \rangle$$
$$+ \langle C_z'^l - C_z' \rangle + \langle C_z' - C_z'^u \rangle\}$$

The $\langle \ \rangle$ function is defined as follows: $\langle \mathcal{A} \rangle$ equals $\mathcal{A}$ if $\mathcal{A}$ is greater than zero; otherwise $\langle \mathcal{A} \rangle$ equals zero. Thus, any coordinate value beyond the specified bounds is multiplied by the penalty parameter β and added to the potential energy. Consequently, the minimization of the objective function eliminates solutions in which the C' position falls outside the specified bounds.

The global optimization formulation is then as follows:

$$L = E_{\text{Total}} + \alpha\left\{ \sum_{m=1}^{M} \left(\phi^{mL} - \phi^m \right)\left(\phi^{mU} - \phi^m \right) + \left(\psi^{mL} - \psi^m \right)\left(\psi^{mU} - \psi^m \right) \right.$$
$$+ \left(\omega^{mL} - \omega^m \right)\left(\omega^{mU} - \omega^m \right) + \sum_{k=1}^{K} \left(\chi_k^{mL} - \chi_k^m \right)\left(\chi_k^{mU} - \chi_k^m \right)$$
$$+ \left(N_x^{mL} - N_x^m \right)\left(N_x^{mU} - N_x^m \right)$$
$$+ \left(N_y^{mL} - N_y^m \right)\left(N_y^{mU} - N_y^m \right) + \left(N_z^{mL} - N_z^m \right)\left(N_z^{mU} - N_z^m \right)$$
$$+ \left(\varepsilon_1^{mL} - \varepsilon_1^m \right)\left(\varepsilon_1^{mU} - \varepsilon_1^m \right)$$
$$\left. + \left(\varepsilon_2^{mL} - \varepsilon_2^m \right)\left(\varepsilon_2^{mU} - \varepsilon_2^m \right) + \left(\varepsilon_3^{mL} - \varepsilon_3^m \right)\left(\varepsilon_3^{mU} - \varepsilon_3^m \right) \right\}$$

where α is a nonnegative parameter that must be greater or equal to the negative one-half of the minimum eigenvalue of the Hessian of E_{Total} in the considered domain defined by the lower and upper bounds (i.e., $x^L = -\pi, x^U = \pi$) of the dihedral angles, translation variables, and Euler angles. This parameter can be rigorously calculated based on the techniques introduced by Adjiman and Floudas [14] and Adjiman et al. [16,17].

For the problem of determining the binding sites of the unknown HLA molecules, the global variable set includes the ϕ, ψ, and χ_k variables. All of the dihedral angles of the substituted residues, as well as their translation vectors and Euler angles, are continuous variables in the problem and are treated as local variables.

4. Deterministic Global Optimization

The implementation of the approach involves the connection of the conformational energy program PACK [74], which allows the evaluation of all energy interactions when more than one protein chain is involved in the system, to the deterministic global optimization framework αBB. PACK evaluates all energy components through repeated calls to the ECEPP/3 potential energy function program. The local optimization solver NPSOL is used for the minimization of the overall potential energy provided by PACK and for the minimization of the convexified potential function (L) provided by αBB. MSEED [52], the program for the determination of solvation energy, is interfaced to αBB to allow the consideration of the solvation energy at the local minima. The algorithmic procedure is represented graphically in Fig. 53.

The implementation of the proposed approach is illustrated in Fig. 54 and involves the following steps:

1. The Program for Pocket Definition (*PPD*) uses the input files *residue.pdb* and *pocket.pdb* to generate the coordinates of the residues involved in the considered pocket.

2. The program *ARAS* is used to determine the translation vectors, Euler angles and dihedral angles of the residues in the pocket given their (x, y, z) coordinates. This information and the initial values for the translation vector, Euler angles, and dihedral angles of the substituted residues are incorporated within the input file *name.input*.

3. The program *prePACK* utilizes the *residue.data* file (a set of initial atomic coordinates that are based on fixed bond lengths, fixed bond angles, and each variable dihedral angle initially set to $180°$), the *mol.in* file for each one of the amino acids involved in the pocket, and the *prep.name.abb* file (which specifies the fixed and substituted residues) to create a *name.date* file. The *name.date* file is the standard input for the potential function program, *PACK*.

4. The global optimization program αBB requires a *name.abb* file that defines the optimization problem, including the variable bounds. αBB also uses the *name.input* file and the *name.bounds* file, which contains the C' bounds used to evaluate the coordinates of C' as a function of the independent variables.

5. The program *PACK* uses the *name.date* file and is connected with *ECEPP/3* in order to evaluate the potential function, which is minimized by the local optimization solver *NPSOL*.

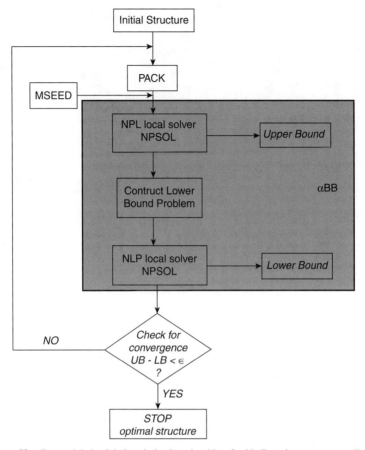

Figure 53.   Deterministic global optimization algorithm for binding site structure prediction.

6. The *MSEED* solvation energy program uses the *JRF.dat* file, which defines the solvation parameters σ_i and evaluates the solvation energy at the current local minimum structure.

5.   Computational Studies

Comparison with Crystallographic Data. To compare the predicted structure of the pockets accurately with the crystallographic data, the best rotation and translation to relate the two different sets of atomic positions must be obtained. Given two proteins A and B with N_{atom} atoms, the best superposition is the one that minimizes the sum of squared distances between each B atom and the

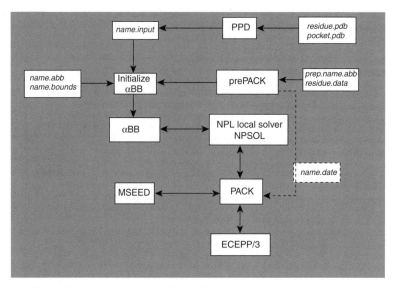

Figure 54.   Implementation of the binding site structure prediction approach.

corresponding A atom. Existing approaches to this problem are based on the following:

   (i) Iterative minimization using rotation angles [239,240].

  (ii) The use of decomposition approaches, where the transformation matrix L is determined by calculating the best unrestricted linear transformation that converts A into B using the least-squares matrix method [241]; or the formation of a generalized inverse of the molecular structure [242], and then the decomposition $L = RS$ where R is a rotation matrix and S is a symmetric distortion matrix.

 (iii) The construction of a matrix U which yields an orthogonal rotation directly [243–246].

As pointed out by McLachlan [246], the rotation angles method is very slow, while the rotation matrix methods depend on whether A is fitted to B or vice versa and do not minimize the RMS distance. McLachlan [246] proposed an approach to improve the speed and accuracy of determining the matrix U and moreover to cover all special cases which arise when U is degenerate or singular.

We formulated and solved the problem of obtaining the best fit of two protein structures as a global optimization problem. The best rotation and translation matrices that minimize the "fitting distance" for the two protein structures are guaranteed to be found in all special cases without having to perform any additional tests and calculations.

Consider two protein structures A and B, with A obtained from the crystallographic data and B determined from our methodology. Both structures involve N_{atom} atoms with Cartesian coordinates $(x_c(i), y_c(i), z_c(i))$ for the crystal and $(x_p(i), y_p(i), z_p(i))$ for the predicted structure. The mathematical formulation of the best-fitting problem can then be posed as follows:

$$\min \quad \text{RMS} = (1/N_{atom}) \sqrt{\sum_{i=1}^{N_{atom}} (x_c(i) - x'(i))^2 + (y_c(i) - y'(i))^2 + (z_c(i) - z'(i))^2}$$

subject to

$$\begin{bmatrix} x'(i) \\ y'(i) \\ z'(i) \end{bmatrix} = R \begin{bmatrix} x_p(i) \\ y_p(i) \\ z_p(i) \end{bmatrix} + T \tag{111}$$

$$R = \begin{bmatrix} r_{11} & r_{12} & r_{13} \\ r_{22} & r_{22} & r_{22} \\ r_{31} & r_{32} & r_{33} \end{bmatrix}, \qquad T = \begin{bmatrix} t_1 \\ t_2 \\ t_3 \end{bmatrix}$$

$$RR^\top = I$$

where R and T are the required rotation and translation vectors that translate the predicted binding sites that correspond to $(x_p(i), y_p(i), z_p(i))$ coordinates to the Cartesian system of the crystal $(x_c(i), y_c(i), z_c(i))$. The coordinates $(x'(i), y'(i), z'(i))$ correspond to the transformed system following the rotation and translation.

Formulation (111) constitutes a special case of global optimization problems because it involves the minimization of a convex function subject to a set of linear equality and nonconvex equality constraints $RR^\top = I$. The deterministic global optimization algorithm αBB [14–18], presented briefly in previous sections, is used for the solution of this global optimization problem. The results obtained for the superposition of the predicted HLA-DR3 and I-E^k binding sites with the crystallographic data are presented in the following sections. Four tests are performed in order to evaluate the prediction accuracy of our methodology.

(i) For each predicted binding site the root-mean-square deviations of Cartesian coordinates of all the atoms (cRMSD) and the C^α atoms are evaluated.

(ii) For each one of the substituted residues, the cRMSD is evaluated considering all the atoms.

(iii)   For each one of the substituted residues, a relative cRMSD is evaluated
        based on the following formula:

$$R - cRMSD$$

$$= \frac{1}{N_{atom}} \sqrt{\sum_{i \in N_{atom}} \frac{1}{3} \left[\frac{(x_p(i) - x_c(i))^2}{x_c(i)} + \frac{(y_p(i) - y_c(i))^2}{y_c(i)} + \frac{(z_p(i) - z_c(i))^2}{z_c(i)} \right]}$$

to measure the relative predictive error of the procedure.

(iv)   Computational binding studies are performed to compare the
        energetic-based rank ordering of the amino acids in the predicted
        binding site versus the rank ordering of the amino acids in the binding
        site based on the crystallographic data, as discussed in later sections.

Prediction of HLA-DR3 Binding Sites. We applied our approach to the prediction of the three-dimensional structure of HLA-DR3 binding sites.

As presented in Table XXXVII, by substituting Gly to Val in position β86
in pocket 1 of HLA-DR1, the pocket 1 of HLA-DR3 is formulated. The
predicted pocket of HLA-DR3 is shown in Fig. 55 with the crystallographically
obtained pocket superposition. The cRMSD difference between these two
pockets is found to be 1.09 Å based on the differences of the coordinates of

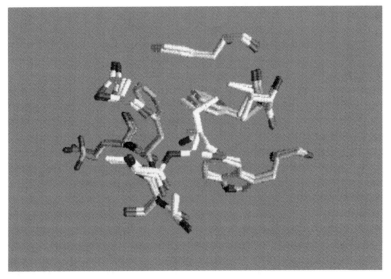

Figure 55.   Superposition of the predicted pocket 1 of HLA-DR3 versus crystallographic data.

all the atoms involved in the pocket. The relative cRMSD for the whole binding site is 0.0425, which corresponds to 4.25% difference of the predicted Cartesian coordinates of the binding site and the crystallographic data. The cRMSD difference based on the α carbons is 0.55 Å. The cRMSD for the substituted residue (Val) is 1.584 Å and the relative-cRMSD is 0.04601, which indicates a 4.6% difference between the predicted valine and the valine determined based on the crystallographic data of the HLA-DR3 molecule [236].

To generate pocket 4 of HLA-DR3, three substitutions are made on the composition of the pockets of HLA-DR1 at the positions β13: Phe → Ser; β26: Leu → Tyr; and β74: Ala → Arg. The cRMSD difference for all the residues in the pocket is 1.11 Å, and the overall relative difference of the predicted pocket compared to the crystallographic data is 2.08%. The cRMSD difference based on the α carbons is 0.49 Å. The cRMSD for each one of the substituted residues is 1.67 Å for Ser, 0.83 Å for Tyr, and 1.46 Å for Arg and correspond to relative differences of 3.2%, 1.2% and 2.3%, respectively.

For pocket 6 of HLA-DR3, the substitutions are at positions β11: Leu to Ser; β13: Phe to Ser; and β71: Arg to Lys. The cRMSD difference for this pocket is 1.22 Å based on all atom deviations, which corresponds to a relative cRMSD of 4.9%. The cRMSD difference based on the α carbons is 0.61 Å. The individual cRMSD for Serβ11 is 1.26 Å, for Serβ13 it is 1.62 Å, and for Lys β71 it is 1.82 Å, which correspond to relative predictive errors of 7.4%, 3.7% and 3.2%, respectively.

For pocket 7 of HLA-DR3, three substitutions are made at the positions β28: Glu to Asp; β47: Tyr to Phe, and β71: Arg to Lys. The cRMSD difference for this pocket is 1.94 Å based on all atom deviations, which corresponds to a 4.69% deviation. The cRMSD difference based on the α carbons is 0.71 Å. The cRMSD for each one of the substituted residues are 1.08 Å for Phe, 3.08 Å for Asp, and 3.4 Å for Arg and correspond to relative differences of 1.4%, 5.1%, and 4.7%, respectively.

Finally, for pocket 9 only one substitution is needed, namely Trp to Glu in position β9 to obtain pocket 9 of HLA-DR3 from pocket 9 of HLA-DR1. The resulting pocket is found to have a cRMSD difference of 1.03 Å based on all atoms and 0.56 Å based on the C^α atoms. The relative cRMSD based on all atom deviations is 37.2%. Considering only the substituted residue, the cRMSD is 1.67 Å. The large predictive deviation in this pocket is due to the large inherent deviation between the HLA-DR1 and the HLA-DR3 crystallographic data. This cRMSD difference for pocket 9 is 1.05 Å, which corresponds to an inherent relative cRMSD of 20.7%.

The results of our prediction approach for all the pockets are summarized in Table XXXVIII. Note that the percentage predictive error is less than 5%, except for pocket 9 where the large inherent deviation between the two crystals prohibits a more accurate prediction.

TABLE XXXVIII
Results for HLA-DR3 Prediction

Pocket	Pocket		Substituted Residues	
	cRMSD (Å)		Relative cRMSD (%)	
Pocket	All Atoms	C^α	All Atoms	cRMSD (Å)
1	1.09	0.55	4.6	Val: 1.58
				Ser: 1.67
4	1.11	0.49	2.1	Tyr: 0.83
				Arg: 1.46
				Ser: 1.26
6	1.22	0.61	4.9	Ser: 1.62
				Lys: 1.82
				Asp: 3.08
7	1.94	0.71	4.7	Phe: 1.08
				Lys: 3.40
9	1.32	0.56	37.2	Glu: 1.67

The coordinates of N and C' are variables in this formulation with bounded ranges for their values around the corresponding atoms in HLA-DR1. Based on the differences observed in the N and C' (x, y, z) coordinates of the HLA-DR1, HLA-DR3, and I-E^k crystals [171,236,237] after superposition, tight bounds in the range of 0.3–1.0 suffice. To study further the effect of the bounds, we considered bound variations of $(\pm 0.5), (\pm 0.7)$ and (± 1.0). The predicted structures of pocket 1 exhibited small cRMSD differences of 1.18, 1.11, and 1.09 Å, respectively, calculated based on all atoms.

Our prediction approach considers the simultaneous substitution of all amino acids responsible for the differences of MHC class II molecules. The required substitutions usually involve 2, 3, or 4 residues and give rise to a global optimization problem that includes as variables the dihedral angles of each residue as well as the translation vector and Euler angles defining the relative position of each residue. In order to reduce the size of the resulting global optimization problem, the following two simplifying alternative procedures also were explored. The first approach is sequential in nature. Instead of considering all amino acids substitutions simultaneously, we considered them sequentially. In particular, the conformation of the first considered substituted amino acid was determined by minimizing the intra- and intermolecular interactions between the specific amino acid and the other residues of the HLA-DR1 binding site. Then, this residue was considered as part of the pocket, and the structure of the second substituted residue was determined. In the second alternative approach we considered each of the substituted amino acids independently and determined their conformations based on minimized energy interactions with the rest of amino acids involved in the pocket of HLA-DR1 molecule. The results obtained for the case of pocket 1 of HLA-DR3 are better than that of the sequential approach having an cRMSD of 2.17 Å compared to 2.51 Å of the sequential procedure but worse than that of the simultaneous approach (cRMSD = 1.09 Å). The reason is that in the sequential approach the error from the first determined amino acid conformation is accumulated as its conformation affects greatly the conformation of the other sequentially considered amino acids.

Prediction of I-E^k Binding Sites. Pocket 1 of I-E^k requires three substitutions: β85: Val → Ile; β86: Gly → Phe, and β90: Thr → Leu. The predicted pocket is illustrated in Fig. 56 with the crystallographic data of I-E^k [78]. The cRMSD difference based on all atoms deviations is 1.67 Å and corresponds to 9.2% relative predictive error. The cRMSD differences for the individual substituted residues are 2.45, 3.36, and 1.76 Å, for Ile, Phe, and Leu, respectively.

For pocket 4 of I-E^k there are four substitutions needed, as shown in Table XXXVII (β13: Phe to Ser; β74: Ala to Glu; β78: Tyr to Val; and β:71 Arg to Lys). The cRMSD difference is 1.58 Å, which corresponds to 3.49% predictive error. For the individual substituted residues the cRMSD differences

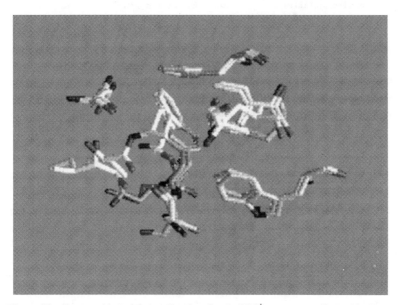

Figure 56.   Superposition of the predicted pocket 1 of I-E^k versus crystallographic data.

are 0.78, 1.35, 2.88, and 1.61 Å, for Ser, Glu, Val, and Lys, respectively. These individual differences correspond to relative predictive errors of 1.59%, 2.16%, 4.48%, and 2.03%.

For pocket 6 of I-E^k, three substitutions are required at the positions β11: Leu → Ser, β13: Phe → Cys; and β71: Arg → Lys. The cRMSD difference is 1.28 Å based on all atoms, which corresponds to 5.19% relative predictive error. For the individual substituted residues, the cRMSD differences are 1.89, 2.67, and 1.64 Å for Ser, Cys, and Lys, respectively. These differences correspond to 4.41%, 14.06%, and 2.82% relative predictive error.

Pocket 7 of I-E^k requires four substitutions, as shown in Table XXXVII (β28: Glu to Val; β47: Tyr to Phe; β67: Leu to Phe; and β71: Arg to Lys). The cRMSD difference is 2.03 Å and corresponds to 4.33% relative predictive deviation. For the individual residues the cRMSD differences are 2.89, 2.15, 2.20, and 3.23 Å for Val, Pheβ47, Pheβ67, and Lys, respectively, and correspond to 3.95%, 3.1%, 5.28%, and 4.41% relative predictive deviation.

Finally, pocket 9 of I-E^k features three substitutions: α72: Ile to Val; β9: Trp to Glu; and β60: Tyr to Asn. The cRMSD difference is 1.35 Å, which corresponds to 23.3% relative predictive deviation. For the individual residues the cRMSD differences are 1.56, 2.46, and 1.56 Å for Val, Glu, and Asn, respectively. The larger relative predictive deviation for this pocket is mainly due to the large relative error for Val at position α72, and the large deviation between the crystals HLA-DR1 and HLA-DR3 gives a cRMSD of 1.09 Å and a 21.4% relative deviation. The results for all the pockets are summarized in Table XXXIX.

In order to study the effect of considering different bounds on N and C' coordinates, the proposed approach was applied to all the pockets for ±0.5 and ±0.3 Å bounds around the coordinates of the corresponding atoms of HLA-DR1 molecule. The results are shown in Table XL. As was found from the crystallographic data of the I-E^k molecule binding with different peptides (i.e., a peptide derived from murine hemoglobin Hb(64–76), or a peptide from murine heat shock protein 70 Hsp(236–248)), there is some inherent variability in the range of 0.01–0.4 Å for N and C' coordinates. These differences correspond to pocket flexibility to accommodate different peptides.

The obtained cRMSD data for all predicted pockets show good agreement with the crystallographic data considering that there is an inherent difference between the crystals, as shown in Table XLI. The cRMSD differences shown in Table XLI represent the differences in the common atoms of the pockets of HLA-DR1 and HLA-DR3 crystals, as well as the differences between HLA-DR1 and I-E^k crystals. These cRMSD differences serve as a reference point in the evaluation of the predicted pockets. For instance, for pocket 1 of HLA-DR3 the predicted structure via the proposed approach has a cRMSD difference of 1.09 Å, whereas the crystallographic data of pocket 1 for HLA-DR1 and

TABLE XXXIX
Results for I-E^k Prediction

Pocket	Pocket cRMSD (Å)		Substituted Residues Relative cRMSD (%)	
Pocket	All Atoms	C^α	All Atoms	cRMSD (Å)
1	1.67	0.47	9.2	Ile: 2.45 Phe: 3.36 Leu: 1.76
4	1.58	0.83	3.5	Ser: 0.78 Glu: 1.35 Val: 2.88 Lys: 1.61
6	1.28	0.65	5.2	Ser: 1.89 Cys: 2.67 Lys: 1.64 Val: 2.89
7	2.03	0.93	4.3	Phe: 2.15 Phe: 2.20 Lys: 3.23 Val: 1.56
9	1.35	0.63	23.3	Glu: 2.46 Asn: 1.56

pocket 1 of HLA-DR3 exhibit a cRMSD of 1.03 Å among their common atoms. Comparing the results shown in Tables XXXVIII, XXXIX, and XLI, it is evident that the predicted structures are close to their reference points, even for pocket 9.

TABLE XL
Effect of Different Bounds on N and C′ Coordinates (I-E^k)

Pocket	Bounds	cRMSD (Å)
1	±0.5	2.26
	±0.3	1.67
4	±0.5	1.81
	±0.3	1.58
6	±0.5	1.28
	±0.3	1.44
7	±0.5	3.17
	±0.3	2.41
9	±0.5	1.84
	±0.3	1.77

TABLE XLI
cRMSD Differences Between HLA-DR1, HLA-DR3, and I-E^k Crystals

Pocket	HLA-DR1 vs. HLA-DR3 Crystals—All Atoms cRMSD (Å)	HLA-DR1 vs. I-E^k-HB Crystals—All Atoms cRMSD (Å)
1	1.03	1.24
4	0.84	1.23
6	0.84	0.84
7	0.996	0.997
9	1.05	1.092

Our approach couples the modeling of energetic interactions and deterministic global optimization approaches and can predict the pockets of HLA-DR3 and I-E^k with small (RMS) differences.

C. Prediction of Relative Binding Affinities

We have developed a theoretical approach that determines the relative binding affinities of amino acids binding to the five pockets of the MHC II molecule HLA-DR1. MHC II molecules such as HLA-DR1 are cell surface glycoproteins that play a pivotal role in the development of an effective immune response. An important function of HLA molecules is to bind and present antigen peptides to T cells. Presently there is no comprehensive way of predicting and energetically evaluating peptide binding for HLA molecules.

To determine quantitatively the relative binding affinities of different peptides for HLA molecules, we developed a decomposition approach based on deterministic global optimization that takes advantage of the topography of the HLA binding groove. Our computational results for binding the 20 naturally occurring amino acids in the five pockets of the HLA allele HLA-DRB1*0101 are in excellent agreement with experimental binding assays and with X-ray crystallography data.

1. Definition of Problem

Class II histocompatibility molecules are cell surface molecules that form complexes with self and nonself peptides and then present them to T cells as part of the immune response. MHC II molecules are important in autoimmune diseases such as diabetes and rheumatoid arthritis, and being able to predict and design the sequences and affinities of peptides which bind to MHC II molecules would increase our understanding of these diseases as well as our ability to design drugs to treat them.

The question that is addressed is stated as follows: *Given the (x,y,z) coordinates of the atoms in HLA-DR1 [171], can we predict the affinity and conformation of the peptides which bind to it?*

2. Approach

We have developed a decomposition approach for predicting the binding affinity and conformation of peptides binding to HLA-DR1. Our decomposition approach takes advantage of the fact that the binding affinity of a peptide for HLA-DR1 molecules is determined primarily by the binding affinity of individual amino acid residues for HLA-DR1's five binding pockets. Our approach uses a sequence of three steps [234]:

I: Consideration of each binding pocket individually

II: Evaluation of the binding of one amino acid at a time to a given pocket

III: Creation of a rank-ordered list of strong, average, and weak amino acid binders for each pocket

Step I involves determining which residues of the HLA-DR1 molecule compose a given pocket. This process is discussed in Section V.C.3 below. Step II involves formulating a mathematical model for the potential and solvation energy of the pocket and the binding amino acid and then using this model to predict the amino acid conformation which corresponds to the global minimum potential and solvation energy state of the system. This global minimum energy state is considered to be the system's most stable state. The mathematical model used to describe the energy of the HLA-DR1/peptide system is discussed in Section V.C.3 below, while the global minimization algorithm used to find the peptide conformation corresponding to the global minimum energy is discussed in Section V.C.4 below. Step III involves comparison of the amino acids binding to a given pocket. The comparison standard used is the change in potential and solvation energy of an amino acid on binding, ΔE. This quantity is defined as the difference between the global minimum potential and solvation energy of an amino acid when it is bound in the pocket (E_{Total}) and the global minimum potential and solvation energy of a free amino acid far from the pocket or any other interactions (E_{Res}^0):

$$\Delta E = E_{\text{Total}} - E_{\text{Res}}^0 \tag{112}$$

The quantity ΔE can be thought of as the difference between the final (bound) and initial (free) states of an amino acid. Thermodynamics predicts that events will proceed in the direction that lowers the total energy of their components. Thus ΔE is a measure of the tendency of an amino acid to bind with the pocket of the HLA-DR1 molecule. A very negative ΔE corresponds to very strong binding.

3. Modeling

Pocket Definition. Consideration of each of HLA-DR1's five binding pockets independently, which corresponds to Step I in Section V.C.2 above, involves determining which residues of the HLA-DR1 molecule compose a given pocket. X-ray crystallography data are available that provide the (x,y,z) Cartesian coordinates of the atoms in the complex of HLA-DRB1*0101 and the influenza peptide HA [171]. The Program for Pocket Definition (PPD) is able to define a given HLA-DR1 pocket from this crystallographic data by calculating which HLA-DR1 amino acids have atoms within a radius $\mathscr{R}$ of the atoms of the influenza peptide amino acid bound to the pocket [234]. The HLA-DR1 residues that do have atoms within this radius constitute the pocket. The inputs required for PPD operation are the value of $\mathscr{R}$, the crystallographic data for the entire HLA-DR1/peptide complex [171], and the crystallographic data for the peptide amino acid bound in the HLA-DR1 pocket. The crystallographic coordinates of the pocket residues are given in an output file. A range of $\mathscr{R}$ values has been evaluated [234] in order to determine an appropriate radius which represents a pocket realistically but which does not include so many residues in the pocket that energy minimization is computationally intractable. Table XLII presents the residues defining each of HLA-DR1's five pockets at different radii. The general trends of this table include increased pocket complexity with increased radius (such as in Pocket 1), constant pocket complexity despite increased radius (such as in Pocket 7), and the much larger number of amino acids in Pocket 1 in comparison to the other four pockets [234]. Based on the results in Table XLII, a radius of 5 Å was used to define Pockets 1, 4, 6, and 7 of HLA-DR1, whereas a radius of 4.5 Å was used to define Pocket 9.

Problem Formulation. The position of a particular peptide or amino acid chain in space can be described completely by a translation vector, a rotation matrix, and a set of dihedral angles. The translation vector is defined as the coordinates of the backbone nitrogen atom on the first residue of a chain. The rotation matrix is defined by the Euler angles of the first chain residue. In our work, the HLA-DR1 pockets are considered rigid and fixed. Thus the variables are the nitrogen coordinates, Euler angles, and dihedral angles of the amino acid which is attempting to bind to a pocket.

Because the decomposition approach described in Section V.C.2 above implicitly assumes that the binding residue is part of a longer peptide, the Cartesian coordinates of the carboxyl carbon atom (C') must be constrained. The decomposition approach is based on the assumption that the rest of the peptide, although not explicitly modeled, is binding normally, and thus that the backbone atoms of the binding peptide do not vary greatly from their crystallographic positions. Because the optimization problem is formulated on internal coordinates, the Cartesian coordinates of C' are implicit variables defined as functions of the translation vector, Euler angles, and dihedral angles of the peptide [234].

TABLE XLII
PPD Pocket Compositions for $\mathscr{R} = 4.0$–5.0 Å

Pocket	$\mathscr{R} =$	4.0	$\mathscr{R} =$	4.5	$\mathscr{R} =$	5.0
1	ileα31	pheα32	ileα31	pheα32	ileα31	pheα32
	trpα43	alaα52	trpα43	alaα52	trpα43	alaα52
	serα53	pheα54	serα53	pheα54	serα53	pheα54
	valβ85	glyβ86	valβ85	glyβ86	valβ85	glyβ86
	pheβ89		pheβ89	pheα24	pheβ89	pheα24
			asnβ82		asnβ82	gluα55
					thrβ90	
4	glnα09	asnα62	glnα09	asnα62	glnα09	asnα62
	pheβ13	glnβ70	pheβ13	glnβ70	pheβ13	glnβ70
	argβ71	alaβ74	argβ71	alaβ74	argβ71	alaβ74
	tyrβ78		tyrβ78	gluα11	tyrβ78	gluα11
			leuβ26		leuβ26	
6	gluα11	asnα62	gluα11	asnα62	gluα11	asnα62
	valα65	aspα66	valα65	aspα66	valα65	aspα66
	leuβ11		leuβ11		leuβ11	pheβ13
					argβ71	
7	valα65	asnα69	valα65	asnα69	valα65	asnα69
	gluβ28	tyrβ47	gluβ28	tyrβ47	gluβ28	tyrβ47
	trpβ61	leuβ67	trpβ61	leuβ67	trpβ61	leuβ67
	argβ71		argβ71		argβ71	
9	ileα72	asnα69	ileα72	asnα69	ileα72	asnα69
	metα73	argα76	metα73	argα76	metα73	argα76
	trpβ09	aspβ57	trpβ09	aspβ57	trpβ09	aspβ57
	tyrβ60		tyrβ60	trpβ61	tyrβ60	trpβ61
					leuα70	

With these variables in mind, formulation of the energy minimization problem proceeds as follows [234]. Let E be the function which calculates the potential and solvation energy of the HLA-DR1 pocket/binder system. Let the Cartesian coordinates of the nitrogen translation vector be defined by the variables N_x, N_y, and N_z. Let the Euler angles be represented by ε_1, ε_2, and ε_3. Let $k = 1, \ldots, K$, where K is the total number of side-chain dihedral angles of the amino acid residue binding to a pocket. The set of variable dihedral angles then includes the backbone dihedral angles ϕ, ψ, and ω, and the side chain angles χ_k. The Cartesian coordinates of the backbone carboxyl carbon (C') are defined by C'_x, C'_y, and C'_z. Utilizing these variable definitions, the potential energy minimization problem can be formulated as follows:

$$\min \quad E(\phi, \psi, \omega, \chi^k, N_x, N_y, N_z, \varepsilon_1, \varepsilon_2, \varepsilon_3) \tag{113}$$

$$\text{subject to} \quad -\pi \le \phi \le \pi \tag{114}$$

$$-\pi \le \psi \le \pi \tag{115}$$

$$-\pi \le \omega \le \pi \tag{116}$$

$$-\pi \le \chi^k \le \pi, \qquad k = 1, \ldots, K \tag{117}$$

$$-\pi \le \varepsilon_1 \le \pi \tag{118}$$

$$-\pi \le \varepsilon_2 \le \pi \tag{119}$$

$$-\pi \le \varepsilon_3 \le \pi \tag{120}$$

$$N_x^l \le N_x \le N_x^u \tag{121}$$

$$N_y^l \le N_y \le N_y^u \tag{122}$$

$$N_z^l \le N_z \le N_z^u \tag{123}$$

$$C_x^{\prime l} \le C_x^{\prime}(\phi, \psi, \omega, \chi^k, N_x, N_y, N_z, \varepsilon_1, \varepsilon_2, \varepsilon_3) \le C_x^{\prime u} \tag{124}$$

$$C_y^{\prime l} \le C_y^{\prime}(\phi, \psi, \omega, \chi^k, N_x, N_y, N_z, \varepsilon_1, \varepsilon_2, \varepsilon_3) \le C_y^{\prime u} \tag{125}$$

$$C_z^{\prime l} \le C_z^{\prime}(\phi, \psi, \omega, \chi^k, N_x, N_y, N_z, \varepsilon_1, \varepsilon_2, \varepsilon_3) \le C_z^{\prime u} \tag{126}$$

In the formulation above, the superscripts u and l denote upper and lower bounds, respectively, for the Cartesian coordinates of both the amino nitrogen and the carboxyl carbon.

Although the constraints on the amino nitrogen in the formulation above can be considered directly as problem variables, the C' coordinates are not explicit variables and consequently must be defined as a function of the other variables [234]. Because the energy minimization problem described above involves these implicit constraints on the location of C', a penalty function must be added to the function E in order to implement these constraints. The modified form of the function E is then [234]:

$$E' = E + \beta\{\langle C_x^{\prime l} - C_x^{\prime}\rangle + \langle C_x^{\prime} - C_x^{\prime u}\rangle \tag{127}$$

$$+ \langle C_y^{\prime l} - C_y^{\prime}\rangle + \langle C_y^{\prime} - C_y^{\prime u}\rangle \tag{128}$$

$$+ \langle C_z^{\prime l} - C_z^{\prime}\rangle + \langle C_z^{\prime} - C_z^{\prime u}\rangle\} \tag{129}$$

The $\langle\ \rangle$ function has the following definition: $\langle \mathscr{A}\rangle$ equals $\mathscr{A}$ if $\mathscr{A}$ is greater than zero; otherwise $\langle \mathscr{A}\rangle$ equals zero. Therefore, if the coordinates of C' are within their respective bounds, the function E will not be modified. If, however, a particular coordinate falls outside of its bounds, the function will be increased by an arbitrarily large constant β. The conformation's energy then would be arbitrarily large, and the conformation would be discarded as a choice for the minimum energy conformation.

Note that E in the formulation above is a nonconvex function involving numerous local minima that correspond to metastable states of the specific amino acid binding to the pocket. A single global minimum defines the energetically most favorable peptide conformation. In establishing a ranked-order list of binding peptides, one needs to identify rigorously the best conformation of (i) the binding residue far from the pocket and (ii) the complex of Pocket 1 with the binding residue. Consequently, there is a need for a method that can guarantee convergence to the global minimum potential energy conformation and which is capable of solving large-scale constrained optimization problems. The global optimization approach αBB [18,20,247] is one such method.

GLO-DOCK. The αBB algorithm is interfaced and supported with several other programs in the overall energy minimization scheme, and the entire collection of programs is known as GLO-DOCK. The additional programs include MSEED, RRIGS, NPSOL, and PACK. The MSEED program is discussed in Section III.A.2 above and calculates solvent-accessible surface areas, and the RRIGS program is discussed in Section III.A.2 above and calculates solvent-accessible volumes. Only one of these programs is utilized for calculating solvent energies during a given peptide docking optimization. The program NPSOL [28] is a nonlinear local optimization solver used in the calculation of upper and lower bounds for αBB. The PACK program [74], and its associated program prePACK, is a peptide calculation program. The prePACK program initializes PACK by converting the amino acid residue data supplied by the program's user into the format required by the ECEPP/3 potential energy model. The prePACK program also generates the parameter values used by PACK in calculating energy potentials. The PACK program transforms Cartesian coordinates into internal (dihedral angle) coordinates and uses the ECEPP/3 potential energy model to provide function and gradient evaluations to αBB. The PACK program is able to keep track of data for several peptides and make appropriate calls to ECEPP/3 for calculation of their interaction energies [74]. As discussed below, solvation contributions based on solvent-accessible area are added only at local minima, so the program MSEED is called from αBB through PACK once a local minimum has been found. The program RRIGS is called from αBB though PACK at each local minimization step.

Several supporting programs generate the input files used in this overall minimization scheme. These programs include PPD and ARAS. PPD, the Program for Pocket Definition, was discussed above and defines a given pocket from crystallographic data. The output file from PPD is then used as an input file for the program ARAS, the Amino acid Residue Angle Solver. ARAS converts the crystallographic data from the PPD output file into translation vectors, Euler angles, and dihedral angles for each amino acid in the file. An ARAS output file (*name.input* in Figure 57) is then used as an input file for PACK. Three other input files are required for peptide docking optimizations: *name.abb* and *prep.name.abb*, which provide the bounds on the initial nitrogen atom and other information needed by αBB; and *name.bounds*, which provides the bounds on C' for the penalty function. The (x,y,z) nitrogen and C' bounds for each pocket binder are determined by examining the crystallographic data for the corresponding peptides in the HLA-DRB1*0101/influenza virus peptide complex presented by Stern et al. [171]. These bounds are set at ± 0.7 Å from the crystallographic coordinates. A schematic diagram for the overall global optimization scheme is given in Fig. 57.

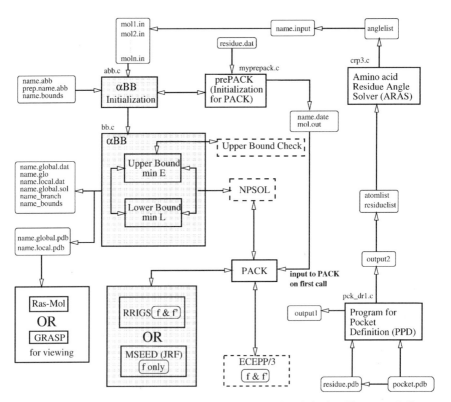

Figure 57. Schematic diagram for peptide docking global optimization. The arrows indicate the direction of information flow. The names of input, output, and source code files are indicated. References to "f & f'" and "f only" describe whether gradient evaluations or only function evaluations are used.

Solvation Methods.   Because the polar, cohesive nature of water profoundly affects all molecular interactions in biological systems [248], the effects of solvation on the conformation of a protein must be included in an accurate protein model.

There are many types of solvation models. Explicit solvation models arrange individual water molecules around peptides and calculate solvent–peptide interactions with potential models similar to those discussed in Section III.A.1 above. These explicit models are prohibitively expensive computationally because of the large number of water molecules involved and because a given peptide conformation has a large number of equivalent possible water molecule arrangements, making it necessary to calculate the energy of many solvent arrangements and average them together [83]. Neglecting the molecular nature of water molecules yields much simpler, implicit solvation models. Models of this type often estimate energies of solvation as functions of solvent-accessible surface areas or volumes.

Our work is based on two separate implicit methods of determining solvation potentials. One method involves solvent-accessible area calculations, and the other involves solvent-accessible volume calculations. These models are based on two assumptions: that a solvation energy can be calculated for each functional group of a peptide by calculating an averaged energy of interaction between the group and a layer of solvent known as the solvation shell, and that these solvation energies are additive. Thus the model assumes that the total energy of solvation of a peptide can be expressed as the sum of the energies of solvation for each of the functional groups of the peptide [83].

The solvent-accessible area solvation model used in our work is based on a program called MSEED [52]. This model assumes that the energy of solvation is proportional to the solvent-accessible surface area of the peptide, as discussed in Section III.A.2 above. MSEED area calculations have some limitations, however. First, MSEED does not always search effectively for the peptide's surface areas. The error incurred by this ineffective search, however, has been shown to be less than 2% for a number of test problems [52]. Second, changes in peptide conformations produced by minimization of the total energy of the peptide proceed continuously but not necessarily smoothly, and surface area gradients may thus have discontinuities. Large discontinuities may cause minimization techniques that require calculation of first derivatives to fail to converge. This problem is avoided in our work because gradients for area-accessible solvation contributions are not calculated and surface-accessible solvation energies are included in the total energy only at local minimum energy conformations and are not part of local minimization processes.

The solvent-accessible volume model used in our work is based on a program called RRIGS, which stands for Reduced Radius Independent Gaussian Sphere [53]. This model assumes that the energy of solvation of a peptide is proportional to the solvent-accessible volume of a solvation layer or shell around the peptide, as discussed in Section III.A.2 above. This method provides continuous derivatives of the solvation potential, so solvation contributions to total energy can be added at every step of local minimizations and not just at the local minimum itself. Thus the RRIGS solvation model interfaces well with the ECEPP/3 potential energy model [83].

Combining the ECEPP/3 potential energy model with a solvation model creates an expression for the total potential and solvation energy (E_{Total}) of the system: $E_{Total} = E_{Potential} + E_{Solvation}$, where $E_{Potential}$ is calculated from ECEPP/3 and $E_{Solvation}$ is calculated from either the MSEED or RIGGS solvation models. With this mathematical model for the potential and solvation energy of the pocket and the binding amino acid in place, the next step in evaluating the binding of one amino acid at a time to a given HLA-DR1 pocket is finding the amino acid conformation that corresponds to the global minimum potential and solvation energy of the system.

4. Deterministic Global Optimization

The first step in implementing a global optimization algorithm like αBB is the formulation of the optimization problem. This involves choosing the functions that will be optimized (either minimized or maximized), choosing the variables that will be optimized, and choosing the constraints that will be included in the problem. For the peptide docking prediction problem, implementing a global optimization algorithm also involves deciding whether to minimize the total energy function based on the Cartesian coordinates of the peptide atoms or based on the dihedral angles of the peptide. Because optimization constraints are more easily applied to internal coordinates like dihedral angles than to Cartesian coordinates [20], we used internal coordinates for our work. The problem formulation is developed in Section V.C.3 above. The function E shown in Section V.C.3 is difficult to minimize because it is nonlinear and nonconvex and has multiple local minima. These local minima correspond to metastable states of the amino acid binder being modeled, but the single global minimum is the minimum that defines the energetically most stable peptide conformation.

Our minimization scheme determines the peptide conformation that corresponds to the global minimum total potential and solvation energy through a series of steps [83,234]:

1. Upper bound calculation: The local solver NPSOL identifies a local minimum of the potential energy function supplied by PACK in a region (rectangle) defined by the lower and upper bounds of the variables. These bounds are supplied by αBB. If solvent-accessible volume is being considered, potential energy evaluations during local minimization are made

using the ECEPP/3 model and RRIGS. If solvent-accessible surface area is being considered, potential energy evaluations are made using only the ECEPP/3 model and the solvation energy is calculated by MSEED and added only at the local minimum.

2. The current best upper bound is updated to be the minimum of those stored thus far.

3. The current rectangle (region) is partitioned by bisecting its longest side.

4. Lower bound calculation: The convex underestimator function L is minimized in each new rectangle using NPSOL and PACK (with ECEPP/3). If solvent-accessible volume is being considered, potential energy evaluations are also made using RRIGS. If solvent-accessible surface area is being considered, potential energy evaluations are not made with MSEED, and the solvation contributions are added only at the local minimum. If a minimum is greater than the best upper bound, the corresponding rectangle will be eliminated from the search. Otherwise, the local minimum value is stored.

5. The rectangle with the current minimum value for L is selected for further partitioning.

6. If the best upper and lower bounds are within the user-specified tolerance ϵ, the program will finish; otherwise it will proceed back to Step 1.

We then introduced an energetic-based criterion to evaluate the energy of interaction between a given pocket and each naturally occurring amino acid. This measure, which we denote as ΔE, corresponds to the difference between (i) the global minimum total potential and solvation energy that considers all the energetic atom-to-atom interactions—classified as inter-interactions between the atoms of the residues that define the pocket of HLA-DR1 protein and the atoms of the considered naturally occurring amino acid, and classified as intra-interactions among the atoms of the considered naturally occurring amino acid—and (ii) the global minimum potential and solvation energy of the considered naturally occurring amino acid when it is far away from the pocket. Equation (130) illustrates this criterion:

$$\Delta E = E^0_{\text{Total}} - E^0_{\text{Res}} \tag{130}$$

where E^0_{Total} is the global minimum of the potential energy of the complex of the pocket and the binding peptide or amino acid, and E^0_{Res} is the global minimum of the potential energy of the peptide or amino acid away from the pocket. Note that ΔE does not represent a difference in the free energies of the complex and isolated amino acids. Instead, it denotes the difference between potential and solvation energy for the complex and the isolated amino acids.

Repeating this optimization scheme for each naturally occurring amino acid in each of HLA-DR1's five binding pockets and then listing each pocket's amino acid binders in order of increasing global minimum potential and solvation energy (decreasing binding affinity) creates a rank-ordered list of strong, average, and weak amino acid binders for each pocket.

5. Computational Studies

Binding Affinity Evaluation in HLA-DR1 Pockets. The area and volume solvation methods correctly predict the binding affinity and conformations of the strongest binders to Pockets 1, 4, and 6. Neither the area nor the volume solvation methods correctly predict the crystallographic binder conformations for Pocket 7 and 9, perhaps due the pockets' incomplete definition.

The volume solvation method appears to be a stronger method for considering solvation than the area solvation method. The area solvation method does not use separate parameters for charged and neutral atoms, and its structure does not permit consideration of area contributions at each step of local minimizations of the total energy. Solvation energy contributions in the volume method are of the same order of magnitude as nonbonded contributions, whereas solvation energy contributions are an order of magnitude larger than nonbonded energy contributions in the area solvation method. The domination of total energy values by solvation in the area solvation method may not distinguish amino acid binders sufficiently from one another in rank-ordered binding lists.

Our global optimization results are in excellent agreement with available experimental data. Experimental data [234] for amino acids binding to Pocket 1 are shown in Fig. 58. We were able to reproduce the relative binding affinities shown in the figure, and all of our other relative binding affinity results agree with literature data. The results for Pocket 1 and Pocket 4 are especially encouraging, because Pocket 1 is considered to be the most discriminating and most important pocket for successful peptide binding [249] and Pocket 4 is considered to be one of the most important pockets in T-cell recognition interactions [250]. This agreement indicates that our approach is an accurate, effective tool for approaching the peptide docking problem.

The need for determining the conformation of a binding amino acid which corresponds to its global minimum total energy instead of to a local minimum total energy is illustrated in Fig. 59. Figure 59 shows a local minimum conformation and the global minimum conformation of tyrosine in Pocket 1 for the volume solvation method. The volume solvation method's local minimum conformation of tyrosine has a ΔE of -17.349 kcal/mol and is shown in a lighter shade, whereas the global minimum conformation of tyrosine has a ΔE of -20.155 and is shown in darker shade. There is only a 13.9% difference between these two ΔE values, but there is a significant difference in their conformations. The global minimum conformation corresponds closely to the

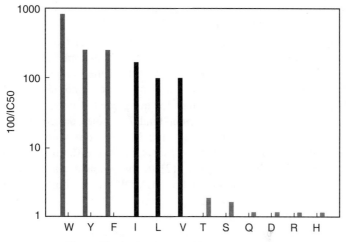

Figure 58.   Pocket 1 competitive binding assays.

crystallographically determined conformation, highlighting the necessity of not mistaking a local solution for the global solution. This comparison also highlights the need for global optimization methods in approaches to the peptide docking prediction challenge.

Binding Affinity Evaluation After Structure Prediction.   Our prediction of the structures of MHC class II binding sites has significant implications for the evaluation of peptide binding to HLA molecules. We applied our binding affinity prediction methodology to the binding pocket structures we predicted in

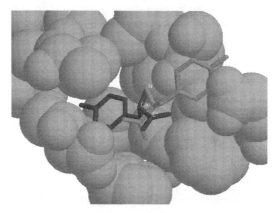

Figure 59.   Global (darker) versus local (lighter) tyrosine conformations in pocket 1, volume solvation.

Section V.B.5. We then compared the results of predicting binding affinities for predicted versus crystallographic pockets.

We applied our methodology to the pocket we predicted for pocket 1 of HLA-DR3 and to the pocket obtained from crystallographic data [221] for the binding amino acids Phe, Ile, and Met. Based on the energy differences we found that Phe is a better binder than Met by 1.1 kcal/mol and that Met is a better binder than Ile by 3.9 kcal/mol for the predicted pocket. For pocket 1 based on the crystallographic data, our binding studies determined the same sequence (i.e., Phe followed by Met and Met by Ile) with corresponding differences of 2.37 kcal/mol for Phe to Met and 2.06 for Met to Ile. Application of our predictive binding approach [12] to the predicted, as well as to the crystallographicaly obtained, pocket 4 of HLA-DR3 for the binding amino acids Asp, Glu, Ile, and Phe showed that the negatively charged Asp and Glu are very strong binders. In contrast, Ile and Phe were weaker binders than Asp and Glu.

We also applied our predictive binding approach to the predicted pocket 1 of I-E^k for the binding amino acids Ile, Val, and Phe. Our results showed that Phe is a better binder than Ile by an energy difference of 6.1 kcal/mol, and that Ile binds better than Val by an energy difference of 2.8 kcal/mol. We obtained similar results from the crystallographic data, with Phe being a better binder than Ile and Ile being a better binder than Val with energy differences of 4.4 and 0.7 kcal/mol, respectively.

In order to verify further the correct prediction of the binding sites of HLA molecules, we used the crystal of HLA-DR3 [78] to predict pocket 1 of HLA-DR1. We compared the results obtained using the predicted pocket to those found with the crystallographically obtained pocket [221]. As shown in Table XLIII, the binding studies using the predicted pocket illustrate the same trends as the binding studies using the crystallographic pocket. Therefore, our approach not only predicts the binding site structure of class II HLA molecules, but also provides results consistent with the binding studies of individual amino acids based on the crystallographic data.

D. Perspectives and Future Work

We currently are expanding and extending our binding site structure and binding affinity prediction methods. We are expanding our methods to incorporate rigorous calculation of free energies. Our approach to these free energy calculations involves the terms in the following equation:

$$E_{\text{Total}} = E_{\text{Vacuum}} - TS_{\text{Vacuum}} + E_{\text{Cavity}} + E_{\text{Solvation}} + E_{\text{Ionize}} \qquad (131)$$

where E_{Total} is the total free energy of a protein–protein system. In this approach, as in our earlier approach discussed above, E_{Vacuum} is the potential energy of a protein system conformation in a vacuum calculated from the ECEPP/3 force

TABLE XLIII
Comparison of Binding Studies in Predicted Binding Sites Versus Crystallographic Binding Sites in
Pocket 1 of HLA-DRI ($\mathscr{R} = 5.0$ Å), Area Solvation

Residue	ΔE Crystal (kcal/mol)	ΔE Prediction (kcal/mol)	Difference (kcal/mol)	Difference (%)
Tyr	− 20.000	− 18.850	− 1.15	5.75
Phe	− 19.625	− 18.040	− 1.58	2.95
Trp	− 16.950	− 17.754	0.80	4.72
Gln	− 15.396	− 15.916	0.52	3.37
Met	− 13.943	− 13.928	− 0.02	0.14
Asn	− 13.784	− 14.644	0.86	6.24
Thr	− 13.297	− 13.297	0.00	0.00
Leu	− 12.481	− 12.399	− 0.08	0.64
Ile	− 12.465	− 12.486	0.02	0.16
Ser	− 11.557	− 11.187	− 0.37	3.20
Cys	− 11.280	− 11.087	− 0.19	1.68
Val	− 11.209	− 11.324	0.12	1.07
Ala	− 10.355	− 10.338	− 0.02	0.19
Gly	− 10.091	− 9.996	− 0.09	0.89
Glu-	− 7.744	− 6.891	− 0.85	10.97
Asp-	− 2.431	− 2.594	0.16	6.58

field. In our expanded approach, S_{Vacuum} is the entropy of a protein conformation in a vacuum. In order to calculate this term, we generate a large set of unique conformers and then apply a harmonic approximation to obtain the entropy of each conformation. The E_{Cavity} term in this approach is the energy required to form a protein conformation's cavity in aqueous solvent. This cavity energy is estimated to be proportional to the surface area of the protein system exposed to water. We calculate the $E_{\text{Solvation}}$ term in this expanded approach with Poisson–Boltzmann electrostatics by using the DELPHI software package [251–253]. The $E_{\text{Solvation}}$ term is the difference in a protein system conformation's polarization energy in solvent (dielectric constant $\varepsilon = 80$) and in a vacuum (dielectric constant $\varepsilon = 1$), as shown in the equation below:

$$E_{\text{Solvation}} = \frac{1}{2}\sum_i \sum_s \frac{q_i \sigma_{s,\varepsilon=80}}{|r_i - r_s|} - \frac{1}{2}\sum_i \sum_s \frac{q_i \sigma_{s,\varepsilon=1}}{|r_i - r_s|} \tag{132}$$

where q_i is the charge associated with atom i, and σ_s is the surface charge induced by each charge s other than i. The E_{Ionize} term is the energy due to the ionization state of a protein system at a given pH. These expansions to our binding site structure and binding affinity prediction methods will allow us to model solvent effects more rigorously and more accurately, as well as allowing the study of ionization effects. We then will have the ability to calculate and predict not only relative binding affinities, but accurate, quantitative binding affinites. The drawback of employing the entropic and Poisson–Boltzmann calculations discussed above is the large increase in computational time they require. We are exploring ways of parallelizing our algorithm to address this issue.

In addition to expanding our methods, we are extending them to the investigation of larger systems. We are examining the role of the peptide residues intermediate to the pocket-binding residues in peptides that bind to HLA molecules, as well as modeling the docking of entire peptides to HLA molecules. Our future plans include extending our methods to the examination of T-cell interactions with HLA molecules and bound peptides.

Our computational and experimental results demonstrate that applying atomistic level modeling and deterministic global optimization is a promising approach to a systematic framework for peptide docking prediction. The strengths of our peptide docking prediction model are its guaranteed convergence to the global minimum energy, its detailed modeling of entropic, electrostatic, and other energetic interactions, and its quantitative prediction of binding free energy.

The predictive power of protein–protein interaction and peptide docking models is of significant and increasing importance. Accurate prediction will lead to the more efficient and effective design of drugs and devices. Peptide docking and protein–protein interaction prediction thus will play a valuable role in capitalizing on the data provided by the mapping of human and other genomes.

VI. CONCLUSIONS

The intense worldwide experimental and theoretical research effort directed toward solving the protein folding and peptide docking problems underscores their importance. The ability to predict computationally the folding of proteins and the formation of protein–protein complexes would support and help direct experimental work in biology, chemistry, biophysics, and pharmaceutical development. In this review we have shown that molecular modeling and global optimization are the dominant factors that will provide solutions to these problems.

In particular, this review has focused on the use of *ab initio* models, which give rise to a series of complex mathematical problems. A second important component has been the application of deterministic global optimization, namely the αBB algorithm, for solving the resulting problems. In this review we have analyzed and discussed many issues related to the modeling of protein folding and peptide docking systems. These observations have highlighted the extraordinary difficulty of these problems and the crucial interdependence of *ab initio* modeling and deterministic global optimization approaches.

Acknowledgments

The authors gratefully acknowledge financial support from the National Science Foundation and the National Institutes of Health (R01 GM52032, 1 F32 GM20007).

References

1. C. B. Anfinsen, E. Haber, M. Sela, and F. H. White, Jr., The kinetics of formation of native ribonuclease during oxidation of the reduced polypeptide chain. *Proc. Natl. Acad. Sci. USA* **47**, 1309–1314 (1961).

2. P. S. Kim and R. L. Baldwin, Intermediates in the folding reactions of small proteins. *Annu. Rev. Biochem.* **59**, 631–660 (1990).

3. C. Levinthal, Are there pathways to protein folding? *J. Chem. Phys.* **65**, 44–45 (1968).

4. O. M. Becker and M. Karplus, The topology of multidimensional potential energy surfaces: Theory and application to peptide structure and kinetics. *J. Chem. Phys.* **106**(4), 1495–1517 (1997).

5. R. Czerminski and R. Elber, Reaction path study of conformational transitions in flexible systems: Applications to peptides. *J. Chem. Phys.* **92**(9), 5580–5601 (1990).

6. B. W. Church, M. Orešič, and D. Shalloway, Tracking metastable states to free-energy global minima, in *DIMACS Series in Discrete Mathematics and Theoretical Computer Science*, Vol. 23, American Mathematical Society, 1996, pp. 41–64.

7. P. Leopold, M. Montal, and J. Onuchic, Protein folding funnels: A kinetic approach to the sequence-structure relationship. *Proc. Natl. Acad. Sci. USA* **89**, 8721–8725 (1992).

8. A. Šali, E. Shaknovich, and M. Karplus, Thermodynamics and kinetics of protein folding, in *DIMACS Series in Discrete Mathematics and Theoretical Computer Science*, Vol. 23, American Mathematical Society, 1996, pp. 199–213.

9. D. Goldberg, *Genetic algorithms in search, optimization and machine learning*. Addison-Wesley, Reading, MA, (1989).

10. S. Kirkpatrick, C. D. Gelatt, Jr., and M. P. Vecchi, Optimization by simulated annealing. *Science* **220**, 671–680 (1983).

11. C. A. Floudas, *Deterministic Global Optimization: Theory, Methods and Applications. Nonconvex Optimization and Its Applications*, Kluwer Academic Publishers, Hingham, MA, 2000.

12. R. Horst and P. M. Pardalos, eds., *Handbook of Global Optimization*, Kluwer Academic Publishers, Hingham, MA, 1995.

13. R. Horst and H. Tuy, *Global Optimization: Deterministic Approaches*, 2nd revised edition, Springer-Verlag, Berlin, 1993.

14. C. S. Adjiman and C. A. Floudas, Rigorous convex underestimators for general twice–differentiable problems. *J. Glob. Opt.* **9**, 23–40 (1996).

15. C. S. Adjiman, I. P. Androulakis, C. D. Maranas, and C. A. Floudas, A global optimization method, αBB, for process design. *Comput. Chem. Eng.* **20**, S419–S424 (1996).

16. C. S. Adjiman, S. Dallwig, C. A. Floudas, and A. Neumaier, A global optimization method for general twice-differentiable nlps—i. theoretical advances. *Comput. Chem. Eng.* **22**, 1137–1158 (1998).

17. C. S. Adjiman, I. P. Androulakis, and C. A. Floudas, A global optimization method for general twice-differentiable nlps—ii. implementation and computational results. *Comput. Chem. Eng.* **22**, 1159–1179 (1998).

18. I. P. Androulakis, C. D. Maranas, and C. A. Floudas. αbb: A global optimization method for general constrained nonconvex problems. *J. Glob. Opt.* **7**, 337–363 (1995).

19. C. D. Maranas and C. A. Floudas, A deterministic global optimization approach for molecular structure determination. *J. Chem. Phys.* **100**(2), 1247–1261 (1994).

20. C. D. Maranas and C. A. Floudas, Global minimum potential energy conformations of small molecules. *J. Glob. Opt.* **4**, 135–170 (1994).

21. F. A. Al-Khayyal and J. E. Falk, Jointly constrained biconvex programming. *Math. Ops. Res.* **8**, 273–286 (1983).

22. G. P. McCormick, Computability of global solutions to factorable nonconvex programs: Part I—Convex underestimating problems. *Math. Programming* **10**, 147–175 (1976).

23. C. D. Maranas and C. A. Floudas, Finding all solutions of non-linearly constrained systems of equations. *J. Glob. Opt.* **7**(2), 143–182 (1995).

24. H. Ratschek and J. Rokne, *Computer Methods for the Range of Functions*, Ellis Horwood Series in Mathematics and Its Applications, Halsted Press, 1988.

25. A. Neumaier, *Interval Methods for Systems of Equations*. Encyclopedia of Mathematics and Its Applications, Cambridge University Press, New York, 1990.

26. S. Gerschgorin, Über die abgrenzung der eigenwerte einer matrix. *Izv. Akad. Nauk SSSR, Ser. Fiz. Mat.* **6**, 749–754 (1931).

27. Bruce A. Murtagh and Michael A. Saunders, *MINOS 5.4 User's Guide*. Systems Optimization Laboratory, Department of Operations Research, Stanford University, 1993. Technical Report SOL 83-20R.

28. P. E. Gill, W. Murray, M. A. Saunders, and M. H. Wright, *NPSOL 4.0 User's Guide*, Systems Optimization Laboratory, Department of Operations Research, Stanford University, CA, (1986).

29. W. L. Jorgensen and J. Tirado-Rives, The opls potential functions for proteins. Energy minimizations for crystals of cyclic peptides and crambin. *J. Am. Chem. Soc.* **110**, 1657–1666 (1988).

30. S. Weiner, P. Kollman, D. A. Case, U. C. Singh, C. Ghio, G. Alagona, S. Profeta, and P. Weiner, A new force field for molecular mechanical simulation of nucleic acids and proteins. *J. Am. Chem. Soc.* **106**(3), 765–784 (1984).

31. S. Weiner, P. Kollman, D. Nguyen, and D. Case, An all atom force field for simulations of proteins and nucleic acids. *J. Comp. Chem.* **7**, 230–252 (1986).

32. B. R. Brooks, R. E. Bruccoleri, B. D. Olafson, D. J. States, S. Swaminathan, and M. Karplus, Charmm: A program for macromolecular energy, minimization, and dynamics calculations. *J. Comp. Chem.* **4**(2), 187–217 (1983).

33. P. Dauber-Osguthorpe, V. A. Roberts, D. J. Osguthorpe, J. Wolff, M. Genest, and A. T. Hagler, Structure and energetics of ligand binding to peptides: *Escherichia coli* dihydrofolate reductase–trimethoprim, a drug receptor system. *Proteins* **4**, 31 (1988).

34. F. A. Momany, L. M. Carruthers, R. F. McGuire, and H. A. Scheraga, Intermolecular potential from crystal data. III. *J. Phys. Chem.* **78**, 1595–1620 (1974).

35. F. A. Momany, L. M. Carruthers, and H. A. Scheraga, Intermolecular potential from crystal data. IV. *J. Phys. Chem.* **78**, 1621–1630 (1974).

36. F. A. Momany, R. F. McGuire, A. W. Burgess, and H. A. Scheraga, Energy parameters in polypeptides. VII. *J. Phys. Chem.* **79**, 2361–2381 (1975).

37. G. Némethy, M. S. Pottle, and H. A. Scheraga, Energy parameters in polypeptides. 9. *J. Phys. Chem.* **87**, 1883–1887 (1983).

38. G. Némethy, K. D. Gibson, K. A. Palmer, C. N. Yoon, G. Paterlini, A. Zagari, S. Rumsey, and H. A. Scheraga, Energy parameters in polypeptides: X. improved geometrical parameters and nonbonded interactions for use in the ecepp/3 algorithm, with application to proline-containing peptides. *J. Phys. Chem.* **96**(15), 6472–6484 (1992).

39. V. Daggett and M. Levitt, Realistic simulations of native-protein dynamics in solution and beyond. *Annu. Rev. Biophys. Biomol. Struct.* **22**, 353–380 (1993).

40. M. Levitt, Protein folding by restrained energy minimization and molecular dynamics. *J. Mol. Biol.* **170**, 723–764 (1983).

41. W. F. van Gunsteren and H. J. C. Berendsen, *GROMOS.* Groningen Molecular Simulation, Groningen, The Netherlands, 1987.

42. N. L. Allinger, Conformational analysis. A 130 mm^2 hydrocarbon force field utilizing v_1 and v_2 torsional terms. *J. Am. Chem. Soc.* **99**, 8127–8134 (1977).

43. N. L. Allinger, Y. H. Yuh, and J. H. Lii, Molecular mechanics. The mm^3 force field for hydrocarbons. *J. Am. Chem. Soc.* **111**(23), 8551–8565 (1989).

44. J-H. Lii and N. L. Allinger, Molecular mechanics the mm3 force field for hydrocarbons. 2. Vibrational frequencies and thermodynamics. *J. Am. Chem. Soc.* **111**, 8566–8575 (1989).

45. J-H. Lii and N. L. Allinger, Molecular mechanics. The mm3 force field for hydrocarbons. 3. The van der waals' potentials and crystal data for aliphatic and aromatic hydrocarbons. *J. Am. Chem. Soc.* **111**, 8576–8582 (1989).

46. R. F. McGuire, F. A. Momany, and H. A. Scheraga, Energy parameters in polypeptides. v. An empirical hydrogen bond potential function based on molecular orbital calculations. *J. Phys. Chem.* **76**, 375–393 (1972).

47. A. Dejaegere and M. Karplus, Analysis of coupling schemes in free energy simulations: A unified description of nonbonded contributions to solvation free energies. *J. Phys. Chem.* **100**, 11148–11164 (1996).

48. P. Kollman, Free energy calculations: Applications to chemical and biochemical phenomena. *Chem. Rev.* **93**, 2395–2417 (1993).

49. T. P. Straatsma and J. A. McCammon, Computational alchemy. *Annu. Rev. Phys. Chem.* **43**, 407–435 (1992).

50. A. Kitao, F. Hirata, and N. Go, Effects of solvent on the conformation and the collective motions of a protein. 2. Structure of hydration in melittin. *J. Phys. Chem.* **97**, 10223–10230 (1993).

51. B. Honig, K. Sharp, and A. Yang, Macroscopic models of aqueous solutions: Biological and chemical applications. *J. Phys. Chem.* **97**, 1101–1109 (1993).

52. G. Perrot, B. Cheng, K. D. Gibson, K. A. Palmer, J. Vila, A. Nayeem, B. Maigret, and H. A. Scheraga, Mseed: A program for the rapid analytical determination of accessible surface areas and their derivatives. *J. Comp. Chem.* **13**, 1–11 (1992).

53. J. D. Augspurger and H. A. Scheraga, An efficient, differentiable hydration potential for peptides and proteins. *J. Comp. Chem.* **17**, 1549–1558 (1996).

54. M. L. Connolly, Analytical molecular surface calculation. *J. Appl. Cryst.* **16**, 548–558 (1983).

55. B. von Freyberg and W. Braun, Minimization of empirical energy functions in proteins including hydrophobic surface area effects. *J. Comp. Chem.* **14**, 510–521 (1993).

56. F. Eisenhaber, P. Lijnzaad, P. Argos, C. Sander, and M. Scharf, The double cubic lattice method: Efficient approaches to numerical integration of surface area and volume and to dot surface contouring of molecular assemblies. *J. Comp. Chem.* **16**, 273–284 (1995).

57. F. Eisenhaber and P. Argos, Improved strategy in analytic surface calculation for molecular systems: Handling of singularities and computational efficiency. *J. Comp. Chem.* **14**, 1272–1280 (1993).

58. R. J. Wawak, K. D. Gibson, and H. A. Scheraga, Gradient discontinuities in calculations involving molecular surface area. *J. Math. Chem.* **15**, 207–232 (1994).

59. L. Wesson and D. Eisenberg, Atomic solvation parameters applied to molecular dynamics of proteins in solution. *Protein Sci.* **1**, 227 (1992).

60. R. Wolfenden, L. Andersson, P. M. Cullis, and C. C. B. Southgate, Affinities of amino acid side chains for solvent water. *Biochemistry* **20**, 849 (1981).

61. J. Kyte and R. F. Doolittle, A simple method for displaying the hydropathic character of a protein. *J. Mol. Biol.* **157**, 105 (1982).

62. R. Friedman, K. A. Sharp, A. Nicholls, and B. Honig, Reconciling the magnitude of the microscopic and macroscopic hydrophobic effects. *Science* **252**, 106 (1991).

63. A. H. Juffer, F. Eisenhaber, S. J. Hubbard, D. Walther, and P. Argos, Comparison of atomic solvation parametric sets: Applicability and limitations in protein folding and binding. *Protein Sci.* **4**, 2499 (1995).

64. A. Ben-Naim and R. M. Mazo, Size dependence of the solvation free energies of large solutes. *J. Phys. Chem.* **97**, 10829 (1993).

65. T. Ooi, M. Oobatake, G. Némethy, and H. A. Scheraga, Accessible surface areas as a measure of the thermodynamic parameters of hydration of peptides. *Proc. Natl. Acad. Sci. USA* **84**, 3086 (1987).

66. C. A. Schiffer, J. W. Caldwell, P. A. Kollman, and R. M. Stroud, Protein structure prediction with a combined solvation free energy-molecular mechanics force field. *Mol. Sim.* **10**, 121 (1993).

67. R. L. Williams, J. Vila, G. Perrot, and H. A. Scheraga, Empirical solvation models in the context of conformational energy searches: Application to bovine pancreatic trypsin inhibitor. *Proteins* **14**, 110–119 (1992).

68. A. J. Hopfinger, Polymer–solvent interactions for homopolypeptides in aqueous solution. *Macromolecules* **4**, 731–737 (1971).

69. Y. K. Kang, G. Némethy, and H. A. Scheraga, Free energies of hydration of solute molecules. 1. Improvement of hydration shell model by exact computations of overlapping volumes. *J. Phys. Chem.* **91**, 4105 (1987).

70. Y. K. Kang, G. Némethy, and H. A. Scheraga, Free energies of hydration of solute molecules 2. Application of the hydration shell model to nonionic organic molecules. *J. Phys. Chem.* **91**, 4109 (1987).

71. Y. K. Kang, G. Némethy, and H. A. Scheraga, Free energies of hydration of solute molecules 3. Application of the hydration shell model to charged organic molecules. *J. Phys. Chem.* **91**, 4118 (1987).

72. Y. K. Kang, K. D. Gibson, G. Némethy, and H. A. Scheraga, Free energies of hydration of solute molecules. 4. Revised treatment of the hydration shell model. *J. Phys. Chem.* **92**, 4739 (1988).

73. C. S. Adjiman, I. P. Androulakis, and C. A. Floudas, Global optimization of minlp problems in process synthesis and design. *Comput. Chem. Eng.* **21**, S445–S450 (1997).

74. H. A. Scheraga, *PACK: Programs for Packing Polypeptide Chains*, 1996, online documentation.

75. T. Noguti and N. Go, A method of rapid calculation of a second derivative matrix of conformational energy for large molecules. *J. Phys. Soc. Japan* **52**(10), 3685–3690 (1983).

76. V. Madison and K. D. Kopple, Solvent-dependent conformational distributions of some dipeptides. *J. Am. Chem. Soc.* **102**(15), 4855–4863 (1980).

77. Z. Li and H. A. Scheraga, Structure and free energy of complex thermodynamic systems. *J. Mol. Struct. (Theochem.)* **179**, 333–352 (1988).

78. I. P. Androulakis, C. D. Maranas, and C. A. Floudas, Global minimum potential energy conformation of oligopeptides. *J. Glob. Opt.* **11**(1), 1–34 (1997).

79. W. H. Graham, E. S. Carter II, and R. P. Hicks, Conformational analysis of met-enkephalin in both aqueous solution and in the presence of sodium dodecyl sulfate micelles using multidimensional nmr and molecular modeling. *Biopolymers* **32**, 1755–1764 (1992).

80. S. K. Burley and G. A. Petsko, Aromatic–aromatic interaction: A mechanism of protein structure stabilization. *Science* **229**, 23–28 (1985).

81. J. L. Klepeis and C. A. Floudas, Comparative study of global minimum energy conformations of hydrated peptides. *J. Comput. Chem.* **20**, 636 (1999).

82. F. H. Stillinger and T. A. Weber, Inherent pair correlation in simple liquids. *J. Chem. Phys.* **80**(9), 4434–4437 (1984).

83. J. L. Klepeis, I. P. Androulakis, M. G. Ierapetritou, and C. A. Floudas, Predicting solvated peptide conformations via global minimization of energetic atom-to-atom interactions. *Comput. Chem. Eng.* **22**, 765–788 (1998).

84. N. Gō and H. A. Scheraga, Analysis of the contribution of internal vibrations to the statistical weights of equilibrium conformations of macromolecules. *J. Chem. Phys.* **51**(11), 4751–4767 (1969).

85. N. Gō and H. A. Scheraga, On the use of classical statistical mechanics in the treatment of polymer chain conformations. *Macromolecules* **9**(4), 535–542 (1976).

86. P. J. Flory, Foundations of rotational isomeric state theory and general methods for generating configurational averages. *Macromolecules* **7**(3), 381–392 (1974).

87. M. Vásquez, G. Némethy, and H. A. Scheraga, Conformational energy calculations on polypeptides and proteins. *Chem. Rev.* **94**, 2183–2239 (1994).

88. H. Meirovitch and E. Meirovitch, Efficiency of monte carlo minimization procedures and their use in analysis of nmr data obtained from flexible peptides. *J. Comput. Chem.* **18**, 240–253 (1997).

89. H. Meirovitch and M. Vásquez, Efficiency of simulated annealing and the monte carlo minimization method for generating a set of low energy structures of peptides. *J. Mol. Struct. (Theochem.)* **398–399**, 517–522 (1997).

90. S. S. Zimmerman, M. S. Pottle, G. Némethy, and H. A. Scheraga, Conformational analysis of the 20 naturally occurring amino acid residues using ecepp. *Macromolecules* **10**, 1–9 (1977).

91. U. H. Hansmann, M. Masuya, and Y. Okamoto, *Proc. Natl. Acad. Sci. USA* **94**, 10652–10656 (1997).

92. J. L. Klepeis, C. A. Floudas, D. Morikis, and J. D. Lambris, Predicting peptide structures using nmr data and deterministic global optimization. *J Comp Chem.* **20**, 1354–1370 (1999).

93. D. M. Standley, V. A. Eyrich, A. K. Felts, R. A. Friesner, and A. E. McDermott, A branch and bound algorithm for protein structure refinement from sparse nmr data sets. *J Mol Biol*, **285**, 1691–1710 (1999).

94. P. Güntert, C. Mumenthaler, and K. Wüthrich, Torsion angle dynamics for nmr structure calculation with the new program dyana. *J. Mol. Biol.* **273**, 283–298 (1997).

95. L. M. Rice and A. T. Brünger, Torsion angle dynamics: Reduced variable conformational sampling enhances crystallographic structure refinement. *Proteins* **19**, 277–290 (1994).

96. A. T. Brünger, *X-PLOR, Version 3.1: A System for X-ray Crystallography and nmr*, Yale University Press, New Haven, CT, 1992.

97. Y. Duan and P. A. Kollman, Pathways to a protein folding intermediate observed in a 1-microsecond simulation in aqueous solution. *Science* **282**, 740–744 (1998).

98. V. Daggett, A. J. Li, and A. R. Fersht, Combined molecular dynamics and phi-value analysis of structure-reactivity relationships in the transition state and unfolding pathway of barnase: Structural basis of Hammond and anti-Hammond effects. *J. Am. Chem. Soc.* **120**, 12740–12754 (1998).

99. L. S. D. Caves, J. D. Evanseck, and M. Karplus, Locally accessible conformations of proteins: Multiple molecular dynamics simulations of cramb in. *Protein Sci.* **7**, 649–666 (1998).

100. A. Jain, N. Vaidehi, and G. Rodriguez, A fast recursive algorithm for molecular dynamics simulation. *J. Comp. Phys.* **106**, 258–268 (1993).

101. H. J. C. Berendsen, J. P. M. Postma, W. F. van Gunsteren, A. DiNola, and J. R. Haak, Molecular dynamics with coupling to an external bath. *J. Chem. Phys.* **81**, 3684–3690 (1984).

102. A. Sahu, B. K. Kay, and J. D. Lambris, Inhibition of human complement by a c3-binding peptide isolated from a phage-displayed random peptide library. *J. Immunol.* **157**, 884–891 (1996).

103. D. Morikis, N. Assa-Munt, A. Sahu, and J. D. Lambris, Solution structure of compstatin, a potent complement inhibitor. *Protein Sci.* **7**, 619–627 (1998).

104. J. L. Klepeis and C. A. Floudas, Deterministic global optimization and torsion angle dynamics for molecular structure prediction. *Comp. Chem. Eng.* **24**, 1761–1766 (2000).

105. D. Hinds and M. Levitt, Exploring conformational space with a simple lattice model for protein structure. *J. Mol. Biol.* **243**, 668 (1994).

106. A. R. Ortiz, A. Kolinski, and J. Skolnick, Native like topology assembly of small proteins using predicted restraints in Monte Carlo folding simulations. *Proc. Natl. Acad. Sci. USA* **95**, 1020–1025 (1998a).

107. J. Skolnick, A. Kolinski, and A. R. Ortiz, Monsster: A method for folding globular proteins with a small number of distance restraints. *J. Mol. Biol.* **265**, 217 (1997).

108. K. T. Simons, I. Ruczinki, C. Kooperberg, B. A. Fox, C. Bystroff, and D. Baker, Improved recognition of native like structures using a combination of sequence dependent and sequence independent features of proteins. *Proteins* **34**, 82 (1999).

109. D. Shortle, K. T. Simons, and D. Baker, Clustering of low energy conformations near the native structure of small proteins. *Proc. Natl. Acad. Sci. USA* **95**, 11158 (1998).

110. S. Sun, P. D. Thomas, and K. A. Dill, A simple protein folding algorithm using a binary code and secondary structure constraints. *Protein Eng.* **8**, 769 (1995).

111. A. Monge, R. A. Friesner, and B. Honig, An algorithm to generate low resolution protein tertiary structures from knowledge of secondary structure, *Proc. Natl. Acad. Sci. USA* **91**, 5027 (1994).

112. A. Monge, E. J. P. Lathrop, J. R. Gunn, P. S. Shenkin, and R. A. Friesner, Computer modeling of protein folding: Conformational and energetic analysis of reduced and detailed models. *J. Mol. Biol.* **247**, 995, (1995).

113. H. A. Scheraga, J. Lee, J. Pillardy, Y. J. Lee, A. Liwo, and D. Ripoll, Surmounting the multiple minima problem in protein folding. *J. Glob. Opt.* **15**, 235 (1999).

114. A. Liwo, J. Lee, D. Ripoll, J. Pillardy, and H. A. Scheraga, Surmounting the multiple minima problem in protein folding. *Proc. Natl. Acad. Sci. USA* **96**, 5482 (1999).

115. J. Y. Lee, H. A. Scheraga, and S. Rackovsky, Conformational analysis of the 20-residue membrane-bound portion of melittin by conformational space annealing. *Biopolymers* **46**, 103–115 (1998).

116. R. Srinivasan and G. D. Rose, A physical basis for protein secondary structure. *PNAS* **96**, 14258–14263 (1999).

117. K. Yue and K. A. Dill, Folding proteins with a simple energy function and extensive conformational searching. *Protein Sci.* **5**, 254 (1996).

118. K. A. Dill, A. T. Phillips, and J. B. Rosen, Protein structure and energy landscape dependence of sequence using a continuous energy function. *J. Comput. Biol.* **4**, 227 (1997).

119. C. A. Orengo, J. E. Bray, T. Hubbard, L. LoConte, and I. Sillitoe, Analysis and assessment of *ab initio* three dimensional prediction, secondary structure and contacts prediction. *Proteins Suppl.* **3**, 149 (1999).

120. J. L. Klepeis and C. A. Floudas, *Ab-initio* structure prediction in protein folding. In preparation, 2000.

121. C. A. Floudas, *Nonlinear and Mixed-Integer Optimization*, Oxford University Press, New York, (1995).

122. R. L. Baldwin and G. D. Rose, Is protein folding hierarchic? I. Local structure and peptide folding. *Trends Biochem. Sci.* **24**(1), 26–33 (1999).

123. R. L. Baldwin and G. D. Rose, Is protein folding hierarchic? II. Folding intermediates and transition states. *Trends Biochem. Sci.* **24**(2) 77–83 (1999).

124. A. Chakrabartty and R. L. Baldwin, Stability of α-helices. *Adv. Protein Chem.* **46**, 141–175 (1995).

125. H. J. Dyson and P. E. Wright, Defining solution conformations of small linear peptides. *Annu. Rev. Biophys. Biophys. Chem.* **20**, 519–538 (1991).

126. U. H. E. Hansmann and Y. Okamoto, Finite-size scaling of helix-coil transitions in polyalanine studied by multicannonical simulations. *J. Chem. Phys.* **110**(2), 1267–1276 (1999).

127. D. T. Clarke, A. J. Doig, B. J. Stapley, and G. R. Jones, The α-helix folds on the millisecond time scale. *Proc. Natl. Acad. Sci. USA* **96**(13), 7232–7237 (1999).

128. S. Marqusee and R. L. Baldwin, Helix stabilization by glu⁻ ··· lys⁺ salt bridges in short peptides of *de novo* design. *Proc. Natl. Acad. Sci. USA* **84**, 8898–8902 (1987).

129. K. R. Shoemaker, P. S. Kim, E. J. York, J. M. Stewart, and R. L. Baldwin, Tests of the helix dipole model for stabilization of α-helices. *Nature* **326**, 563–567 (1987).

130. K. M. Westerberg and C. A. Floudas, Locating all transition states and studying the reaction pathways of potential energy surfaces. *J. Chem. Phys.* **110**(18), 9259–9295 (1999).

131. K. M. Westerberg and C. A. Floudas, Dynamics of peptide folding: Transition states and reaction pathways of solvated and unsolvated tetra-alanine. *J. Glob. Opt.* **15**(3), 261–297 (1999).

132. T. Kortemme, M. Ramirez-Alvarado, and L. Serrano, Design of a 20-amino acid, three-stranded beta-sheet protein. *Science* **281**, 253–256 (1998).

133. M. Ramirez-Alvarado, F. J. Blanco, and L. Serrano, *De novo* design and structural analysis of a model beta-hairpin peptide system. *Natl. Struct. Biol.* **3**(7), 604–612 (1996).

134. E. de Alba, M. A. Jiménez, M. Rico, and J. L. Nieto, Conformational investigation of designed short linear peptides able to fold into β-hairpin structures in aqueous solution. *Fold. Des.* **1**(2), 133–144 (1996).

135. F. J. Blanco, G. Rivas, and L. Serrano, A short linear peptide that folds into a native stable beta-hairpin in aqueous solution. *Nat. Struct. Biol.* **1**(9), 584–590 (1994).

136. F. B. Sheinerman and C. L. Brooks III, Calculations on folding of segment b1 of streptococcal protein g. *J. Mol. Biol.* **278**(2), 439–456 (1998).

137. V. S. Pande and D. S. Rokhsar, Molecular dynamics simulations of unfolding and refolding of a β-hairpin fragment of protein g. *Proc. Natl. Acad. Sci. USA* **96**(16), 9062–9067 (1999).

138. A. R. Dinner, T. Lazaridis, and M. Karplus, Understanding beta-hairpin formation. *Proc. Natl. Acad. Sci. USA* **96**, 9068–9073 (1999).

139. S. Honda, N. Kobayashi, and E. Munekata, Theormodynamics of a β-hairpin structure: Evidence for cooperative formation of folding nucleus. *J. Mol. Biol.* **295**(2), 269–278 (2000).

140. B. Ma and R. Nussinov, Molecular dynamics simulations of a β-hairpin fragment of protein g: Balance between side-chain and backbone forces. *J. Mol. Biol.* **296**(4), 1091–1104 (2000).

141. V. Munoz, E. R. Henry, J. Hofrichter, and W. A. Eaton, A statistical mechanical model for β-hairpin kinetics. *Proc. Natl. Acad. Sci. USA* **95**(11), 5872–5879 (1998).

142. W. A. Eaton, V. Munoz, P. A. Thompson, E. R. Henry, and J. Hofrichter, Kinetics and dynamics of loops, α-helices, β-hairpins and fast-folding proteins, *Acc. Chem. Res.*, **31**(11), 745–753 (1998).

143. B. D. Bursulaya and C. L. Brooks III, Folding free energy surface of a three-stranded β-sheet protein. *J. Am. Chem. Soc.* **121**(43), 9947–9951 (1999).

144. A. M. J. J. Bonvin and W. F. van Gunsteren, β-Hairpin stability and folding: Molecular dynamics studies of the first β-hairpin of tendamistat. *J. Mol. Biol.* **296**(1), 255–268 (2000).

145. C. J. Tsai and K. D. Jordan, Use of an eigenmode method to locate the stationary points on the potential-energy surfaces of selected argon and water clusters. *J. Phys. Chem.* **97**(43), 11227–11237 (1993).

146. J. Simons, P. Jorgensen, H. Taylor, and J. Ozment, Walking on potential energy surfaces. *J. Phys. Chem.* **87**(15), 2745–2753 (1983).

147. A. Banerjee, N. Adams, J. Simons, and R. Shepard, Search for stationary points on surface. *J. Phys. Chem.* **89**(1), 52–57 (1985).

148. C. J. Cerjan and W. H. Miller, On finding transition states. *J. Chem. Phys.* **75**(6), 2800–2806 (1981).

149. D. O'Neal, H. Taylor, and J. Simons, Potential surface walking and reaction paths for $be + h_2 \rightarrow beh_2 \rightarrow be + 2h$. *J. Phys. Chem.* **88**(8), 1510–1513 (1984).

150. P. Culot, G. Dive, V. H. Nguyen, and J. M. Ghuysen, A quasi-newton algorithm for first-order saddle-point location. *Theor. Chim. Acta.* **82**(3–4), 189–205 (1992).

151. R. S. Berry, Potential surfaces and dynamics: What clusters tell us. *Chem. Rev.* **93**(7), 2379–2394 (1993).

152. R. S. Berry, H. L. Davis, and T. L. Beck, Finding saddles on multidimensional potential surfaces. *Chem. Phys. Lett.* **147**(1), 13–17 (1988).

153. J. Y. Lee, H. A. Scheraga, and S. Backovsky, New optimization method for conformational energy calculations on polypeptides: Conformational space annealing. *J. Comp. Chem.* **18**(9), 1222–1232 (1997).

154. J. Y. Lee and H. A. Scheraga, Conformational space annealing by parallel computations: Extensive conformational search of met-enkephalin and of the 20-residue membrane-bound portion of melittin. *Int. J. Quant. Chem.* **75**(3), 255–265 (1999).

155. R. J. Wawak, J. Pillardy, A. Liwo, K. D. Gibson, and H. A. Scheraga, Diffusion equation and distance scaling methods of global optimization: Applications to crystal structure prediction. *J. Phys. Chem. A* **102**(17), 2904–2918 (1998).

156. K. A. Dill, A. T. Phillips, and J. B. Rosen, Cgu: An algorithm for molecular structure prediction, in *IMA Volumes in Mathematics and Its Applications*, Vol. 94, Springer-Verlag, Berlin, 1997, pp. 1–21.

157. M. F. Jarrold, Introduction to statistical reaction rate theories, in *Clusters of Atoms and Molecules*, H. Haberland, ed., Springer, Berlin, 1994, pp. 163–186.

158. R. E. Kunz and R. S. Berry, Statistical interpretation of topographies and dynamics of multidimensional potentials. *J. Chem. Phys.* **103**(5), 1904–1912 (1995).

159. H. B. Schlegel, Geometry optimization on potential energy surfaces, in *Modern Electronic Structure Theory*, D. R. Yarkony, ed., World Scientific Publishing, Singapore, (1995), pp. 459–500.

160. R. Fletcher and M. J. D. Powell, A rapidly convergent descent method for minimization. *Comput. J.* **6**(2), 163–168 (1963).

161. J. Greenstadt, Variations on variable-metric methods. *Math. Comp.* **24**(109), 1–22 (1970).

162. D. M. Gay, Sumsl, 1980 (FORTRAN source code).

163. K. D. Ball and R. S. Berry, Realistic master equation modeling of relaxation on complete potential energy surfaces: Partition function and equilibrium results. *J. Chem. Phys.* **109**(19), 8541–8556 (1998).

164. K. D. Ball and R. S. Berry, Realistic master equation modeling of relaxation on complete potential energy surfaces: Kinetic results. *J. Chem. Phys.* **109**(19), 8557–8572 (1998).

165. T. Lazaridis and M. Karplus, Effective energy function for proteins in solution. *Proteins: Struct. Funct. Genet.* **35**(2), 133–152 (1999).

166. A. D. MacKerell, Jr., D. Bashford, M. Bellott, R. L. Dunbrack, Jr., J. D. Evanseck, M. J. Field, S. Fischer, J. Gao, H. Guo, S. Ha, D. Joseph-McCarthy, L. Kuchnir, K. Kuczera, F. T. K. Lau, C. Mattos, S. Michnick, T. Ngo, D. T. Nguyen, B. Prodhom, W. E. Reiher III, B. Roux, M. Schlenkrich, J. C. Smith, R. Stote, J. Straub, M. Watanabe, J. Wiorkiewicz-Kuczera, D. Yin, and M. Karplus, All-atom empirical potential for molecular modeling and dynamics studies of proteins. *J. Phys. Chem. B* **102**(18), 3586–3616 (1998).

167. J. W. Ponder, *TINKER, Software Tools for Molecular Design, Version 3.6*, Department of Biochemistry and Molecular Biophysics; Washington University School of Medicine, St. Louis, MO, 1998.

168. R. G. Urban and R. M. Chicz, *MHC Molecules: Expression, Assembly and Function*. R. G. Landes Company and Chapman & Hall, London, (1996).

169. D. H. Fremont, M. Matsumura, E. A. Stura, P. A. Peterson, and I. A. Wilson, Crystal structures of two viral peptides in complex with murine mhc class I h-2 k^b. *Science* **257**, 919–927 (1992).

170. M. L. Silver, H. C. Guo, J. L. Strominger, and D. Wiley, Atomic structure of a human mhc molecule presenting an influenza virus peptide. *Nature* **360**, 367–368 (1992).

171. L. Stern, J. Brown, T. Jardetzky, J. Gorga, R. Urban, L. Strominger, and D. Wiley, Crystal structure of the human class II mhc protein hla-dr1 complexed with an influenza virus peptide. *Nature* **368**, 215–221 (1994).

172. T. L. Blundell, B. L. Sibanda, M. J. E. Sternberg, and J. M. Thornton, Knowledge-based prediction of protein structures and the design of novel molecules. *Nature* **326**, 347 (1987).

173. M. J. Sutcliffe, I. Haneef, D. Carney, and T. L. Blundell, Knowledge-based modeling of homologous proteins, part I: Three dimensional frameworks derived from the simultaneous superposition of multiple structures. *Protein Eng.* **1**, 377, (1987).

174. R. Chandrasekaran and G. N. Ramachandran, Studies on the conformation of amino acids. xi. Analysis of the observed side group conformations in proteins. *Int. J. Protein Res.* **2**, 223 (1970).

175. R. L. Dunbrack and M. Karplus, Backbone-dependent rotamer library for proteins: Application to side-chain prediction. *J. Mol. Biol.* **230**, 543 (1993).

176. H. Schauber, F. Eisenhaber, and P. Argos, Rotamers: To be or not to be? An analysis of amino acid side-chain conformations in globular proteins. *J. Mol. Biol.* **230**, 592 (1993).

177. M. Vasquez, An evaluation of discrete and continuous search techniques for conformational analysis of side-chains in proteins. *Biopolymers* **36**, 53 (1995).

178. S. Y. Chung and S. Subbiah, A structural explanation for the twilight zone of protein sequence homology. *Structure* **4**, 1123 (1996).

179. P. Koehl and M. Delarue, Application of a self-consistent mean field theory to predict protein side-chains conformation and estimate their conformational entropy. *J. Mol. Biol.* **239**, 249 (1994).

180. L. Holm and C. Sander, Fast and simple Monte-Carlo algorithm for side-chain optimization in proteins: Application to model building by homology. *Proteins: Struct. Funct. Genet.* **14**, 213 (1994).

181. P. Tuffery, C. Etchebest, S. Hazout, and R. Lavery, A new approach to the rabid determination of protein side-chain conformations. *J. Biomol. Struct. Dynam.* **8**, 1267 (1991).

182. J. K. Hwang and W. F. Liao, Side-chain prediction by neural networks and simulated annealing optimization. *Protein Eng.* **8**, 363 (1995).

183. M. G. Ierapetritou, I. P. Androulakis, D. S. Monos, and C. A. Floudas, Structure prediction of binding sites of MHC Class II molecules based on the crystal of HLA–DRB1 and global optimization, in *Optimization in Computational Chemistry and Molecular Biology: Local and Global Approaches*, Kluwer Academic Publishers, Hingham, MA, 2000, pp. 157–189.

184. I. D. Kuntz, J. M. Blaney, S. J. Oatley, R. Landgridge, and T. E. Ferrin, A geometric approach to macromolecule–ligand interactions. *J. Mol. Biol.* **161**, 269–288 (1982).

185. M. L. Connolly, Solvent accessible surfaces of proteins and nucleic acids. *Science* **221**, 709–713 (1983).

186. B. Lee and F. M. Richards, The interpretation of protein structures: Estimation of static accessibility. *J. Mol. Biol.* **55**, 379–400 (1971).

187. P. D. J. Grootenhuis and P. A. Kollman, Crown ether–neutral molecule interactions studied by molecular mechanics and free energy perturbation calculations. near quantitative agreement between theory and experimental binding free energies. *J. Am. Chem. Soc.* **111**, 4046–4051 (1989).

188. J. Shen and F. A. Quiocho, Calculation of binding energy differences for receptor–ligand systems using the Poisson–Boltzmann methods. *J. Comput. Chem.* **16**, 445–448 (1995).

189. S. Miyamoto and P. A. Kollman, What determines the strength of noncovalent association of ligands to proteins in aqueous solutions? *Proc. Natl. Acad. Sci. USA.* **90**, 8402–8406 (1993).

190. C. A. Reynolds, P. M. King, and W. G. Richards, Free energy calculations in molecular biophysics. *Mol. Phys.* **76**, 251–275 (1992).

191. E. Di Cera, *Thermodynamic Theory of Site-Specific Binding Processes in Biological Macromolecules*, Cambridge University Press, New York, 1995.

192. A. Wallquist, R. L. Jernigan, and D. G. Covell, A preference-based free energy parameterization of enzyme-inhibitor binding. applications to hiv-1 protease inhibitor design. *Protein Sci.* **4**, 1881–1903 (1995).

193. R. D. Head, M. L. Smyte, T. I. Oprea, C. L. Waller, S. M. Green, and G. R. Marshall, Validate: A new method for receptor-based prediction of binding affinities of novel ligands. *J. Am. Chem. Soc.* **118**, 3959–3969 (1996).

194. G. Verkhivker, K. Appelt, S. T. Freer, and J. E. Vilafranca, Empirical free energy calculations of ligand–protein crystallographic complexes. 1. knowledge based ligand–protein interaction potentials applied to the prediction of human immunodeficiency virus 1 protease binding affinity. *Protein Eng.* **8**, 677–691 (1995).

195. A. N. Jain and M. A. Murcko, Computational methods to predict binding free energy in ligand–receptor complexes. *J. Med. Chem.* **38**, 4953–4967 (1995).

196. S. Vajda, M. Sippl, and J. Novotny, Empirical potentials and functions for protein folding and binding. *Curr. Opin. Struct. Biol.* **7**, 222–228 (1997).

197. J. Janin and S. J. Wodak, Reaction pathway for the quartenary structure change in hemoglobin. *Biopolymers* **24**, 509–526 (1985).

198. S. J. Wodak and J. Janin, Computer analysis of protein–protein interactions. *J. Mol. Biol.* **124**, 323–342 (1978).

199. S. J. Wodak, M. De Crombrugghe, and J. Janin, Computer studies of interactions between macromolecules. *Prog. Biophys. Mol. Biol.* **49**, 29–63 (1987).

200. J. Cherfils, S. Duquerroy, and J. Janin, Protein–protein recognition analyzed by docking simulation, *Proteins* **11**, 271–280 (1991).

201. R. H. Lee and G. D. Rose, Molecular recognition. i. Automatic identification of topographic surface features. *Biopolymers* **24**, 1613–1627 (1985).

202. M. L. Connolly, Shape complementarity at the hemoglobin $\alpha_1\beta_1$ subunit interface. *Biopolymers* **25**, 1229–1247 (1986).

203. D. J. Bacon and J. Moult, Docking by least-squares fitting of molecular surface patterns. *J. Mol. Biol.* **225**, 849–858 (1992).

204. R. L. DesJarlais, G. L. Seibel, I. D. Kuntz, P. S. Furth, J. C. Alvarez, P. R. Ortiz de Montellano, D. L. Decamp, L. M. Babe, and C. S. Craik, Structure based design of nonpeptide inhibitors specific for the human immunodeficiency virus-1 protease. *Proc. Natl. Acad. Sci. USA* **87**, 6644–6648 (1990).

205. R. L DesJarlais, R. P. Sheridan, G. L. Seibel, J. S. Dixon, I. D. Kuntz, and R. Venkataraghavan, Using shape complementarity as an initial screen in designing ligands for a receptor binding site of known three-dimensional structure. *J. Med. Chem.* **31**, 722–729 (1988).

206. A. R. Leach and I. D. Kuntz, Conformational analysis of flexible ligands in macromolecular receptor sites. *J. Comput. Chem.* **13**, 733–748 (1992).

207. B. K. Shoichet and I. D. Kuntz, Protein docking and complementarity. *J. Mol. Biol.* **221**, 327–346 (1991).

208. F. Jiang and S-H. Kim, "Soft docking": Matching of molecular surface cubes. *J. Mol. Biol.* **219**, 79–102 (1991).

209. P. J. Goodford, A computational procedure for determining energetically favorable binding sites on biologically important molecules. *J. Med. Chem.* **28**, 849–857 (1985).

210. P. M. Pardalos, X. Liu, and G. L. Xue, Protein conformation of a lattice model using tabu search. *J. Glob. Optim.* **11**, 55–68 (1997).

211. E. C. Meng, B. K. Shoichet, and I. D. Kuntz, Automated docking with grid-based energy evaluation. *J. Comput. Chem.* **13**, 505–524 (1992).

212. H. Wang, Grid-search molecular accessible surface algorithm for solving the protein docking problem. *J. Comput. Chem.* **12**, 746–750 (1991).

213. S. K. Kearsley, D. J. Underwood, R. P. Sheridan, and M. D. Miller, Flexibases: A way to enhance the use of molecular docking methods. *J. Comput. Aided Mol. Design* **8**, 565–582 (1994).

214. M. D. Miller, S. K. Kearsley, D. J. Underwood, and R. P. Sheridan, Flog: A system to select "quasi-flexible" ligand complementary to a receptor of known three-dimensional structure. *J. Comput. Aided Mol. Design* **8**, 153–174 (1994).

215. D. S. Goodsell and A. J. Olson, Automated docking of substrates to proteins by simulated annealing. *Proteins* **8**, 195–202 (1990).

216. A. Calfisch, P. Niederer, and M. Anliker, Monte Carlo docking of oligopeptides to proteins. *Proteins* **13**, 223–230 (1992).

217. T. N. Hart and R. J. Read, A multiple-start Monte Carlo docking method. *Proteins* **13**, 206–222 (1992).

218. T. N. Hart and R. J. Read, Multiple-start Monte Carlo docking of flexible ligands, in *The Protein Folding Problem and Tertiary Structure Prediction*, Birkhäuser, 1994, pp. 71–108.

219. Jean-Yves Trosset and Harold A. Scheraga, Prodock: Software package for protein modeling and docking. *J. Comput. Chem.* **20**, 412–427 (1999).

220. C. M. Oshiro, I. D. Kuntz, and J. S. Dixon, Flexible ligand docking using a genetic algorithm. *J. Comput. Aided Mol. Design* **9**, 113–130 (1995).

221. G. Jones, P. Willett, and R. C. Glen, Molecular recognition of receptor sites using a genetic algorithm with a description of desolvation. *J. Mol. Biol.* **245**, 43–53 (1995).

222. R. S. Judson, Y. T. Tan, E. Mori, C. Melius, E. P. Jaeger, A. M. Treasurywala, and A. Mathiowetz, Docking flexible molecules: A case study of three proteins. *J. Comput. Chem.* **16**, 1405–1419 (1995).

223. J. B. Moon and J. Howe, Computer design of bioactive molecules: A method for receptor-based *de novo* ligand design. *Proteins* **11**, 314–328 (1991).

224. S. H. Rotstein and M. A. Murcko, Groupbuild: A fragment-based method for *de novo* drug design. *J. Med. Chem.* **36**, 1700–1710 (1993).

225. S. H. Rotstein and M. A. Murcko, Gensstar: A program for *de novo* drug design. *J. Comput. Aided Mol. Design* **7**, 23–43 (1993).

226. M. C. Lawrence and P. C. Davis, Clix: A search algorithm for finding novel ligands capable of binding proteins of known three-dimensional structure. *Proteins* **12**, 31–41 (1992).

227. H. J. Böhm, The computer program ludi: A new method for the *de novo* design of enzyme inhibitors. *J. Comput. Aided Mol. Design* **6**, 61–78 (1992).

228. H. J. Böhm, Ludi: Rule-based automatic design of new substituents for enzyme inhibitor leads. *J. Comput. Aided Mol. Design* **6**, 593–606 (1992).

229. A. Miranker and M. Karplus, Functionality maps of binding sites: A multicopy simultaneous search method. *Proteins* **1**, 29–34 (1991).

230. R. Rosenfeld, Q. Zheng, S. Vajda, and C. DeLisi, Computing the structure of bound peptides. application to antigen recognition by class I major histocompatibility complex receptors. *J. Mol. Biol.* **234**, 515–521 (1993).

231. A. Calfisch, A. Miranker, and M. Karplus, Multiple copy simultaneous search and construction of ligands in binding sites: Application of inhibitors of hiv-1 aspartic proteinase. *J. Med. Chem.* **36**, 2142–2164 (1993).

232. K. Gulukota, S. Vajda, and C. DeLisi, Peptide docking using dynamic programming. *J. Comput. Chem.* **17**, 418–428 (1996).

233. D. Rogman, L. Scapozza, G. Folkers, and A. Daser, Molecular dynamics simulation oof mhc-peptide complexes as a tool for predicting potential T cell epitopes. *Biochemistry* **33**, 11476–11485 (1994).

234. I. P. Androulakis, N. N. Nayak, M. G. Ierapetritou, D. S. Monos, and C. A. Floudas, A predictive method for the evaluation of peptide binding in pocket 1 of hla-drb1 via global minimization of energy interactions. *Proteins* **29**, 87–102 (1997).

235. C. A. Floudas, J. L. Klepeis, and P. M. Pardalos, Global optimization approaches in protein folding and peptide docking. *DIMACS Ser. Discrete Math. Theor. Comput. Sci.* **47**, 141–171 (1999).

236. P. Ghosh, M. Amaya, E. Mellins, and D. C. Wiley, The structure of an intermediate in class II mhc maturation: Clip bound to hla-dr3. *Nature* **378**, 457–462 (1995).

237. D. H. Fremont, W. A. Hendrickson, P. Marrack, and J. Kappler, Structures of an mhc class ii molecule with covalently bound single peptides. *Science* **272**, 1001–1004 (1996).

238. J. Vila, R. L. Williams, M. Vasquez, and H. A. Scheraga, Empirical solvation models can be used to differentiate native from non-native conformations of bovine pancreatic trypsin inhibitor. *Proteins* 199–218 (1991).

239. S. J. Remington and B. W. Matthews, *Proc. Natl. Acad. Sci. USA* **75**, 2180 (1978).

240. S. T. Rao and M. G. Rossmann, *J. Mol. Biol.* **76**, 241 (1973).

241. R. Diamond, On the comparison of conformations using linear and quadratic transformations. *Acta Cryst.* 1 (1976).

242. A. L. Mackay, The generalized inverse and inverse structure. *Acta Cryst.* 212 (1977).

243. W. Kabsh, A solution for the best rotation to relate two sets of vectors. *Acta Cryst.* 922 (1976).

244. W. Kabsh, A discussion of the solution for the best rotation to relate two sets of vectors. *Acta Cryst.* 827 (1978).

245. A. D. McLachlan, A mathematical procedure for superimposing atomic coordinates of proteins. *ACTA Cryst.* **A28**, 656 (1972).

246. A. D. McLachlan, Gene duplications in the structural evolution of chymotrypsin. *J. Mol. Biol.* **128**, 49 (1979).

247. I. P. Androulakis, C. D. Maranas, and C. A. Floudas, Prediction of oligopeptide conformations via deterministic global optimization. *J. Glob. Opt.* **11**, 1–34 (1997).

248. L. Stryer, *Biochemistry*, 4th ed., W. H. Freeman, New York, 1995.

249. J. Hammer, C. Bolin, D. Papadopoulos, J. Walsky, J. Higelin, W. Danho, F. Sinigaglia, and Z. A. Nagy, High-affinity binding of short peptides to major histocompatibility complex class ii molecules by anchor combinations. *Proc. Natl. Acad. Sci.* **91**, 4456 (1994).

250. X. Fu, C. Bono, S. Woulfe, C. Swearingen, N. Summers, F. Sinigaglia, A. Sette, B. Schwartz, and R. W. Carr, Pocket 4 of the hla-dr molecule is a major determinant of T cell recognition of peptide. *J. Exp. Med.* **181**, 915–926 (1995).

251. M. K. Gilson and B. H. Honig, Calculation of the total electrostatic energy of a macromolecular system: Solvation energies, binding energies, and conformational analysis. *Proteins Struct. Funct. Genet.* **4**, 7–18 (1988).

252. M. K. Gilson, K. A. Sharp, and B. H. Honig, Calculating the electrostatic potential of molecules in solution: Method and error assessment. *J. Comput. Chem.* **9**, 327–335 (1988).

253. K. A. Sharp and B. H. Honig, Electrostatic interactions in macromolecules: Theory and applications. *Ann. Rev. Biophys. Biophys. Chem.* **19**, 301–332 (1990).

DETECTING NATIVE PROTEIN FOLDS AMONG LARGE DECOY SETS WITH THE OPLS ALL-ATOM POTENTIAL AND THE SURFACE GENERALIZED BORN SOLVENT MODEL

ANDERS WALLQVIST, EMILIO GALLICCHIO, ANTHONY K. FELTS, AND RONALD M. LEVY

Department of Chemistry, Rutgers University, Wright-Rieman Laboratories, Piscataway, NJ, U.S.A.

CONTENTS

Computational Methods for Protein Folding: A Special Volume of Advances in Chemical Physics, Volume 120, Edited by Richard A. Friesner. Series Editors I. Prigogine and Stuart A. Rice.
ISBN 0-471-20955-4. © 2002 John Wiley & Sons, Inc.

I. INTRODUCTION

The ability to distinguish native protein conformations from misfolded ones is a problem of fundamental importance in the development of methods designed to predict protein structure. To this end, several empirical functions for scoring protein conformations have been proposed. [1–6]. Some of these empirical scoring functions implement knowledge-based statistical potentials that are "trained" to recognize native conformations. Knowledge-based potentials are best suited for "threading" applications where the best conformation of a protein is selected from a database of known protein conformations. Scoring functions applicable to *ab initio* folding studies, which require differentiable potentials and the inclusion of excluded volume terms, have also been developed. These are based on combinations of knowledge-based potentials and reduced atomic models sometimes augmented by simplified solvation models based on hydrophobic or hydrophilic exposure [7].

Physics-based all-atom molecular mechanics force fields have not been generally considered practical for fold detection because they are parameterized on small molecule data rather than on proteins directly; the level of atomic detail contained in these models is considered poorly matched to the fold detection problem with respect to both accuracy and computational cost. Recent studies have shown, however, that a scoring function based on the potential energy from an all-atom molecular mechanics force field can recognize native protein conformations among a set of decoys as well as the best available knowledge-based scoring functions [6].

The use of an all-atom force-field minimizes the assumptions that are inherent in an empirical scoring function; and, as will be shown, the inclusion of more refined solvation models enhances our ability to discriminate native folds. An additional value of the all-atom potential lies in its suitability for modeling proteins at higher resolution. This is an important feature for applications in structure–function relation studies such as homology modeling, drug design, and protein–protein recognition.

Although all-atom force fields allow for explicit simulations of solvent, the cost required to appropriately sample solvent configurations rapidly becomes prohibitive. Simplified solvation models are more computationally efficient while preserving a reasonably accurate representation of the interactions between the protein and the water solvent. Although no continuum model can wholly account for the explicit inclusion of solvation [8,9], free energies of solvation of small molecules have been obtained accurately to within a fraction of a kcal/mol relative to experiments using these methods [10–15].

Solvation effects have been included using a variety of simple models [16–23]. These models have been based on exposed surface area, dielectric continuum methods, and screened or modified Coulomb interactions. The validity of a continuum representation of the solvent based on the Poisson–Boltzmann equation has been studied extensively for small and large molecules [24–30]. Continuum solvation models that treat solute and solvent as two dielectric regions with different dielectric constants have been used successfully to account for solute free energies of hydration [11,31–34]. Dielectric models based on the Born model [35] have been developed for which the free energies of hydration are comparable to the predictions of Poisson–Boltzmann and explicit solvent models [36–42].

The inclusion of solvation effects with an all-atom molecular mechanics force field has been shown to be important for the recognition of the native state [16,17,43–45]. Scheraga and co-workers [46,47] used explicit all-atom protein models in conjunction with solvation models based on the molecular exposed surface area. A similar approach by Wang et al.[48,49] showed that inclusion of solvation effects can be successful in discriminating native from non-native structures. Vieth et al. [50] generated structures of the small 33-residue GCN4 leucine zipper proteins using a simplified lattice model; promising structures were then converted to all-atom models and evaluated using a molecular mechanics force field. A hierarchical method of generating large numbers of protein folds was also employed by Monge et al. [20] to select and evaluate structures using the AMBER all-atom force field model [51]. The generalized Born continuum solvent model of Still et al. [37] has been used in this context to represent the aqueous environment. For decoy sets of three different proteins the protocol performed reasonably well in distinguishing the native structure. All-atom models with continuum solvent were also used as the basis for discrimination of non-native states for a small set of 12 deliberately misfolded proteins studied by Vorobjev et al. [52]. In their protocol, conformations for each protein are first sampled from a molecular dynamics trajectory in order to capture micro-states of the protein; this is followed by an evaluation using a dielectric continuum model. Lazaridis and Karplus [22] used the CHARMM19 protein force field together with a Gaussian solvation shell model for the solvation free energy to distinguish deliberately misfolded from native conformations considered on a pairwise basis and in large decoy sets.

Given the complexity of the protein potential surface, it is virtually impossible to consistently find the global minimum starting from an arbitrary point on the surface. Instead, tests have been designed whereby the scoring function is "challenged" to find the native conformation among an ensemble of conformations, most of which are compact but non-native. Many empirical energy functions have been used to identify the correct native structure among a collection of known protein structures using threading techniques [1,53–58]. Scoring functions are also used to identify native-like conformations from a large set containing native and decoy non-native conformers [22,59–63]. Due to the large ensemble of conformations available, the use of large decoy sets to evaluate scoring functions is a more demanding test than threading and is well-suited for the evaluation of scoring functions based on an all-atom force field.

In this work we show that the all atom (OPLS-AA) force field for proteins [64] together with a surface integral formulation of the generalized Born model (SGB) [40,42] is capable of discriminating between native and non-native folds among large sets of compact decoy structures. Validation of the scoring protocol is performed on a large database of well-packed misfolded and near-native protein conformations generated by an algorithm designed to cover exhaustively the relevant parts of conformational space [65,60,66]. The inclusion of near-native decoys in these sets is important in determining whether the scoring function is well-behaved in the vicinity of an idealized native conformation, because it is unlikely that any *ab initio* method of generating conformations will generate that state exactly. In any case, the native state actually represents an ensemble of closely related conformations.

Two additional decoy data sets of misfolded proteins [17] and of predicted protein structures from the Critical Assessment of Techniques for Protein Structure Prediction (CASP) [67] are also used to illustrate the method and its utility. Individual components of the energy perform worse than the total energy; for example, for the bulk of the well-packed decoys, the van der Waals energy provides very little information about structural similarity between a well-packed non-native structure and the native state. It is also shown that some aspects of the SGB model results can be mimicked by a screened electrostatic energy, although the SGB approximation provides a better discriminatory measure between non-native and native states.

II. METHODS

A. Details of the Calculations

The energy of each protein structure investigated was calculated using the OPLS-AA/SGB force field implemented in the IMPACT modeling program (Schrödinger, Inc.) [68]. Initial structures were first minimized in order to remove any artifacts that result from the coordinates being generated with a different energy function; only minimized energies are reported here. All non-native coordinates were taken from independently generated data sets as described below; native protein coordinates were obtained from the Protein Data Bank (PDB) [69]. The force field employed in the calculation of the atomic interactions was the OPLS all-atom force field [64], including parameters for all intramolecular degrees of freedom. The surface formulation of the generalized Born model [37,39] (SGB) as coded in IMPACT was used to estimate the solvation energy [40,70].

The total energy for a protein in vacuum is given by

$$U_{\text{tot}}^{\text{vac}} = U_{\text{bond}} + U_{\text{angle}} + U_{\text{torsion}} + U_{\text{Coulomb}} + U_{\text{vdW}} \tag{1}$$

where the first three terms refer to intramolecular interactions arising from the connectivity of the molecule, and the last terms reflect nonlocal interactions within the protein. The van der Waals energy, U_{vdW}, is modeled by the standard 6-12 Lennard-Jones interaction. The energy of the protein in water calculated according to the SGB continuum solvent model is

$$U_{\text{tot}}^{\text{con}} = U_{\text{tot}}^{\text{vac}} + U_{\text{SGB}} + U_{\text{cav}} \tag{2}$$

where U_{SGB} denotes the electrostatic contribution to the solvation energy calculated using the SGB method, and the cavity term U_{cav} is taken as γA where A is the accessible surface area of the molecule and $\gamma = 5$ cal/(Å^2 mole) [40].

The SGB model is the surface implementation [40,42] of the generalized Born model [37]. The generalized Born equation

$$U_{\text{SGB}} = -\frac{1}{2}\left(\frac{1}{\epsilon_{\text{in}}} - \frac{1}{\epsilon_{\text{w}}}\right)\sum_{ij}\frac{q_i q_j}{f_{ij}(r_{ij})} \tag{3}$$

(where q_i is the charge of atom i, and r_{ij} is the distance between atoms i and j) gives the electrostatic component of the free energy of transfer of a molecule with interior dielectric ϵ_{in} from vacuum to a continuum medium of dielectric constant ϵ_{w}, by interpolating between the two extreme cases that can be solved analytically: one in which the atoms are infinitely separated and the other in which the atoms are completely overlapped. The interpolation function f_{ij} in Eq. (3) is defined as

$$f_{ij} = [r_{ij}^2 + \alpha_i \alpha_j \exp(-r_{ij}^2/4\alpha_i\alpha_j)]^{1/2} \tag{4}$$

where α_i is the Born radius of atom i defined as the effective radius that reproduces through the Born equation

$$U_{\text{single}}^i = -\frac{1}{2}\left(\frac{1}{\epsilon_{\text{in}}} - \frac{1}{\epsilon_{\text{w}}}\right)\frac{q_i^2}{\alpha_i} \tag{5}$$

the electrostatic free energy, U_{single}^i, of the molecule when only the charge of atom i is turned on. The SGB method estimates U_{single}^i by integrating the interaction between atom i and the charge induced on the molecular surface by the Coulomb field of this atom:

$$U_{\text{single}}^i = -\frac{1}{8\pi}\left(\frac{1}{\epsilon_{\text{in}}} - \frac{1}{\epsilon_{\text{w}}}\right)\int_S \frac{(\mathbf{r} - \mathbf{r}_i)}{|\mathbf{r} - \mathbf{r}_i|^4}\cdot\mathbf{n}(\mathbf{r})d^2\mathbf{r} \tag{6}$$

The SGB method has been shown to compare well with the exact solution of the Poisson–Boltzmann (PB) equation. The SGB implementation used in this work includes further correction terms that bring the SGB reaction field energy even closer in agreement with exact PB results [40].

To help assess the ability of the energy function to discriminate between non-native and native protein conformations, the energy gaps between the decoy conformations and the native are evaluated:

$$\Delta U = U_{\text{tot}}^{\text{decoy}} - U_{\text{tot}}^{\text{native}} \tag{7}$$

Energy gaps of individual energy terms have also been examined [see Eqs. (1) and (2)]. Unless explicitly noted, all results presented below were performed without energy cutoffs; that is, all possible non-bonded interactions are included in the total energy. The structural similarity between two protein conformations is expressed as a root mean square deviation (RMSD) between the best overlap of the alpha-carbon (C_α) atoms of the two conformations.

B. Data Sets of Decoys

Although we are probing various energy functions for their ability to differentiate between native and non-native structures, none of the coordinate sets were originally generated by these functions. The vastness of the conformational space and the complexity of an all-atom potential energy function effectively hinders the full sampling of the appropriate degrees of freedom. Scoring conformations with the OPLS/SGB potential may be considered as a last step in the process of generating protein folds; that is, only at the end would it be appropriate to spend the time and effort to evaluate a complex all-atom potential energy function. For this study we focus on existing decoy data sets as our conformational space. These data sets have proven to be highly nontrivial to score correctly.

The first data set contains structure decoys for seven small proteins compiled by Park and Levitt [60]. The protein structures were generated by exhaustively enumerating the backbone rotamers states of 10 selected residues in each protein using an off-lattice model with four discrete dihedral angle states per rotatable bond. From this data set, containing hundreds of thousand of conformations, the authors selected for further evaluation only compact structures that scored well using a variety of scoring functions as well as those having a reasonable RMSD from the native [60]. The coordinates, available on the internet (http://dd.stanford.edu), are all-atom models built from the C_α atoms with the program SEGMOD [71]. No further refinement of these coordinates was done except for minimizing the structures using our energy function (see Eqs. 1 and 2). The decoy data sets are summarized in Table I and encompass a range of small proteins from 54 to 75 residues with varying topological folds. The number of

TABLE I

The Sequence Length, N_{res}, the Number of Decoys, N_{decoy}, and Total Charge of the Seven Proteins of the Park and Levitt Set [60]

PDB Name	N_{res}	N_{decoy}	q (e)
1ctf	68	630	-2
1r69	63	675	$+4$
1sn3	65	660	$+1$
2cro	65	674	$+6$
3icb	75	653	-7
4pti	58	687	$+6$
4rxn	54	677	-12

decoys in these sets ranged from 630 for 1ctf (the carboxy-terminal domain of L7/L12 50s ribosomal protein from *Escherichia coli*) to 687 for 4pti (bovine pancreatic trypsin inhibitor).

An extended data set for the calcium-binding protein calbindin D9K from bovine intestine (4icb) was also investigated using 2000 best-scoring conformations constructed using an *ab initio* procedure [72]. These structures were generated from an exhaustive enumeration on a tetrahedral lattice [73,74] and selected using a combination of scoring functions.

A third data set consists of 26 misfolded protein coordinates constructed by threading the original sequence on to non-native folds with the same number of residues [17]. These structures were generated by swapping main chains between folds and placing the side chains using an annealing protocol. From this data set we selected 25 misfolded structures with continuous backbone coordinates for analysis. These latter coordinate sets were also taken from the internet site listed above.

A fourth data set derived from the CASP3 [67] targets and model submissions was also investigated. CASP3 is the third experiment run by the Protein Structure Prediction Center at Lawrence Livermore National Laboratory to test how well protein structures can be predicted from amino acid sequence. Results are available on the internet at http://predictioncenter.llnl.gov/casp3/Casp3.html. For our calculations, submitted targets were chosen for which coordinates of the native structure were available from the PDB. For each target, models were chosen which had predictions over all residues given in the PDB file. We selected 11 targets and a total of 167 models, with RMS deviations ranging from 1.3 Å to 22.9 Å. The target structures investigated are given in Table IV.

The energy of each native and model structure was minimized using the full atomic model with and without the SGB dielectric continuum solvation energy term.

III.   RESULTS AND DISCUSSION

The problem of differentiating non-native states from native-like states can be expressed as the ability of a scoring function, depending only on the coordinates of each structure, to score the native states better than any other structures. If such a scoring function were used also to generate structures, a further desirable property would be that in the vicinity of the native state the structural similarity to the native state would be a monotonically increasing function of improved scores.

A.   Park and Levitt Decoys

Examination of minimized energies for the seven extensive data sets of protein decoys (see Fig. 1) shows that using the OPLS-AA/SGB potential, no decoy scores better than the X-ray structure. The correlation between structural similarity and score is strong only for structures with low RMSD. For RMSD > 4 Å this correlation breaks down. Native-like states appear around 2 Å at low energies, with the bulk of the decoys being in non-native-like conformations with RMSD above 4 Å.

In Table II we report the statistical indicators of the quality of the scoring function. Some of the indicators depend on defining the reference structure as the native X-ray structure. It has been verified that similar results are obtained by selecting any native-like decoy as the reference structure. A global view of the results for the Park and Levitt sets is given in Fig. 2. The fraction, $P(\Delta U)$, of native-like decoys with an energy gap from the native less than ΔU is shown. A decoy conformation with an RMS less than 3 Å is considered native-like. Figure 2 indicates, for example, that structures with an energy gap from the native less than 100 kcal/mol have a ~90% chance of being native-like, whereas a decoy with a +200 kcal/mol energy gap from the native has only a 20% chance of being native-like. For these data sets there are no decoy structures with a total energy, U_{tot}^{con}, below that of the native state (i.e., energy-minimized X-ray coordinates; see Fig. 1). This suggests that if a fold prediction program can generate protein structures within 100 kcal/mol of the native state, there should be a high (>90%) chance of finding native-like states in this data set.

Another measure of the fitness of the scoring functions is to evaluate the RMSD of the lowest-energy structure in each decoy set. The results are summarized in Table II. The RMSD of the lowest-energy decoy range from 0.94 Å to 2.20 Å with an average RMSD of 1.9 Å. These decoys fall within the native-like designation. The average energy deviation from the native energy is +79.5 kcal/mol, which represents an average deviation of +2% from the native total energy values. As we shall see below, not all scoring functions examined yield decoy energies consistently higher than the native energy.

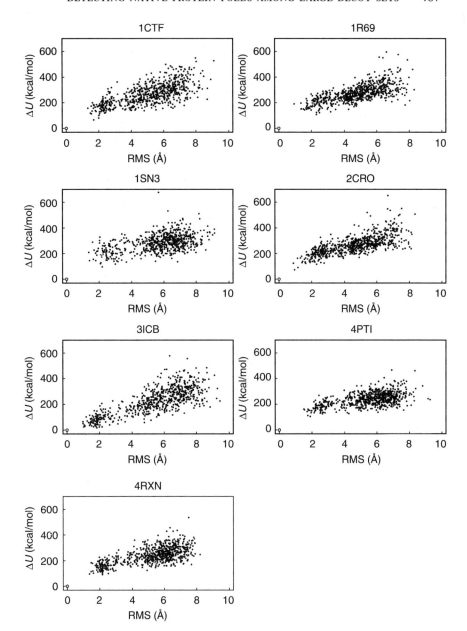

Figure 1. OPLS-AA/SGB: Energy gap/RMS correlation plots for the Park and Levitt decoy sets.

TABLE II

OPLS-AA/SGB Results: The Minimized energy, U_{native}, of the Native Conformation; the Energy Gap, min (ΔU), and the RMS Devition Between the Best-Scoring Decoy and the Native Conformation; the Native Z-Score Z_{nat} and the Average Z-Score $\bar{Z}_{nat-like}$ of the Native-like Conformations of the Park and Levitt Decoy Sets [60]

PDB Name	U_{native}	min(ΔU)	RMSD	Z_{nat}	$\bar{Z}_{nat-like}$
1ctf	−4213.92	+65.55	1.69	−3.24	−1.08
1r69	−3499.46	+107.16	2.30	−4.03	−1.01
1sn3	−3467.53	+96.08	2.19	−4.22	−1.04
2cro	−3628.30	+72.55	0.94	−3.69	−0.95
3icb	−4694.45	+18.08	1.84	−2.18	−1.34
4pti	−3055.04	+105.07	1.89	−4.53	−1.15
4rxn	−3363.51	+92.06	2.16	−3.76	−1.29

In Table II we also report the native Z score, Z_{nat}, and the average Z score of the native-like decoys, $\bar{Z}_{nat-like}$. The Z score of conformation i is defined as

$$Z_i = \frac{E_i - \bar{E}}{\sigma} \tag{8}$$

where E_i is the energy of the particular conformation, $\bar{E}$ is the average score and σ is the standard deviation of the distribution of scores in the set. The average Z score, $\bar{Z}_{nat-like}$, is obtained by averaging the Z scores of the native-like decoys. A decoy is defined as native-like if its RMSD with respect to the native is less than

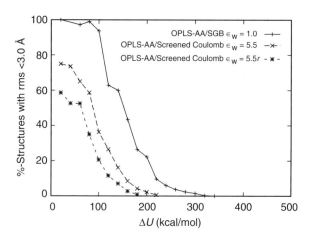

Figure 2. Fraction of the Park and Levitt decoys with energy gap from the native less than ΔU which are native-like (RMSD from native < 3 Å), using the OPLS-AA/SGB potential function and the vacuum OPLS-AA potential with screened Coulomb interactions.

3 Å. The Z score measures the ability of the scoring function to recognize native conformations. Assuming the distribution of scores is approximately Gaussian, a native Z score of, say, -2 indicates that the native structure is ranked in the best 1% in the decoy set. In general, the more negative the Z score, the better. The values of the native Z scores range from -3.2 to -4.5, indicating that the scoring function is extremely successful in finding the native structure among the decoys. The native-like average Z score represents the ability of the scoring function to discriminate the native-like conformations from the non-native conformations. The more negative the average native-like Z score, the larger the probability that a low-energy conformation is a conformation structurally similar to the native. The calculated values of the Z scores ranging from -0.95 to -1.34 indicate that, although on average the native-like conformations have lower energies than the non-native conformations, a significant number of native-like structures have a favorably low Z score. This can also be seen from Fig. 1 by looking at the vertical position of the low-RMSD structures with respect to the bulk of the decoys. This does not necessarily indicate a deficiency of the energy function but rather that for native-like conformations (i.e., those with the correct fold) the energy is also sensitive to the position and orientation of the amino acid side chains. An incorrect placement of a side chain may be enough to increase the energy of a native-like fold to the level of the misfolded conformations. A native-like energy is achieved only when all of the structural elements of the protein are placed correctly [22].

Park and Levitt [60] have evaluated six simple empirical scoring functions using the same decoy sets examined in this work. A comparison between the native and native-like Z scores calculated here with those obtained by Park and Levitt shows that the OPLS-AA/SGB energy model clearly outperforms the six empirical scoring functions examined in the Park and Levitt work. Moreover, none of the empirical scoring functions examined by Park and Levitt was able to consistently rank first the native conformation, whereas the OPLS/SGB model does.

It is instructive to evaluate the importance of each component of the OPLS-AA/SGB energy function in recognizing native conformations. Because all the decoys are well-packed, there is very little discrimination based on packing (as measured by the van der Waals energies) of the non-native states from the near-native conformations. In order to establish the role of intramolecular and solvent electrostatic interactions, we have calculated the energy scores in vacuum, U_{tot}^{vac}, using the same protocol used for the calculations in continuum solvent. The results are summarized in Table III. For several proteins the native conformation does not correspond to the minimum energy, and decoys with large RMSD from the native have very favorable scores. The native Z score and the near-native average Z scores have also significantly degraded (compare Tables II and III). This can be clearly seen in Fig. 3 showing the energy RMSD correlation plots

TABLE III
Vacuum OPLS-AA Results: The Minimized Energy, U_{native}, of the Native Conformation; the Energy
Gap, min (ΔU), and the RMS Devition Between the Best-Scoring Decoy and the Native
Conformation; the Native Z-Score Z_{nat} and the Average Z-Score $\bar{Z}_{nat\text{-}like}$ of the
Native-like Conformations of the Park and Levitt Decoy Sets [60]

PDB Name	U_{native}	$\min(\Delta U)$	RMSD	Z_{nat}	$\bar{Z}_{nat\text{-}like}$
1ctf	− 2795.74	+ 43.68	6.49	− 2.62	− 0.51
1r69	− 2489.72	+ 76.49	1.65	− 3.03	− 0.42
1sn3	− 2495.10	+ 0.04	1.42	− 3.10	− 0.59
2cro	− 1122.06	− 35.12	0.93	− 2.37	− 0.68
3icb	− 2795.74	− 282.69	1.19	− 0.63	− 0.84
4pti	− 1324.06	+ 37.53	6.21	− 2.97	− 0.71
4rxn	− 3581.88	− 8.95	1.60	− 2.47	− 1.13

for the seven proteins studied. The gain achieved by including the solvation term is particularly noticeable for the 3icb data set. Figure 4 shows the distribution of energy gaps from the native for the 3icb decoys using either the vacuum OPLS-AA energy or the OPLS-AA/SGB energy. A shift of the distribution to positive values indicates that no decoy structures have energies lower than the native structure. Vacuum energies are scattered above and below the native state energy with little correlation between energy and structural similarity. The OPLS-AA/SGB energies produce a sharper distribution than the vacuum energies. It is clear that for this decoy set the vacuum energy is significantly poorer than the energy in solution in discriminating native folds.

An important contribution to protein stability arises from the tendency for packing nonpolar side-chains in the interior of the proteins and placing polar residues on the solvent exposed surface of the protein [75,76]. These tendencies are not represented well by the intramolecular potential in vacuum, which in general is equal to the strength of interaction between two nonpolar residues and between a nonpolar residue and polar residue and does not particularly favor the placement of a polar residue on the protein surface. The solvation energy calculated using the SGB model, however, reproduces hydrophobic interactions and favors the placement of polar residues on the protein surface where they can interact strongly with the solvent. The presence of a hydrophobic core and a polar surface is a key feature of the native protein conformation in solution. Several empirical scoring function have been designed to recognize these features [20,60,65,66,62]. A model that does not take into account solvation effects is likely to perform poorly in native fold recognition among large numbers of compact decoys.

Another important function of dielectric continuum models is to dampen the strength of the electrostatic interactions between polar and charged residues. Conformations having salt bridges and intramolecular hydrogen bonds are

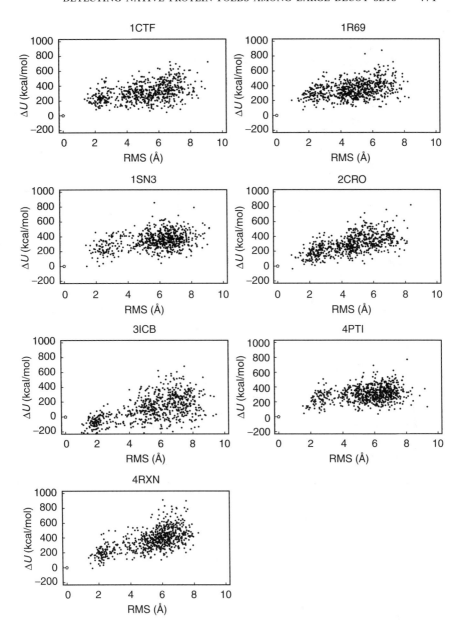

Figure 3.   Vacuum OPLS-AA: Energy gap/RMS correlation plots for the Park and Levitt decoy sets.

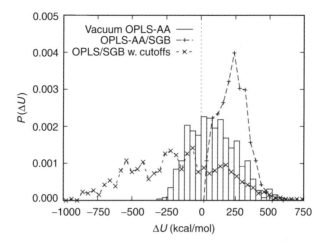

Figure 4. The distribution of energy gaps from the native for the 3icb data set of the Park and Levitt decoys using various energy functions.

strongly favored in vacuum, but much less so in solution. The SGB implicit solvent model provides a mechanism to filter out non-native conformations with artificially low intramolecular electrostatic energies that would be otherwise given a favorable score.

In these calculations, all charged interactions are included in the total energy; employing a cutoff for atom–atom interactions destroys the correlation between low energy values and native-like structures. Figure 4 shows that the proper evaluation of the long-range Coulomb interactions is crucial in selecting native conformations. If the electrostatic interactions are spatially truncated, many non-native structures assume lower total energies than do the native structure. As shown in Fig. 4, the correlation between energy and structural similarity using the OPLS-AA/SGB force field with a nonbonded cutoff of 9 Å is poor. This is a direct consequence of neglecting the long-range part of Coulomb interactions and is aggravated by the highly charged nature of some of the proteins examined (see Table I).

B. Holm and Sander Single Decoys

Recognizing single misfolded structures that have been carefully selected or devised as possible alternate folds poses a different challenge than distinguishing native-like states in large decoy data sets. Instead of picking native-like conformations among a large set of decoys, the challenge is to differentiate between two well-folded proteins, one of which corresponds to the native state. In the decoy set of Holm and Sander [17], misfolded conformations were constructed by swapping parts of the polypeptide chains with segments from

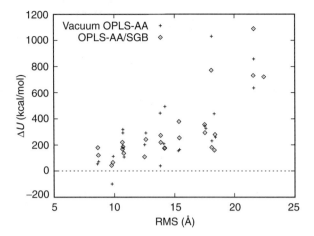

Figure 5. Energy gaps between the Holm and Sander [17] misfolded decoys and the corresponding native conformations, using the vacuum OPLS-AA and the OPLS-AA/SGB potentials.

known crystal structures. The proteins in the Holm and Sander set cover a wide range of sizes, from 36 residues for the smallest protein to over 300 residues for the largest protein. Figure 5 reports the energy gaps from the native of the misfolded proteins using the vacuum OPLS-AA energy and the OPLS-AA/SGB energy. The misfolded conformations are compact and have RMSDs from the native of 8 Å or more. Both the vacuum OPLS-AA and the OPLS-AA/SGB models are successful in ranking the native structures higher then the corresponding misfolded decoys; the only exception is for the avian pancreatic polypeptide (1ppt), a small 36 residue polypeptide, using the vacuum OPLS-AA model. Although smaller energy differences are generally correlated with higher structural similarity (see Fig. 5), the smallest (∼8 Å) RMSD structure in this data set is well above the RMSD threshold of ∼4 Å, above which energy and structural similarity were no longer correlated for the proteins in the Park and Levitt set.

The apparent correlation between RMSD and energy gap visible in Fig. 5 is mostly due to the fact that the RMSDs and the energy gaps increase with increasing protein size. As shown in Fig. 6, the energy gaps grow roughly linearly with the sequence length of the protein (a slightly better correlation is observed when using the OPLS-AA/SGB model). The energy gaps calculated using the OPLS-AA/SGB model are generally of the same relative magnitude, when normalized by size, as the energy gaps calculated for the Park and Levitt set. This confirms that the energy function used here can discriminate between native and misfolded structures over a wide range of protein sizes.

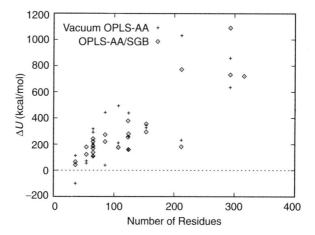

Figure 6. Protein size dependence of the energy gaps from the native of the misfolded protein structures from the Holm and Sander [17] data set.

C. CASP3 Targets

We have also analyzed some of the structures submitted to the CASP3 competition [67]. The target proteins are listed in Table IV. Our results are shown in Fig. 7, which shows the differences between the energy of each predicted structure and the energy of the corresponding native conformation. The targets can be divided into two groups: the "easy" targets for which the majority of the predicted models have an RMS deviation from the native of 3 Å or less, and the

TABLE IV
A Summary of the CASP3 Target Evaluated in this Study[a]

Target	N_{res}	Resolution (Å)	N_{res} Predicted	Models	Class	RMS (Å)	PDB
T0043	158	1.5	158	8	α/β	14.2–16.8	1hka
T0047	162	2.5	158	14	mostly β	1.3–1.9	1a2u
T0052	101	NMR	101	8	all β	13.7–17.1	2ezm
T0055	125	2.0	123	17	mostly β	2.8–7.4	1byf a
T0058	229	1.6	225	10	α/β	1.6–3.3	1eug
T0060	117	1.54	117	17	α/β	1.3–5.2	1dpt
T0064	111	1.9	103	22	All α	7.8–19.1	1b0n a
T0065	57	1.9	31	49	All α	2.7–10.1	1b0n b
T0068	376	1.9	376	4	Mainly β	8.9–18.5	1bhe
T0082	190	1.75	190	12	α + β	4.6–19.3	1bk7
T0085	211	2.6	211	6	Mostly α	17.8–22.9	1bvb

[a]Out of the structures predicted by the participants in CASP, we have selected those that have near- or full-length predictions only and whose PDB coordinates were available at the time of this study.

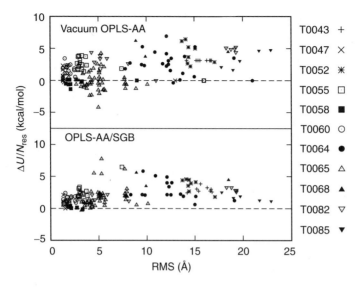

Figure 7. Energy difference per residue between native and predicted structures for a selection of targets from the CASP3 competition: T0043 (1hka), T0047 (2a2u), T0052 (2ezm), T0055 (1byf), T0058 (1eug), T0060 (1dpt), T0064 (1b0n_a), T0065 (1b0n_b), T0068 (1bhe), T0082 (1bk7), and T0085 (1bvb).

difficult targets in which none of the predicted models is native-like (RMS deviations from the native of 10 Å or more). For a few of the targets the predictions ranged from near-native (<3 Å) to non-native (>3 Å).

As shown in Fig. 7, the OPLS-AA/SGB model achieves nearly 100% discrimination of the native conformations. Only a few predictions, structurally similar to the native, score slightly better than the native. The vacuum OPLS-AA energy function does not perform as well as the OPLS-AA/SGB energy function; several high-RMS predictions for the T0055, T0058, T0064, and T0065 targets have scores significantly lower than the native. As observed for the Park and Levitt [60] decoy set, neither the vacuum OPLS-AA nor OPLS-AA/SGB energy functions are able to differentiate between models with large RMS deviations from the native; that is, a 15 Å structure can easily score better than a 10 Å structure.

D. Energy Components

The ability of a scoring function to discriminate between native and non-native conformations depends on the delicate balance between the components of the scoring function [1,20,60,66,62]. As described in this section, we find that, although some combinations of energy components show improvement over

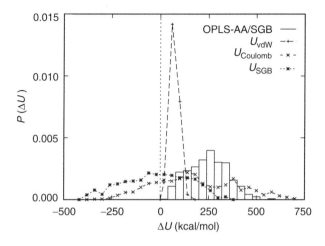

Figure 8. Distribution of energy gaps from the native of the 3icb Park and Levitt decoys for the total OPLS-AA/SGB energy and for the van der Waals, U_{vdW}, intramolecular Coulomb, $U_{Coulomb}$, and solvation, U_{SGB}, energy components.

each individual component, the total OPLS-AA/SGB energy is the best scoring function overall.

An analysis of the energy components of Eqs. (1) and (2) presented in Fig. 8 shows that for the Park and Levitt data set (Table I), containing only well-packed structures, the van der Waals energy difference with respect to the native is positive for most of the decoys. The van der Waals energy, however, does not strongly correlate with structural similarity to the native. This point is illustrated in Fig. 9, which shows the distribution of energy gaps from the native of both the native-like (RMSD <3 Å) and misfolded (RMSD >3 Å) 3icb decoys. In contrast, the discriminating power of the total OPLS-AA/SGB energy is indicated by the relatively small overlap between the native-like and misfolded distributions of energy gaps (see Fig. 9). A similar separation is not achieved with the van der Waals energy, indicating that the van der Waals energy alone does not provide good discrimination when used as a scoring function.

The electrostatic energy components, the intramolecular Coulomb energy, and the solvation energy, taken individually, are not effective scoring functions; the sum of the two, however, is significantly better as indicated in Figs. 10 and 11 ($\epsilon_w = 1$ distribution). As shown in Fig. 10, the solvation energy is strongly anticorrelated with the electrostatic energy. A positive intramolecular electrostatic energy gap from the native is counteracted by a negative solvation energy gap, and vice versa. Because the solvation energy does not completely offset the intramolecular electrostatic energy, decoys having an intramolecular electrostatic energy less favorable than the native will generally continue to

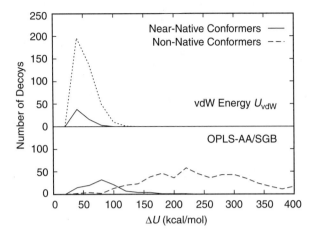

Figure 9. Near-native and non-native distributions of the OPLS-AA/SGB and van der Waals energy gaps from the native for the Park and Levitt 3icb decoys.

have a less favorable total electrostatic energy (intramolecular + solvation) with respect to the native. The contribution of the solvation energy term, however, is large enough to reverse the sign of the energy gap for those decoys having an intramolecular energy more favorable than the native, for which there are many examples in the Park and Levitt set (see Fig. 11). The native state

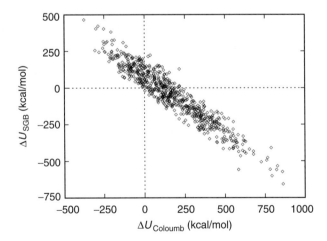

Figure 10. Correlation plot between the intramolecular Coulomb energy gap $\Delta U_{\text{Coulomb}}$ and the solvation energy gap ΔU_{SGB} for the 3icb decoys.

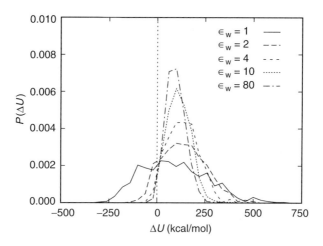

Figure 11. The distributions of the screened Coulomb OPLS-AA energy gaps from the native for the 3icb decoys as a function of dielectric constant.

corresponds to a balance between optimizing the intramolecular Coulomb interactions and the intermolecular protein–solvent interactions.

Monge et al. [20] have also studied various energy decompositions of an all-atom force field supplemented by a continuum solvation model. They analyzed a decoy data set generated by a simplified model employing a fixed, known secondary structure. The authors observe that the relative differences of both van der Waals and Coulomb energies are about 1–2% above the native values, but the total electrostatic component is the dominant factor in distinguishing non-native states from the native ones. They found that a fraction of the decoys had vdW energies lower than that of the native. Their model performed reasonably well, though some non-native conformations had better scores than the native state. This was not observed in the data sets we studied using the OPLS-AA/SGB scoring function.

E. Approximate Effective Dielectric Models

1. Screened Coulomb Approximation

As shown in Fig. 10, the solvation energy gaps with respect to the native are strongly correlated with the intramolecular Coulomb energy gaps. The equation

$$\Delta U_{\text{SGB}} = \alpha + \beta \Delta U_{\text{Coulomb}} \tag{9}$$

can be fitted obtaining $\beta = -0.82$ with a regression coefficient of 0.94. If we collate the total electrostatic interaction energy ΔU_{ele} as the sum of the Coulomb and solvation energies, we find

$$\Delta U_{\text{ele}} \equiv \Delta U_{\text{Coulomb}} + \Delta U_{\text{SGB}} \cong 0.18 \Delta U_{\text{Coulomb}} \qquad (10)$$

This suggests that it might be possible to employ a screened Coulomb model to account for solvation effects.

The screened Coulomb effective electrostatic interaction between two charges q a distance r apart is

$$\frac{U_{\text{Coulomb}}(r)}{\epsilon_w} = \frac{q^2}{\epsilon_w r} \qquad (11)$$

The effect of the surrounding medium is accounted for by the value of ϵ_w, usually taken as 80 for water. Figure 11 shows the energy distributions for the 3icb decoy set relative to the native state for the vacuum case and for various values of the effective dielectric constant. A good energy function should only produce energy gap values in the positive range. It is clear that for this decoy set, a simple electrostatic energy evaluation in vacuum ($\epsilon_w = 1$) results in many decoy structures with energies substantially below the native values. Moreover, no correlation between the RMSD from the native and the energy is observed. Increasing the value of the effective dielectric constant removes some of the negative energy gaps and increases the propensity for the low-energy decoy structures to have low RMSD (not shown). None of the effective dielectric constants used, however, was able to differentiate all of the decoys from the native structure. This point is also illustrated in Fig. 2, which depicts the fraction of native-like structures with energy gaps from the native less than ΔU using $\epsilon_w = 5.5$ as suggested by the relation in Eq. (10). It is clear that the screened Coulomb scoring function provides less discrimination between decoys and native structures than does the SGB solvation model.

If a simple relationship between the reaction field energy calculated via the SGB model and the Coulomb energy as in Eq. (11) could be found, there would be no need to employ more complicated continuum models. Although the bulk of the correlation between these two terms can be explained by a screened Coulomb interaction, the discrimination between native and non-native states is degraded by such an approximation. The dispersion in the reaction field energy versus the Coulomb energy, which is not contained in the screened Coulomb model, provides a more detailed description of solvation effects which aids the discrimination of native-like conformations from misfolded ones.

Although the SGB solvation energy is correlated with the intramolecular Coulomb energy, it is not clear that the best values to use for an effective dielectric constant is given by Eq. (10). The fraction of native-like structures with energy gap less than a given energy difference calculated over all the data

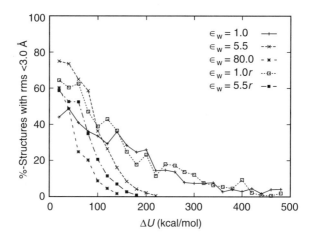

Figure 12. Fraction of the Park and Levitt decoys with energy gap from the native less than ΔU which are native-like (RMSD from native < 3 Å), using the vacuum OPLS-AA potential with screened Coulomb interactions.

sets in Table I, reported in Fig. 12, shows the efficiency achieved using different values of ϵ_w. None of the effective dielectric models achieves 100% discrimination for energy values within 20 kcal/mol of the native state energy. Using $\epsilon_w = 1$ yields a broad range of energies for both native-like and non-native states as discussed above. In comparison, using a value of ϵ_w either 5.5 or 80.0 yields distributions of energies that are like those given in Fig. 11 for the calbindin data set. The fraction of native-like structures with energies similar to the native state is around 60% for an effective dielectric constant of 80.0. This fraction increases to about 75% for an ϵ_w value of 5.5.

2. Distance-Dependent Dielectric Approximation

An alternative to the simple screened Coulomb interaction in protein modeling is the distance-dependent dielectric function [51]. In this approximation the effective electrostatic interaction between two partial charges q at distance r is written as

$$\frac{U_{\text{Coulomb}}(r)}{\epsilon_w r} = \frac{q^2}{\epsilon_w r^2} \tag{12}$$

Although unphysical in nature, it has been suggested that the extra screening afforded by the $1/r^2$ function can capture some of the additional polarization effects contained in higher-level implicit solvent models [51]. By calculating the energies of the decoy conformers in Table I using the distance-dependent dielectric approximation, we obtain energy distributions similar to those obtained using the simple screened Coulomb model. Moreover, as shown in Fig. 2, both effective dielectric models produce qualitatively similar results. For both values of ϵ_w studied, 1.0 and 5.5, the fraction of native-like structures with energy similar to the native energy, is significantly less than 100%. Comparison between the distance-dependent dielectric and the non-distance-dependent dielectric function in Figs. 12 and 2 demonstrate that the distance-dependent function is less discriminatory for the decoy data sets studied here. While the distance-dependent dielectric constant has been successfully employed in some cases [77], we find that, though it is better than the vacuum Coulomb potential, a simple non-distance-dependent screened Coulomb model is more effective (Fig. 12). None of the screened Coulomb models are as effective as the SGB solvation potential for the protein decoy data sets investigated here.

F. Dependence on the Interior Dielectric Constant

The SGB solvent model requires the separation of space into an exterior region containing the solvent medium and an interior region containing the protein charge distribution. In the current implementation of the SGB model, the van der Waals surface of the protein is used to define the dividing surface. The default value for the dielectric constant of the solvent is 80, corresponding to pure water at room temperature. Up to this point, the dielectric constant of the interior region, ϵ_{in}, has been set at the value of 1, corresponding to the vacuum dielectric constant. We have also examined the cases $\epsilon_{in} = 2$ and 5.5 to see whether the OPLS-AA/SGB results can be further improved. The energy components obtained for the native conformations contained in the Park and Levitt set are given in Table V. A larger interior dielectric constant results in a lower total energy of the system due to the increase of the dielectric shielding inside the protein. The Coulomb energy and the reaction field contributions are both reduced in an amount roughly proportional to the interior dielectric constant. The van der Waals energy partly compensates for the reduction in electrostatic energy, but the variation in U_{vdW}^{native} is relatively small.

The fraction of native-like decoys of the Park and Levitt set as a function of energy gap is shown in Fig. 13 for the values of ϵ_{in} examined. The number of native-like conformations (RMSD <3 Å) with an energy score similar to the native increases as we decrease the dielectric constant of the interior region. It is only with an interior dielectric of 1.0 that all misfolded conformations can be eliminated based on energy alone. The discriminatory power of the OPLS-AA/SGB energy model in this fold recognition test is optimal for this choice of the internal dielectric, though it may not be optimal in other modeling contexts.

TABLE V
Selected Energy Components from Eqs. (1) and (2) for the Native State Using the Continuum Model
($\epsilon_w = 80.0$) as a Function of Interior Dielectric Constant, ϵ_{in}

PDB	ϵ_{in}	U_{total}^{native} (kcal/mol)	U_{vdW}^{native} (kcal/mol)	$U_{Coulomb}^{native}$ (kcal/mol)	U_{SGB}^{native} (kcal/mol)	U_{cav}^{native} (kcal/mol)
1ctf	1.0	−4213.9	−475.5	−5340.3	−1367.6	+37.9
	2.0	−2065.9	−519.7	−2595.2	−688.3	+38.4
	5.5	−730.6	−532.8	−925.5	−244.0	+38.7
1r69	1.0	−3499.5	−497.2	−3722.9	−1168.9	+37.2
	2.0	−1709.9	−539.0	−1781.7	−593.3	+37.7
	5.5	−599.5	−554.3	−627.9	−210.8	+38.1
1sn3	1.0	−3467.5	−465.1	−4784.2	−972.8	+36.3
	2.0	−1688.1	−499.8	−2315.2	−500.3	+36.8
	5.5	−585.3	−511.8	−821.5	−180.1	+37.1
2cro	1.0	−3628.3	−522.4	−3514.8	−1462.2	+40.4
	2.0	−1763.1	−567.2	−1662.8	−749.7	+41.0
	5.5	−604.8	−578.9	−585.2	−264.8	+41.4
3icb	1.0	−4694.5	−587.3	−5163.5	−2350.6	+45.4
	2.0	−2271.4	−641.0	−2466.5	−1195.6	+46.1
	5.5	−766.8	−656.8	−865.7	−427.2	+46.4
4pti	1.0	−3055.0	−423.9	−2542.0	−1366.9	+34.1
	2.0	−1464.2	−448.4	−1208.6	−686.6	+34.6
	5.5	−473.2	−455.1	−425.0	−240.9	+34.8
4rxn	1.0	−3363.5	−373.6	−2496.6	−2791.5	+31.3
	2.0	−1598.8	−399.3	−1190.1	−1389.9	+31.6
	5.5	−498.1	−407.6	−410.9	−489.1	+31.8

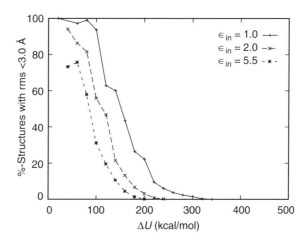

Figure 13. Fraction of the Park and Levitt decoys with energy gap from the native less than
ΔU which are native-like (RMSD from native < 3 Å), using the OPLS-AA/SGB potential with
various values of the interior dielectric constant.

IV. CONCLUSIONS

The OPLS-AA molecular mechanics energy function coupled with the surface generalized Born solvation model is found to be able to discriminate the native structures of several proteins from their decoys. The results show that for a number of cleverly constructed decoys the OPLS-AA/SGB scoring function correctly singles out native-like states from the bulk of the non-native conformations. Not all of the native-like structures were clearly separated in the data sets; indeed some distant non-native conformations score better than some native-like (RMSD <3 Å) conformations. This suggests that if the current scoring method is to be applied to a set of *ab initio* generated structures, it is critical that the algorithm for constructing native-like structures be such that a broad range of the relevant parts of the native-like conformational space are sampled.

The ability of the OPLS-AA/SGB model to recognize native conformations is found to be comparable, and in many cases superior, to the best knowledge-based scoring functions. Other studies have shown the usefulness of molecular mechanics force fields augmented by implicit solvation models in this area [6]. Lazaridis and Karplus [22] have shown that the CHARMM protein force field combined with their EEF1 effective solvation free energy model [78] is able to achieve 100% discrimination of the native conformations in a large decoy data set and in the single decoy data set they examined. They also observe, in agreement with our findings, that significantly poorer results are obtained by omitting the solvation free energy term. They obtain these results despite the use of a computationally fast solvation model which has the form of an effective pair potential and is simpler than the SGB solvation model. Recently, Petrey and Honig [79] have applied the CHARMM protein force field, together with a dielectric continuum model based on the Poisson–Boltzmann equation, to the problem of native fold recognition in the single decoy data set [17] (also examined in this work) achieving a discrimination level close to 100%. They also applied a simplified solvation model containing only the intramolecular electrostatic energy and a hydrophobic residue burial estimator to evaluate the Park and Levitt decoy sets. In two cases (3icb and 4rxn) their method does not clearly rank the X-ray conformation favorably. Petrey and Honig observe that the solvation energy often favors the misfolded conformation in the single-decoy sets, concluding that the solvation energy is not useful in recognizing the native conformation. However, even though the solvation energy generally favors misfolded conformations, these structures tend to be disfavored relative to the native conformation when the total electrostatic energy (sum of the direct Coulomb and solvation term) is considered. In contrast, the SGB solvation term is essential for destabilizing the relatively large number of Park and Levitt decoys for which the direct Coulomb energy is more favorable than the corresponding value for the native.

The OPLS-AA/SGB scoring function was also compared with the screened Coulomb OPLS-AA scoring function. Whereas a significant fraction of the decoys with scores within 100 kcal/mol from the native are misfolded using a screened Coulomb potential, essentially all of the decoys within 100 kcal/mol from the native are native-like using the OPLS-AA/SGB scoring function.

The ability to discriminate native-like protein conformations from non-native conformations is one of the fundamental problems in theoretical protein structure prediction. The use of knowledge-based scoring potentials, derived from a combination of structural and thermodynamic data, is currently the most widely used method. It is often assumed that such models are inherently better than all-atom force fields. This work shows the importance of correctly modeling the physical forces underlying protein folding. Thanks to their simplicity, knowledge-based scoring schemes are less costly to evaluate compared to all-atom models. In the future it should be possible to combine the best features of the two approaches to rapidly generate plausible protein conformations using knowledge-based potentials more reliably, and then discriminate between conformers using all-atom scoring functions.

Acknowledgments

This project has been supported by the National Institutes of Health Grant GM-30580, the Center for Biomolecular Simulations at Columbia University, and the High Performance Computing Project at Rutgers University. The authors thank Dr. Lynne Reed Murphy for help with some of the calculations.

References

1. S. J. Wodak and M. J. Rooman, *Curr. Opin. Struct. Biol.* **3**, 247–259 (1993).

2. D. T. Jones and J. M. Thornton, *Curr. Opin. Struct. Biol.* **6**, 210–216 (1996).

3. K. W. Plaxco, D. S. Riddle, V. Grantcharova, and D. Baker, *Curr. Opin. Struct. Biol.* **8**, 80–85 (1988).

4. M. Hao and H. A. Scheraga, *Curr. Opin. Struct. Biol.* **9**, 184–188 (1999).

5. D. J. Osguthorpe, *Curr. Opin. Struct. Biol.* **10**, 146–152 (2000).

6. T. Lazaridis and M. Karplus, *Curr. Opin. Struct. Biol.* **10**, 139–145 (2000).

7. V. Eyrich, D. Standley, A. Felts, and R. Friesner, *Proteins* **35**, 41–57 (1999).

8. S. W. Rick and B. J. Berne, *J. Am. Chem. Soc.* **116**, 3949–3954 (1994).

9. R. M. Levy and E. Gallicchio, *Annu. Rev. Phys. Chem.* **49**, 531–567 (1998).

10. A. A. Rashin and M. A. Bukatin, *J. Phys. Chem.* **98**, 386–389 (1994).

11. D. Sitkoff, K. A. Sharp, and B. Honig, *J. Phys. Chem.* **98**, 1978–1988 (1994).

12. D. J. Tannor, B. Marten, R. Murphy, R. A. Friesner, D. Sitkoff, A. Nicholls, M. Ringnalda, W. A. Goddard III, and B. Honig, *J. Am. Chem. Soc.* **116**, 11875–11882 (1994).

13. D. Sitkoff, N. Ben-Tal, and B. Honig, *J. Phys. Chem.* **100**, 2744–2752 (1996).

14. G. Hawkins, C. Cramer, and D. Truhlar, *J. Phys. Chem.* **100**, 19824–19839 (1996).

15. E. Gallicchio, L. Zhang, and R. M. Levy, submitted (2001).

16. D. Eisenberg and A. D. McLachlan, *Nature* **319**, 199–203 (1986).

17. L. Holm and C. Sander, *J. Mol. Biol.* **225**, 93–105 (1992).

18. W. F. van Gunsteren, F. J. Luque, D. Timms, and A. E. Torda, *Annu. Rev. Biophys. Biomol. Struct.* **23**, 847–863 (1994).

19. P. E. Smith and B. M. Pettitt, *J. Phys. Chem.* **98**, 9700–9711 (1998).

20. A. Monge, E. J. P. Lathrop, J. R. Gunn, P. S. Shenkin, and R. A. Friesner, *J. Mol. Biol.* **247**, 995–1012 (1995).

21. M. Schaefer, H. W. van Vlijmen, and M. Karplus, *Adv. Protein Chem.* **51**, 1–57 (1998).

22. T. Lazaridis and M. Karplus, *J. Mol. Biol.* **288**, 477–487 (1999).

23. Y. N. Vorobjev and J. Hermans, *Biophys. Chem.* **78**, 195–205 (1999).

24. M. K. Gilson and B. Honig, *Proteins Struct., Funct., Genet.* **4**, 7–18 (1988).

25. D. Bashford and M. Karplus, *Biochemistry* **29**, 10219–10225 (1990).

26. A. A. Rashin, *J. Phys. Chem.* **94**, 1725–1733 (1990).

27. K. A. Sharp and B.Honig, *Annu. Rev. Biophys. Chem.* **19**, 301–332 (1990).

28. A. Warshel and J. Åqvist, *Annu. Rev. Biophys. Chem.* **20**, 267–298 (1991).

29. M. K. Gilson, M. E. Davis, B. A. Luty, and J. A. McCammon, *J. Phys. Chem.* **97**, 3591–3600 (1993).

30. B. Honig, K. Sharp, and A.-S. Yang, *J. Phys. Chem.* **97**, 1101–1109 (1993).

31. V. Mohan, M. E. Davis, J. A. McCammon, and B. M. Pettitt, *J. Phys. Chem.* **96**, 6428–6431 (1992).

32. T. Simonson and A. T. Brünger, *J. Phys. Chem.* **98**, 4683–4694 (1994).

33. K. Ösapay, W. S. Young, D. Bashford, C. L. Brooks III, and D. A. Case, *J. Phys. Chem.* **100**, 2698–2705 (1996).

34. S. R. Edinger, C. Cortis, P. S. Shenkin, and R. A. Friesner, *J. Phys. Chem. B* **101**, 1190–1197 (1997).

35. M. Born, *Z. Physik* **1**, 45–48 (1920).

36. F. Hirata, P. Rejfern, and R. Levy, *J. Quantum Chem.* **15**, 179–188 (1988).

37. W. C. Still, A. Tempczyk, R. C. Hawley, and T. Hendrickson, *J. Am. Chem. Soc.* **112**, 6127–6129 (1990).

38. A. Jean-Charles, A. Nichols, K. Sharp, B. Honing, A. Tempczyk, T. F. Hendrickson, and W. C. Still, *J. Am. Chem. Soc.* **113**, 1454–1455 (1991).

39. D. Qiu, P. S. Shenkin, F. P. Hollinger, and W. C. Still, *J. Phys. Chem. A* **101**, 3005–3014 (1997).

40. A. Ghosh, C. S. Rapp, and R. A. Friesner, *J. Phys. Chem. B* **102**, 10983–10990 (1998).

41. B. Roux and T. Simonson, *Biophys. Chem.* **78**, 1–20 (1999).

42. L. Zhang, E. Gallicchio, R. Friesner, and R. M. Levy, *J. Comp. Chem.* **22**, 591–607 (2001).

43. J. Novotny, R. Bruccoleri, and M. Karplus, *J. Mol. Biol.* **177**, 787–818 (1984).

44. J. Novotny, A. A. Rashin, and R. Bruccoleri, *Proteins Struct. Funct. Genet.* **4**, 19–30 (1988).

45. L. Chiche, L. M. Gregoret, F. E. Cohen, and P. A. Kollman, *Proc. Natl. Acad. Sci. USA* **87**, 3240–3243 (1990).

46. J. Vila, R. L. Williams, M. Vasquez, and H. A. Scheraga, *Proteins Struct. Funct. Genet.* **10**, 199–218 (1991).

47. R. L. Williams, J. Vila, G. Perrot, and H. A. Scheraga, *Proteins Struct. Funct. Genet.* **14**, 110–119 (1992).

48. Y. Wang, H. Zhang, W. Li, and R. A. Scott, *Proc. Natl. Acad. Sci. USA* **92**, 709–713 (1995).

49. Y. Wang, H. Zhang, and R. A. Scott, *Protein Sci.* **4**, 1402–1411 (1995).

50. M. Vieth, A. Kolinski, C. L. Brooks III, and J. Skolnick, *J. Mol. Biol.* **237**, 361–367 (1994).

51. S. J. Weiner, P. A. Kollman, D. A. Case, U. C. Singh, C. Ghio, G. Alagone, S. Profeta, and P. Weiner, *J. Am. Chem. Soc.* **106**, 765–784 (1984).

52. Y. N. Vorobjev, J. C. Almagro, and J. Hermans, *Proteins Struct. Funct. Genet.* **32**, 399–413 (1998).

53. M. Hendlich, P. Lackner, S. Weitckus, H. Floeckner, R. Froschauer, K. Gottsbacher, G. Casari, and M. J. Sippl, *J. Mol. Biol.* **216**, 167–180 (1990).

54. M. J. Sippl, *Curr. Opin. Struct. Biol.* **5**, 229–235 (1995).

55. R. L. Jernigan and I. Bahar, *Curr. Opin. Struct. Biol.* **6**, 195–209 (1996).

56. S. Miyazawa and R. J. Jernigan, *J. Mol. Biol.* **256**, 623–644 (1996).

57. A. Wallqvist, G. W. Smythers, and D. G. Covell, *Protein Sci.* **6**, 1627–1642 (1997).

58. S. Miyazawa and R. L. Jernigan, *Proteins Struct. Funct. Genet.* **36**, 357–369 (1999).

59. D. Covell and R. Jernigan, *Biochemistry* **29**, 3287–3294 (1990).

60. B. Park and M. Levitt, *J. Mol. Biol.* **258**, 367–392 (1996).

61. B. Ozkan and I. Bahar, *Proteins Struct. Funct., Genet.* **32**, 211–222 (1998).

62. R. Samudrala and J. Moult, *J. Mol. Biol.* **275**, 895–916 (1998).

63. K. T. Simons, I. Ruczinski, C. Kooperberg, B. A. Fox, C. Bystroff, and D. Baker, *Proteins Struct. Funct. and Genet.* **34**, 82–95 (1999).

64. W. L. Jorgensen, D. S. Maxwell, and J. Tirado-Rives, *J. Am. Chem. Soc.* **118**, 11225–11236 (1996).

65. E. S. Huang, S. Subbiah, J. Tsai, and M. Levitt, *J. Mol. Biol.* **257**, 716–725 (1996).

66. B. H. Park, E. S. Huang, and M. Levitt, *J. Mol. Biol.* **266**, 831–846 (1997).

67. J. Moult, T. Hubbard, K. Fidelis, and J. T. Pedersen, *Proteins Struct. Funct. Genet. Suppl.* **3**, 2–6 (1999).

68. D. B. Kitchen, F. Hirata, J. D. Westbrook, R. Levy, D. Kofke, and M. Yarmush, *J. Comp. Chem.* **11**, 1169–1180 (1990).

69. E. E. Abola, F. C. Bernstein, S. H. Bryant, T. F. Koetzle, and J. Weng, Protein data bank, in *Crystallographic Databases—Information Content, Software Systems, Scientific Applications*, F. H. Allen, G. Bergerhoff, and R. Sievers, eds., Data Commission of the International Union of Crystallography, Bonn/Cambridge/Chester, 1987.

70. L. Zhang, E. Gallicchio, and R. M. Levy, Implicit solvent models for protein–ligand binding: Insights based on explicit solvent simulations, in *Simulation and Theory of Electrostatic Interactions in Solution, AIP Conference Proceedings 492*, L. R. Pratt and G. Hummer, eds., American Institute of Physics, New York, 1999.

71. M. Levitt, *J. Mol. Biol.* **226**, 507–533 (1992).

72. R. Samudrala, Y. Xia, M. Levitt, and E. S. Huang, *Proc. Pacific Symp. Biocomput.* **4**, 505–516 (1999).

73. D. A. Hinds and M. Levitt, *Proc. Natl. Acad. Sci. USA* **89**, 2536–2540 (1992).

74. D. A. Hinds and M. Levitt, *J. Mol. Biol.* **243**, 668–682 (1994).

75. K. A. Dill, *Biochemistry* **29**, 7133–7155 (1990).

76. K. A. Dill, *Curr. Opin. Struct. Biol.* **3**, 99–103 (1993).

77. M. Schaefer, C. Bartels, and M. Karplus, *Theor. Chem. Acc.* **101**, 194–204 (1998).

78. T. Lazaridis and M. Karplus, *Proteins* **35**, 133–152 (1999).

79. D. Petrey and B. Honig, *Protein Sci.* **9**, 2181–2191 (2000).

AUTHOR INDEX

Numbers in parentheses are reference numbers and indicate that the author's work is referred to although his name is not mentioned in the text. Numbers in *italic* show the pages on which the complete references are listed.

Santiago, J. V.: 28(57), *34*; 37(35), 53-55(35), 57-58(35), 61-62(35), *71*
Sato, S., *34*(65)
Saunders, J. A., 195(9), 199(9), *220*
Saunders, M. A., 277(27-28), 299(28), 345(28), 435(28), *446*
Sayle, R., 50(126), 54(125), 60(126), *76*
Scapozza, L., 414(233), 419(233), *457*
Schaefer, M., 460(21), 481(77), *485–486*
Schaffer, A. A.: 125(59), *130*; 151(182), 183(247), *191–192*
Scharf, M.: 78(5), *127*; 148(174), *190*; 293(56), *448*
Schauber, H., 407(176), *454*
Schellman, J. A., 10(46), *33*
Scheraga, H. A.: 79(12), 83(42), 93(12), *127, 129*; 134(16), 139(108-110), 143(147,149), 144(108-110), 145(16,108-110,159-163), *186, 189–190*; 195(9), 199(9,30), 200(46), *220–221*; 225(9), 233(20), 239(23), 241(25), 244(25), 258(25,45), *262–263*; 288(34-38), 290(38,46), 291(38), 292(52-53), 293(52,58,65), 294(67), 296(52,69-72), 299(52-53,74), 304(77), 314(84-85), 318(77), 321(87), 325(90), 340(38), 345(74), 360(113-115), 365(115,153-154), 366(155), 369(38), 393(38,115,153-154), 411(84), 413(219), 416(38), 417(52,238), 420(52,74) 435(74), 437(52-53), *446–449, 451–453, 456–457*; 460(4), 461(46-47), *484–485*
Schiffer, C. A., 294(66), *448*
Schindler, T., 36(2), 58(2), 61(2), *70*
Schlegel, H. B., 392(159), *453*
Schlenkrich, M., 404(166), *453*
Schmid, F. X., 36(2), 58(2), 61(2), *70*
Schneider, R.: 78(5), *127*; 79(25), *128*; 247-248(34), *263*
Schoonman, M. J., 196(11), *220*
Schwartz, B., 440(250), *457*
Scott, R. A.: 196(15), *220*; 461(48-49), *485*
Scott, W., 211(76), *221*
Seebach, D., 197(24), *220*
Seibel, G. L., 411(204-205), 413(205), *455*
Sela, M., 267(1), 331(1), *445*
Selbig, J., 136(46), *187*
Semenza, G., 68(124), *76*
Seno, F.: 196(12), 212(12), *220*; 201(49), *221*
Serrano, L.: 28(58,60), *34*; 364(132-133,135), 404(135), *451–452*

Sette, A., 440(250), *457*
Sfatos, C. D., 64(101), *74*
Shafran, G., 79-80(32), 87(32), 93(32), *128*
Shaklınovich, E.: 2(2,6,9), 3(2,6,9,18), 4(9), 6(6,29), 7(2,6,9,35), 8(29,38-39), 29(9), 30(2,6,9,18), *31–33*; 37(30), 39(52), 40(56), 53-55(30,52), 57-58(30,52), 61-62(30), 64(101), *71–72, 74*; 78(8), *127*; 139(97-105), *188–189*; 268(8), *445*
Shalloway, D., 268(6), *445*
Sharp, K. A.: 292(51), 293(62), 443(252-253), *447–448, 457*; 460(11), 461(11,27,30,38), *484–485*
Shastry, R., 36(19), 53-55(19), 57-58(19), 61-62(19), *71*
Shats, O., 201(53), *221*
Sheinerman, F. B., 364(136), *452*
Shen, J., 408(188), *454*
Shenkin, P. S.: 217(85), *222*; 360(112), *450*; 460(20), 461(20,34,39), 462(39), 470(20), 475(20), 478(20), *485*
Shepard, R., 365(147), *452*
Sheridan, R. P., 411(205,213-214), 413(205), *455–456*
Sherman, S., 201(53), *221*
Shimanouchi, T., 224(6), *262*
Shoemaker, K. R., 364(129), *451*
Shoichet, B. K., 411(211-212), *455*
Shortle, D., 360(109), *450*
Shtilerman, M., 67-68(120), *75*
Sibanda, B. L., 406(172), *453*
Siddiqui, A. S., 247(35), *263*
Siew, N., 133(12-13), 173-174(12), *186*
Sigler, P. B., 60(112), 64(100,105,107-108,112), 65(100,107-108), 67(100), 68(107), *74–75*
Sikorski, A., 139(85-88), *188*
Sillitoe, I.: 225(11), 258(11), 261(11), *262*; 360(119), *451*
Silver, M. L., 406(170), *453*
Simmerling, C., 171(208), *191*
Simons, J., 365(146-147,149), *452*
Simons, K. T.: 3(12,14), 4(12), 7(12,14), 9-10(12,14), 14-15(12), 16(12,14), 24(12), 26(14), 29(14), *32*; 37(37), 53-55(37), 57-58(37), 61-62(37), *71*; 79(26), *128*; 141(128), 184(128), 180(243), *189, 192*; 200(44), *221*; 246(30-31), *262–263*; 360(108-109), *450*; 461(63), *486*
Simonson, T., 461(32,40), *485*
Singh, J. P., 205(71), *221*

SUBJECT INDEX